2009 IEEE 18th Conference on Electrical Performance of Electronic Packaging and Systems

(EPEPS 2009)

Portland, Oregon, USA
19 – 21 October 2009

IEEE Catalog Number: CFP09EPP-PRT
ISBN: 978-1-4244-4447-2

Copyright © 2009 by the Institute of Electrical and Electronic Engineers, Inc
All Rights Reserved

Copyright and Reprint Permissions: Abstracting is permitted with credit to the source. Libraries are permitted to photocopy beyond the limit of U.S. copyright law for private use of patrons those articles in this volume that carry a code at the bottom of the first page, provided the per-copy fee indicated in the code is paid through Copyright Clearance Center, 222 Rosewood Drive, Danvers, MA 01923.

For other copying, reprint or republication permission, write to IEEE Copyrights Manager, IEEE Service Center, 445 Hoes Lane, Piscataway, NJ 08854. All rights reserved.

***This publication is a representation of what appears in the IEEE Digital Libraries. Some format issues inherent in the e-media version may also appear in this print version.**

IEEE Catalog Number:	CFP09EPP-PRT
ISBN 13:	978-1-4244-4447-2
Library of Congress No.:	2009903119

Additional Copies of This Publication Are Available From:

Curran Associates, Inc
57 Morehouse Lane
Red Hook, NY 12571 USA
Phone: (845) 758-0400
Fax: (845) 758-2633
E-mail: curran@proceedings.com
Web: www.proceedings.com

TABLE OF CONTENTS

SIGNAL INTEGRITY I

Improving Single-Ended Channel Performance with a Dynamic Reference Voltage Scheme 1
Tingdong Zhou, Daniel M. Dreps, Wiren D. Becker

Designing a ZXnoise Pseudo-Differential Link 5
Frédéric Broydé, Bernard Démoulin

Extraction of Via and Trace Model from PCB Channel S-Parameter Data by Stochastic Optimization 9
Juyoung Lee, Drew Doblar

Through Silicon Via (TSV) Equalizer 13
Joohee Kim, Eakhwan Song, Jeonghyeon Cho, Jun So Pak, Junho Lee, Hyungdong Lee, Kunwoo Park, Joungho Kim

POWER INTEGRITY I

Achieving Near Zero SSN Power Delivery Networks by Eliminating Power Planes and Using Constant Current Power Transmission Lines 17
Suzanne Huh, Daehyun Chung, Madhavan Swaminathan

Statistical Simulation of SSO Noise in Multi-Gigabit Systems 21
Wendemagegnehu T. Beyene, Amir Amirkhany, Ali Abbasfar

Low-Impedance Power Distribution Network of Decoupling Capacitor Embedded Interposers for 3-D Integrated LSI System 25
Katsuya Kikuchi, Koichi Takemura, Chihiro Ueda, Osamu Shimada, Toshio Gomyo, Yukiharu Takeuchi, Toshikazu Okubo, Kazuhiro Baba, Masahiro Aoyagi, Toshio Sudo, Kanji Otsuka

Resonance-aware Methodology for System Level Power Distribution Network Co-design 29
Amirali Shayan, Kevin Bowles, Sorin Dobre, Mikhail Popovich, Xiaoming Chen, Christopher Pan

HIGH SPEED LINKS

Design and Characterization of a 12.8GB/s Low Power Differential Memory System for Mobile Applications 33
Dan Oh, Sam Chang, Chris Madden, Joong-Ho Kim, Ralf Schmitt, Ming Li, Chuck Yuan, Fred Ware, Brian Leibowitz, Yohan Frans, Nhat Nguyen

Low-power Self-equalizing Driver for Silicon Carrier Interconnects with Low Bit Error Rate 37
Peter Gadfort, Paul D. Franzon

Energy-Efficient Performance Budgeting in FEC-Based High-Speed I/O Links 41
Rajan Lakshmi Narasimha, Naresh Shanbhag

DSP-based Multimode Signaling for FEXT Reduction in Multi-Gbps Links 45
Pavle Milosevic, José E. Schutt-Ainé, Naresh R. Shanbhag

SIGNAL INTEGRITY II

Clock Jitter Modeling in Statistical Link Simulation 49
Dan Oh, Sam Chang

Hybrid Equalizer Design for 12.5 Gbps Serial Data Transmission 53
Eakhwan Song, Jeonghyeon Cho, Joungho Kim

The Effects of Time Windowing on the Accuracy of the Short-Pulse Propagation Technique 57
Lionelle F. Wells, Alina Deutsch, Zhen Zhou, Kathleen L. Melde

MACROMODELING

On the Performance of Weighting Schemes for Passivity Enforcement of Delayed Rational Macromodels of Long Interconnects 61
A. Chinea, S. Grivet-Talocia, P. Triverio

Passivity Verification and Enforcement of Delayed Rational Approximations from Scattering Parameter Based Tabulated Data 65
Andrew Charest, Michel Nakhla, Ram Achar

Order Estimation for Time-Domain Vector Fitting 69
Se-Jung Moon, A. C. Cangellaris

Least Squares Convolution: A Method to Improve the Fidelity of Convolution in Transient Circuit Simulation 73
Michael Tsuk, Subramanian Lalgudi

Application of Surrogate Modeling to Generate Compact and PVT-sensitive IBIS Models 77
Ting Zhu, Paul D Franzon

SIGNAL INTEGRITY III

Generalized Leapfrog Scheme for Large-Scale Circuit Simulation 81
Tadatoshi Sekine, Hideki Asai

Modeling of IC Power Supply and I/O Ports from Measurements 85
I. S. Stievano, L. Rigazio, I. A. Maio, A. Girardi, R. Izzi, F. Vitale, T. Lessio

Solver for Current Source Type Driver(s) and Interconnect with Linear or Nonlinear Loads 89
Albert E. Ruehli, Jerry Hayes

SYSTEM AND ON-CHIP ISSUES

Challenges and Solutions for Next Generation Main Memory Systems 93
Joong-Ho Kim, Dan Oh, Ravi Kollipara, John Wilson, Scott Best, Thomas Giovannini, Ian Shaeffer, Michael Ching, Chuck Yuan

Active Circuit to Through Silicon Via (TSV) Noise Coupling 97
Jonghyun Cho, Jongjoo Shim, Eakhwan Song, Jun So Pak, Junho Lee, Hyungdong Lee, Kunwoo Park, Joungho Kim

Experimental Characterization of Metal Fill Placement and Size Impact on Spiral Inductors 101
Vikas S. Shilimkar, Steven G. Gaskill, Andreas Weisshaar

POWER INTEGRITY II

Power Integrity Optimization of 3D Chips Stacked Through TSVs 105
Waqar Ahmad, Li-Rong Zheng, Roshan Weerasekera, Qiang Chen, Awet Yemane Weldezion, Hannu Tenhunen

Extraction of Equivalent Inductance in Package-PCB Hierarchical Power Distribution Network 109
Jingook Kim , Jaemin Kim , Liehui Ren, Jun Fan , Joungho Kim, James L. Drewniak

Effect of System Components on Electrical and Thermal Characteristics for Power Delivery Networks in 3D System Integration 113
Jianyong Xie, Daehyun Chung, Madhavan Swaminathan, Michael McAllister, Alina Deutsch, Lijun Jiang, Barry J Rubin

INNOVATIVE PACKAGING SOLUTIONS

Electrical Modeling of Annular and Co-axial TSVs Considering MOS Capacitance Effects 117
Tapobrata Bandyopadhyay, Ritwik Chatterjee, Daehyun Chung, Madhavan Swaminathan, Rao Tummala

Multi-Bit Fractional Equalization for Multi-Gb/s Inductively Coupled Connectors 121
Evan Erickson, John Wilson, Karthik Chandrasekar, Paul D. Franzon

Hybrid Substrate Integrated Waveguides Developed Using Flexible Substrates 125
Mohammad S. Mahani, Asanee Suntives, Ramesh Abhari

Frequency-Dependent Circuit Models of Carbon Nanotube Networks 129
Mahmoud A. El Sabbagh, Samir M. El-Ghazaly

EM MODELING AND SIMULATION

A Two-Level Optimization Scheme for Bandwidth Optimization of a Microprocessor Vertical Interconnect 133
Arun V. Sathanur, Vikram Jandhyala, Henning Braunisch

Defining a MultiChannel Infrastructure to Enable MNA Formulation of Opto-Electronic Circuits.......137
  P. Gunupudi, T. Smy, J. Klein, J. Jakubczyk

Accelerated Frequency Domain Analysis by Susceptance-Element Based Model Order Reduction of 3D Full-wave Equations.......141
  Narayanan T. V., Sung-Hwan Min, Madhavan Swaminathan

An Unconditionally Stable Time-Domain Finite Element Method of Significantly Reduced Computational Complexity for Large-Scale Simulation of IC and Package Problems.......145
  Houle Gan, Dan Jiao

POSTERS

Design of a SIW-Based Data Communication System Using a SIW Six-Port Receiver.......149
  Abdulhadi E. Abdulhadi, Asanee Suntives, Ramesh Abhari

Model-to-Hardware Correlation of Disk Resonators for Via-Array Modeling in High-Speed PCBs.......153
  Arun Reddy Chada, Young H. Kwark, Xiaoxiong Gu, Jun Fan

A Precise Analytical Eye-diagram Estimation Method for Non-ideal High-Speed Channels.......157
  Jeonghyeon Cho, Eakhwan Song, Jongjoo Shim, Jiseong Kim, Joungho Kim

A Fast Methodology for the Synthesis of Dispersive Multi-Port Equivalent Circuit Model of Multiple Coupled Bond Wires.......161
  J. H. Chung, V. Okhmatovski, A. C. Cangellaris

A New EBG Structure for Low Frequency Power Plane Noise Mitigation.......165
  A. Ciccomancini Scogna, G. Romo

Sensitivity Analysis of Lossy Transmission Lines based on the Passive Method of Characteristics.......169
  Amir Beygi, Anestis Dounavis

Fast Full Wave Analysis of PCB Via Arrays with Model-to-Hardware Correlation.......173
  Xiaoxiong Gu, Boping Wu, Christian Baks, Leung Tsang

Efficient Capacitance Solver for 3D Interconnect Based on Template-Instantiated Basis Functions.......177
  Yu-Chung Hsiao, Tarek El-Moselhy, Luca Daniel

An Ultra Compact Electromagnetic Band Gap Filter for GHz Power Noise Suppression Using LTCC Technology.......181
  Yu-Wen Huang, Ting-Kuang Wang, Tzong-Lin Wu

A New Extraction Method of Characteristic Parameters of a Coupled Transmission Line.......185
  Minwoo Kang, Daehoon Jang, Kwangsik Park, Chilhyeun Gwon, Kwisoo Kim, Jongsik Lim, Kwansun Choi, Dal Ahn

Design and Testing of a High Speed Module Based Memory System.......189
  Ravi Kollipara, Ming Li, Don Mullen, Wendemagegnehu Beyene, Chris Madden, Chuck Yuan, Hideki Kusamitsu, Toshiyasu Ito

The Extraction and Measurement of On-Die Impedance for Power Delivery Analysis.......193
  Xiaoping Liu, Yi-Feng Liu

Bit-Pattern Sensitivity Analysis and Optimal On-Die-Termination for High-Speed Memory Bus Design.......197
  Evelyn Mintarno, Steven Yun Ji

An LCP Package Model for Use in Chip/Package Co-Design of an X-band SiGe Low Noise Amplifier.......201
  Chung Hang John Poh, Tushar K. Thrivikraman, Swapan K. Bhattacharya, Chad E. Patterson, John D. Cressler, John Papapolymerou

On Adding Metalization to Improve Via Performance on PCBs.......205
  Albert E. Ruehli, Xiaoxiong Gu, Mark B. Ritter

Next Generation I/O Power Delivery Design through SIPD Co-Analysis and Comprehensive Platform Validation.......209
  Yee Hung See Tau, Marcus Chan

Perturbation Based Modeling Strategy for Weakly Coupled Interconnects.......213
  Hao Shi

The Impact of Guard Trace with Open Stub on Time-Domain Waveform in High-Speed Digital Circuits.......217
  Po-Wei Chiu, Guang-Hwa Shiue

Numerical Acceleration of Spectral Domain Approach for Shielded Microstrip Lines by Approximating Summation with Corrected Integral.......221
  Sidharath Jain, Jiming Song

A New Isolation Structure for Crosstalk Reduction for Pogo Pins in a Test Socket.......225
  Ruey-Bo Sun, Chang-Yi Wen, Yen-Chih Chang, Ruey-Beei Wu

Chip-Package Codesign with Redistribution Layer.......229
  Mahadevan Suryakumar, Yidnek Mekonnen, Ananda Sarangi

An Approach for Quantifying the Conductor and Dielectric Losses in PCB Transmission Lines 233
Reydezel Torres-Torres, Víctor H. Vega-González

GPGPU-FDTD Method for 2-Dimensional Electromagnetic Field Simulation and Its Estimation 237
Masaki Unno Yuta Inoue, Hideki Asai

Distributed Via Connectivity in High Resolution Package Power Delivery Modeling 241
Omer Vikinski

Design of Shorting Vias in Alternative PCB Planes for Suppressing Ground-Bounce Induced Electromagnetic Emission 245
Kai-Bin Wu, Fu-Sheng Chang, Ruey-Beei Wu

On-Chip Global Clock Distribution Using Directional Rotary Traveling-Wave Oscillator 249
Yulei Zhang, James F. Buckwalter, Chung-Kuan Cheng

Minimizing Crosstalk in High-Speed Differential Buses by Optimizing Power/Ground and Signal Assignment 253
Yang Yi, Yifang Liu, Yaping Zhou, Wiren Dale Becker

Author Index

INTRODUCTION

The 18th Conference on Electrical Performance of Electronic Packaging and Systems (EPEPS) provides a forum for the presentation and discussion of the latest advances in the electrical design, analysis, modeling and characterization of interconnections and packaging structures of electronic systems covering all the application families and frequency ranges namely, digital, RF, microwave and mm-wave applications. One of the key objectives of this meeting is to bring together researchers and practicing engineers from industry, universities, and government laboratories from around the world to address all current and future issues affecting the electrical performance of high speed electronic systems.

The conference is organized into ten sessions of oral presentations and one open forum (poster) session for one-to-one discussions. The meeting begins with a keynote speech by Dr. Greg Taylor of Intel Corporation and an Intel Fellow. This year we continue a recently introduced special feature offering three embedded tutorials and a panel discussion giving deep background on several topical areas of the conference. The three embedded tutorials are entitled "Fundamentals and Advances in Jitter Analysis for High-Speed Links," "Fundamentals of Macromodeling for Signal Integrity Analysis," and "Challenges in High-Frequency Measurement and Simulation Comparison" and are presented by experts in their fields. The panel discussion is on "Grand Challenges in Signal and Power Integrity."

The following ten sessions are dedicated to Signal Integrity I, Power Integrity I, High Speed Links, Signal Integrity II, Macromodeling, Signal Integrity III, System and On-Chip Issues, Power Integrity II, Innovative Packaging Solutions, and EM Modeling and Simulation.

This year EPEPS will continue with the tradition to offer a best student paper award. A total of 25 student papers will be competing for this award. Last year the Texas Instruments best student paper award was presented to Krishna Bharath, Georgia Institute of Technology and the Intel best student paper award was given to Zhi Guo Qian from the University of Illinois at Urbana Champaign, USA. The student paper award adds a special dimension and fulfills another important objective of this meeting. This is the third year that EPEPS offers a best conference paper award. Last year this award was given to A. Chinea, P. Triverio, and S. Grivet-Talocia from the Politecnico di Torino, Italy. Both of this year's awards will be presented at the last day of the conference.

A further objective of the meeting is the encouragement of informal interaction among the participants during the time between sessions and evening forum. In our experience, such interaction has resulted in the most productive industry-academia collaboration and leads to significant advances in the area of electronic packaging.

Last, but not least, the chairs of the 18th EPEPS conference wish to thank the invited speakers, authors, presenters, tutorial instructors, and the members of the Technical Program Committee for their contributions in creating this year's outstanding technical program. The sponsorship of the IEEE Microwave Theory and Techniques Society, and the IEEE Components, Packaging and Manufacturing Society is acknowledged and greatly appreciated. We would like also to acknowledge the continuing effort and dedication of the office of Engineering Professional Development at the University of Arizona in the planning and organization of the EPEPS Meetings. And finally, we want to thank the session chairs and our paper award committee members for their contributions.

Andreas Weisshaar   Dale Becker

2009 Meeting Co-Chairs

Embedded Tutorial

Challenges in High Frequency Measurement and Simulation Comparisons

Dr. Kathleen L. Melde and Terry Burcham

Speakers: Dr. Kathleen L. Melde and Mr. Terry Burcham

High frequency modeling and simulation tools can be used to rapidly develop new packaging structures, yet the models must often be verified by comparison to broadband measurements. This tutorial will discuss some of the key issues involved in an accurate comparison of electromagnetic simulations with measurements. The important factors to consider include which measurement calibration method to use, how to fabricate the necessary calibration artifacts, and the impact that the calibration method may have on the measurement results. The increase in clock speeds and the push to higher frequencies of operation mean that many of the measurement approaches typically used at lower frequencies must be re-evaluated. The verification of accurate high frequency design and analysis tools mean that the measured results on the test structures must also be very accurate at increasingly higher frequencies.

Biography:

Kathleen Melde (Senior member, IEEE) is an Associate Professor in the Electrical and Computer Engineering Department at the University of Arizona. She has more than 20 years industrial and academic experience in the design of antennas and high frequency electrical packaging. She has over 75 publications and 4 US patents.

Terry Burcham entered the microwave industry in 1968, responsible for amplifier R&D until 1974. From 1974 to1982, he was co-founder and vice president of Aercom Industries, producer of GaAs FET amplifiers and microwave components. Burcham joined Tektronix in 1982 as marketing manager for the Frequency Domain Instruments Division. At Cascade Microtech, from 1987 to 1997, he held positions of marketing manager and sales vice president. In 2002, Burcham came out of retirement to be worldwide applications development manager.

Embedded Tutorial

Fundamentals and Advances in Jitter Analysis of High-Speed Links

Speaker: Dr. Lei Luo, Rambus

Abstract: For high speed links, it is essential to meet the voltage and timing margin to achieve the Bit-Error-Rate (BER) requirement. With ever increasing data rate and lower BER requirement, meeting the timing margin, or limiting the timing jitter, becomes more and more critical. This tutorial will analyze the major jitter components in high speed links; explain how these jitter components are generated from circuitry and channel; demonstrate how the jitter components are measured in the lab; illustrate methods to reduce the jitter components; and finally review several system level clock architectures and analyze their jitter performance."

Biography: Lei Luo is a senior circuit design engineer with Rambus working on high speed serial link and memory bus design. He received the Ph.D degree from NC State University in electrical and computer engineering in 2005. His research interests include mixed-signal circuit design for high speed chip-to-chip communications, related signal integrity analysis and signal processing techniques for communications.

Embedded Tutorial

Fundamentals of Macromodeling for Signal Integrity Analysis

Dr. Piero Triverio and Dr. Michel Nakhla

Speaker: Dr. Piero Triverio

Abstract: Computer aided design tools are an integral part of the design flow of modern electronic systems. However, lack of physical consistency of the models that are used in the simulation and optimization cycle can cause the design tools to slow down unpredictably or even fail.

This tutorial provides a comprehensive analysis of this important topic, with emphasis on relevant properties of the macromodel and associated data such as causality, stability and passivity and their interdependency. These properties are crucial for the transient simulation of high-speed modules. Several design scenarios will be examined to show the effects of consistency violations on real design tasks, while outlining suitable best-practice rules to avoid them. Finally, an overview of the best numerical techniques for checking and enforcing physical consistency during measurement, modeling and simulation tasks will be presented.

Biography:

Piero Triverio received the Ph.D. degree in 2009 in Electronics Engineering from Politecnico di Torino, where He is currently a research assistant with the Electromagnetic Compatibility group. His research interests include macromodeling, numerical algorithms for Signal Integrity and Electromagnetic Compatibility analyses, and parallel computing. He has received several awards, including the 2007 Best Paper Award of the IEEE Transactions on Advanced Packaging, the EPEP 2008 Best Paper Award and the EPEP 2006 Best Student Paper Award.

Michel Nakhla *(Fellow IEEE)* is a Chancellor's Professor and founder of the high-speed CAD research group at Carleton University. He has more than 30 years industrial and academic experience in the design automation area. He has authored more than 280 technical papers in this subject.

Improving Single-Ended Channel Performance
With a Dynamic Reference Voltage Scheme

Tingdong Zhou, Daniel M. Dreps, and Wiren D. Becker
IBM Corporation
11410 Burnet Road, Austin, 78758
Phone: 512-2869400
Email: tingdong@us.ibm.com, drepsdm@us.ibm.com, and wbecker@us.ibm.com

Abstract

For high-speed signaling, differential channels are now commonly used because of the improved noise immunity and receiver sensitivity that can be achieved compared to the single-ended channels that had been widely used in electronic systems until recently. However, single-ended channels have the distinct advantage of only needing one signal trace per signal, which in a well-controlled impedance, low-noise environment can provide equivalent bandwidth per pin at a lower silicon area and lower power demand than differential signals.

To provide an effective low-noise environment for the receiver threshold tolerance, a power noise compensation scheme consisting of a dynamic forwarded reference voltage, is proposed. This improves the overall performance of the Elastic Interface (EI) bus [1], a single-ended high speed signaling technology used in IBM systems. We explore the range of voltage differences that could be observed at the transmit and receive circuits and how those differences can be mitigated to enhance the performance of the system. Simulations for channels with varying voltage characteristics confirm the anticipated performance improvement.

Introduction

As integrated circuit (IC) technology advances, the IC clock frequency and chip power density increase. In addition, the voltage levels decrease. Power supply noise and DC power supply variation become increasingly more important to the performance of an IC. For a processor chip, the noise not only impacts the core/nest performance [2], but also affects the performance of the input/output (I/O) buffer circuit and link.

From a circuit point of view, power supply noise is caused by resistance and inductance of a power delivery network (PDN) and circuit current demand variations. In addition, we also have the regulation tolerance of the active voltage regulation to take into consideration. The PDN resistance is a spatial resistance distribution, which produces an IR drop [3] variation when a chip has a change in load current. The loop inductance of a PDN causes high, middle, and low frequency noises [4, 5] with a change in the load current through the PDN.

Consider the case of two ICs on a first-level package whose transmit and receive circuits share a voltage domain on a printed circuit board with the voltage regulator setpoint between the ICs as represented in Fig. 1. We constructed a test case using created a model of the package and printed circuit board and applied the currents as shown in Table 1.

Table 1 Voltage overview of a common voltage domain for the two IC system .

	domain current for the chips (A)	DC drop/gradient and AC noise component						DC gradient + AC noise		setpoint		onchip IR	DC @ CKT		voltage tolerance @ CKT	
		module (IRmax +IRmin)/2 (mv)	module (IRmax - IRmin)/2 (mv)	Card (IRmax +IRmin)/2 (mv)	Card (IRmax - IRmin)/2 (mv)	middle frequency noise (mv)	SSN (peak to peak) (mv)	voltage droop (mv)	voltage overshoot (mv)	set point level (V)	regulation tolerance (+- V)	average onchip IR drop (mv)	when 100% chip power (V)	when 50% GND and domain power (V)	CKT MIN (V)	CKT MAX (V)
IC1	72.000	67.600	7.100	68.500	8.000	36.000	4.500	48.640	25.825	1.200	0.012	18.000	1.060	1.130	1.015	1.151
IC2	76.700	34.000	11.000	54.000	5.000	13.000	4.500	27.650	14.713	1.200	0.012	15.000	1.106	1.153	1.077	1.169

We show the voltage drops due to the resistance and inductance of the PDN under the assumption that the transmit circuits would be required to run at speed when the chips are at full power and in a reduced power mode drawing one-half the current and the maximum current delta that produces the inductive mid-frequency noise is 30% of maximum current. A common voltage domain from a single voltage regulator is used in this example. The IC to IC interconnect is a high speed bus using the Elastic Interface (EI) [1] technology that has been used for several generations of IBM servers. A reference voltage (V_{ref}) is compared with the data voltage level at the receiver. Note that the loop IR drop from the voltage regulator set point to IC1 is higher than the loop IR drop from the set point to the IC2. We, therefore, see a DC offset between the driver, V_D, and the receiver, V_R, at ICs at nominal power. As more clock gating or power saving is designed into the chips for power savings, the voltage variations at the ICs will be more independent because of the independent stimulus during system operation. Previous implementations of EI used a forwarded clock to derive V_{ref}, but this has the shortcoming of being able to compensate for the transient on V_D but not V_R. Similarly, using a local receiver voltage to derive V_{ref} would compensate for the transient on V_R but not on V_D. In either case, the signal voltages at the receiver as shown in an eye diagram will shift relative to V_{ref} as V_D and V_R. The undesirable result is the frequency of operation of the channel is reduced as compared with an ideally generated V_{ref}.

In this paper, we will investigate a power-noise-aware EI I/O circuit design using a dynamic forwarded reference voltage scheme. This paper is organized as follows. Section I, will introduce the EI topology. The dynamic forward reference voltage scheme follows in Section II. Simulation results and discussions are contained in Section III.

I. Elastic interface topology

The EI is a high-speed source-synchronous interface used to transfer address, controls, and data between CPUs, L2 caches, memory subsystems, switches, and I/O hubs. An EI uses single-ended data lines. For each data link, there is a driver and a receiver connected to the channel at the ends. Figure 2 and 3 show the topology and the DC circuit of an EI

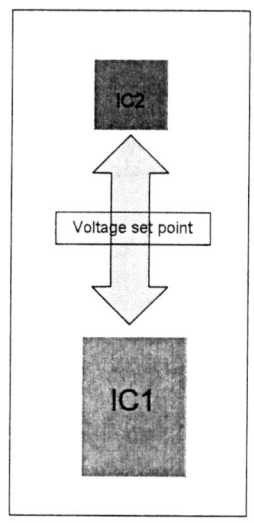

Fig 1, A high performance computer system card placement plane.

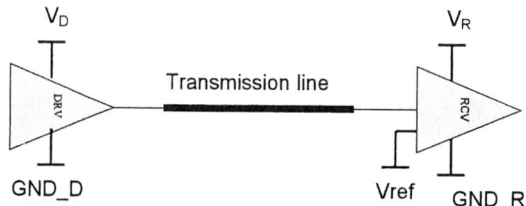

Fig 2, The EI3 topology

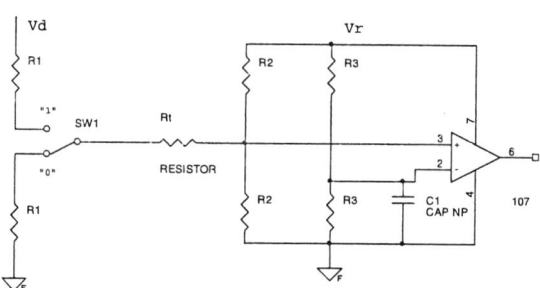

Fig 3, DC circuit of the EI3 interface.

interface. The termination depends on the specifics of the implementation including the frequency of operation, the length of the interconnect and the nature of the discontinuities. A common element of operation among all the generations of EI is that the transmitted signal at the receiver end is compared to a reference voltage to determine if the data bit is "1" or "0". Therefore, the level of the V_{ref} is critical. It directly impacts how fast the interface can operate. Ideally, the V_{ref} should be set at the center of the eye diagrams voltage so that we have a balanced eye opening for "1" and "0" bits. Setting the V_{ref} to a value that is too high or too low would reduce the performance of the interface because a smaller "1" or "0" eye would result. In Fig. 3, the V_{ref} is derived from receiver local power supply with a balanced resistor voltage divider. However, the voltage supply at the driver and receiver ends can fluctuate due to power noise. EI power supply is usually common for both drivers and receivers and can easily be separated by more than 30 inches as illustrated in Fig 1.

II. Dynamic forwarded reference voltage scheme

In order to derive the V_{ref} so that it is always near the center of the eye diagram, both the power supply transients at V_D and V_R need to be factored into to the voltage level of V_{ref}. A DC circuit analysis for the circuit in Fig. 3 for "1" and "0" levels (DC analysis only) at the receiver is given in Equations (1) and (2).

$$V^{high} = \frac{V_D R_2 + V_R R_C + V_{GND}^R R_C}{2R_C + R_2} \qquad (1) \qquad V^{Low} = \frac{V_{GND}^D R_2 + V_R R_C + V_{GND}^R R_C}{2R_C + R_2} \qquad (2)$$

Where, V_D is the voltage at the driver side of the power supply relative to power supply GND. V_R is the voltage at the receiver side power supply relative to power supply GND and V_{GND}^D is the voltage at the DRV side GND relative to power supply GND. V_{GND}^R is voltage at the RCV side GND relative to power supply GND. R_C is the channel resistance, including the output resistance of the driver (about 40Ω or 15 Ω if pre-compensation is on) and path resistance from the driver output to the receiver input (last metal, C4, module, card, module, C4, last metal, range from 5 Ω to 10 Ω). R_2 is the pull up or pull down termination resistor at the receiver input (assuming balanced termination). In order to track the level shifting of the eye diagram, the V_{ref} need to be shifted by the same amount to compensate the eye diagram shifting.

$$V_{ref} = \frac{V^{high} + V^{low}}{2} = \frac{V_D + V_{GND}^D}{2} \frac{R_2}{2R_C + R_2} + \frac{V_R + V_{GND}^R}{2} \frac{2R_C}{2R_C + R_2} \qquad (3)$$

Ideally the V_{ref} should be set at the level as in Equation (3). The level can be derived using the circuit in Fig 4, where R7/R6 = 2Rc/R2. The differential clock nets are part of the existing EI interface. The differential common mode voltage, i.e., $(V_D + V_{GND}^D)/2$ can be derived with a voltage divider. C5 is part of the high frequency pass filter to clean

978-1-4244-4447-2/09 $25.00 © 2009 IEEE

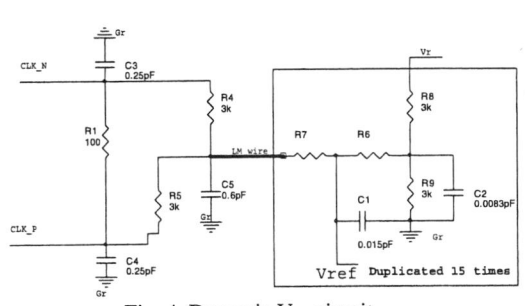

Fig. 4, Dynamic V_{ref} circuit

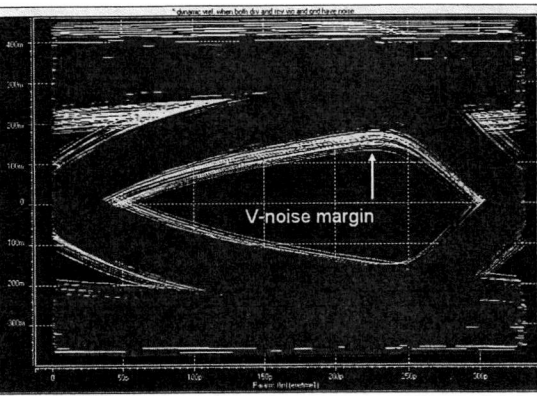

Fig. 5, Typical eye-diagram of an off-module EI net using dynamic V_{ref} and receiver local V_{ref}. (3Gbps, RLGC line, 6ns filter, Random data)

out the high-frequency transient due to skewed clock nets. The C3 and C4 are input parasitic capacitors of the differential clock receivers. Resistor R1 is the differential termination of the differential clocks. The C1 is the receiver parasitic capacitance of the V_{ref} pin. Resistors R6 and R7 need to be much higher than R4 and R5 or R8 and R9 so that the current flowing through R6 and R7 is small and does not affect the voltage level on both ends. Also, resistors R4, R5, R6, R7, R8, and R9 can be designed as a mix of fixed resistors and variable resistors to allow the circuit to be tuned during power on and compensate for variations due to manufacturing tolerances. R7 and R6 can be implemented as a series of resistors. The junction of these resistors can be programmed to be wired to the V_{ref} pin.

The dynamic V_{ref} circuit above can track DC to middle frequency transients of the V_D as well as the V_R so that the V_{ref} level is always near the center of the eye diagram. That will allow a balanced eye opening for "1" and "0" bits. The impact of DC offset to middle frequency power transients at both driver and receiver ends can be reduced and therefore increase the speed of the EI interface. A large amount of on-chip capacitance will remove the first droop in the power noise and the second and third droop power noises would become the dominant power transient harmonics and are lower in frequency. The differential common mode voltage derived from forwarded clock can track the V_D transient even after filtering of the high frequency noise caused by crosstalk or skewed clocked nets.

III. Simulation

The EI data channel of different bus types or representation forms are simulated using a spice simulator. The EI driver is represented with the Mπlog [6] model that can handle the impact of power supply noise to driver output. We considered two buses: a bus including 25 inches of printed circuit board wiring between IC1 and IC2 and the potential case of a multi-chip module where two ICs share the same package. This analysis is performed with the interconnections operating at 3 Gigabits per second. The channels are represented with a tabular RLGC transmission line model and S-parameters to model all the components of the channel. Fig. 5 shows a typical eye-diagram comparison when using a dynamic V_{ref} versus a local V_{ref}. Both the horizontal and vertical eye openings are improved when using the dynamic V_{ref}. Table 2 summarizes the improvement of the dynamic V_{ref} scheme comparing to the local V_{ref} scheme, where the data are either a clock pattern of a repeating "01" or a Pseudo-Random pattern. The time

Table 2, Eye summary for 3Gbps simulations with +-100mv noise at drv and rcv ends

Sim ID	Eye using dynamic Vref (ps)	Eye using local Vref (ps)	Vertical eye margin dynamic Vref (mv)	Vertical eye margin local Vref (mv)	Eye improvement (ps)	Eye improvement (%)	Vertical eye improvement (mv)	Vertical eye improvement (%)
W bus, RLGC card line, 01 data (1.8ns filter)	299.4	260.7	196	147	38.7	14.8%	49	33%
W bus, RLGC card line, random data (1.8ns filter)	258.8	229.7	152	133	29.1	12.7%	19	14.3%
W bus, RLGC card line, random data (6ns filter)	262.8	223.8	157	133	39	17.4%	24	18%
W bus, s-parameter, 01 data (1.8ns filter)	305	245.7	197	138	59.3	24.1%	59	42.7%
W bus, s-parameter, random data (1.8ns filter)	223.6	175.9	132	79.2	47.7	27.1%	52.8	66.7%
W bus, s-parameter, random data (6ns filter)	225	180.4	133	79.2	44.6	24.7%	53.8	68%
On module bus, RLGC line, 01 data (1.8ns filter)	286	268.7	212.6	186.3	17.3	6.3%	26.3	14.1%
On module bus, RLGC line, random data (1.8ns filter)	297	291	204	193	6	2.1%	11	5.7%
On module bus, RLGC line, random data (6ns filter)	297	292	212	193	5	1.7%	19	9.8%

978-1-4244-4447-2/09 $25.00 © 2009 IEEE

constant of the front end filter is 1.8ns or 6ns, the power noise is assumed to be ±100mv sinusoid with different frequencies used for driver and receiver ends.

For the short on-module nets, the pre-compensation of the EI driver is not turned on. The driver has close to constant output impedance. For long off-module nets, pre-compensation of the EI is turned on to over-drive the channel when the bit pattern is changed from 0 to 1 or 1 to 0. In that case, the output impedance will vary from bit to bit. This will have impact on the resistance required for R6 and R7 in Fig. 4. Therefore the fully designed circuit will need to have a variable resistance for R6 and R7 that is configurable in the system. We chose to implement a series of resistors where a tap point can be selected with switch to be used as the V_{ref} input. This enables the tuning of the V_{ref} for each instantiation of the interface and allows one to characterize the effectiveness in the lab just by selecting the tap point. Fig. 6 shows the eye improvement in an off-module bus simulation environment (pre-compensation is turned on) when R6 and R7 are replaced with five 3Kohm and nine 5Kohm resistors.

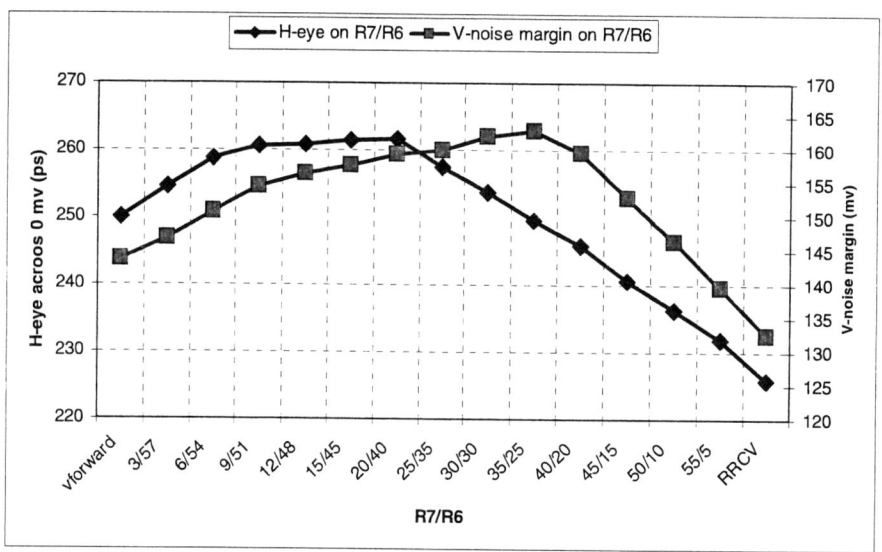

Fig. 6, Sensitivity of eye on R7/R6 for a off-module net

Conclusions

A dynamic forward reference voltage scheme is introduced to the EI I/O design. The noise-aware I/O design compensates for the DC and AC voltage differences between driver power supply and receiver power supply and therefore significantly improves the bus performance in a noisy environment. Applying this to a 3 Gigabit per second interface (333 ps unit interval), we can effectively open the eye by about 60 ps in the horizontal opening and 60 mV in the vertical opening for a representative application in a server. Using this technology, we are able to have an option of using a single-ended interface which uses less power and less silicon area than a differential interface of the same effective bandwidth per pin.

References

1. D. M. Berger, J. Y. Chen, F. D. Ferraiolo, J. A. Magee, and G. A. Van Huben, "High-speed source-synchronous interface for the IBM system Z9 processor," *IBM J. Res. & Dev.* 51, No. ½, pp. 53-64, 2007.
2. M. Saint-Laurent, M. Swaminathan, "Impact of power supply noise on timing in high performance microprocessors," in *Proc. IEEE 11th EPEP Topical Meeting*, pp.261-264, Monterey CA, 2002.
3. A. Huber, T. Zhou, W. D. Becker, R. Weekly, and E. Klink, "Power distribution analysis for IBM eServer system integration optimization," in *Proc. IEEE 13th EPE Topical Meeting*, pp. 189-192, Portland, OR, Oct. 2004.
4. L. D. Smith *et al.*, "Power distribution system design methodology and capacitor selection for modern CMOS technology," *IEEE Trans. Adv. Packag.*, vol. 22, pp. 284–290, Aug. 1999.
5. K. L. Wong, T. Rahal-Arabi, M. Ma, and G. Tayler, "Enhancing microprocessor immunity to power supply noise with clock-data compensation," *IEEE Journal of Solid-State Circuits*, vol. 41, no. 4, Apr. 2006.
6. I. S. Stievano et al., "Mπlog, Macromodeling via parametric identification of logic gates", *IEEE Trans. Adv. Packag.*, vol. 27, no. 1, pp. 15–23, Feb. 2004.

IBM is a trademark of International Business Machines Corp..

Designing a ZXnoise Pseudo-Differential Link

Frédéric Broydé
Excem,
12, chemin des Hauts de Clairefontaine, 78580 Maule, France
e-mail: fredbroyde@eurexcem.com
Tel.: + 33 1 34 75 13 65   Fax.: + 33 1 34 75 13 66

Bernard Démoulin
Université des Sciences et Technologies de Lille,
Bâtiment P3, 59655 Villeneuve d'Ascq Cedex, France
e-mail: bernard.demoulin@univ-lille1.fr
Tel.: + 33 3 20 43 48 56   Fax.: + 33 3 20 43 65 23

Abstract — An interconnection-ground structure used according to the ZXnoise method may be modeled using a $(n+2)$-conductor multiconductor transmission line (MTL) model. However, a $(n+1)$-conductor MTL model is used to design the link. We validate this design procedure.

I. Introduction

Pseudo-differential links (PDLs) providing several channels may be used to convey analog or digital signals between a transmitting circuit (TX circuit) and a receiving circuit (RX circuit). We shall refer to crosstalk between the different channels as *internal crosstalk* and to crosstalk with other circuits as *external crosstalk*. A multichannel PDL may provide a reduced external crosstalk compared to multiple single-ended links, using fewer conductors than multiple differential links. A low external crosstalk makes low-swing transmission possible, and low-swing is the key to high-speed transmission. A low external crosstalk is also necessary to implement multilevel digital signaling, which, for a given bit rate, uses a reduced bandwidth compared to binary signaling.

A wide variety of pseudo-differential transmission schemes use an interconnection having $n \geq 2$ transmission conductors (TCs) and one common conductor distinct from the reference conductor (ground):

■ some schemes use no termination circuit, so that their bandwidth is limited by reflections, for a given length;

■ some schemes use type 1 termination circuits, for which an impedance matrix with respect to ground is defined, this impedance matrix being diagonal, such termination circuits creating an undesirable coupling with other circuits using ground as return path;

■ some schemes use type 2 termination circuits or type 3 termination circuits, such termination circuits being characterized by an impedance matrix with respect to the return conductor, denoted by $\mathbf{Z}_{RL}$, $\mathbf{Z}_{RL}$ being a matrix of size $n \times n$, $\mathbf{Z}_{RL}$ being a diagonal matrix in the case of a type 2 termination circuit, or a non-diagonal matrix in the case of a type 3 termination circuit.

The combination of floating termination circuits (i.e., type 2 termination circuits or type 3 termination circuits) with appropriate interconnection-ground structures is referred to as *ZXnoise method* [1] [2]. In the ZXnoise method, the common conductor is usually referred to as *return conductor* because it is used as a return path for the return current produced by the current flowing in the TCs. In Section II, we present new theoretical results on propagation in an interconnection such as the one shown in Fig. 1, used for implementing the ZXnoise method. A design procedure is presented in Section III, and Section IV provides a design example.

II. Propagation in the Interconnection

Each interconnection-ground structure shown in Fig. 1 may be modeled as a $(n+2)$-conductor multiconductor transmission line (MTL), this MTL using natural voltages referenced to ground and natural currents as variables. For such a model, it is necessary to consider, at a given abscissa z along the interconnection:

■ for any integer α such that $1 \leq \alpha \leq n$, the natural current i_α ;

■ the current flowing on the RC, denoted by i_{n+1} ;

■ for any integer α such that $1 \leq \alpha \leq n$, the voltage between the TC number α and the reference conductor, denoted by $v_{G\alpha}$;

■ the voltage between the RC and the reference conductor, denoted by $v_{G\,n+1}$.

We define the column-vector $\mathbf{I}_G$ of the currents $i_1,..., i_{n+1}$ and the column-vector $\mathbf{V}_G$ of the natural voltages referenced to ground $v_{G\,1},..., v_{G\,n+1}$. For the $(n + 2)$-conductor MTL, the telegrapher's equations are:

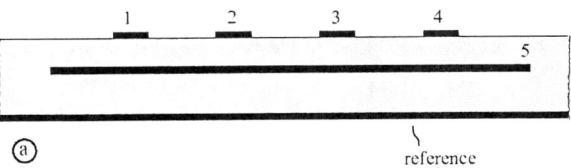

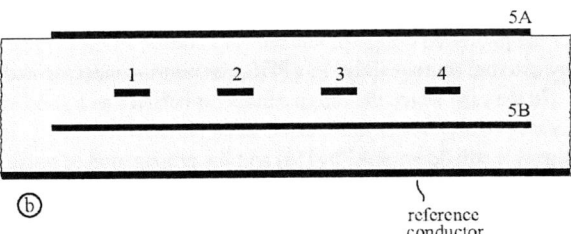

Fig. 1. Two possible cross-sections for an interconnection-ground structure used with floating termination circuits, where 1 to 4 are the TCs, where 5 is the return conductor in *a* and where the return conductor in *b* is made of 5A and 5B.

$$\begin{cases} \dfrac{d\mathbf{V}_G}{dz} = -\mathbf{Z}_G\,\mathbf{I}_G \\[2mm] \dfrac{d\mathbf{I}_G}{dz} = -\mathbf{Y}_G\,\mathbf{V}_G \end{cases} \quad (1)$$

978-1-4244-4447-2/09 $25.00 © 2009 IEEE

where $\mathbf{Z}_G$ and $\mathbf{Y}_G$ are the per-unit-length (p.u.l.) impedance matrix with respect to ground, and the p.u.l. admittance matrix with respect to ground, respectively. $\mathbf{Z}_G$ and $\mathbf{Y}_G$ are symmetric matrices of size $(n+1) \times (n+1)$.

These equations and appropriate boundary conditions provide a full description of all propagation phenomena in the interconnection-ground structure, including signal propagation, echo and internal crosstalk. They are also suitable for describing external crosstalk phenomena which do not involve a capacitive or inductive coupling with the conductors of the interconnection, for instance the very important problem of simultaneous switching output (SSO) noise produced within IC packages when the interconnection is built in the substrate of a multi-chip module (MCM) or in a printed circuit board (PCB). However these equations are not very convenient for the investigation of signal propagation since a receiving circuit (RX circuit) used in a PDL senses the natural voltages referenced to the return conductor, denoted by $v_{R1},..., v_{Rn}$, where $v_{R\alpha}$ denotes the voltage between the TC number α and the return conductor. We may define the column-vector $\mathbf{I}_R$ of the natural currents $i_1,..., i_n$ and the column-vector $\mathbf{V}_R$ of the natural voltages referenced to the return conductor $v_{R1},..., v_{Rn}$, such that $v_{R\alpha} = v_{G\alpha} - v_{Gn+1}$ for $1 \le \alpha \le n$.

The equations involving $d\mathbf{V}_G/dz$ and $d\mathbf{V}_R/dz$ have been compared in the case $n = 1$, in a problem where the interconnection was a coaxial cable [3]. In order to write the telegrapher's equations applicable to $\mathbf{I}_R$ and $\mathbf{V}_R$ in the case $n \ge 2$ relevant to PDLs, we must use 2 additional variables and 6 new p.u.l. quantities. The additional variables are the common-mode current $i_{MC} = i_1 + ... + i_{n+1}$ and the common-mode voltage $v_{MC} = v_{Gn+1}$. The new p.u.l. quantities are [4]:

■ the p.u.l. impedance matrix with respect to the return conductor, a symmetric matrix of size $n \times n$ denoted by $\mathbf{Z}_R$, the p.u.l. transfer impedance vector, of size $n \times 1$ and denoted by $\mathbf{Z}_E$, and the external impedance, denoted by Z_{EE}, defined by

$$Z_{G\alpha\beta} = Z_{R\alpha\beta} + Z_{EE} - Z_{E\alpha} - Z_{E\beta} \quad \text{and} \quad Z_{Gn+1\alpha} = Z_{G\alpha n+1} = Z_{EE} - Z_{E\alpha} \quad \text{and} \quad Z_{Gn+1n+1} = Z_{EE} \qquad (2)$$

■ the p.u.l. admittance matrix with respect to the return conductor, a symmetric matrix of size $n \times n$ denoted by $\mathbf{Y}_R$, the p.u.l. transfer admittance vector, of size $n \times 1$ and denoted by $\mathbf{Y}_E$, and the external admittance, denoted by Y_{EE}, defined by

$$Y_{G\alpha\beta} = Y_{R\alpha\beta} \quad \text{and} \quad Y_{Gn+1\alpha} = Y_{G\alpha n+1} = Y_{E\alpha} - \sum_{\beta=1}^{n} Y_{R\alpha\beta} \quad \text{and} \quad Y_{Gn+1n+1} = Y_{EE} + \sum_{\alpha=1}^{n}\sum_{\beta=1}^{n} Y_{R\alpha\beta} - 2\sum_{\alpha=1}^{n} Y_{E\alpha} \qquad (3)$$

where α and β are integers such that $1 \le \alpha \le n$ and $1 \le \beta \le n$ and where indices have been used to denote the entries of matrices and vectors. Using $^t\mathbf{X}$ to denote the transpose of $\mathbf{X}$, it can been shown that (1) is exactly equivalent to

$$\begin{cases} \dfrac{d\mathbf{V}_R}{dz} = -\mathbf{Z}_R \mathbf{I}_R + i_{MC}\mathbf{Z}_E \\[2mm] \dfrac{d\mathbf{I}_R}{dz} = -\mathbf{Y}_R \mathbf{V}_R - v_{MC}\mathbf{Y}_E \end{cases} \qquad (4a) \qquad \text{and} \qquad \begin{cases} \dfrac{dv_{MC}}{dz} = {}^t\mathbf{Z}_E \mathbf{I}_R - i_{MC} Z_{EE} \\[2mm] \dfrac{di_{MC}}{dz} = -{}^t\mathbf{Y}_E \mathbf{V}_R - v_{MC} Y_{EE} \end{cases} \qquad (4b)$$

The equations (4a) and (4b) are not based on any approximation and are therefore valid for any pseudo-differential transmission scheme. Also, they clarify and improve the existing theory for the ZXnoise method.

In the case where the return conductor behaves as an ideal screen, the terms containing the p.u.l. transfer impedance vector $\mathbf{Z}_E$ or the p.u.l. transfer admittance vector $\mathbf{Y}_E$ vanish in (4a) and (4b), so that

$$\begin{cases} \dfrac{d\mathbf{V}_R}{dz} = -\mathbf{Z}_R \mathbf{I}_R \\[2mm] \dfrac{d\mathbf{I}_R}{dz} = -\mathbf{Y}_R \mathbf{V}_R \end{cases} \qquad (5a) \qquad \text{and} \qquad \begin{cases} \dfrac{dv_{MC}}{dz} = -i_{MC} Z_{EE} \\[2mm] \dfrac{di_{MC}}{dz} = -v_{MC} Y_{EE} \end{cases} \qquad (5b)$$

Thus, the reference conductor being not regarded as a part of the interconnection, (5a) means that the interconnection may be modeled as a $(n + 1)$-conductor MTL using the natural voltages referenced to the return conductor and the natural currents $i_1,..., i_n$ as natural electrical variables. According to (5b), the propagation in the return-conductor-and-ground circuit may be modeled as a 2-conductor MTL using the common-mode voltage and the common-mode current as natural electrical variables. The equation (5a) is the basis of the published design procedure for ZXnoise PDLs [1] [2]. The equation (5b) describes only the propagation of noise produced by external sources since, in a PDL, the return-conductor-and-ground circuit is not used for signals.

In the case where the return conductor behaves as a good screen, the terms containing $\mathbf{Z}_E$ or $\mathbf{Y}_E$ in (4a) and (4b) may be regarded as small disturbances and treated using perturbation theory. In the framework of first order perturbation theory, the propagation of signal is still determined by (5a) and the propagation of noise produced by external sources by (5b), these results being used in (4a) to obtain the external crosstalk received by the PDL and in (4b) to obtain the external crosstalk produced by the PDL.

If the electric and magnetic fields of the signals are mainly confined between the TCs and the return conductor, we can say that the return conductor behaves as a good screen for signals. Thus, (5a) may be used to design the PDL so as to obtain a suitable propagation of signals. In particular, a termination circuit designed in this manner is a floating termination circuit, since it creates a boundary condition for the natural voltages referenced to the return conductor and the natural currents $i_1,..., i_n$. At a later stage, an analysis of the PDL may be directly based on (1), or on (4a) and (4b).

978-1-4244-4447-2/09 $25.00 © 2009 IEEE

III. DESIGN FOR THE ZXNOISE METHOD

Based on (5a) alone, we shall use $\mathbf{T}_R$ and $\mathbf{S}_R$ to denote two regular matrices such that [5]:

$$\begin{cases} \mathbf{T}_R^{-1}\mathbf{Y}_R\mathbf{Z}_R\mathbf{T}_R = \Gamma_R^2 \\ \mathbf{S}_R^{-1}\mathbf{Z}_R\mathbf{Y}_R\mathbf{S}_R = \Gamma_R^2 \end{cases} \qquad \text{where} \qquad \Gamma_R = \mathrm{diag}_n\big(\gamma_{R1},\ldots,\gamma_{Rn}\big) \qquad (6)$$

is the diagonal matrix of order n of the propagation constants for the different propagation modes of the $(n+1)$-conductor MTL, for waves propagating toward the far-end (that is to say in the direction of increasing z). The squares of the propagation constants are the eigenvalues of $\mathbf{Y}_R\mathbf{Z}_R$, which are also the eigenvalues of $\mathbf{Z}_R\mathbf{Y}_R$.

Any matrices $\mathbf{T}_R$ and $\mathbf{S}_R$ satisfying (6) define a "modal transform" for the natural currents and for the natural voltages referenced to the return conductor, and the results of this transform are called modal currents and modal voltages, respectively. If we use $\mathbf{I}_{RM}$ to denote the column-vector of the n modal currents $i_{RM\,1},\ldots,i_{RM\,n}$ and $\mathbf{V}_{RM}$ to denote the column-vector of the n modal voltages $v_{RM\,1},\ldots,v_{RM\,n}$, we get:

$$\begin{cases} \mathbf{V}_R = \mathbf{S}_R\mathbf{V}_{RM} \\ \mathbf{I}_R = \mathbf{T}_R\mathbf{I}_{RM} \end{cases} \qquad (7)$$

The characteristic impedance matrix of the $(n+1)$-conductor MTL, denoted by $\mathbf{Z}_{RC}$, and the matrix of the voltage reflection coefficients of a floating termination circuit with respect to the return conductor, denoted by $\mathbf{P}_R$, are given by:

$$\begin{aligned} \mathbf{Z}_{RC} &= \mathbf{S}_R\Gamma_R^{-1}\mathbf{S}_R^{-1}\mathbf{Z}_R = \mathbf{S}_R\Gamma_R\mathbf{S}_R^{-1}\mathbf{Y}_R^{-1} \\ &= \mathbf{Y}_R^{-1}\mathbf{T}_R\Gamma_R\mathbf{T}_R^{-1} = \mathbf{Z}_R\mathbf{T}_R\Gamma_R^{-1}\mathbf{T}_R^{-1} \end{aligned} \qquad (8a) \qquad \text{and} \qquad \mathbf{P}_R = \big(\mathbf{Z}_{RL}-\mathbf{Z}_{RC}\big)\big(\mathbf{Z}_{RL}+\mathbf{Z}_{RC}\big)^{-1} \qquad (8b)$$

where $\mathbf{Z}_{RL}$ is the impedance matrix of the floating termination circuit with respect to the return conductor. $\mathbf{Z}_{RC}$ is a matrix of size $n \times n$ referred to as the "characteristic impedance matrix with respect to the return conductor".

Let us consider a given interconnection-ground structure meeting our requirements regarding external crosstalk, in which the propagation of signals may be modeled using (5a). If the desired reduction of reflection, expressed as a maximum value for a norm of $\mathbf{P}_R$, can be obtained with a diagonal matrix $\mathbf{Z}_{RL}$, and if the resulting internal crosstalk is acceptable, then a PDL implementing a type 2 termination circuit can be designed. Such a PDL uses one TC for each channel. In the opposite case, since $\mathbf{P}_R = \mathbf{0}$ for $\mathbf{Z}_{RL} = \mathbf{Z}_{RC}$, we may design a type 3 termination circuit producing very low echo. It will also provide a low internal crosstalk if we use one modal current or one modal voltage defined by (7) for each channel, so as to transpose to PDLs the ZXtalk method [5] or the special ZXtalk method for completely degenerate interconnection (CDI) [6]. In general, analog and/or digital processing are needed in the TX circuits and in the RX circuits, to perform linear combinations defined by $\mathbf{S}_R$ or $\mathbf{T}_R$. However, in the special case where the propagation constants of the $(n+1)$-conductor MTL model may be regarded as equal, linear combinations are not needed in the TX circuits and/or in the RX circuits, since $\mathbf{S}_R$ or $\mathbf{T}_R$ may be chosen equal to the identity matrix. Nearly equal propagation constants may for instance be obtained using the structure shown in Fig. 1b.

IV. EXAMPLE

We now discuss the design of a unidirectional PDL providing $m = 4$ channels, between two chips of a MCM. The PDL uses the ZXnoise method, according to the voltage-driven common conductor architecture. We assume that the PDL is equivalent to the schematic diagram shown in Fig. 2. At the near-end, the return conductor is connected to a node intended to present a fixed voltage e_{CC} and a low impedance Z_{CC} with respect to a ground plane of the substrate of the MCM. The TX circuit, being connected to this node, produces accurate open-circuit output voltages with respect to it. This node is also subject to a noise voltage e_N caused by other circuits within the chip of the TX circuit (e.g., SSO noise). At the far-end, the return conductor is connected to the termination circuit and to a damping resistor R_D. We assume an ideal RX circuit that is undisturbed by noise source inside its chip.

The PDL uses a 30-mm-long interconnection having the cross-section shown in Fig. 1a. The parameters defined in Fig. 2 have the following nominal values: conductor spacing $s \approx 1.2\,w$, distance between conducting layers $h \approx H \approx 0.64\,w$ and a return conductor overhang $v \approx 1.2\,w$. The values of $\mathbf{Z}_G$ (at 50 MHz) and $\mathbf{Y}_G$ measured on a scaled-up model are given by $\mathbf{Z}_G = j\omega\,\mathbf{L}_G$ and $\mathbf{Y}_R = j\omega\,\mathbf{C}_R$, with

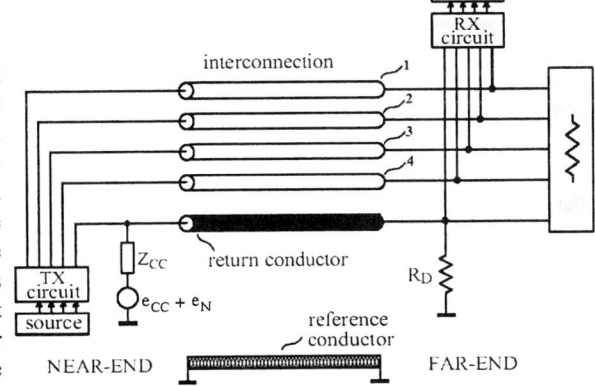

Fig. 2. A PDL with voltage-driven common conductor. The block containing the resistor symbol is a termination circuit.

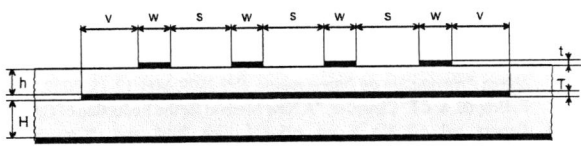

Fig. 3. Cross-section of an interconnection-ground structure built in a printed circuit board.

$$\mathbf{L}_G \approx \begin{pmatrix} 407 & 92 & 69 & 61 & 66 \\ 92 & 414 & 94 & 69 & 70 \\ 69 & 94 & 410 & 93 & 69 \\ 61 & 69 & 93 & 404 & 66 \\ 66 & 70 & 69 & 66 & 85 \end{pmatrix} \text{nH/m}$$

and

$$\mathbf{C}_G \approx \begin{pmatrix} 116.6 & -3.8 & -0.4 & -0.2 & -108.4 \\ -3.8 & 115.9 & -3.7 & -0.4 & -106.4 \\ -0.4 & -3.7 & 116.9 & -3.9 & -107.6 \\ -0.2 & -0.4 & -3.9 & 117.3 & -109.4 \\ -108.4 & -106.4 & -107.6 & -109.4 & 1098 \end{pmatrix} \text{pF/m}$$

For the p.u.l. impedance parameters of the $(n+1)$-conductor MTL model, using (2), we obtain $\mathbf{Z}_R = j\omega\, \mathbf{L}_R$ and $\mathbf{Z}_E = j\omega\, \mathbf{L}_E$ with

$$\mathbf{L}_R \approx \begin{pmatrix} 359 & 41 & 17 & 13 \\ 41 & 359 & 39 & 17 \\ 17 & 39 & 356 & 42 \\ 13 & 17 & 42 & 356 \end{pmatrix} \text{nH/m}, \quad \mathbf{L}_E \approx \begin{pmatrix} 18 \\ 15 \\ 15 \\ 18 \end{pmatrix} \text{nH/m}$$

and $Z_{EE} = j\omega\, L_{EE}$ with $L_{EE} \approx 85$ nH/m.

For the p.u.l. admittance parameters of the $(n+1)$-conductor MTL model, using (3), we obtain $\mathbf{Y}_R = j\omega\, \mathbf{C}_R$ and $\mathbf{Y}_E = j\omega\, \mathbf{C}_E$ with

$$\mathbf{C}_R \approx \begin{pmatrix} 116.6 & -3.8 & -0.4 & -0.2 \\ -3.8 & 115.9 & -3.7 & -0.4 \\ -0.4 & -3.7 & 116.9 & -3.9 \\ -0.2 & -0.4 & -3.9 & 117.3 \end{pmatrix} \text{pF/m}, \quad \mathbf{C}_E \approx \begin{pmatrix} 3.8 \\ 1.6 \\ 1.4 \\ 3.5 \end{pmatrix} \text{pF/m}$$

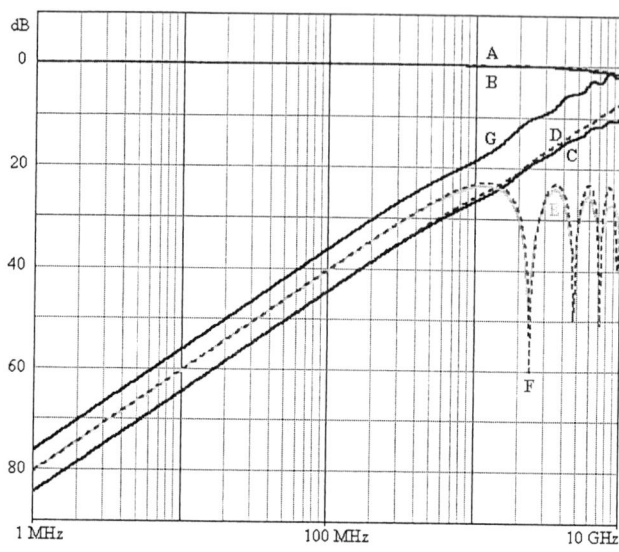

Fig. 4. Some frequency domain simulation results. Attenuation of transmitted signal when the TC number 1 is excited: curve A (blue, solid) according to (1) and curve B (red, dash) according to (5a). Far-end crosstalk loss on the TC number 2 when the TC number 1 is excited: curve C (blue, solid) according to (1) and curve D (red, dash) according to (5a). Near-end crosstalk loss on the TC number 2 when the TC number 1 is excited: curve E (cyan, solid) according to (1) and curve F (magenta, dash) according to (5a). Far-end external crosstalk loss on the TC number 1 when all conductors are excited at the near-end: curve G (blue, solid) according to (1).

and $Y_{EE} = j\omega\, C_{EE}$ with $C_{EE} \approx 1520$ pF/m. As expected, the entries of $\mathbf{L}_E$ and $\mathbf{C}_E$ are much smaller than the diagonal entries of $\mathbf{L}_R$ and $\mathbf{C}_R$, respectively, so that we may plan to use $\mathbf{L}_R$ and $\mathbf{C}_R$ for designing a ZXnoise PDL, according to Section III.

For a type 2 termination circuit, the minimum value of the matrix norm $\| \mathbf{P}_R \|_\infty$ is 0.083, which may be obtained with a termination circuit comprising two 57.7 Ω resistors and two 55.0 Ω resistors. However, our simulations use a termination circuit comprising one 50.0 Ω resistor connected between each TC and the return conductor, for which $\| \mathbf{P}_R \|_\infty = 0.131$, and a damping resistor of $R_D = 10\ \Omega$. Using the approximate $(n+1)$-conductor MTL model corresponding to (5a), we obtain the curves B, D and F of Fig. 4 when a signal is sent through the channel 1. Using the exact $(n+2)$-conductor MTL model for the same configuration, we obtain the curves A, C and E of Fig. 4. The curve G of Fig. 4 represents the rejection of external crosstalk, which can only be computed using the $(n+2)$-conductor MTL model.

V. Conclusion

Up to 3 GHz, the rejection of external crosstalk exceeds 10 dB and the agreement between the $(n+1)$-conductor MTL model and the $(n+2)$-conductor MTL model is excellent. This validates (2), (3) and (4), and also the use of the design procedure outlined in Section III, which produces floating termination circuits because it is based on the $(n+1)$-conductor MTL model. We have also computed the rejection of external crosstalk, which compares the performance of the PDL to that of multiple single-ended links. Our simple PDL design provides an effective reduction of external crosstalk, for instance 36 dB at 100 MHz and 19 dB at 1 GHz.

References

[1] F. Broydé, and E. Clavelier, "Pseudo-differential links using a wide return conductor and a floating termination circuit", *Proc. of the 2008 IEEE International Midwest Symposium on Circuits and Systems (MWSCAS)*, August 10-13, 2008, pp. 586-589.

[2] F. Broydé, E. Clavelier, "A new pseudo-differential transmission scheme for on-chip and on-board interconnections", *Proc. of the CEM 08 Int. Symp. on Electromagnetic Compatibility*, Paris, France, May 2008, Available: http://www.eurexcem.com/zxnoisedef.htm.

[3] B. Démoulin, A.P.J. van Deursen, "Deux approches pour établir le lien entre la notion usuelle d'impédance de transfert et le formalisme des lignes couplées", *Proc. of the 10e Colloque International & Exposition sur la Compatibilité Électromagnétique, CEM 2000*, Clermont-Ferrand, France, March 2000, pp. 98-103.

[4] F. Broydé, and E. Clavelier, "Modeling the interconnection of a pseudo-differential link using a wide return conductor", *Proc. of the 13th IEEE Workshop on Signal Propagation on Interconnects, SPI 2009*, May 12-15, 2009.

[5] F. Broydé, and E. Clavelier, "A New Method for the Reduction of Crosstalk and Echo in Multiconductor Interconnections", *IEEE Trans. Circuits Syst. I: Regular Papers*, vol. 52, No. 2, pp. 405-416, Feb. 2005. and "Corrections to «A New Method for the Reduction of Crosstalk and Echo in Multiconductor Interconnections»", *IEEE Trans. Circuits Syst. I: Regular Papers*, vol. 53, No. 8, p. 1851, Aug. 2006.

[6] F. Broydé, and E. Clavelier, "Echo-Free and Crosstalk-Free Transmission in Particular Interconnections", *IEEE Microwave and Wireless Components Letters*, Vol. 19, No. 4, April 2009, pp. 209-211.

978-1-4244-4447-2/09 $25.00 © 2009 IEEE

Extraction of Via and Trace Model from PCB Channel S-Parameter Data by Stochastic Optimization

Juyoung Lee, Drew Doblar
Sun Microsystems Inc.
4120 Network Circle, Santa Clara, CA 95054
Phone: (408) 276-6878, Fax: (408) 276-4550
juyoung.lee@sun.com

Abstract

A new method to extract via and trace model from given s-parameter data of PCB channels with symmetrical ends is presented. By applying stochastic optimization to two channels of different channel lengths, the two-port s-matrix of via and trace can be accurately determined by searching the multi-dimensional parameter space.

Introduction

The building blocks of multi-layer PCB channels such as via and trace play an important role in the design and the analysis of high bandwidth serial link channels. Via and trace models are often prepared by simulation method such as field solver and behavioral model, and their correlation with measurements are commonly pursued. [1], [2] However, PCB via and trace are not readily accessible as isolated entities to testing equipment such as VNA (vector network analyzer). [3] The only directly observable entity is often the s-parameters of the whole end-to-end channel. If the via and trace can be accurately separated from the measured s-parameter of the whole channel, it will be beneficial for multiple reasons. Analysis of isolated via or trace will be easier than that of the whole channel. A reasonably accurate via model library, which is based on measurements, will be available. The same method can also apply to the de-embedding of coaxial launcher on a test fixture in place of via and potentially can become an alternative to the existing de-embedding procedures. [3] - [6]

In this paper, a method to extract via and trace s-matrix from a given set of s-matrix data of whole channel with symmetrical ends is presented. A pair of single ended channels of different channel length leads to a unique solution for via and trace. Stochastic optimization is used to find the solution. Then, extension to differential pairs is made when the mode conversion is small.

Method

A symmetric single ended via-trace-via PCB channel is the simplest form and is represented by a two-port 2 x 2 s-matrix S_{ch}. Since the channel is symmetric ($S_{ch}11=S_{ch}22$), there are only two distinct matrix elements, $S_{ch}11$, $S_{ch}12$, which are the return loss and insertion loss of the channel, respectively. Then the measurement of a channel provides 4 observables as each matrix element is a complex number. The same channel can be specified by total of 8 unknown parameters: the via s-matrix has 6 unknowns from the three distinct matrix elements S_v11, S_v12, and S_v22, while the trace s-matrix S_{tr} has two unknowns, as a transmission line is specified by insertion loss $S_{tr}12$. The characteristic impedance Z_0 of the trace is considered known as it can be determined by inspecting low frequency $S_{ch}11$ with renormalized port impedance. Measurements of the two PCB channels of identical composition but different channel lengths provide effectively 8 observables. The two PCB channels are given by two sets of matrix equation

$$S_{ch\_1} = S_v \ (cascade) \ S_{tr\_1} \ (cascade) \ S_{v\_r} \tag{1}$$

$$S_{ch\_2} = S_v \ (cascade) \ S_{tr\_2} \ (cascade) \ S_{v\_r} \tag{2}$$

where S_{ch_1}, and S_{ch_2} are the cascaded s-matrix for channel 1, and channel 2, respectively. S_{v_r} represents the identical via as S_v but is connected in reverse direction as in Fig. 1. S_{tr_1} is for the trace of channel 1. The exponent of S_{tr_1} insertion loss is scaled to S_{tr_2} linearly by trace length due to the translational symmetry of the transmission line. The operator *(cascade)* represents the cascade operation of two s-matrices. It is implemented by multiplication of two abcd-matrices after converting the s-matrix into abcd-matrix. Then the resultant abcd-matrix is converted back to s-matrix after the multiplication. [5] Eq. (1) and (2) pose multiple cubic equations with 8 unknowns. Instead of pursuing algebraic approach, we used stochastic optimization to find the solution in 8-dimensional parameter space. The objective function for the optimization is given by

$$norm(S_{ch\_m1} - S_{ch\_1}) + norm(S_{ch\_m2} - S_{ch\_2}) \tag{3}$$

978-1-4244-4447-2/09 $25.00 © 2009 IEEE

where S_{ch_m1} is the given data of channel 1 such as data from measurement. *norm(A)* is the magnitude of the vector which is composed of the elements of matrix A. The global minimum of Eq. (3) coincides with the solution S_v and S_{tr} of Eq. (1) and (2). The steepest descent minimization algorithm, which is available as a sub-module in Matlab, is used. [7] The minimization process is prone to get into local minima of the objective function. Stochastic initialization of S_v and S_{tr} is applied with repetition in order to discard the local minima by selecting a smaller objective function in the search of the global minimum.

Simulations

The simulated example that emulates VNA measured data is prepared with 4" and 6" long channels. The channel s-matrix data are prepared by cascading the known via and trace s-matrices. The 4" long channel can be divided into two halves S_{vt}, and S_{vt_r} as in Fig. 1 for more efficient process. Eq. (1) and (2) are modified as

$$S_{ch\_1} = S_{vt} \, (cascade) \, S_{vt\_r} \tag{4}$$

$$S_{ch\_2} = S_{vt} \, (cascade) \, S_{tr\_2in} \, (cascade) \, S_{vt\_r} \tag{5}$$

$$S_{vt} = S_v \, (cascade) \, S_{tr\_2in} \tag{6}$$

where S_{tr_2in} is the 2" trace s-matrix. Then, from the solution, S_{vt} and S_{tr_2in} of Eq. (4) and (5), the s-matrix of via S_v is obtained by de-embedding S_{tr_2in} from S_{vt}. [5]

The solved s-matrix and the pristine s-matrix are compared to confirm effectiveness of the method. Fig. 2 shows insertion loss of trace, via, and channel 1 and 2 in the order from top to bottom. The two curves of S_{tr_2in} agree so well that they appear as one on top part of Fig. 2. The two S_v curves also show good agreement in about 0.2% error. The solved via model shows stub resonance dip at 4.8 GHz and one at 16 GHz. At the bottom part of Fig. 2, the gap between S_{ch_1} and S_{ch_2} coincides with more attenuation in the two inch longer channel. Both curves show periodic wiggles that is resulted from the reflection between the two discontinuities at the via and trace junction. It is noted that the smoothness of S_v and S_{tr_2in} is recovered out of the wiggles in S_{ch_1} and S_{ch_2}. Fig. 3 is the comparison of the phase of S_v21. Because Eq. (4) is ambiguous with the sign of S_{vt} insertion loss, stochastic search leads to random distribution between two branches of phase that are different by 180 degrees. The one with zero phase at low frequency is the physical one. Fig. 4 compares the return loss of S_v confirming that all elements of s-matrix are recovered.

In Fig. 5, random noise is injected into S_{ch_1} and S_{ch_2} to emulate for the noise in measured data. Noise level of 0.5% (-46dB) is applied uniformly across the whole frequency range, which is worse than the typical VNA noise floor. The propagated noise in the solved S_{tr_2in} appears largest in the vicinity of the stub resonance dips where the signal to noise ratio for the channels is smallest. The similar level of s/n ratio is transferred to S_{tr} and S_v curves.

The differential via model is of practical interest for the high bandwidth channels. Four-port mixed mode s-matrix is effectively decomposed into both differential and common mode two port sub-matrices, which become independent of each other when the mode conversion is small. The 4" and 6" long differential channels are prepared and the same extraction method is applied to each mode of the differential pair. In Fig. 6, differential mode shows good agreement in most frequency ranges with exception at the stub resonance region near 10GHz. Any unaccounted contributions such as mode conversion effect (less than -40dB in this example), and trace characteristic impedance mismatch to 50 ohm should appear as error. It is assumed to be compounded with the near mathematical singularity associated with the sharp resonance dip of about -70 dB. It is compared with the common mode results in Fig. 7, which is relatively smooth near the stub resonance at 6.8GHz.

Conclusion

A new method to simultaneously extract PCB via and trace model from symmetric single ended channel s-parameter data is presented. The two s-matrices from channels of different lengths constitute a set of equations that can be solved for its building block, via and trace, uniquely. Stochastic initialization, combined with the steepest descent optimization method, demonstrates to be very effective in solving for via and trace s-matrix. Extension to differential channel can be made when mode conversion is small. The method can be applied to constructing via model library that is based on measurements. It can also potentially be used for de-embedding of the coaxial to non-coaxial transition structure in test fixtures.

References

1. E. Bogatin, L. Simonovich, S. Gupta, M. Resso, "Practical Analysis of Backplane Vias," Designcon 2009.
2. V. Balasubramanian, S. Smith, S. Agili, "Comparison of S-parameter Concatenation to Full-Wave Simulation for High-Speed Interconnect Analysis," Designcon 2007.
3. G. Antonini, A. Scogna, , A. Orlandi, "De-Embeddig Procedure Based on Computed/Measured Data Set for PCB Structures Characterization," *IEEE trans. on advanced packaging*, vol. 27, no. 4, pp. 597-602, Nov. 2004.
4. S. Agili, V. Balasubramanian, A. Morales, "De-embedding techniques in signal integrity: a comparison study," 2005 Conference on Information Sciences and Systems, The Johns Hopkins University, March 2005.
5. G. Hernandez-Sosa, G. Romo, R. Torres-Torres, "Characterization and Modeling of Electronic Packages Using S-parameters," Proceedings of the International Caribbean Conference on Devices, Circuits and Systems, Mexico, Apr. 2008.
6. E. McGibney, J. Barrett, "An overview of electrical characterization techniques and theory for IC packages and interconnects," *IEEE trans. on advanced packaging*, vol.29, no.1, pp. 131-139, Feb. 2006
7. J. Snyman, *Practical Mathematical Optimization: An Introduction to Basic Optimization Theory and Classical and New Gradient-Based Algorithms*. Springer Publishing, 2005.

Figures

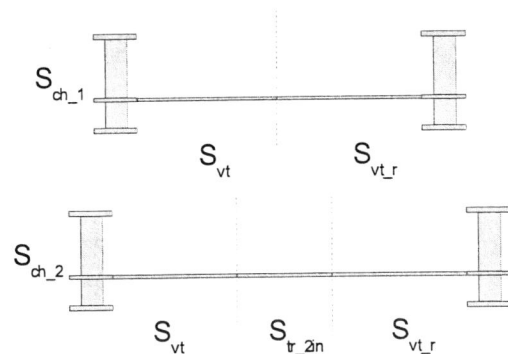

Fig. 1. Two symmetric PCB channels of via-trace-via configuration

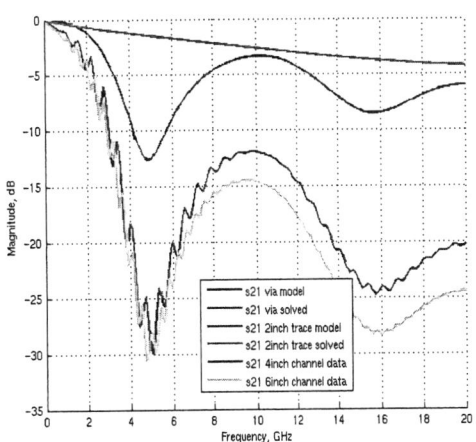

Fig. 2. Insertion loss of trace, via, and channels.

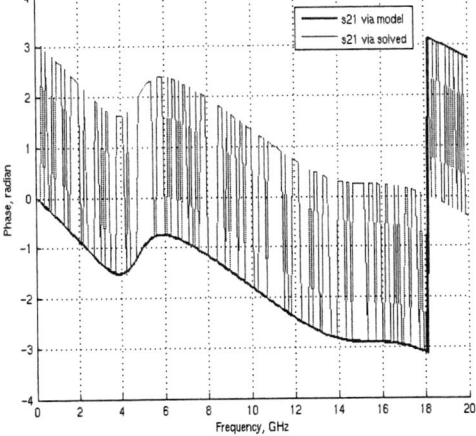

Fig. 3. Phase of via S21.

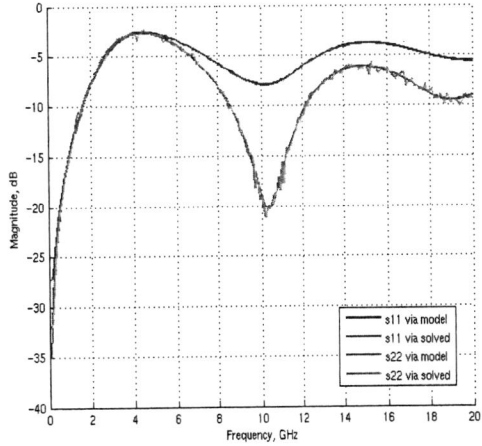

Fig. 4. Return loss of via.

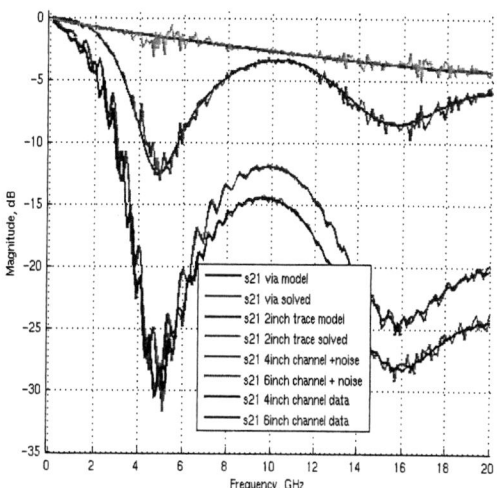

Fig. 5. Propagated noise of via and trace.

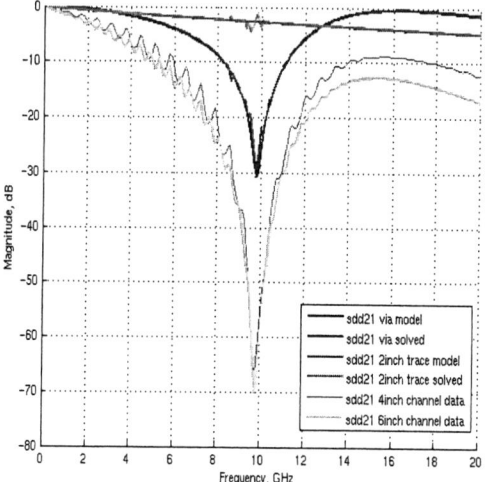

Fig. 6. Differential mode insertion loss of trace, via, and channels.

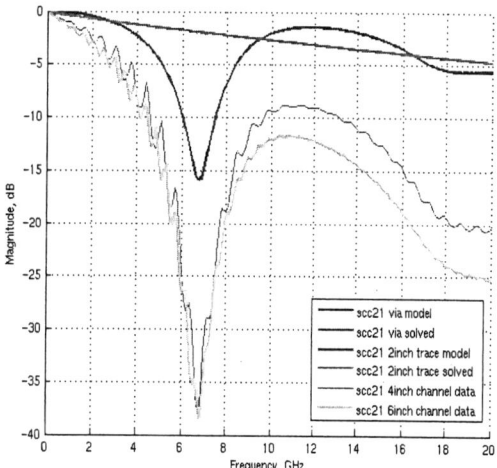

Fig. 7. Common mode insertion loss of trace, via, and channels.

Through Silicon Via (TSV) Equalizer

[1]Joohee Kim, [1]Eakhwan Song, [1]Jeonghyeon Cho, [1]Jun So Pak, [2]Junho Lee, [2]Hyungdong Lee, [2]Kunwoo Park
and [1]Joungho Kim

[1]Korea Advanced Institute of Science and Technology, Daejeon, South Korea
Tel) +82-42-869-9869, Fax) +82-42-869-8058, E-mail) joohee@eeinfo.kaist.ac.kr; teralab@ee.kaist.ac.kr
[2]Advanced Design Team, Hynix Semiconductor Inc., Icheon-si, Kyoungki-do, Korea

Abstract — **Through silicon via (TSV) is a promising vertical interconnection method to achieve a 3-dimensional integrated circuit (3D IC) system. However, high-speed digital signals suffer from severe distortions induced by TSV interconnects. In this paper, we propose a TSV equalizer using an ohmic contact on a double-sided silicon interposer to reduce the inter-symbol interference (ISI) of the TSV interconnects in a 3D IC system.**

I. INTRODUCTION

In order to realize highly-dense packaging and to improve channel bandwidth in high-speed integrated circuit systems, 3-dimensional integrated circuit (3D IC) technology using through silicon via (TSV) has become one of the key solutions. In 3D IC technology, TSV is a core technology for vertical interconnection between multiple chips and has a greatly reduced interconnection length [1]. In addition, for the external interconnection of the stacked 3D IC, a silicon interposer has been presented as another essential technology, which provides not only global 2D interconnect layers but also 3D interconnects between stacked chips with TSVs. In high-speed digital device 3D IC applications, the TSV channel and the interposer must be designed with a wide-band and a small form-factor for further miniaturization and higher performance [2]-[4].

However, TSV has to be electrically isolated from the silicon substrate with a thin silicon dioxide film, as shown in Fig. 1-(a), which causes a capacitive effect. Therefore, high-speed digital I/O signals suffer from the capacitive loading as well as the frequency-dependent loss from the lossy silicon substrate, which results in degradation of the eye opening and timing jitter in the high-speed digital system. In order to compensate for the frequency-dependent losses in a high-speed I/O channel, as shown in Fig 1-(b), equalization methods have been exploited as practical loss compensation techniques. Among various equalization techniques on silicon, on-chip passive equalizers have been introduced with wide-band achievement, but they consume significantly larger area on-chip compared to other active circuitries [5], [6]. To achieve small size, we used an ohmic contact to mitigate the capacitive effect of the TSV interconnect.

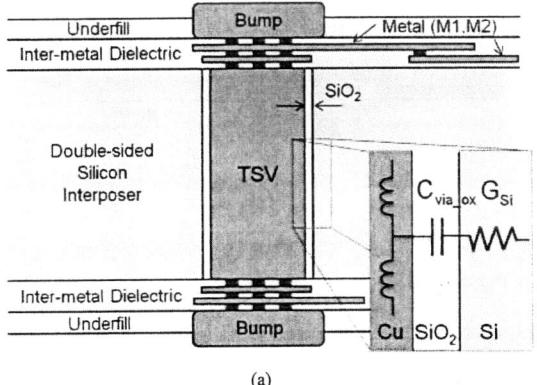
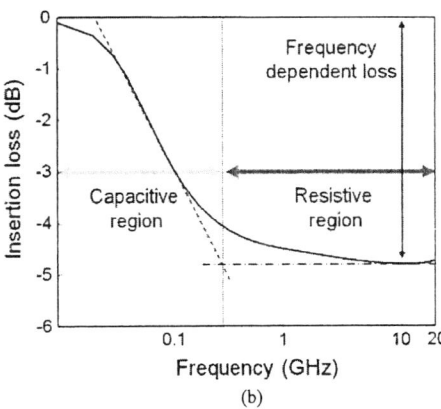

(a) (b)

Fig. 1. (a) A Through Silicon Via (TSV) structure with via-first process and (b) Insertion loss for 8 stacked GSG (Ground-Signal-Ground) TSV up to 20 GHz.

In this paper, we propose a novel TSV equalizer using intentional DC attenuation caused by an ohmic contact. An ohmic contact is a metal-semiconductor junction that provides current conduction between TSV and the silicon substrate with low resistance. By using an ohmic contact, we intentionally cause DC attenuation to compensate for the capacitive effect of the TSV interconnects. The proposed structure was implemented on a double-sided silicon interposer, which offers global 2D interconnect layers on both sides of the silicon wafer and with a via-first TSV process. A noticeable improvement in voltage and timing margin was demonstrated for the 8 stacked GSG (Ground-Signal-Ground) TSV equalizer using an ohmic contact of Al/n+-type with a data rate of 20 Gbps. The newly proposed TSV equalizer successfully achieved normalized eye opening and timing jitter, 20% and 32%, even though the unequalized eye was completely closed.

II. THE PROPOSED SIGNAL TSV EQUALIZER

To reduce ISI, we intentionally open a loss path for DC attenuation by using an ohmic contact, which is a metal-semiconductor junction

978-1-4244-4447-2/09 $25.00 © 2009 IEEE

with low resistance. The structures of a conventional TSV structure and the newly proposed TSV equalizer structure are shown in Fig 2-(a) and (b). As shown in Fig. 2-(b), with a contact width, $w_{contact}$, of 22.5 μm from TSV to the contact edge, the ohmic contact area is 6892 μm² per one TSV, which has a doughnut shape. Therefore, we can achieve significant area reduction (92%) with the proposed TSV equalizer compared to conventional on-chip passive equalizer with on-chip passive components [6]. TSV is considered to be fabricated throughout the via-first TSV process so that TSV passes through the silicon substrate but not the inter-metal dielectric (IMD). In addition, we selected a GSG signal TSV with a via hole that is filled with copper and surrounded by silicon dioxide, SiO₂, as an insulator. We assumed that the underfill and IMD are also formed with SiO₂. For the TSV dimensions, we adopted via holes of 75 μm diameter, 90 μm height, 150 μm pitch between TSVs, and 0.1 μm-thick SiO₂ surrounding the via [9]. The ohmic contact was formed with an Al/n+ metal-semiconductor junction that has a contact resistivity of 0.008 Ω-cm² under 200□ on an n-type silicon substrate [8]. The n+ doped junction depth is 1 μm. Because n+ well resistance is inversely proportional to the area of the contact, we can control the amount of DC attenuation by adjusting the contact area. The gray-colored doughnut shape region in Fig. 2-(b) represents the ohmic contact area of the proposed TSV equalizer. The proposed TSV equalizer was implemented on a double-sided silicon interposer, which provides twice the available are for the 2D interconnects compared to that of the one-sided silicon interposer. Therefore, this TSV equalizer enables vertical chip-to-chip interconnection and external interconnection through 2D interconnects as well as compensates for the degraded signal caused by the capacitive effect of the TSV interconnect. In addition, the proposed signal TSV equalizer has an advantage of using an ohmic contact because it is very popular and easy to be fabricated among silicon wafer processes.

However, if a contact is formed far from TSV, this DC control technique may not be very effective because of an undesired effect. Therefore, when ohmic contact is directly connected to TSV, we obtain the best equalization performance from the proposed equalization method. In addition, we formed an ohmic contact on both sides of the silicon interposer in order to generate more intentional DC attenuation by reducing effective n+ well resistance caused by an ohmic contact, which resulted in better ISI reduction of the TSV interconnects. Thus, we implemented a doughnut-shaped ohmic contact (Al/n+-type) with a contact width of 22.5 μm, which is formed adjacent to TSV, which is a GSG signal TSV, on both sides of the double-sided silicon interposer.

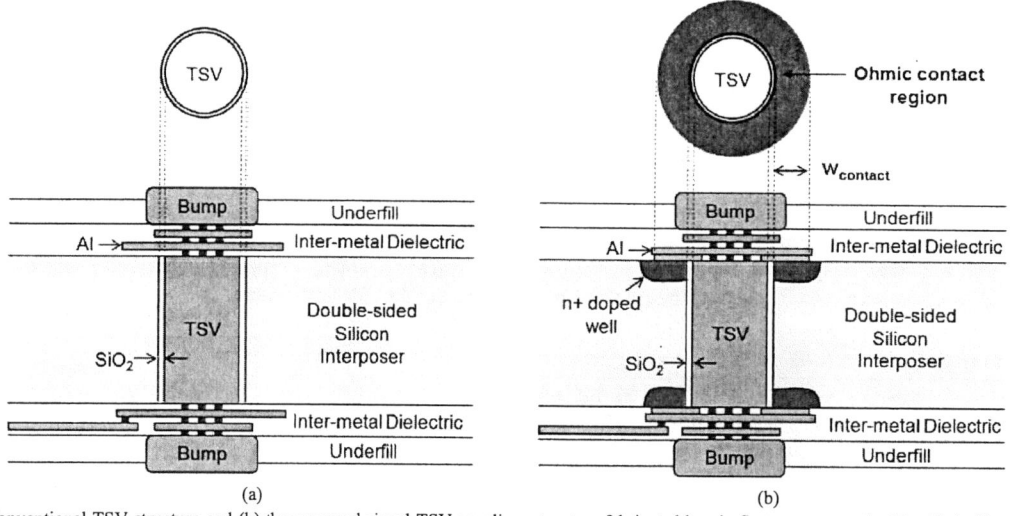

Fig. 2. (a) Conventional TSV structure and (b) the proposed signal TSV equalizer structure fabricated by via-first process on double-sided silicon interposer. The top view and cross-sectional view of each structure are shown.

III. EQUIVALENT CIRCUIT MODEL FOR THE PROPOSED SIGNAL TSV EQUALIZER

In Fig. 3-(a), we propose the equivalent circuit model and empirical equations for the proposed GSG signal TSV equalizer which is verified with the 3D field solver, HFSS. Since ideal ohmic contact has a straight line I-V characteristic that is voltage independent, we added n+ well resistance, R_{well}, to the equivalent circuit model of the TSV, which was verified by measurement [9]. With the proposed TSV equalizer, we can adjust the DC attenuation with n+ well resistance, which is determined by the contact area, $A_{contact}$, and specific contact resistivity, $\rho_{contact}$, which is determined by material properties [8]. Therefore, we can express n+ well resistance, R_{well}, as a closed-form equation including $A_{contact}$ and $\rho_{contact}$, as shown in Eq. (1). In addition, the capacitance and conductance of the silicon substrate increases as the contact width increases. Consequently, this causes the insertion loss to increase throughout the frequency range. Therefore, we propose empirical closed form equations of C_{Si} and G_{Si} in Eq. (2) and (3), respectively, with contact width, $w_{contact}$, as a variable and the fixed structural dimensions of TSV as constant values: d is the via hole diameter (75 μm); h is the via height (90 μm); p is the TSV-to-TSV pitch (150 μm); t_{ox} is the thickness of SiO₂ surrounding the via (0.1 μm). In addition, other RLC parameters in the proposed equivalent circuit model are set with the equation or constant values from the verified TSV equivalent circuit model, as shown in Eq.(4a)- (4d) [9]. Therefore, we propose empirical closed-form equations that can successfully estimate the equalization performance of the proposed signal TSV equalizer, including the effect of contact width variation.

$$R_{well} = \frac{\rho_{contact}}{A_{contact}} = \frac{8 \times 10^{-7}}{\pi((d/2 + w_{contact})^2 - (d/2 + t_{ox,via})^2)} \quad [\Omega] \quad (1)$$

$$C_{Si} = \varepsilon_0 \varepsilon_{r,Si} \cdot \frac{\pi(h - w_{contact})}{\ln\left(\frac{2 \cdot p}{d}\right)} + \varepsilon_0 \varepsilon_{r,Si} \cdot \frac{1.6 \times w_{contact} \times d}{p - w_{contact}} \quad [F] \quad (2)$$

978-1-4244-4447-2/09 $25.00 © 2009 IEEE

$$G_{Si} = 10 \times \frac{10 \times w_{contact} \times d}{p - 2 \cdot w_{contact}} + 10 \times \frac{(h - w_{contact}) \times 7 \times d}{0.95 \times p} \quad [1/\Omega] \tag{3}$$

where $d = 75um$, $h = 90um$, $p = 150um$ and $t_{ox,via} = 0.1um$

$$R_{via} = 4 \times 10^{-3} \times \sqrt{1 + \frac{frequency}{1 \times 10^9}} \ [\Omega] \tag{4a}$$

$$L_{via} = 15 \ [pH] \tag{4b}$$

$$C_{via\_ox} = 880 \ [fF] \tag{4c}$$

$$C_{ox} = 3 \ [fF] \tag{4d}$$

Fig. 3-(b) shows simulation results of the proposed signal TSV equalizer from the equivalent circuit model using ADS with contact width, $w_{contact}$, variation from 2.5 μm to 22.5 μm with 10 μm steps. With comparing to the insertion loss of a GSG-type signal TSV without an ohmic contact, the amount of DC attenuation of the proposed TSV equalizer increases from 0.25 dB to 0.8 dB as contact width increases from 2.5 μm to 22.5 μm, as we expected.

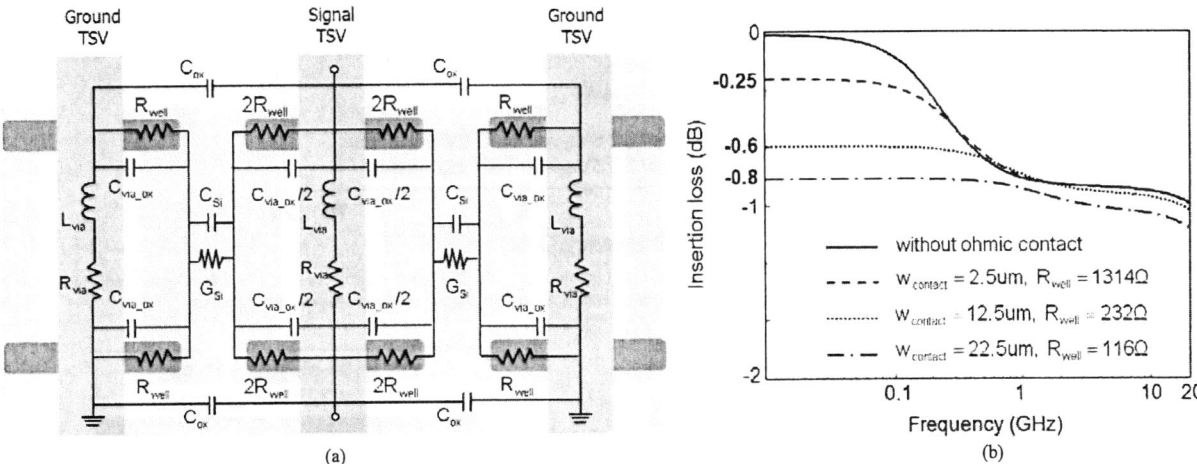

(a)   (b)

Fig. 3. (a) The proposed equivalent circuit model for the GSG type signal TSV equalizer. (b) Simulated insertion losses of the proposed signal TSV equalizer from the equivalent circuit model with variation of contact width.

In addition, we note that the insertion loss of TSV interconnect is frequency dependent based on the equivalent circuit model of the signal TSV [9]. The insertion loss of the TSV interconnects is mainly characterized by the capacitance of the thin SiO$_2$ film and silicon substrate, C_{ox} and C_{Si}, and the conductance of the lossy silicon substrate between the signal and ground TSV, G_{Si}. In a low frequency range, the impedance of the capacitance between TSV and silicon substrate, C_{via_ox}, is very high and rapidly decreases as frequency increases. If that impedance becomes lower, the signal experiences more leakage through C_{ox} and starts to feel the conductance of the silicon substrate, G_{Si}, resulting in insertion loss. With this analysis of capacitive and resistive insertion loss characteristics of the TSV interconnects, we present a low impedance signal loss path through ohmic contact, represented by R_{well} in the proposed equivalent circuit model, as an equalization method for the signal TSV. DC attenuation caused by the R_{well} mitigates rapid changes in not only impedance but also insertion loss, as shown in Fig. 3-(b). Therefore, we can achieve ISI reduction by flattening the frequency response using ohmic contact.

As a result, we proposed the equivalent circuit model of the proposed TSV equalizer with help of newly proposed empirical equations of R_{well}, C_{Si} and G_{Si} and analyzed the insertion loss of the TSV with the contact width variation.

IV. SIMULATION RESULTS

In order to verify the equalization performance of the proposed signal TSV equalizer, we designed a GSG signal TSV equalizer and performed insertion loss simulations with 3D field solver, HFSS. We performed simulations for 8 stacked GSG signal TSV with and without ohmic contacts.

In Fig. 4, we successfully validated the equalization performance of the proposed signal TSV equalizer from the simulation results of the 8 stacked signal TSV and 8 stacked TSV equalizer, which are presented by a solid line and dashed line. The insertion loss of the 8-stacked signal TSV equalizer is noticeably flattened from DC to Nyquist frequency, 10GHz, by 3.8 dB. In addition, as we expected through analysis of the proposed equivalent circuit model, there is an additional loss throughout the frequency range after equalization using an ohmic contact. This loss is caused by increased capacitance and conductance of the silicon substrate due to a shortened distance between the signal TSV and ground TSV. As a result, the overall loss level is slightly lowered by 0.7 dB after equalization, as shown in Fig. 4. However, the ohmic contact still effectively concentrates insertion loss under the Nyquist frequency, which is 10 GHz; and as a result, we successfully achieved a flattened frequency response throughout the frequency range.

In order to verify the equalization performance of the proposed TSV equalizer in the time domain, we simulated eye diagrams with a data rate of 20 Gbps. The input signal was a 2^9-1 pseudo-random bit sequence with pk-pk amplitude of 500 mV. Fig. 5 shows eye diagrams of an

978-1-4244-4447-2/09 $25.00 © 2009 IEEE

8 stacked signal TSV and an 8 stacked TSV equalizer. The newly proposed signal TSV equalizer successfully achieves normalized eye opening and timing jitter, 20% and 32%, while the unequalized eye is seriously distorted.

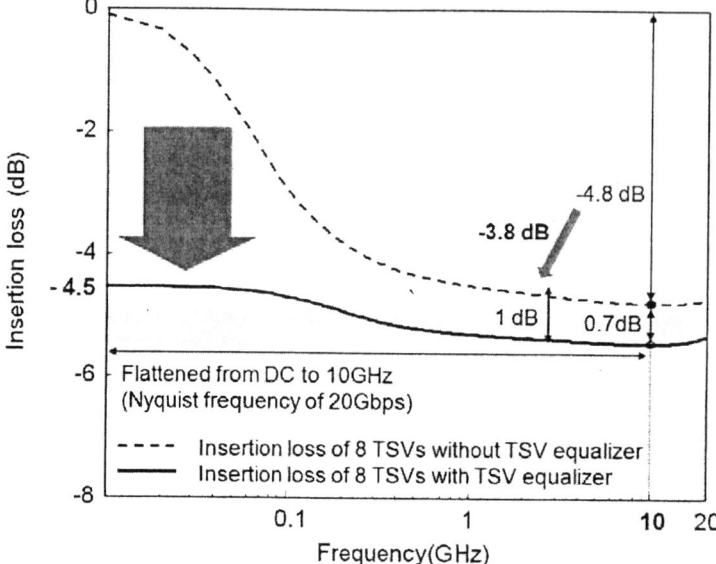

Fig. 4. Insertion losses for 8 stacked signal TSV and 8 stacked TSV equalizer. Insertion loss of 8 TSVs with ohmic contact, which is an equalized one from the proposed signal TSV equalizer, is successfully flattened due to DC attenuation caused by ohmic contact of the proposed TSV equalizer.

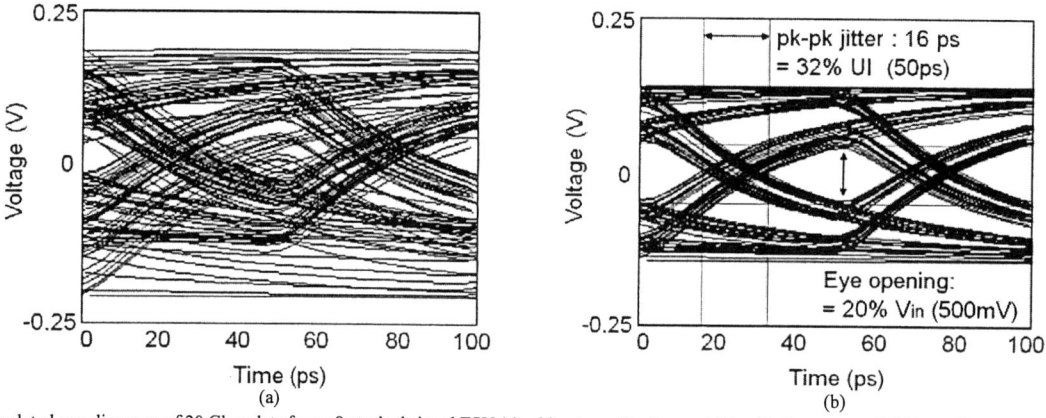

Fig. 5. Simulated eye-diagrams of 20 Gbps data for an 8 stacked signal TSV (a) without equalization and (b) with the proposed TSV equalizer.

V. CONCLUSIONS

In this paper, we proposed a novel signal TSV equalizer using an ohmic contact on a double-sided silicon interposer in order to achieve wide-band throughput of the TSV interconnects. Then, we proposed an equivalent circuit model for the equalizer. Based on the model, we analyzed the insertion loss of the TSV with an ohmic contact. The proposed model was verified with insertion loss simulations in the frequency-domain using a 3D field solver. In addition, eye-diagram simulations were conducted with the proposed signal TSV equalizer; it successfully achieves normalized eye-opening and pk-pk jitter, 20% and 32%, respectively, while eye of the unequalized signal was completely closed.

REFERENCES

[1] Knickerbocker, J. U., Andry, P. S., Dang, B. and Horton, R.R., "Three-dimensional Silicon Integration", IBM Journal of Research and Development, 2008.
[2] Black, B., Nelson, D. W., Webb, C. and Samra, N., "3D Processing Technology and Its Impact on iA32 Microprocessors," 22nd IEEE Int. Conf. on Computer Design Proc., 2004.
[3] Sasaki, K., Matsuo, M., Hayasaka, N. and Okumuram K., "128Mbit NAND Flash Memory by Chip-on-Chip Technology with Cu Through Plug," 2001 Int. Conf. Electron. Packaging Proc., 2001.
[4] Takahashi, K. and Sekiguchi, M., "Through Silicon Via and 3-D Wafer/Chip Stacking Technology," Digest of Technical Papers, 2006.
[5] Sun R., Park, J., Mahony, F. O. and Yue, C. P., "A tunable passive filter for low-power high-speed equalizers," Digest of Technical Papers, 2006.
[6] R. Sun et al., "A low-power, 20-Gb/s continuous-time adaptive passive equalizer," in Proc. ISCAS, ,2005.
[7] X. Wang and R. R. Spencer, "A low power 170MHz discrete-time analog FIR filter," IEEE J. Solid-State Circuits, 1998.
[8] Richard C. Jaeger, Introduction to Microelectronic Fabrication, Prentice Hall
[9] Ryu, C., Lee, J., Lee, H., Lee, K., Oh, T. and Kim, J., "High Frequency Electrical Model of Through Wafer Via for 3-D Stacked Chip Packaging," Electronics System Integration Technology Conference, 2006.

978-1-4244-4447-2/09 $25.00 © 2009 IEEE

Achieving Near Zero SSN Power Delivery Networks by Eliminating Power Planes and Using Constant Current Power Transmission Lines

Suzanne Huh, Daehyun Chung, and Madhavan Swaminathan
Interconnect and Packaging Center, SRC Center of Excellence at GT
School of Electrical and Computer Engineering, Georgia Institute of Technology, Atlanta, GA 30332
Email: shuh3@mail.gatech.edu, d.chung@gatech.edu, madhavan.swaminathan@ece.gatech.edu

Abstract – *Enhanced data rate requires the management of the voltage and timing margins in eye diagrams. This lays a great emphasis on controlling SSN in the power supply network. Thus, the ability of the power delivery network to convey clean power has become more important. In this paper, a practical scheme using power transmission lines is presented. Several accompanying issues are addressed to enable this novel approach for managing noise in the power delivery network. The simulation results demonstrate that with the proposed CC-PTL scheme, SSN-free eye opening can be achieved.*

I. INTRODUCTION

The performance of a system depends highly on the communication speed between the processing unit and memory. Here, one of the major bottlenecks in communication speed is power supply noise [1]. The coupling between signal lines and power delivery network (PDN) in off-chip signaling induces simultaneous switching noise (SSN). Using power/ground planes, SSN is mainly induced by the return current path discontinuity (RPD) and the plane cavity resonance. The return current path in a power plane-based environment is usually disturbed by several kinds of discontinuities such as a power plane split or multiple via holes. Even with a solid return current path, the cavity resonance from the power/ground planes will deteriorate signal transmission. These noise sources constrain the timing margin, and reduce the operating frequency. Therefore, a new type of PDN is required to improve the chip-to-chip communication speed.

The concept of conveying power supply using a transmission line has been suggested previously using power transmission line (PTL) [2]. In that approach, a transmission line replaces the power plane, which eliminates the source of SSN. This can be the most effective method to remove SSN and to increase the per-pin data rate. However, to enable power transmissions lines to work in a real environment, several accompanying issues need to be addressed, namely: 1) DC drop on power supply network caused by DC resistance, 2) Mismatch effect caused by the power transmission line, 3) Line congestion effect due to using lines instead of planes, and 4) Power consumption. First of all, data transition causes the supply current to switch: the high-state of data draws current from the power supply network, but the low-state of data interrupts the current flow. Due to the terminating resistor of the power transmission line, the current through the power transmission line induces DC drop on the power supply network. The resulting DC drop on the power supply network is different from SSN. It is data transition-dependent, and does not affect the data itself. However, the problem arises as mismatch occurs or a PTL supports more than 1 bit. Secondly, since power is supplied through the transmission line, an additional mismatch between the power transmission line and its terminating resistor appears beside a mismatch between the signal transmission line and its terminating resistor. This additional mismatch will not only induce additional voltage reflection but also fail to serve as a backward terminator of the signal transmission line to stop the series of voltage reflections. Hence, the power transmission line reduces the design reliability. Moreover, if one PTL is used per I/O driver, the number of lines on PCB doubles. A solution to relieve line congestion is to support more than one I/O driver with one PTL. Then, the power supply node of several I/O drivers will be tied together so that the amount of current through the power transmission line will vary with the data pattern. This will affect both the power supply node and the output nodes of all the transmitters. Finally, the power consumption should be reasonable when compared with that of the conventional power plane scheme. Once these issues are resolved, the power transmission line will serve as a robust method to deliver clean power with minimum SSN.

In this paper, a practical scheme of using power transmission line for high speed signaling is presented, which is called Constant Current Power Transmission Line (CC-PTL). The proposed CC-PTL scheme solves the first two issues: data transition-dependent DC drop will not occur, and additional mismatch will not happen. This scheme is extended to multiple I/O scheme and verified with simulations. For the multiple I/O scheme, the total power consumption is compared to that of the conventional power plane case along with eye height and jitter.

With the newly suggested CC-PTL scheme applied to a real environment, the plane cavity resonance can be removed along with the power plane and decoupling capacitors, as shown in Fig. 1. Also, the return current path loop will be completed without any disturbance, resulting in the absence of RPD.

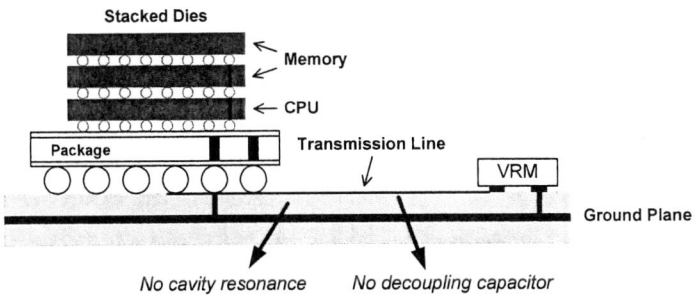

Fig. 1. A system example supplied by a power transmission line

By eliminating the RPD and plane cavity resonance which are the major noise sources in high speed signaling, this new PDN scheme can improve power integrity, and timing and quality of signals.

II. PROPOSED SCHEME AND SIMULATION RESULTS

A. Single-Ended Signaling

Among the accompanying issues of the conceptual power transmission line scheme in Fig. 2(a), the data-dependent DC drop at the power supply node is illustrated in Fig. 2(b). As the data at the output of the driver, data_tx transits from low to high, current flows through the power transmission line yielding DC drop across the terminating resistance. As a result, the voltage at the power supply node of the driver, TxPwr is a fraction of the original supply voltage. On the other hand, as data_tx transitions from high to low, no current flows through TxPwr so that the node voltage equals the original supply voltage. In case of Fig. 2(b), the impedance values are chosen to dual-match the signal transmission line: Z_0, Z and R_{pmos} equals 50ohm, 25ohm and 25ohm, respectively. While data_tx is high, the voltage at TxPwr drops to 1.875V which is ¾ of the original supply voltage, 2.5V. During the low state of data_tx, TxPwr is floating that the voltage stays to be 2.5V. To remove the DC drop on power supply network, a dummy path is connected to TxPwr as shown in Fig. 3(a). The dummy path consists of a switch and an equivalent resistor. When data_tx is 1, the dummy path is disconnected, and the current flows from the power transmission line toward the signal transmission line. When data_tx is 0, the dummy path is activated so that the same amount of current flows through the dummy path. Consequently, constant current flows through the power transmission line regardless of the data state. It then results in a constant DC drop over the PTL terminating resistance. The potential at TxPwr will be less than the original supply voltage, but constant and independent of the data transition. Moreover, such a DC drop does not reduce the eye amplitude of data_tx and data_rx.

The proposed constant current PTL also solves the additional mismatch problem. When the electrical length of the line is short, the transmission line is a mere wire that connects components. Then, the voltages on the wire at a certain time can be assumed to be the same at all points, and the mismatch effect will not appear. The constant current creates such an environment by eliminating the process to repeatedly charge and discharge the power transmission line. Therefore, the additional mismatch effect is removed as shown in Fig. 4. In the conceptual PTL scheme, when the characteristic impedance of the power transmission line

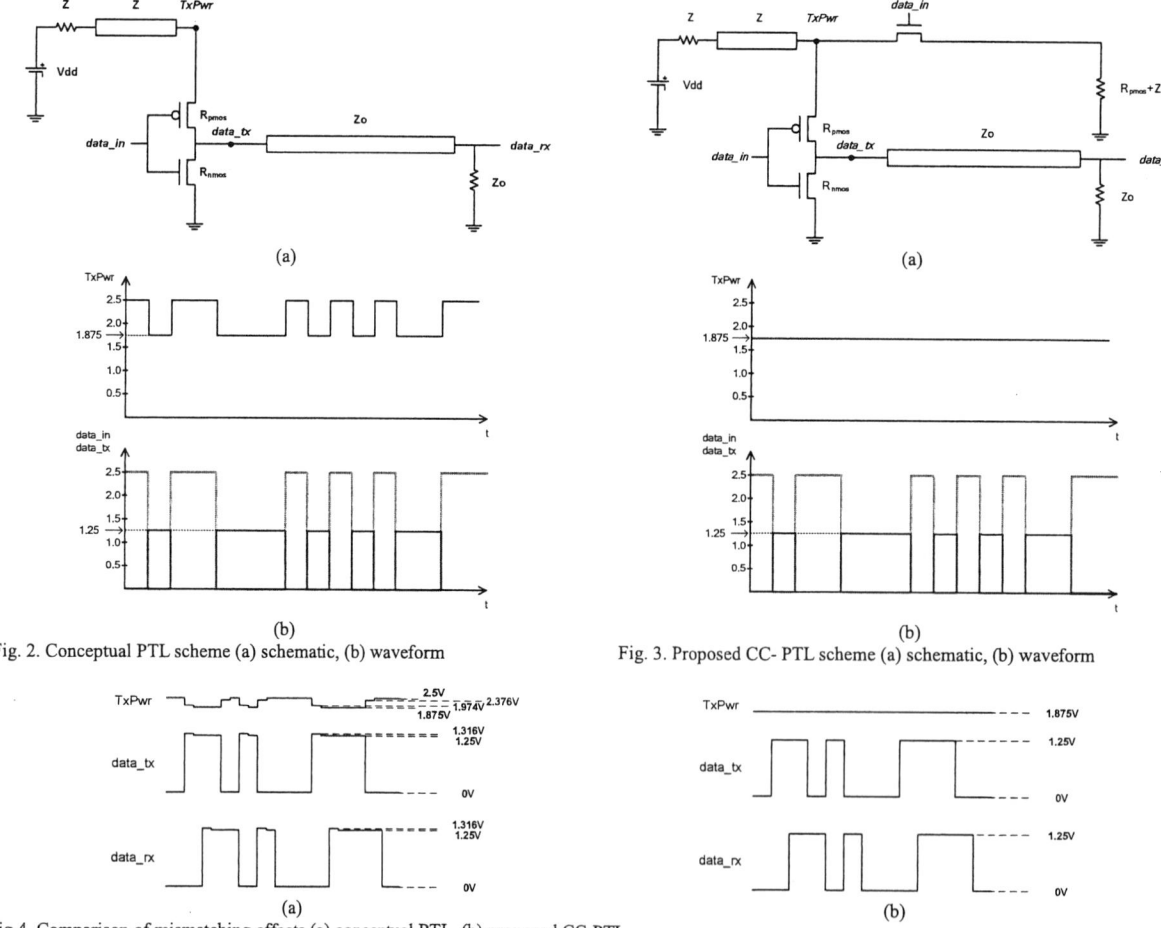

Fig. 2. Conceptual PTL scheme (a) schematic, (b) waveform

Fig. 3. Proposed CC- PTL scheme (a) schematic, (b) waveform

Fig.4. Comparison of mismatching effects (a) conceptual PTL, (b) proposed CC-PTL

is less than the desired value, the distortion due to reflection appears on the power supply node and the data nodes (Fig. 4(a)). However, in the proposed CC-PTL scheme, such a mismatch effect is not found as shown in Fig. 4(b). The eye diagrams of data_rx in the conventional power plane scheme and the proposed CC-PTL scheme are shown in Fig. 5. The eye height of power plane case is 1.085V, which is only 86.8% of the desired height (Fig. 5(a)), while that of CC-PTL case is 1.18V, which is 94.4% of the desired height (Fig. 5(b)). The jitter of the power plane case is 4.43 psec, while that of the CC-PTL case is 0 psec. The small distortion seen on the eye of CC-PTL case is due to port discontinuity in 3D simulation. The CC-PTL scheme offers better jitter performance as well as larger voltage margin.

B. Differential Signaling

In single-ended signaling, the dummy path is attached intentionally to the circuit. However, in differential signaling, the natural dummy path is created. As shown in Fig. 6(a), two kinds of current loops are formed according to the state of the data. During the high-state of data_tx, the current flows along the dotted-line. During the low-state of data_tx, the current flows along the dashed-line. Both the dotted-line path and the dashed-line path consist of the same components and hence are actually the same current loop. Both current loops have the same kind of influence on TxPwr node and output nodes. The eye diagram of the received data (data_rx - data_rx') in differential signaling with CC-PTL is shown in Fig. 6(b). The eye height is 2.316V, which is 92.6% of the desired size. In Fig. 7, the schematic and the resulting eye diagram of the conventional power plane

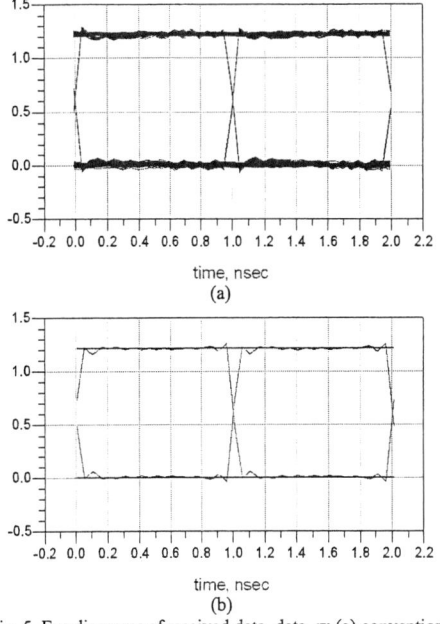

Fig. 5. Eye diagrams of received data, data_rx (a) conventional power plane case, (b) CC-PTL case

case are illustrated. Here, the balance of the differential line is intentionally broken by splitting the power plane. The eye height is 1.991V, which is only 79.6% of the desired size. Although a pair of differential lines affords the return current path by itself, signal integrity still depends on the PDN environment for its balance. Hence, the CC-PTL scheme can also be effective in differential signaling when the balance of symmetry between a pair of coupled lines is upset.

III. EXTENSION TO MULTIPLE I/OS AND REDUCING POWER CONSUMPTION

The only problem of CC-PTL is the increased power consumption in single-ended signaling due to the artificial dummy path and the constant current. Since the current flows during both the low-state and high-state of the data, the power consumption doubles. This problem can be mitigated when the proposed CC-PTL scheme is applied to multi-I/O single-ended signaling.

The total power consumption can be decreased if the target eye height is reduced. It is illustrated in Table 1. Here, it is assumed that the power 'P' is consumed when the eye height is ½ of V_{DD} in the power plane case. The relative amount of power in the CC-PTL case is listed in Table 1. As the eye height reduces from ½ to ⅓ and ¼ of V_{DD}, the power consumption of the CC-PTL

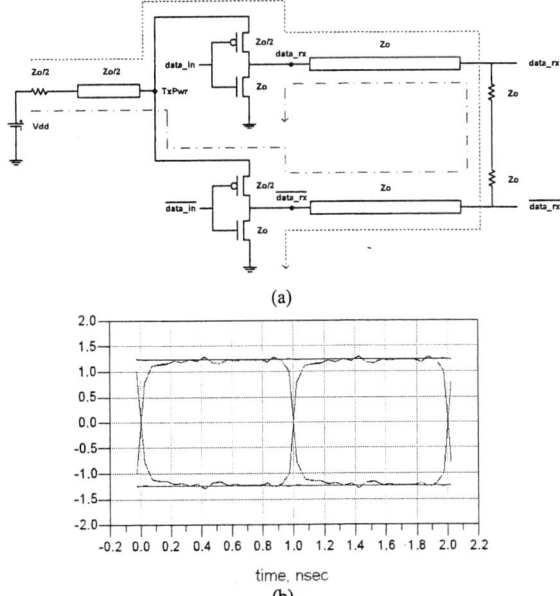

Fig. 6. Differential signaling with CC-PTL (a) schematic, (b) eye

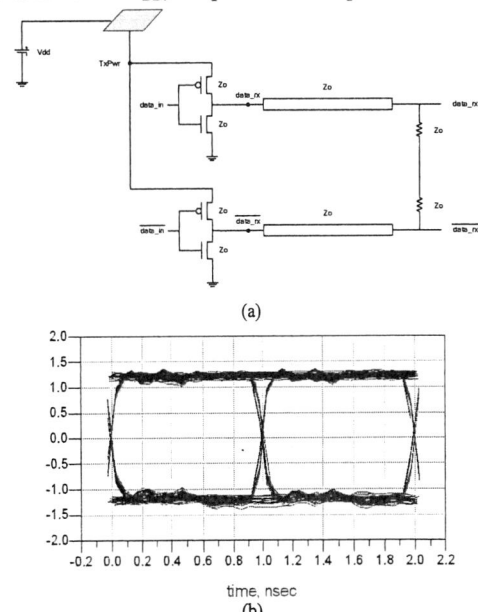

Fig. 7. Differential signaling with power plane (a) schematic, (b) eye

case decreases from 2P to 1.33P and P, respectively. Hence, the trade-off between the target eye height and the power consumption can be used

Table 1. Comparison of power consumptions

eye height	Power consumption	
	Power Plane	CC-PTL
$V_{DD}/2$	P	2P
$V_{DD}/3$	-	1.33P
$V_{DD}/4$	-	P

Table 2. Comparison of eye openings and jitters

bit	eye height (V)		Jitter (ps) @1Gbps	
	Power Plane (Target: 1.25V)	CC-PTL (Target: 0.833V)	Power Plane	CC-PTL
1	1.085	0.816	4.43	0
2	1.062	0.813	4.43	0
3	1.037	0.809	8.87	0
4	1.006	0.803	8.87	0

to save power. This is possible because in the power plane case, the actual eye size at the receiver side grows less than the target eye size as the number of I/O driver connected to a PDN increases.

With the conventional power plane scheme, SSN due to the plane cavity worsens as the number of I/O driver supplied by the power plane increases. It is shown in Table 2: the growing bit number decreases the eye height. The actual eye height is 1.085V in 1-bit case, which is 86.8% of the target height, and even decreases to be 1.006V in 4-bit case, which is only 80.5% of the target height. Although the target eye height of the power plane scheme is ½ of V_{DD}, the actual achieved eye height is reduced to be closer to ⅓ of V_{DD} in 4-bit case. Moreover, the growing bit number increases the peak-to-peak jitter. The jitter is about 4.43psec, which is about 0.5% of the pulse width in 1-bit case, and increases gradually with the bit number. However, with the SSN-free CC-PTL scheme, the increasing bit number causes hardly any harm to signal integrity. Here, the target eye height is reduced to ⅓ of V_{DD} for reasonable power consumption. The actual eye height of the received signal is nearly equal to the target eye height even with the increasing bit numbers. As a result, the actual eye height of the CC-PTL scheme when its target eye height is ⅓ of V_{DD} is comparable to the actual eye height of the power plane scheme when its target height is ½ of V_{DD}. Moreover, the jitter does not appear. Hence, as the bit number connected to the power supply line increases, the trade-off can be used to achieve better eye opening with reasonable amount of power consumption. Furthermore, the line congestion is relived by extending the proposed CC-PTL scheme to multi-I/O single-ended signaling.

In Fig. 8(a), the schematic of 4-bit single-ended signaling with the CC-PTL scheme is shown. The power supply nodes of four I/O drivers are bound together. Four kinds of dummy paths are connected to TxPwr node, and the data pattern detector selectively activates one dummy path at a time. All the component values are selected to generate the target eye height of ⅓ of V_{DD} and to dual-match the signal transmission line at the same time. The eye diagram of data1_rx is shown in Fig. 8(b). The eye amplitude is 0.803V, 96.4% of the desired height, and the jitter is 0 psec.

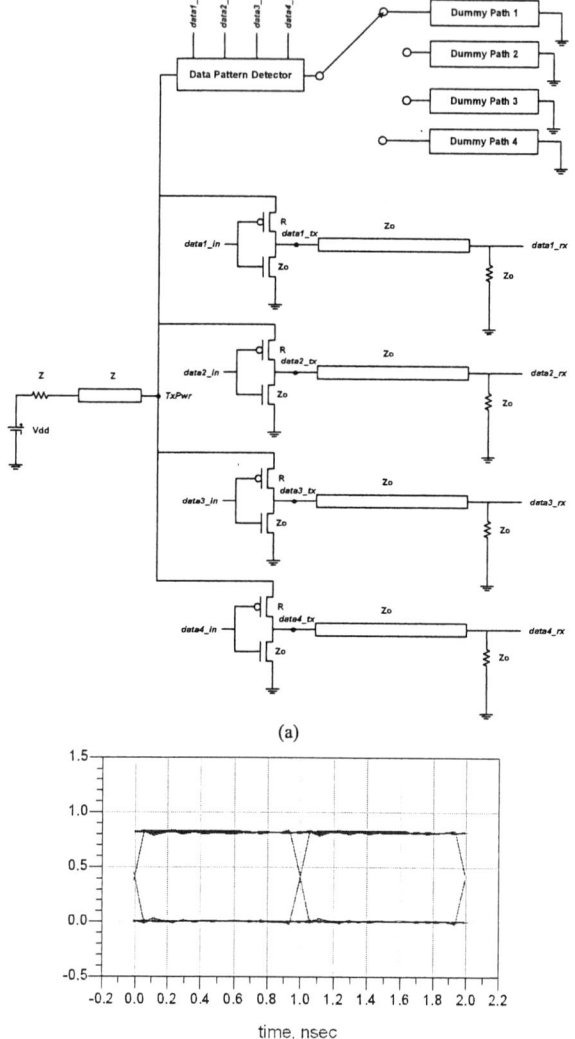

(a)

(b)

Fig. 8. CC-PTL scheme for 4-bit I/O (a) schematic, (b) eye

IV. CONCLUSIONS

In this paper, a practical scheme of using the power transmission line for high speed signaling is presented. The proposed CC-PTL scheme removes the alternating DC drop on power supply network by using dummy path. The dummy path induces constant current through the power transmission line so that it omits the process of charging and discharging the transmission line. Therefore, CC-PTL is free from any mismatch or discontinuity in the power supply path. The simulation result proves that the CC-PTL scheme provides superior jitter performance and larger eye amplitude. The application of the proposed CC-PTL scheme is extended to differential signaling and multi-I/O single-ended signaling. In addition, the total power consumption is comparable to that of the conventional power plane case.

REFERENCES

[1] M. Swaminathan, J. Kim, I. Novak, and J. P. Libous, "Power distribution networks for system-on-package: Status and challenges," IEEE Trans. Advanced Packaging, vol. 27, pp. 286–300, 2004.
[2] A. E. Engin and M. Swaminathan, "Power Transmission Lines: A New Interconnect Design to Eliminate Simultaneous Switching Noise," in Proc. IEEE 58th Electronic Components and Technology Conference, pp. 1139-1143, 2008
[3] M. Swaminathan and A. E. Engin, *Power Integrity Modeling and Design for Semiconductor and Systems*, Prentice Hall, 2007

Statistical Simulation of SSO Noise in Multi-Gigabit Systems

Wendemagegnehu T. Beyene, Amir Amirkhany, Ali Abbasfar

Rambus Inc., 4440 El Camino Real, Los Altos, CA 94022
Tel: 650-947 5000, Fax: 650-947 5001

Abstract

The use of deterministic techniques to evaluate the impact of simultaneous switching output (SSO) noise on the performance of modern high-speed systems with tight timing budget can be pessimistic. These can lead to conservative design, especially, in multi-gigabit systems with embedded coding or scrambling sublayer. To overcome the shortcomings of conventional methodologies, a statistical simulation method of evaluating the impact of SSO noise on high-speed single-ended signaling systems is presented. The method correctly considers the spatial and temporal distributions of switching activities of devices in the system to calculate the performance degradation of the interface due to power supply noise. First, transient behavior that describes the power supply noise coupling to the signal receiver is generated. Then, using the probability distributions of the switching activities of the drivers, the SSO noise distribution is determined. Finally, this distribution is combined with the channel intersymbol interference (ISI) and device noise and jitter distribution of the signals to calculate the bit error rate (BER) of the overall system.

I. INTRODUCTION

AS the signal switching becomes faster and the supply voltage drops with every new process geometry shrink, SSO prediction becomes critical to insure the signal integrity of the system. When a large number of logic gates switches, the voltage supply to the input/output (I/O) circuitry may fluctuate and these disturbances, referred as SSO noise, can cause undesired transient behavior among output drivers, input receivers, or internal logic. Through the coupling between the power and signal distribution systems, SSO noise causes false logic, degrades the signal edge rate, and increases delay skew and signal overshoot or undershoot and jitter [1]. Therefore, accurate determination of SSO noise in gigahertz applications has been of critical importance to maximize bandwidth and minimize I/O power consumption.

The time-domain simulation of the complete system that combines the circuit model of the power delivery network (PDN) and channel is required to evaluate the impact of supply noise on the signal integrity of the high-speed interface. The calculation of BER using time-domain simulation is very time consuming. It is almost impossible to calculate at lower BER with high confidence using conventional circuit simulation techniques. In addition, current high-speed interfaces employ complex circuitries and digital processing components such as a pre-emphasis filter for the transmitter and a decision feedback equalizer for the receiver to mitigate channel impairment such as inter symbol interference (ISI). Since Synchronizing circuits responses to small deviations in voltage and time can be significant, the behaviors of these complex circuitries need to be accurately modeled. Therefore, the time-domain simulation of high-speed channels, PDN with these circuitries is no longer tractable even for BER of 1E-12.

An alternative approach is to use a system simulation approach by modeling the components in a link as blocks whose behaviors are described by higher-level languages such as MATLAB, Verlog-A, or NumPy (Python) [2]. The channel behavior is characterized through an ISI probability density function (*pdf*) or probability mass function (*pmf*) that is analytically derived from the channel characteristic functions. The channel ISI *pdf* is commonly built from the channel pulse or step responses. Then, the ISI eye is constructed by calculating the *pdf* at multiple sampling points of the channel response within one unit interval. The statistical eye is constructed by combining the ISI eye with the device noise. It gives the BER over voltage and time offsets. Thus, instead of performing the time-domain simulation by solving the system equation at each discrete time, the BER is efficiently calculated by convolving together the *pdf*'s or the *pmf*'s of the various blocks in the links. It is important to note that this statistical approach is based on superposition and therefore relies on the linearity of the system. A tutorial and a summary of the recent development in statistical simulation of transmission channels are presented in [2].

Similar statistical simulation approach can be used to analyze the impact of SSO noise in the signal transmission of high-speed bus. The approach can provide a more realistic prediction of SSO noise impact on the performance of high-speed systems with tight voltage and timing margins. It also provides more efficient method of analyzing power supply noise and signal propagation than the conventional and time-intensive deterministic methods. In Section II, the assumption and theory behind method are discussed. Then, a high-speed signaling example is given and the result from deterministic and statistical approaches are compared in Section III.

II. METHOD

The physical implementation of a single-ended signaling, pseudo-open drain logic (PODL) is shown Fig. 1(a). A single-ended signaling bus generates significant SSO noise, if no mitigating factor such as Data Bus Invert (DBI) or other coding is implemented in the design of the interface. Fig. 1(b) shows an interface bus with the PDN. When the transmitters send data, the logic gates switches and the voltage supply to the input/output (I/O) circuitry at locations $(1, 2, \ldots, n)$ on PDN fluctuate. These noise sources at the PDN couple through the termination and R_{on} and generate noise and jitter at the receiver of the devices, (i).

The impact of SSO noise on the signal transmission can be characterized by an impulse response between any arbitrary noise source (location) on the PDN and the receiver. Thus, a family of single-bit (pulse) responses can completely describes the impact of all SSO noise sources on the performance of a signaling system. When a single bit is transmitted from the driver to the receiver, the transfer characteristics between the n noise sources on the power supply network and the m receivers are captured by the bit (pulse) response h_{ij}, where (i, j) are the observation and source locations of the noise waveforms. This relation is described by:

$$
\begin{bmatrix}
y_1 \\
y_2 \\
\vdots \\
y_i \\
\vdots \\
y_m
\end{bmatrix}
=
\begin{bmatrix}
h_{11} & h_{12} & \cdots & h_{1n} \\
h_{21} & h_{22} & \cdots & h_{2n} \\
\vdots & \ddots & \ddots & \vdots \\
h_{i1} & h_{i2} & \cdots & h_{in} \\
\vdots & \ddots & \ddots & \vdots \\
h_{m1} & h_{m2} & \cdots & h_{mn}
\end{bmatrix}
\otimes
\begin{bmatrix}
x_1 \\
x_2 \\
\vdots \\
x_i \\
\vdots \\
x_n
\end{bmatrix},
\tag{1}
$$

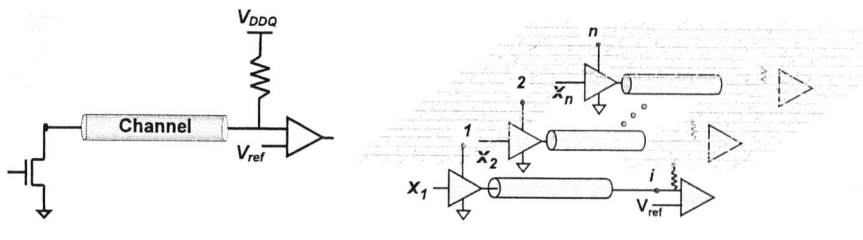

(a) PODL physical signaling system. (b) Power supply network and SSO sources.

Fig. 1. A single-ended signaling system.

where x_j is the transmitted binary data stream and y_i is the noise waveforms at the receiver. To calculate the impact of SSO noise at a receiver, i, Equation (1) reduces to :

$$y_i = \sum_{j=1}^{n} h_{ij} \otimes x_j. \tag{2}$$

In the conventional deterministic (synchronized) SSO noise analysis, all x_j's are switching together. In other words, $x_j = x$ for all j, where x is a binary data stream. This type of analysis assumes a worst case scenario when all the bit slices can switch together over several consecutive bit periods. However, the chances of transmitting identical bit sequence over a 32-bit interface even for as short instance as $3\ UI$ is a very unlikely event with a probability of $p = 1.25e^{-29}$. Thus, this worst case scenario, which is often considered over much larger simulation times, is allowing for a very unlikely event in the analysis that results in pessimistic prediction of the SSO noise impact in high-speed signaling. To be realistic, when designing high-speed links for a certain bit-error rate, the probability of the bit sequences causing eye closure should be included in margining the link, in the same way that ISI and jitter are included in bathtub curves [2].

The impact of the power supply noise at any receiver for an arbitrary bit sequences, x_j's, can be obtained by using the Equation (2). The evaluation of the equation for arbitrary long bit pattern can be computationally intensive. If the noise sources are located very close to each other as compared to the observation point, i, the noise transfer functions of observation point with respect to various noise sources are approximately the same, $\bar{h}_i = h_{ij}, \forall j$. Then, Equation (2) further reduces to :

$$y_i = \bar{h}_i \otimes \bar{x}, \tag{3}$$

where $\bar{x}$ is the set of transmitted random data stream on the bus.

This SSO characteristic function, $\bar{h}_i$, can also be simulated or measured by selecting and holding nearby data bits low or high while the data drivers in the interface switching. The SSO generated on the power supply grid is coupled through the termination and driver R_{on} into the signal nets. Then, the waveform at the receiver of the selected quiet bits is measured.

If the data pattern that each driver transmit is independent and the bit sequences are random, or the probability distribution of the bits switching behavior are known, the calculation of the impact of the SSO noise at the receiver can be dramatically improved. If we assume that each driver switches independently from low to high and from high to low with identical probability, then, the two possible outcomes (0 can represents the switching from low to high and high to low) can be modeled by $Bernoulli$ random variables with $p = 0.5$. If the random variables, $x_1, \ldots, x_n$, are assumed to be both independent and identically distributed (i.i.d.), then the sum of all drivers switching behavior can be represented by the sum of the $Bernoulli(p)$ random variables, $\bar{x} = x_1 + x_2 + \ldots + x_n$. The resulting random variable, $\bar{x}$, is called a $Binomial(n, p)$ random variable and it describes the switching behavior of the n-bit bus. The distributions of each $Bernoulli(p)$ variable, $X_1, \ldots, X_n$, and their sum, $Binomial(n, p)$ variable, $\bar{X}$, are shown in Fig. 2. If the transmitters are driven by independent randomized bit pattern, the resulting switching behavior on n-bit interface can be represented by a $Binomial$ random variables and the pmf is written as,

$$\bar{X}(k) = \binom{n}{k} p^k (1-p)^{n-k}, k = 0, \ldots, n. \tag{4}$$

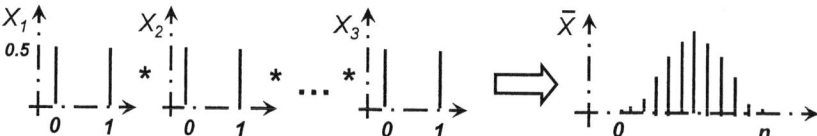

Fig. 2. The pmf distributions of $Bernoulli(p)$ with p=0.5 for each data line and the resulting $Binomial(n, p)$ with $n = 32$ for the interface.

Using the SSO characteristic functions of Equation (1) and the probability distribution of the switching of data or drivers in the interface, accurate prediction of the impact of the SSO noise can be performed. Equation (3) can be further rewritten to show that eacj Y_i sample is a summation of some independent random variables whose pmf is scaled version of $\bar{X}$. First, the convolution in Equation (3) can be written as,

$$y_i(k) = \sum_l \bar{h}_i(l)\bar{x}(k-l). \tag{5}$$

978-1-4244-4447-2/09 $25.00 © 2009 IEEE   22

Then, the distribution of Y_{il} can be rewritten as,

$$Y_{il}(\nu) = \bar{X}\left(\frac{\nu}{\bar{h}_i(l)}\right). \tag{6}$$

Since Y_{il}'s are independent, the Y_i is given by,

$$Y_i = Y_{i1} \otimes Y_{i2} \otimes \ldots \otimes Y_{il} \otimes \ldots. \tag{7}$$

For the high-speed interface described in the next Section, the probability distribution of the SSO noise impact at the receiver is calculated. Fig. 3(a) shows the SSO noise impact using randomized and synchronized switching of data drivers. The SSO impact of the synchronized switching is significantly higher and saturate at relatively higher BER. The assumption of synchronized switching leads to very pessimistic results compared to randomized switching because the former ignores the probability of the synchronizes switching happening at the transmitter. The impact of maximum of 16 and 32 data drivers for randomized switching are shown in Fig. 3(b). It can be gleaned form this graph that the improvement of SSO noise using coding such as DBI and randomizing data switching varies as a function of BER.

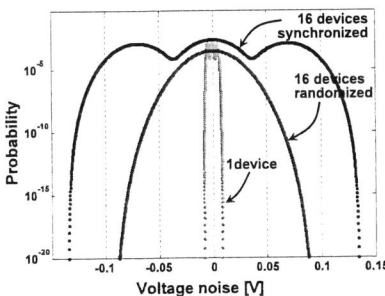

(a) Synchronized and randomized switchings.

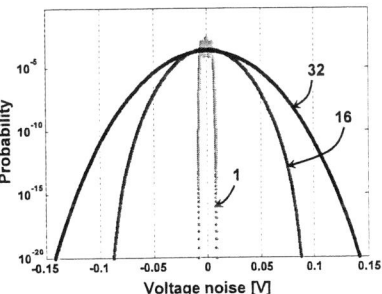

(b) Randomized switching of 16 and 32 devices.

Fig. 3. Probability distributions for SSO noise using synchronized and randomized switchings.

III. RESULTS

To compare the SSO simulation techniques, a multi-gigabit chip-to-chip interconnect system with 32-bit wide bus is analyzed. The system uses single-ended PODL signaling. The power system is modeling using one of the conventional methods described in [1]. The time-domain simulation of the interface is performed by combining the circuit model of the power plane with signaling channel. Since the signal attenuation of the channel is over 15 dB at Nyquist data rate, it is critical for reliable transmissions the use of equalization techniques. The responses of the system to pseudo-random binary sequence excitations are also shown in Fig. 4(a). The eye diagram significantly opens when equalization techniques are applied in series to the channel as shown in in Fig. 4(b). Fig. 5 show the performance predictions using deterministic and statistical simulation techniques. Fig. 5(b) shows significantly wider eye opening than that of the deterministic prediction of Fig. 5(a).

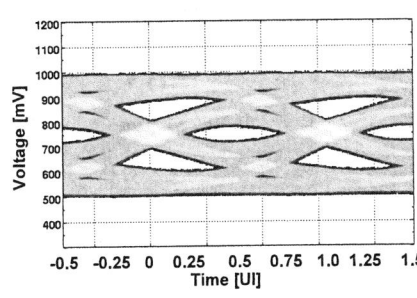

(a) Channel before equalization.

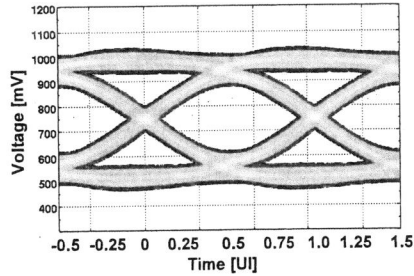

(b) Channel after equalization.

Fig. 4. Received eye diagrams of the channel before and after equalization.

Once the distributions of deterministic and random noises are determined, ths SSO noise distribution is integrated into a behavior simulation tools to perform system-level analysis in similar factions as described in [2]. The bounded distributions of the SSO noise, ISI, and crosstalk (Xtalk), and the bounded and unbounded device noise and jitter are combined to generate the probability distribution or BER of the overall system as a function of time or voltage. The noise and jitter of the devices and sensitivity and bandwidth of the receiver and the channel characteristics are combined with the SSO noise to calculate the BER as shown conceptually in Fig. 6. The BER curves of the complete system as a function of voltage and time are shown is shown in Figs. 7(a) and 7(a), respectively. The curves labeled ① and ② show the system margin degradation due to only SSO noises if the channel were ideal (no attenuation). The SSO noise using randomized switching reduces the voltage margin of the system by 50 mV, while the synchronized switching reduces the margins by more than 150 mV. Similarly, the timing margins reduction for randomized and synchronized switching are almost 0.2 UI and 0.4 UI, respectively. The equalized channel, curves

978-1-4244-4447-2/09 $25.00 © 2009 IEEE

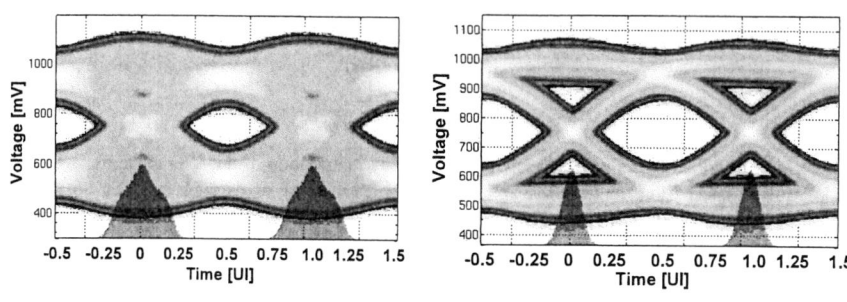

(a) Channel with synchronized switching. (b) Channel with randomized switching.

Fig. 5. Received eye diagrams with SSO noise from synchronized and random switchings.

labled ③, without consider any SSO noise show a voltage and timing margins over 250 mV and 0.5 UI, respectively. When combining the signal channel and SSO noise effects, the overall voltage and timing margins for the randomized switching, curves labled ④, show significantly larger margins for the synchronized switching, curves labled ⑤.

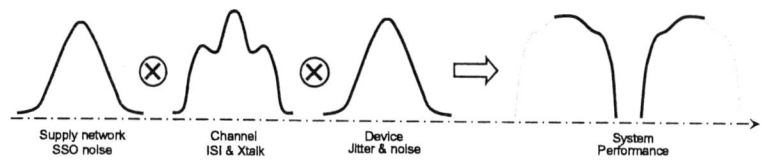

Fig. 6. The conceptual integration of the SSO noise, ISI, and Xtalk, and device noise and jitter distributions to generate system probability or BER.

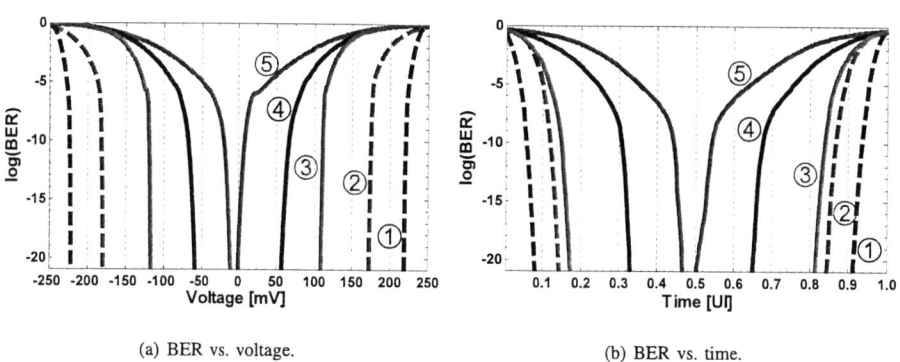

(a) BER vs. voltage. (b) BER vs. time.

Fig. 7. BER of the high-speed interface for ①: Only SSO noise from randomized switching, ②: Only SSO noise from synchronized switching, ③: Only equalized channel, ④: Combined equalized channel with randomized switching, and ⑤: Combined equalized channel with synchronized switching.

IV. CONCLUSIONS

The statistical modeling of SSO noise for high-speed system is presented. The deterministic simulation techniques give worst-case results that often result in very conservative designs as it ignores the probability of the synchronized switching happening at the transmitter. As a result, the voltage and timing margins of gigabit signaling systems using the two approaches give significantly different answers. As current high-speed systems have architectural constraints that minimize or remove the simultaneously switching of large number of drivers for several consecutive switching, the proposed statistical simulation can predict a real impact of SSO noise and give a more practical SSO noise limits to high-speed systems with and without embedded coding or scrambling blocks. The method can easily be integrated to the existing statistical simulation techniques that are currently used to analyze high-speed channels.

REFERENCES

[1] M. Swaminathan and E. Engin, *Power Integrity Modeling and Design for Semiconductors and Systems*, Prentice Hall, Boston, MA 2007.
[2] D A. Sanders, "Statistical simulation of physical transmission media," *IEEE Trans. on Advanced Packaging*, Vol. 32, No. 2, pp. 260-267, May 2009.

Low-Impedance Power Distribution Network
of Decoupling Capacitor Embedded Interposers for 3-D Integrated LSI System

Katsuya Kikuchi[1], Koichi Takemura[2], Chihiro Ueda[3], Osamu Shimada[2], Toshio Gomyo[2], Yukiharu Takeuchi[2],
Toshikazu Okubo[2], Kazuhiro Baba[2], Masahiro Aoyagi[1], Toshio Sudo[4], and Kanji Otsuka[3]

[1]National Institute of Advanced Industrial Science and Technology (AIST)
[2]Association of Super-Advanced Electronics Technologies (ASET)
[3]Faculty of Informatics, Meisei University
[4]Department of Electronic Engineering, Shibaura Institute of Technology
E-mail: k-kikuchi@aist.go.jp

Abstract: We evaluated low-impedance power distribution network (PDN) of decoupling capacitor embedded interposers for 3-D integrated LSI system. Measurements are carried out using the developed impedance analyzer system of a wide frequency range for evaluating ultralow impedance, and calculations are carried out using 2.5-D finite element method (FEM) electromagnetic field simulator. We fabricated various types of capacitor mounted or capacitor embedded interposers test element group (TEG), such as surface-mounted and embedded chip capacitors, and thin film capacitors on silicon interposers using the same simple design to compare measurement results with calculation ones. As a result, the chip capacitor embedded organic interposer TEG and thin film capacitor embedded silicon interposer TEG could provide low PDN impedance at a wide frequency range of up to 10 GHz. In particular, the interposer TEGs of the thin film capacitor embedded interposer that shows a low impedance of approximately 0.001 Ω could be evaluated and calculated accurately. By using chip capacitor embedded or thin film capacitor embedded interposers for 3-D integrated LSI system, it is expected that the PDN of the system can be achieved ultralow PDN impedance.

1. Introduction

Owing to the advancements in the operational speed and integration density of LSI chips, high-performance digital electronic systems have been realized. The clock frequency of a CPU-LSI chip exceeds GHz range. However, to realize a computing system using such a CPU-LSI chip, high- performance LSI packaging and interconnects are required.

Recently, the three-dimensional (3-D) integrated LSI system, which consists of 3-D stacked LSI chips with through silicon vias, a silicon interposer with thin-film decoupling capacitor, and a decoupling chip capacitors embedded organic interposer on a printed circuit board (PCB), has attracted attention for realizing an advanced high-performance system, as shown in Fig. 1 [1]-[6]. The system can provide enormous advantages in terms of achieving multifunctional integration, improving the system speed, and reducing power consumption for next-generation LSIs. Moreover, it is expected that a highly dependable system can be realized owing to the redundancy of multiple LSI chips. However, stacked multiple LSI chips may result in severe power integrity problems. For example, large simultaneous switching noise (SSN) is generated when various stacked chips switch simultaneously. This large SSN may lead to some critical problems, such as the voltage fluctuation on power distribution network (PDN), and cause the system failures. In particular, for realizing ultrahigh-speed signal processing in a 3-D integrated LSI system, it is necessary to suppress the SSN up to a high-frequency range by introducing decoupling capacitor embedded interposers. Such a PDN provides ultralow PDN impedance, as shown in Fig. 2. The required impedance is determined using the operating voltage and average

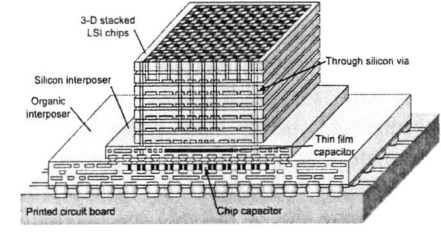

Fig. 1. Schematic of the 3-D integrated LSI system.

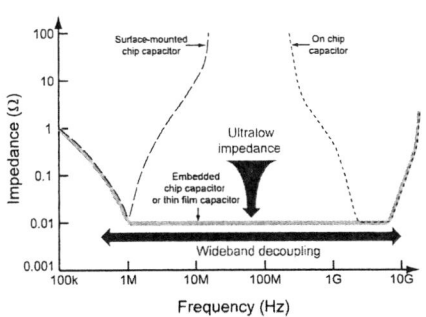

Fig. 2. Schematic of ultralow-impedance PDN using wideband decoupling capacitor of device embedded interposer for 3-D integrated LSI system.

978-1-4244-4447-2/09 $25.00 © 2009 IEEE

current drawn by the component, where the impedance that the PDN must be designed to meet is called the target impedance. The target impedance is defined as the noise voltage tolerance ratio. The target impedance Ztarget is expressed using Eq. (1), where V is supply voltage, I is supply current, and α is maximum voltage ripple tolerance.

$$Z_{target} = \alpha \times V / I \qquad (1)$$

For example, for a supply voltage of 1 V with a 5 % maximum ripple tolerance requirement and a supply current of 5 A, the target impedance should be 10 mΩ. The PDN should have an impedance of less than the target impedance over a frequency range of DC to hundreds of MHz [7][8]. In general, impedance peaks appear in the frequency range from a few MHz to GHz, owing to circuit resonance from the capacitance and the effective inductance of the PDN; thus, a good PDN design must maintain this impedance peak below the target impedance [9][10]. Therefore, it is required that the high-performance interposers that provide ultralow PDN impedance.

In this paper, we have developed various types of capacitor mounted or capacitor embedded interposer test element group (TEG), such as surface-mounted and embedded chip capacitors, and thin film capacitor on silicon interposer using the same design. It is reported that these interposer TEGs were evaluated by using a developed ultralow impedance evaluation system in detail.

2. Impedance Evaluation System

Generally, we use a vector network analyzer (VNA) to measure the PDN impedance. However, there is no VNA that can be used to measure the impedance below 0.01 Ω in the frequency range from DC to tens of MHz. Therefore, the impedance evaluation system realized consists of an impedance analyzer (IMA) and a VNA, as shown in Fig. 3. Low-frequency impedance measurement using the IMA and high-frequency impedance measurement using the VNA are carried out. That is, this system enables the use of the IMA for measuring the impedance below 0.01 Ω in the frequency range from DC to tens of MHz. One impedance measurement using two types of instrument can be seamlessly finished immediately at a wideband frequency. Additionally, the four-point impedance measurement technique is introduced in this system. This impedance measurement can be reliably and accurately carried out by a four-point technique where one pair of contacts is used to inject a signal into the device under test (DUT), and a second pair of contacts is used to sense the resulting signal across the DUT. As a result, an evaluation system for tens of micro-ohms of trans-impedance Z_{21} in the frequency range of 10 Hz to 40 GHz can be realized by introducing an IMA (Ultimetrix P4800K) and a 40 GHz VNA (Agilent 8722ES). To achieve precise contact to the DUT, the measurements were carried out using two 250-μm-pitch microwave contact probes (40-GHz microwave contact probes, Cascade Microtech ACP40-A-GSG-250) on a probe station (Cascade Microtech Summit 9000). By using two microwave contact probes, PDN impedance evaluation with contact pads is realized. Therefore, the PDN impedance can be measured directly without the connectors during the measurement.

3. Measurement Configuration and Simulation

We fabricated various types of capacitor mounted or capacitor embedded interposer test element group (TEG), such as surface-mounted and embedded chip capacitors, and thin film capacitor on silicon interposer using the same

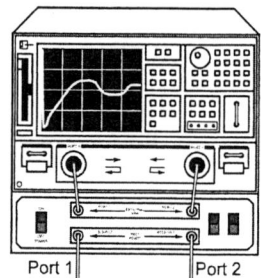

Fig. 3. Schematic of a developed ultralow impedance evaluation system is shown.

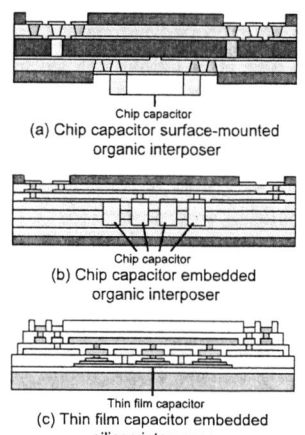

(a) Chip capacitor surface-mounted organic interposer

(b) Chip capacitor embedded organic interposer

(c) Thin film capacitor embedded silicon interposer

Fig. 4. Cross-sectional views of capacitor mounted or capacitor embedded interposer TEGs: (a) chip capacitor surface-mounted organic interposer, (b) chip capacitor embedded organic interposer, and (c) thin film capacitor embedded silicon interposer.

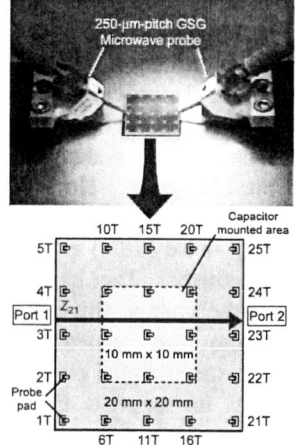

Fig. 5. Measurement system configuration of two 250-μm-pitch microwave contact probes on a probe station.

978-1-4244-4447-2/09 $25.00 © 2009 IEEE   26

design to verify the system, as shown in Fig. 4. The size of this interposer TEG is 20 mm × 20 mm and that of the capacitors mounted area on the interposer is 10 mm × 10 mm, as shown in Fig. 5. The capacitors that show 1.2 - 2.4 μF were mounted on each of these interposer TEGs at a 10 mm × 10 mm part. Twenty-five 0816 chip capacitors of LW reverse type are on the chip capacitor surface-mounted organic interposer. For 0816 chip capacitors, the values are 47 nF. Twenty-four 0603 chip capacitors are inside the chip capacitor embedded interposer TEG. For 0603 chip capacitors, the values are 100 nF. One SrTiO$_3$ thin film capacitor of 1.2 μF is inside the silicon interposer TEG. The PDN trans-impedance Z_{21} was measured at the 3rd and 23rd terminals, as shown in Fig. 5. These interposer TEGs have been evaluated in terms of the PDN impedance using the impedance evaluation system.

We analyzed various types of interposer TEGs using Sigrity PowerSI as 2.5-D finite element method (FEM) electromagnetic field simulator, as shown in Fig. 6. Simulation model are include all element of constructing interposer. For example, probe pads are shown in this figure.

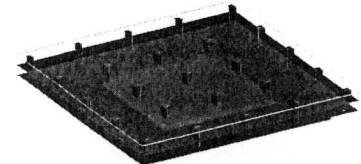

Fig. 6. Schematic of thin film capacitor embedded silicon interposer on the simulator.

4. Measurement Results

We analyzed the PDN impedance of the decoupling capacitor embedded interposer TEGs using the developed ultralow impedance evaluation system and the simulator. The measured PDN trans-impedances Z_{21} are plotted against frequency in Fig. 7. As shown in this figure, measurement results are in good agreement with the calculated ones at a wide frequency range of up to 10 GHz. As a result, the chip capacitor embedded organic interposer TEG and thin film capacitor embedded silicon interposer TEG could provide low PDN impedance at a wide frequency range of up to 10 GHz. In particular, the interposer TEGs like thin film capacitor embedded interposers that show a low impedance of approximately 0.001 Ω could be evaluated and calculated accurately. Therefore, it is considered that the change in the impedance is caused by the small change in the layout of the chip capacitors, and the difference in the electrode structure of the thin film capacitor using high-k dielectric materials can be evaluated and calculated.

Figure 8 is plotted again on the basis of Fig. 7 with the frequency axis on a logarithmic scale. Results show that this measurement system can be used to evaluate the PDN impedance of decoupling capacitor embedded interposer TEGs at a wide frequency range of 1 MHz to 10 GHz. As shown in this figure, measurement results are in good agreement with the calculated ones at a wideband frequency. However, the measured PDN impedance shows a discontinuous change at the frequency between 40 and 50 MHz, as shown in Fig. 8. This discontinuous change is considered to be due to the lack of a measurement dynamic range. Because the dynamic range of the VNA at the frequency between 50 and 840 MHz is 67 dB, this system can be used to measure the impedance of approximately 0.02 Ω at this frequency range. For improvement, we plan to introduce a new VNA with a high measurement dynamic range at the frequency range.

5. Conclusions

We evaluated low-impedance PDN of decoupling capacitor embedded interposers for 3-D integrated LSI system. Measurements are carried out using the developed impedance analyzer system of a wide frequency range for evaluating ultralow impedance, and calculations are carried out 2.5-D FEM electromagnetic field simulator. We fabricated various types of capacitor mounted or capacitor embedded interposers TEG, such as surface-mounted and embedded chip capacitors, and thin film capacitors on silicon interposers using the same simple design to compare measurement results with calculation ones. As a result, the chip capacitor embedded organic interposer TEG and thin film capacitor embedded silicon interposer TEG could provide low PDN impedance at a wide frequency range of up to 10 GHz. In particular, the interposer TEGs of the thin film capacitor embedded interposer that shows a low impedance of approximately 0.001 Ω could be evaluated and calculated accurately. By using chip capacitor embedded or thin film capacitor embedded interposers for 3-D integrated LSI system, it is expected that the PDN of the system can be achieved ultralow PDN impedance.

We plan to introduce a new VNA with a high measurement dynamic range at a frequency between 50 MHz and 1 GHz range. As a result, an ultralow impedance evaluation system that has a high measurement dynamic range at the entire measurable frequency band can be realized. In consequence, it is expected that the accurate and wideband impedance evaluation of the substrate with the EBG structure can be realized.

Acknowledgment

This work was entrusted by NEDO "Development of Functionally Innovative 3D-Integration Circuit (Dream Chip) Technology" project.

References

[1] T. Matsumoto *et al.*, "New three-dimensional wafer bonding technology using the adhesive injection method," *Jpn. J. Appl.Phys.*, vol.37, no.3B, pp.1217-1221, 1998.

[2] M. Koyanagi *et al.*, "Future system-on-silicon LSI chips," *IEEE Micro*, vol.18, no.4, pp.17-22, 1998.

[3] K. Takahashi *et al.*, "Current status of research and development for three-dimensional chip stack technology," *Jpn. J. Appl. Phys.*, vol.40, no.4B, pp.3032-3037, 2001.

[4] K. Tanida *et al.*, "Au bump interconnection in 20 μm pitch on 3D chip stacking technology," *Jpn. J. Appl. Phys.*, vol.42, no.10, pp.6390–6395, 2003.

[5] K. Tanida, M. Umemoto, N. Tanaka, Y. Tomita, and K. Takahashi, "Micro Cu bump interconnection on 3D chip stacking technology," *Jpn. J. Appl. Phys.*, vol.43, no.4B, pp.2264-2270, 2004.

[6] J.U. Knickerbocker *et al.*, "Development of next-generation system- on-package (SOP) technology based on silicon carriers with fine-pitch chip interconnection," *IBM Journal of Research and Development*, vol.49, no.4/5, pp.725-753, 2005.

[7] L.D. Smith, R.E. Anderson, D.W. Forehand, T.J. Pelc, and T. Roy, "Power Distribution System Design Methodology and Capacitor Selection for Modern CMOS Technology," *IEEE Trans. Advanced Packaging*, vol.22, no.3, pp.284-291, 1999.

[8] P. Muthana, M. Swaminathan, E. Engin, P.M. Raj, and R. Tummala, "Mid Frequency Decoupling Using Embedded Decoupling Capacitors," *Proc. 14th Electrical Performance of Electronic Packaging*, pp.271-274, 2005.

[9] B. Garben, R. Frech, J. Supper, and M.F. McAllister, "Frequency Dependencies of Power Noise," *IEEE Trans. Advanced Packaging*, vol.25, no.2, pp.284-291, 2002.

[10] W.D. Becker *et al.*, "Modeling Simulation and Measurement of Mid-Frequency Simultaneous Switching Noise in Computer Systems," *IEEE Trans. Components Packaging and Manufacturing Technology, Part B*, vol.21, no.2, pp.157–162, 1998.

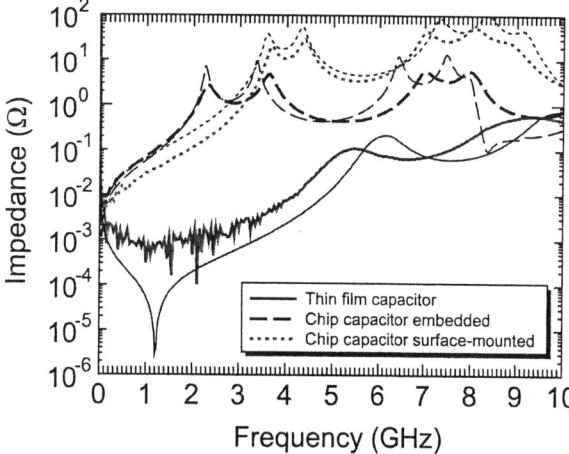

Fig. 7. Measured PDN trans-impedance Z_{21} plotted against frequency, where the solid line indicates the thin film capacitor embedded silicon interposer, the dashed line indicates the chip capacitor embedded organic interposer, and the dotted line indicates the chip capacitor surface-mounted organic interposer. (Horizontal axis: linear frequency; vertical axis: logarithmic PDN impedance. Thin line: simulation; thick line: measurement.)

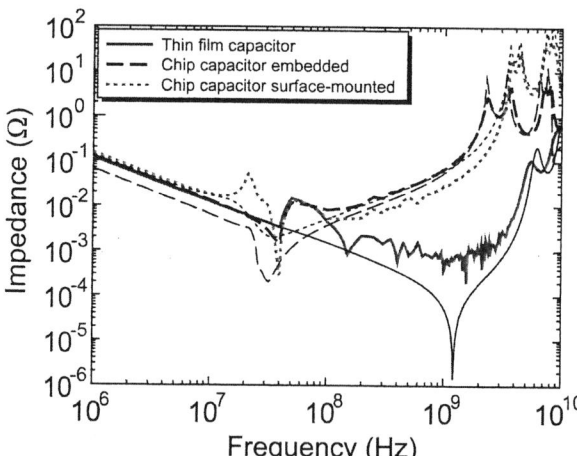

Fig. 8. Logarithmic plot of measured PDN trans-impedance Z_{21} against frequency, where the solid line indicates the thin film capacitor embedded silicon interposer, the dashed line indicates the chip capacitor embedded organic interposer, and the dotted line indicates the chip capacitor surface-mounted organic interposer. (Horizontal axis: logarithmic frequency; vertical axis: logarithmic PDN impedance. Thin line: simulation; thick line: measurement.)

Resonance-aware Methodology for System Level Power Distribution Network Co-design

Amirali Shayan, Kevin Bowles, Sorin Dobre, Mikhail Popovich, Xiaoming Chen, Christopher Pan

Qualcomm, Inc. , San Diego, CA 92121, USA

E-mail: {amiralis, kbowles, sdobre, mikhailp, xiaoming, ycpan} @ qualcomm.com

Abstract- Power delivery network (PDN) design continues to be a major challenge because it demands a good portion of available silicon, package, and board routing resources. In this paper, we outline a frequency and time domain co-design flow that uses frequency domain results to construct time domain input vectors, resulting in a resonance aware time domain analyses flow that can highlight low and mid frequency behaviors dominated by board and package components and parasitics.

I. INTRODUCTION

Power delivery network (PDN) design continues to be a major challenge because it demands a good portion of available silicon, package, and board routing resources. In the past decade, a number of EDA tools came into the market to assist designers optimizing PDN while minimizing its footprint in the overall system. The tools typically are divided into frequency and time domains. The frequency domain tools typically employ specialized fast electromagnetic solvers that take advantage of layered dielectric structures in package and board. Due to much greater complexity and finer feature set of on-die power grids, silicon is usually modeled as a lattice of lumped RLC elements. For the frequency domain tools, the emphasis is therefore on analyzing the low and mid frequency PDN system responses[3]. One shortcoming of frequency domain tools is that the results are not directly expressed in millivolts of voltage drop seen by transistors. The V_{DD} voltage drop and ground bounce behavior could not be distinguished from the frequency domain electromagnetic solver results to reflect modification needed to enhance the power delivery. They also do not analyze on-die power/ground grid structure in minute detail to assist silicon designers detect missing vias and shorts. In the last few years, time domain PDN analysis tools grew out of the static IR drop tools that model full chip power/ground grid structures[2]. In the new time domain tools, die level interconnect is modeled in fine detail, along with on-die decoupling capacitors, power gating transistors, and switching elements. Through proprietary algorithms, time domain tools are now capable of running transient simulations to hundreds of nanoseconds. The results of time domain simulations can be very useful because they contain both current demand and instantaneous voltage information. Current demand, for example, is a critical piece of information when trying to set frequency domain impedance spec, and localized voltage drop hotspots can guide silicon designers to improve power/ground mesh in those regions. One limitation of time domain analyses is the necessity to stimulate a highly complex silicon design that consists of millions of transistors. Often an activity factor (AF) based input vector is used to toggle a set percentage of gates in the design at frequencies of individual functional blocks. At Qualcomm, both frequency domain and time domain PDN analysis flows are used to design the highly integrated system on chip (SoC) mobile phone chipsets. In this paper, we will document a flow improvement to guide time domain analyses with PDN system resonances obtained through frequency domain analyses. In this fashion, the time domain analyses can better highlight PDN system behavior when it is perturbed at or near its resonance frequencies.

The rest of the paper is organized into five sections. Section II consists of a high level review of theoretical background; Section III focuses on frequency co-design of PDN system; Section IV outlines a resonance aware time domain methodology; Section V contains simulation results, and Section VI summarizes our findings and concludes the paper.

II. THEORETICAL BACKGROUND

A. Resonance-aware modulation

The relative importance in creating the right stimulus for PDN analysis can be seen in the following time domain result inFigure1. In this result a simple lumped PDN model having a single resonant frequency at 75 MHz is stimulated by a high speed clock (~1GHz) whose frequency is many times that of the PDN resonance. For the initial portion of the test the circuit draws a fixed charge per instruction cycle which sets up a steady DC current average with no energy near resonance. As a result there is little if any affect since resonance is not disturbed. This is the period prior to 300ns.

At a later time in the test the charge demand per cycle (top pane after 300ns) is made to vary at a rate coincident with PDN resonance frequency. Now an AC current component in the package (middle pane) is developed which perturbs the target frequency and a corresponding low frequency voltage component (lower pane) manifests itself in the power domain. It is of interest to note how the magnitude of this lower frequency component can rival or possibly exceed the localized high speed droop occurring at the clock rate. This result shows how time domain stimulus needs to target certain frequencies identified through frequency domain analysis even though the clock rate is well above PDN resonance.

B. Load current modulation

The notion of load modulation is very applicable for pipelined processors since the executed instructions can vary in both power magnitude as well as completion time relative to issue. The mathematics behind the load modulation properties for both time and frequency domain analysis lend themselves from basic AM radio modulation theory for which there are numerous sources [1].The equation for a generalized sinusoidal carrier can be expressed as the following: $y(t) = a(t) \cdot \cos(\omega_c t)$

978-1-4244-4447-2/09 $25.00 © 2009 IEEE

Where we assume a(t) varies slowly compared to the 'carrier' frequency. The term a(t) is termed the envelope of the carrier frequency. For PDN analysis we would replace the single cosine term representing a 'carrier' with a more appropriate Fourier sum representing the cycle-to-cycle current demand of the on-die circuits whose fundamental clock rate is known. Thus the modulated current demand would take on the following form:

$$y(t) = a(t).\left[\frac{\alpha_0}{2} + \sum_{n=1}^{N}\left(\alpha_n.\cos\left(n\omega_c t\right) + \beta_n.\sin\left(n\omega_c t\right)\right)\right] \quad (1)$$

Even though the modulating carrier a(t) is expressed as a general aperiodic signal the PDN issues which arise occur when enough periodicity exists such that resonance is only momentarily perturbed. This could be for only a few cycles over an extended length of time. In the frequency domain we expect the 'carrier' or high speed current demand which occurs at high frequency to appear as a fundamental clock spur with its associated harmonics implied through Fourier analysis. In addition, we expect to see the modulated energy positioned relative to DC and also positioned around the carrier much like side-band energy with an AM carrier Fig.

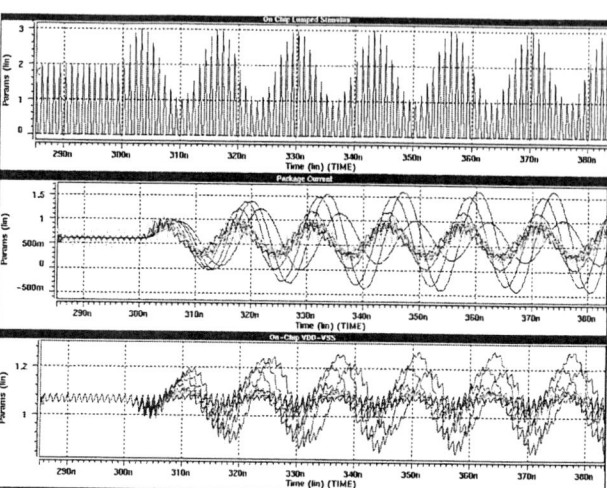

Figure1. Resonance-aware load current modulation impact on voltage variation (Top: modulated current, Middle: package current, Bottom: on-chip voltage variation).

8 and Fig. 9 [1]. For typical PDN systems and the resulting resonant frequencies we only need to concern ourselves with the energy in the 50MHz region for standard cores and somewhat higher frequencies for IO interfaces.

III. BROADBAND FREQUENCY DOMAIN CO-DESIGN

PDN design for highly integrated mobile chipsets spans board, package, and silicon. PDN modeling is complicated by the following unique properties: (i) Large problem size. Board PDN network is electrically large with power shapes ranging in 10's of millimeters; (ii) Wide range of feature sizes. Feature sizes found in a PDN network varies from 10's of millimeter on board, to 10's of microns on package, to micros on silicon; (iii) Fully coupled system. Power delivery shapes and lines are fully coupled in a single system. Accurate modeling of PDN at board/package and package/die resonances is critical to ensure sound design; and (iv) Highly non-uniform die-level capacitive and current loading.

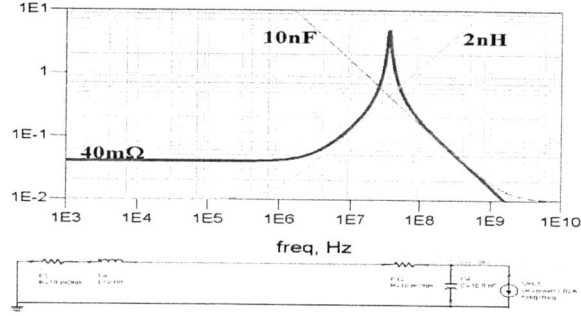

Figure 2. RLC resonance with R=40mΩ, L=2nH, and C=10nF.

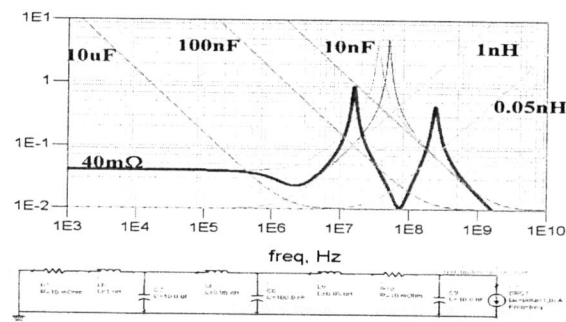

Figure 3. RLC resonance after the 2nH inductance is segmented with shunt capacitances.

PDN design essentially behaves as a RLC circuit. The resistance R is the accumulated board/package routing parasitic and on-die P/G mesh parasitics; the inductance is the accumulation of board and package routing; and the capacitance comes in from silicon intrinsic capacitance and any silicon, package, and board decoupling capacitors.

Figure 2 shows a lumped circuit with 40mΩ of resistance, 2nH of inductance, and 10nF of capacitance. This circuit has a resonance with magnitude of roughly 5Ω at about 30MHz. If the current source on the right hand side perturbs this resonance, a large voltage drop is expected to develop across the inductor. The design of PDN is to minimize this resonance and, if possible, push it outside of current spectral bandwidth expected from the current source. One of the methods for controlling the resonance peak and to shift it higher frequencies is to break the inductance into segments by inserting shunt capacitances. In Figure 3, a 10uF capacitor and a 100nF capacitor are inserted to break the 2nH into 1nH, 0.95nH, and 50pH segments. At about 2.5MHz, the effectiveness of the 10uF is maximized and the 0.95nF inductor after this capacitor starts to dominate the system impedance. The system reaches a resonance at about 20MHz, which is a product of the 0.95nH inductor and 100nF capacitor. After this frequency, the 100nF capacitor dominates the impedance profile until around 100MHz when the 50pH inductor starts to dominate. The system reaches another resonance peak at around 200MHz and after this point the 10nF capacitor becomes the

978-1-4244-4447-2/09 $25.00 © 2009 IEEE   30

dominating factor. This simple RLC example serves as an example of how PDN system impedance can be controlled via insertion of decoupling capacitors. The large capacitors are inserted as close to the silicon package as possible.

IV. RESONANCE-AWARE TIME DOMAIN METHODOLOGY

A. Time domain flow

We have developed a new method based on current demand shaping which can be used in Apache Redhawk® [2] to create a stimulus which will generate a current demand during a VCD analysis run, with important energy content close to the resonant frequency of the PDN as presented in Figure 3. One of the major challenges of PDN sign off is the ability to generate realistic design stimulus which will excite the power delivery network close to the resonant frequency. In other words, the ability to create a current stimulus which is design specific and has enough energy content close to the resonant frequency of the system PDN. We have developed a new method based on current demand shaping which can be used in Apache Redhawk® to create a designer shaped stimulus. This stimulus will generate a current demand during a VCD analysis which has important energy content close to the resonant frequency of the PDN as presented in Figure 3. Current modulation has key parameters: (i) modulation functions (sine, cosine, pulse, etc) (ii) modulation frequency (user defined), and (iii) modulation depth (user defined). The resonant frequency of the PDN is fixed and determined by the physical implementation of the die, package and PCB. Usually the resonant frequency is close to 20-50 MHz as presented in Figure 3. We have extracted the impedance profile for the complete PDN for VDD domains die models, complete layout package and layout PCB. We can create a VCD stimulus from different vectorless runs by knowing the resonant frequency of the PDN which will stimulate the PDN close to the resonant frequency.

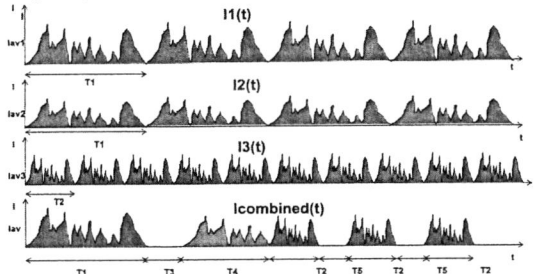

Figure 4. Current waveform generated from individual wavelets.

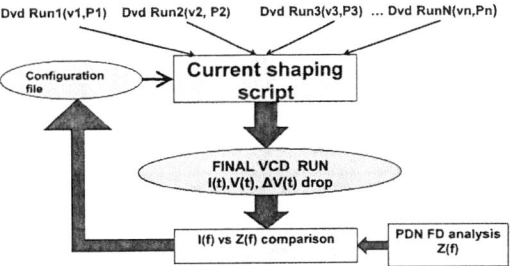

Figure 5. Flowchart of time domain worst case analysis PDN Z(f) frequency aware.

The process of generating the desired current demand is controlled by a configuration file which defines the duration of individual current wavelets. From multiple vectorless runs, we can create in time domain a new VCD run with the desired designed current profile as presented in Figure 4. The combined current shape can be validated using a FFT analysis to ensure that we have maximized the spectral current components in frequency domain close to the resonant frequency of the PDN. The difference in spectral content for the current demand between a vectolress runs and the "designed" VCD run is presented in Figure 7. The basic function for the final DESIGNED current demand is presented below in equation (2) and Figure 4:

$$
I_{com} = \begin{cases}
I_1(t) & (t \le T1) \\
0 & (T1 < t \le T1 + T3) \\
I_2(t) & (T1 + T3 < t \le T1 + T3 + T4) \\
I_3(t) & \{T1 + T3 + T4 + n*(T5+T2) < t \le T1 + T2 + T3 + T4 + n*(T5+T2)\} \\
0 & \{T1 + T2 + T3 + T4 + n*(T5+T2) < t \le T1 + T3 + T4 + (n+1)*(T5+T2)\} \\
& \text{for } n \ge 0
\end{cases}
\tag{2}
$$

Table 1. Setup table for current modulation of the core

Cycle Count	Clock Frequency (MHz)								
	550	575	600	625	650	675	700	725	750
6	91.67	95.83	100.00	104.17	108.33	112.50	116.67	120.83	125.00
8	68.75	71.88	75.00	78.13	81.25	84.38	87.50	90.63	93.75
10	55.00	57.50	60.00	62.50	65.00	67.50	70.00	72.50	75.00
12	45.83	47.92	50.00	52.08	54.17	56.25	58.33	60.42	62.50
14	39.29	41.07	42.86	44.64	46.43	48.21	50.00	51.79	53.57
16	34.38	35.94	37.50	39.06	40.63	42.19	43.75	45.31	46.88
18	30.56	31.94	33.33	34.72	36.11	37.50	38.89	40.28	41.67
20	27.50	28.75	30.00	31.25	32.50	33.75	35.00	36.25	37.50

B. Architectural perspective

Table 1 shows how a processor core can stimulate any given resonant frequency by means of executing finite code loops at various clock frequencies. The table is set up as a guideline for targeted stimulus such that assembly code for either test purposes or VCD generation for dynamic IR drop analysis can be facilitated.

In this context, the cycle count is the implied code loop length which is repeated and the horizontal scale is clock frequency in MHz. The purpose of any code loop is to generate a variation in current demand such that the perceived min and max possible current magnitudes are explored. Generating such stimulus is the essence of exploring what the dynamic range of current variation for any given system component is, during a specific resonance cycle. Only real code running on a design for simulation purposes will yield the true bounds so that effective modulation indexes can then be established for further use.

978-1-4244-4447-2/09 $25.00 © 2009 IEEE

V. EXPERIMENTAL RESULTS

We can generate the worst realistic *di/dt* using proposed method, which drive the worst voltage noise for the complete system. Based on the method presented above using Apache Redhawk®[2], we are creating a system level simulation test bench with package and PCB which will produce the worst realistic voltage noise. This designer driven simulation scenario can be used for the sign off of the PDN, to validate the voltage corners used during time domain analysis and to validate the results of the frequency domain PDN optimization and analysis. As an example of multi resonance PDN design Figure 8 demonstrates time and frequency domain waveforms of Figure 2 with dual resonance of a digital system clocked at 1.0 GHz but having two tone modulations at 20 MHz and 200 MHz. The first pane is the time domain waveform while the later is the frequency domain representation of this modulation scenario. In this example, piece-wise linear current waveform amplitude is 0.5A with rise time and fall time of 0.5ns. Dual resonance frequency (f_1=20MHz and f_2=200MHz) modulation is performed based on: $a_1 \cdot \sin(2\pi f_1 t) + a_2 \cdot \sin(2\pi f_2 t)$ where $a_1 = a_2 = 0.25A$ and the modulation index is 0.5.

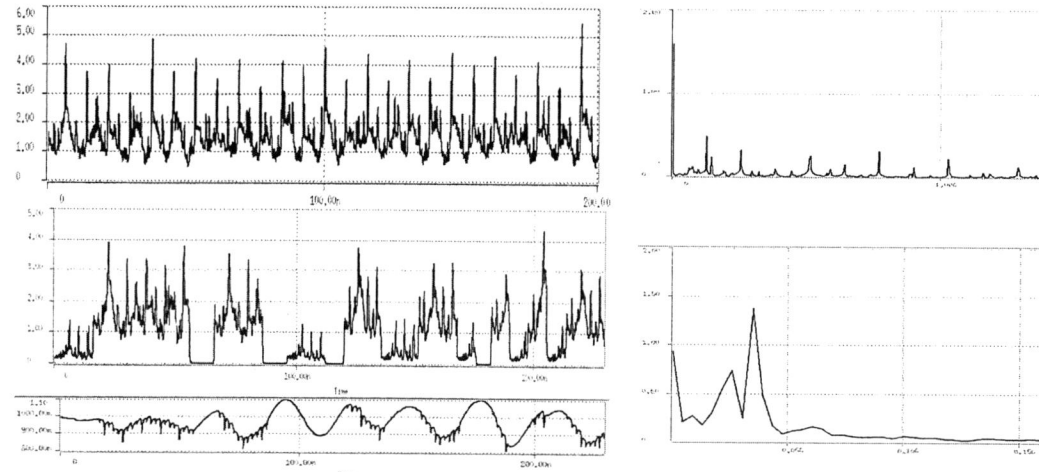

Figure6. Top: Original on-chip current demand from a vectorless run. Middle: Modulated current from the proposed method. Bottom: Voltage waveform for an instance with the modulation where the resonance and low frequency content is more pronounced.

Figure 7. Top: FFT comparison between the two current demands. Bottom: The modulated current has a frequency content in the range of resonance(about 30-40 MHz).

In the provided waveforms one can observe a portion of the 20MHz envelope while multiple cycles of the 200MHz envelope are visible in the time domain pane. The frequency domain pane on the right clearly shows that energy exists at 20MHz and 200MHz and even though this example is contrived, it is the case that pipelined processors which iterate code loops of fixed length can focus energy into a specific frequency while that energy level varies with time. In general, the problem is more aperiodic in nature but we need to be aware of what can or will happen when resonance is perturbed and try to determine the low frequency voltage noise magnitude.

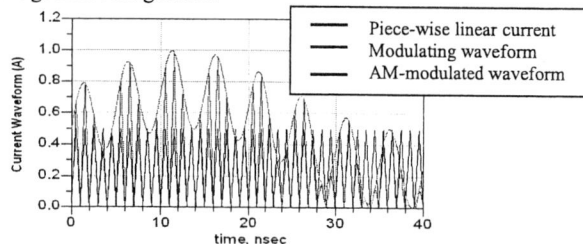

Figure 8. Dual resonance current modulation time domain waveform

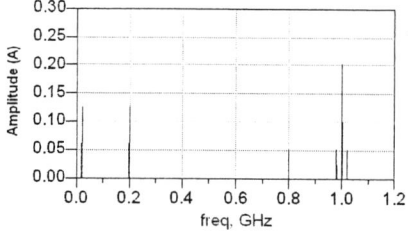

Figure 9. Dual resonance current modulation energy spectrum

VI. CONCLUSIONS

This work, proposes an enhanced flow for the analysis of PDN under worst case realistic load current. The proposed methodology is based on resonance aware modulation of core power activity to perturb the PDN under resonance and mimic core activity realistically under worst case operation mode to guide design of a robust PDN.

REFERENCES

[1] A. Waizman, "CPU power supply impedance profile measurement using FFT and clock gating," EPEP, 2003.

[2] S. Wane, A.Y. Kuo, "Chip-package co-design methodology for global co-simulation of re-distribution layers (RDL)," EPEP, 2008.

[3] A. Shayan, X. Hu, H. Peng, W. Yu, W. Zhang, C.K. Cheng, M. Popovich, X. Chen, L. Chua-Eaon, X. Kong, "Parallel flow to analyze the impact of the voltage regulator model in nanoscale power distribution network," ISQED 2009.

978-1-4244-4447-2/09 $25.00 © 2009 IEEE

Design and Characterization of a 12.8GB/s Low Power Differential Memory System for Mobile Applications

Dan Oh, Sam Chang, Chris Madden, Joong-Ho Kim, Ralf Schmitt, Ming Li, Chuck Yuan
Fred Ware, Brian Leibowitz, Yohan Frans, and Nhat Nguyen

Rambus Inc.
4440 El Camino Real, Los Altos, CA 94022
Tel: 650-947-5363, Fax: 650-947-5001, doh@rambus.com

ABSTRACT

This paper describes the design and characterization of a low power differential memory interface targeted for mobile applications. The initial design of the memory interface achieves 2.7 to 4.3GB/s data bandwidth and consumes 3.3mW/Gb/s at 4.3GB/s operation. The design allows two x16 stacked dies to be fit into a 12mm PoP package, achieving a 12.8GB/s aggregated data bandwidth based on 3.2Gb/s per pin. A low swing signaling based on a voltage-mode differential driver is reviewed and its performance is analyzed. We demonstrate that, compared to LPDDR2 memory interface based on single-ended signaling, the differential memory interface overcomes most of channel related issues such as crosstalk and SSO noise and provides a very clean channel response. Thus, the resulting extra system margin can be used to compensate for extra timing jitter and system noise, enabling lower power and lower system cost. To evaluate the impact of timing jitter and system noise to system performance, a statistical link modeling and simulation methodology is employed. Two test systems are built based on wirebond-based Package-on-Package (PoP) and BGA-based Chip-to-Chip (C2C) module to characterize the memory system performance and to validate the memory statistical link model. The correlation result showed a good agreement in the system bit error rates (BER) between measurement and simulation.

I. INTRODUCTION

Mobile devices, such as smart phones, portable gaming devices, and media players, become an emerging computing platform. Many of these devices not only need to handle rich context web pages but they provide a HD quality video processing power. Providing a required data bandwidth for mobile applications using a small form factor and low power consumption is a quite challenging task. Achieving the bandwidth beyond 6.4GB is very difficult using the current LPDDR2 architecture with cost effective packaging solutions such as PoP. It may require advanced packaging solutions such as Through-Silicon-Via (TSV) or expensive fine ball pitch packaging technologies.

A novel memory architecture based on low-swing differential signaling is reviewed in this paper. Compared to the single-ended signaling which suffers from crosstalk, SSO, EMI issues [1], the differential interface provides a clean channel response enabling a low swing signaling. It is demonstrated that the differential low-swing memory system can provide a 10.8-12.8GB/s aggregated data bandwidth using a low cost PoP with either two DQx16 stacked dies or one DQx32 die. Our estimation shows that our interface consumes only ~40% of an LPDDR2 system at the same data bandwidth.

Small form factor requirement in 3D integration causes additional design challenges. Providing a clean power supply for various interfaces and processor is difficult due to tight integration and power requirement. Hence, it is important to model the impact of various noise sources in 3D packaging. Predicting link performance with noise using traditional SPICE simulation is limited due to computational efficiency [2]. We used the statistical approach described in [3] to predict the link performance. The correlation and simulation result show that the differential memory interface provide a large headroom without using expensive 3D packaging solutions.

II. LOW POWER DIFFERENTIAL MEMORY INTERFACE ARCHITECTURE

The memory interface architecture is shown in Fig. 1 [4]. The link consists of 8 data and 2 command/address (CA) differential signals. These bi-directional links operate from 2.7 to 4.3Gb/s per link based on 8:1 multiplexing suitable for emerging LPDRAM processes. A half-bit-rate clock signal is forwarded to DRAM to avoid any closed-loop timing circuits in the DRAM side; thus, the DRAM side clocking path can be easily shut down and turn on by controller. Separate phase interpolators in the controller are used to adjust both transmit and receive clock phases. The phase adjustment is done per-pin based in order to minimize any timing offset caused by device or a passive channel. This phase adjustment is especially handy for mobile applications with 3D packages as it is very difficult to perform trace length match due to a small form factor.

A low speed CMOS signal (PM) is used to change the power state of the DRAM links. The whole clock distribution on both controller and DRAM can be turned off by gating at the controller during idle (pause) times, saving significant power consumption during idle periods. The interface is implemented using TSMC 40nm LP process and operates from a single 1.1V supply. It consumes 114.7mW (3.3mV/Gb/s) at 4.3GB/s operation.

978-1-4244-4447-2/09 $25.00 © 2009 IEEE

Low-swing near ground signaling is used to reduce the I/O power consumption [5], [6]. The signaling block diagram is shown in Fig. 2. The output driver is an N-over N voltage-mode differential driver with near-ground 100mV common voltage. An on-chip linear regulator is used to provide a 200mV transmitter supply. Since the signaling is reference to ground, the I/O power supply voltage for controller and DRAM does not need to be common, simplifying supply network design. The signal is terminated differentially at the receiver, reducing impact of any power or ground noise at the receiver. The receiver input offset is trimmed with sub-mV resolution by offset calibration DACs in the data slicers. A highly sensitive receiver design is critical in low power signaling interface as it allows a lower transmitter swing which, in turn, reduces the driver size and its parasitic capacitance, allowing further reduction in swing requirement.

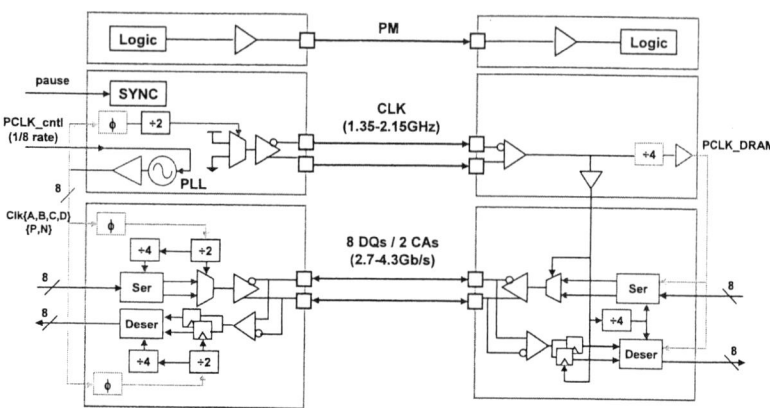

Fig. 1.   Memory interface architecture

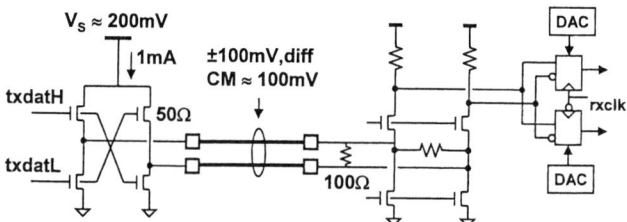

Fig. 2.   Near-ground voltage-mode differential signaling circuitry

III. Designs of PoP and C2C Systems

Two test vehicle systems are built based on BGA-based C2C module and PoP. The total number of 33 pads (23 signal pads, 5 power, and 5 ground pads) are used for each DQx8 interface. For the C2C module, shown in Fig. 3, the die is bonded to a substrate package and the whole package is in a 1mm pitch BGA socket. All signals in the substrate and PCB are routed using microstrip lines. The PoP design, shown in Fig. 4, used the same die design. All DRAM devices are emulated using low power ASIC process. The measured eye diagrams for both cases are also shown in Figs. 3 and 4. As shown in these figures, both systems show a very clean channel response.

A. PoP Design Comparison with LPDDR2 Interface

A multichip package (MCP) system with two stacked dies (Fig. 5) is considered to estimate the potential aggregated system bandwidth. A commonly used 12mm x 12mm (0.4mm pitch, 216 balls) package is considered. Two DQx16 differential memories operating at 3.2Gb/s per pin are used to achieve a total of 12.8GB/s data bandwidth. The ball assignment for our differential interface is shown in Fig. 6. The total of 168 balls (126 IO balls and 42 core power balls) is used out of 216 balls (182 IO balls and 30 core power balls).

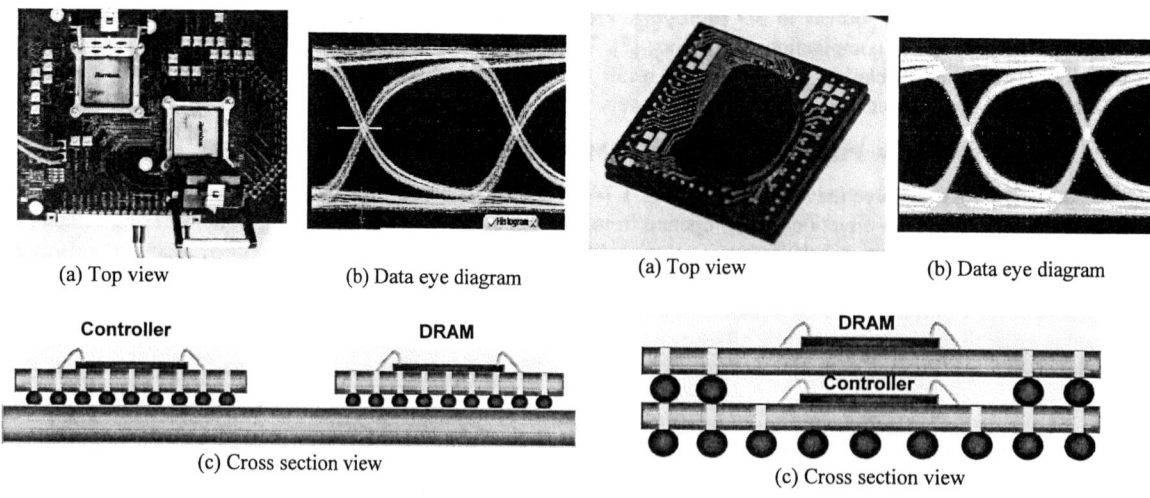

(a) Top view   (b) Data eye diagram

(a) Top view   (b) Data eye diagram

(c) Cross section view

(c) Cross section view

Fig. 3. C2C module based on 35mm BGA

Fig. 4. Package-on-package (PoP) design

As a comparison, we also consider an LPDDR2 case. Two x32 LPDDR2 memories, running at 800Mb/s per pin, are used to achieve a total of 6.4GB/s. The ball assignment is shown in Fig. 7. LPDDR2 interface requires a total of 212 balls leaving virtually no free package balls for core supply. This is unlikely to be an acceptable solution. One either has to use a larger PoP package or use smaller ball pitch. For example, to achieve 12.8GB/s using LPDD2 interface, one has to use four x32 800MHz/link LPDDR2 interface and 20mm x 20mm PoP

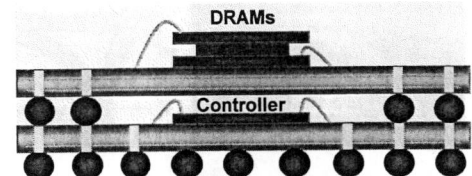

Fig. 5. Two stacked dies MCP system

package which increases system cost dramatically. In addition, the form factor requirement in mobile application such as cell phone, this is again unlikely to be a viable solution. The simulation eye diagram for reading from the top device is shown for LPDDR2 case in Fig. 8. Various simulations are performed to study the impact the individual noise sources: specifically, intersymbol interference (ISI), crosstalk, and SSO. As shown in the figure, the LPDDR2 design suffers significantly from SSO noise in addition to crosstalk, limiting the potential for data rate.[*] Our power estimation shows that the LPDDR2 interface at 6.4GB/s would consume about 2.6 times as much power as the differential case at the same data bandwidth and about 1.3 times of power consumption compared to the differential memory at 12.8GB/s.

IV. STATISTICAL LINK PERFORMANCE MODELING AND CORRELATION

As shown in the previous section, the differential memory interface provides greater channel margin and can run at much higher data rates than the single-ended interface. However, the relatively slow low-power process eventually limits the maximum data rate for efficient operation. Nonetheless, this extra margin is still useful to trade off with more relaxed system noise and timing jitter budgets or cheaper physical designs such as low cost package or PDN network designs. In this section, a statistical link simulation methodology described in [2], [3] is used to predict the overall link performance with system noise and timing jitter. In the clock forwarding architecture, the majority of supply noise at the transmitter can be tracked out and accurate modeling of this tracking effect is critical in determining the system performance. The new statistical formulation described in [7] is used to model this. For convenience, we used the C2C system to perform model to hardware correlation.

Fig. 9 shows the passive channel eye diagram correlation. The measurement is taken with averaging to filter out any random jitter. A linear behavior driver model is used in the statistical simulation. The measurement data shows some nonlinear behavior which cannot be modeled using the linear model. Nonetheless, the comparison shows a fairly good match. In Fig. 10, the measurement data without averaging is also compared with the statistical histogram eye. RJ of 2.3ps is used to match the eye diagram. Using the same system setup, we injected sinusoidal jitter at the transmitter to study jitter tracking at the transmitter. The jitter is excited at 50 and 150MHz and compared with measurement in Figs. 11 and 12.

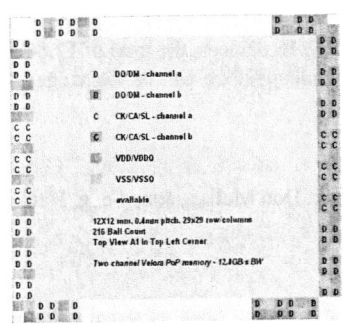

Figure 6. 2 DQx16 ball assignment

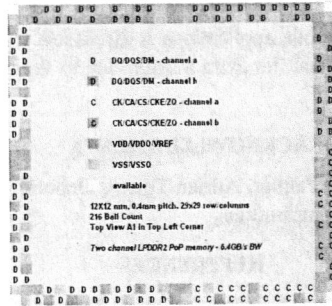

Figure 7. LPDDR2 ball assignment

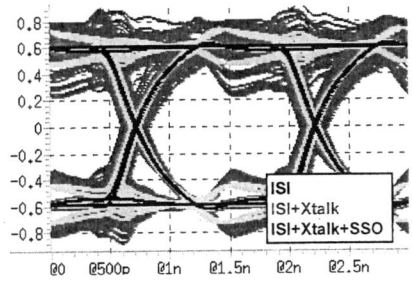

Figure 8. LPDDR2 eye diagrams

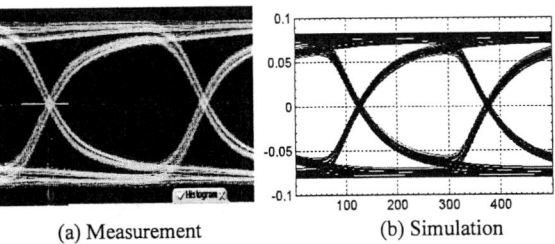

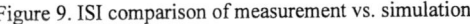

Figure 9. ISI comparison of measurement vs. simulation

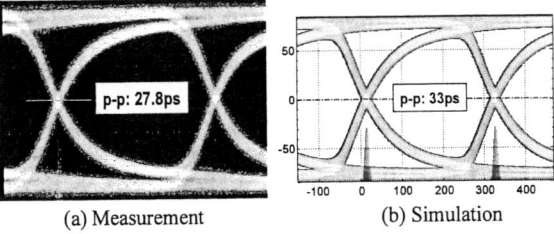

Figure 10. Comparison with random jiiter

---

[*] More detail comparison of single-ended and differential signaling techniques can be found in [4].

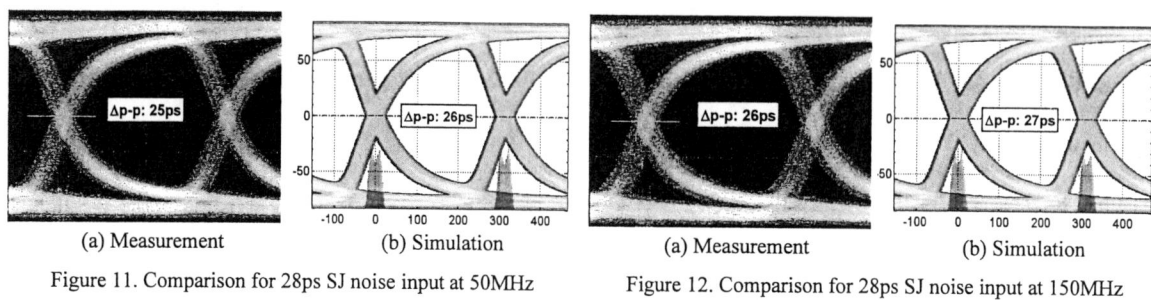

(a) Measurement   (b) Simulation   (a) Measurement   (b) Simulation

Figure 11. Comparison for 28ps SJ noise input at 50MHz   Figure 12. Comparison for 28ps SJ noise input at 150MHz

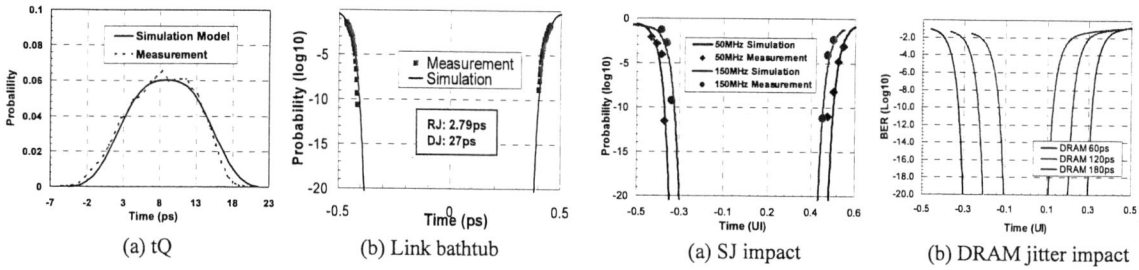

(a) tQ   (b) Link bathtub   (a) SJ impact   (b) DRAM jitter impact

Figure 13. Comparisons of (a) tQ measurement and simulation model, and (b) the measurement and simulated bathtub

Figure 14. (a) Correlation with different SJ frequencies and (b) what-if study of DRAM jitter level

The overall link model is generated based on the component-level measurements in order to compare the link bathtub curves. The transmitter timing error (tQ) is modeled with DCD=3ps, 2σ truncated Gaussian of 11ps p-p jitter, and measured PLL DJ. Fig. 13(a) shows the comparison of this timing model with measurement data after convolving with the ISI due to the transmitter package. The receiver setup and hold times are modeled with 9ps uniform PDF. The measured receiver sensitivity of 2mV p-p of deterministic noise (DN) and 0.8mV random noise (RN) is used in the link simulation. Fig. 13(b) shows the link bathtub correlation. As shown in this figure, an excellent agreement is found. More bathtub correlation is done with additional sinusoidal jitter (SJ) noise and it is shown in Fig. 14(a). The large timing margin in the test system can be used to improve DRAM device yield and to reduce system cost by permitting larger DRAM device jitter. To illustrate this point, Fig. 14(b) shows the link bathtubs of hypothetical cases with relaxed DRAM jitter.

V. CONCLUSION

A low power differential memory system for mobile applications is discussed. Using two x16 devices, the total of 12.8GB/s bandwidth can be achieved. We demonstrated that the extra margin due to the differential signaling can be exchanged for more relaxed timing jitter and system noise.

ACKNOWLEDGMENTS

The authors wish to thank John Poulton, Robert Palmer, Adrian Torres, Deborah Dressler, Don Mullen, June Feng, Hai Lan, Ling Yang, Jun Kim, and Judy Chen for their contributions.

REFERENCES

[1] D. Oh, W. Kim, J.-H. Kim, J. Wilson, R. Schmitt, C. Yuan, L. Luo, J. Kizer, J. Eble, and F. Ware, "Study of signal and power integrity challenges in high-speed memory I/O designs using single-ended signaling schemes", *DesignCon*, Santa Clara, Feb. 2008.

[2] D. Oh, F. Lambrecht, J. Ren, S. Chang, B. Chia, C. Madden, and C. Yuan, "Prediction of system performance based on component jitter and noise budgets," *IEEE 16th Topical Meeting on Electrical Performance of Electronic Packaging, EPEP'07*, pp. 33-36, Oct. 2007.

[3] D. Oh, F. Lambrecht, S. Chang, Q. Lin, J. Ren, C. Yuan, J. Zerbe, and V. Stojanovic, "Accurate system voltage and timing margin simulation in high-speed I/O system designs," *IEEE Trans. Adv. Packag.*, vol. 31, no. 4, pp. 722-730, Nov. 2008.

[4] R. Palmer, J. Poulton, B. Leibowitz, Y. Frans, S. Li, A. Fuller, J Eyles, J. Wilson, M. Aleksic, T. Greer, M. Bucher, N. Nguyen, "A 4.3GB/s mobile memory interface with power-efficient bandwidth scaling," *IEEE Symp. On VLSI Circuits*, pp. 136-137, 2009.

[5] K.-L. J. Wong, H. Hatamkhani, M. Mansuri, and C.-K. K. Yang, "A 27-mW 3.6-Gb/s I/O transceiver," *IEEE J. Solid-State Circuits*, vol. 39, no. 4, pp. 602-612, Apr. 2004.

[6] J. Poulton, R. Palmer, A. M. Fuller, T. Greer, J. Eyles, W. J. Dally, and M. Horowitz, "A 14mV 6.25-Gb/s transceiver in 90nm CMOS," *IEEE J. Solid-State Circuits*, vol. 42, no. 12, pp. 2745-2757, Dec. 2007.

[7] D. Oh and S. Chang, "Clock jitter modeling in statistical link simulation," *IEEE 18th Topical Meeting on Electrical Performance of Electronic Packaging, EPEP'09*, submitted for publication, Oct. 2009.

Low-power Self-equalizing Driver for Silicon Carrier Interconnects with Low Bit Error Rate

Peter Gadfort and Paul D. Franzon
Department of Electrical and Computer Engineering
North Carolina State University, Raleigh, NC 27695
Email: {pgadfor,paulf}@ncsu.edu

Abstract

This paper demonstrates and compares the power efficiency of a standard differential current mode driver operating over an FR-4 channel with an improved driver with pre-emphasis operating over a silicon carrier channel. The drivers were designed for a 45 nm process, and both achieved a bit error rate of 10^{-15} errors per bit while operating at 4 Gbps. The power of the improved driver was reduced to one-fourth that of the standard driver through the utilization of the silicon carrier channels and pre-emphasis.

I. INTRODUCTION

As electronic systems increase in complexity, the ability to integrate all the functionality of a system onto a single chip may not be possible, especially as analog, RF, and digital systems continue to be brought together into a single system [1] [2]. As a result, system on package technologies have become highly desirable as they allow for different technologies to be merged into a single package. However, current packaging technologies are trending towards obsolescence as the data throughput and power limitations requirements increase [3]; therefore, new technologies are required.

Silicon based packages, such as silicon carriers, have been proposed to replace current multichip modules (MCM) by creating "virtual" chips with multiple chips bonded onto a silicon carrier to form the MCM [4]. The use of controlled collapse chip connections (C4) allows for a tenfold increase in the I/O density over that of current wirebonding technologies. Using C4 to provide the physical and electrical connection between the chip and the silicon carrier eliminates the need for large electro-static diodes to protect the I/O circuits of the chips, which removes a huge capacitive load on the channels and allows for higher speed signaling and while reducing the power needed to operate.

Using low swing drivers helps reduce the power requirements of the I/O devices, but they also reduce the noise margins [5]. To improve the noise margins to achieve a low bit error rate, while maintaining the low power requirement, pre-emphasis can be used to improve the frequency response of the channel and driver to help with the higher frequency components of the signal, which have been attenuated by the channel. This work will compare the power and bit error rate performance of a standard differential driver operating over an FR-4 interconnect with the performance of an improved differential driver with pre-emphasis operating over a silicon carrier interconnect [6].

This paper is organized as follows: Section II describes the silicon carrier differential channel. Section III describes the circuits required for chip-to-chip communications: the reference differential driver, and the improved differential driver with pre-emphasis. Section IV presents the simulation results of this channel and the circuits. A summary is given in Section V.

II. CHANNEL CHARACTERIZATION

The channel, shown in Fig. 1, is a differential signaling pair with a ground plane, as reported by IBM for silicon carriers [4]. The figure shows the silicon carrier on the bottom with an oxide layer, where the copper metal traces are placed to provide the interconnects between chips, and a small nitride layer over the oxide. The copper traces make up the differential signaling pair, $S+$ and $S-$, and the ground plane, G. This structure was simulated for various channel lengths ranging from 5 mm to 100 mm using the dimensions shown in the figure in Ansoft HFSS, and the channel S-parameters were extracted and used to simulate the performance of the interconnect in HSPICE.

A differential channel was chosen as opposed to a single-ended channel such as a coplanar waveguide [4], because of the improved performance of the differential channel for high speed signaling and the lower swing requirement on the differential channel which will also help conserve energy.

The performance of the silicon carrier interconnects is shown in Fig. 2. Fig. 2(a) shows the frequency sweep of the channel for a 50 mm interconnect from 1 Hz to 50 GHz. Fig. 2(b) shows the propagation delay of the signals versus the length of the interconnects. The propagation delay is fairly constant over the channel length range. Fig. 2(c) shows the 3 dB frequency of the interconnects versus the length of the interconnects. As evident from the cutoff frequencies, shorter lengths of interconnects will be compatible with high speed signaling due to the high cutoff frequency, whereas longer interconnects will not be able to support high speed signaling.

978-1-4244-4447-2/09 $25.00 © 2009 IEEE

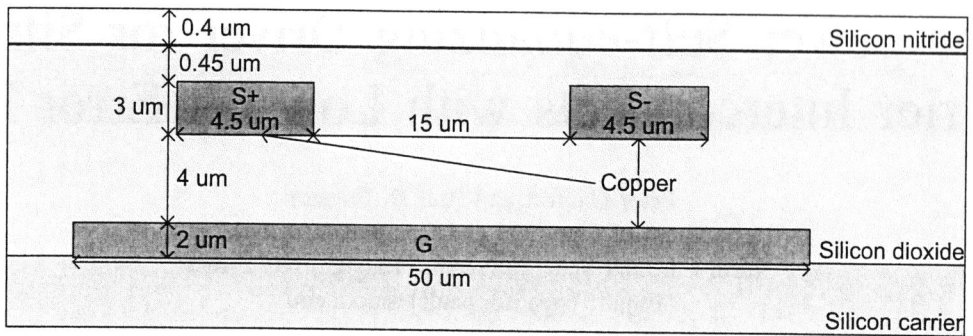

Fig. 1. Differential silicon carrier channel cross-section.

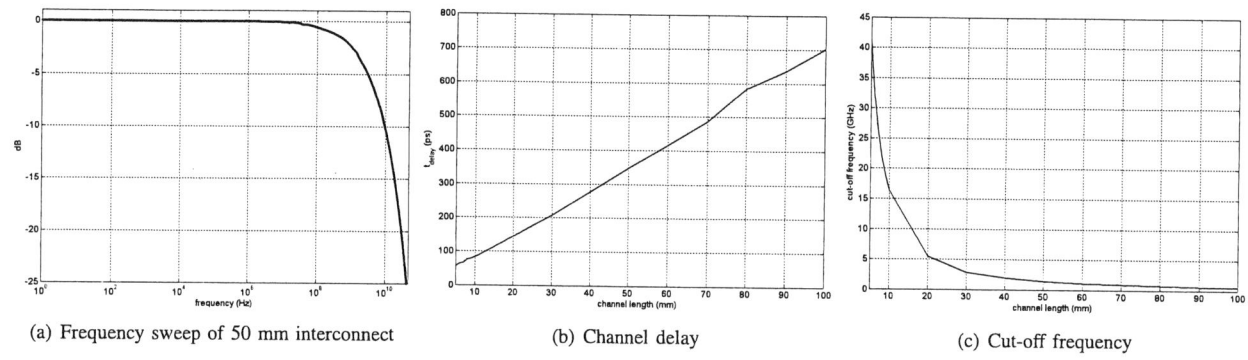

(a) Frequency sweep of 50 mm interconnect

(b) Channel delay

(c) Cut-off frequency

Fig. 2. Differential silicon carrier channel performance.

III. CIRCUIT DETAILS

A. Reference Driver

A reference driver was used to compare the performance of the silicon carrier interconnects and driver. The reference driver was chosen to be a standard differential driver operating on an FR-4 differential interconnect operating at 4 Gbps. The reference driver is shown in Fig. 3(b) and was designed for a 45 nm process [7]. The interconnect that the driver was designed to operate over was designed on a standard FR-4 substrate with copper traces to have a differential impedance of 100 Ω and be 50 mm long.

B. Improved Driver

The driver for the silicon carrier interconnect is based on Zhang's on-chip current mode driver [8] and is shown in Fig. 3(c); it been redesigned for a 45 nm process and for use as an off-chip driver. The driver was designed to operate at 4 Gbps on a 50 mm differential silicon carrier interconnect. Since the cutoff frequency for the 50 mm channel is less than the desired bit rate, pre-emphasis is required to equalize the channel. The pre-emphasis needed to compensate for the channel was determined to be approximately four times the nominal current based on the channel frequency response in Fig. 2(a) at the desired signaling rate.

In order ensure that the pre-emphasis does not apply an unnecessary load on the clock, the pre-emphasis storage cell is provided by long channel transistors instead of a flip-flop. The delay cell provides a delay of approximately one clock period, using several stages of delay. This type of delay cell also compensates for process variations by increasing or decreasing the delay for slow and fast process corners, respectively, as demonstrated by Zhang [8].

The main data path consists of two buffers and the output drivers. The buffers help synchronize the pre-emphasis and the main data paths, which sets both stages to operate in unison. The pre-emphasis data path consists of the delay cell and two tri-state buffers, which are only on during bit transitions (i.e. 0 to 1 or 1 to 0 transitions).

IV. SIMULATION SETUP AND RESULTS

The reference and improved drivers were simulated in HSPICE in a 45 nm predictive development kit [7]. Both drivers were given the same pseudo-random bit stream, and both drivers were driving a sense-amp flip-flop, shown in Fig. 3(a), as the receiver. The sense-amp flip-flop was designed to require a swing of ±50 mV for 100 ps to successfully latch the incoming bit.

978-1-4244-4447-2/09 $25.00 © 2009 IEEE

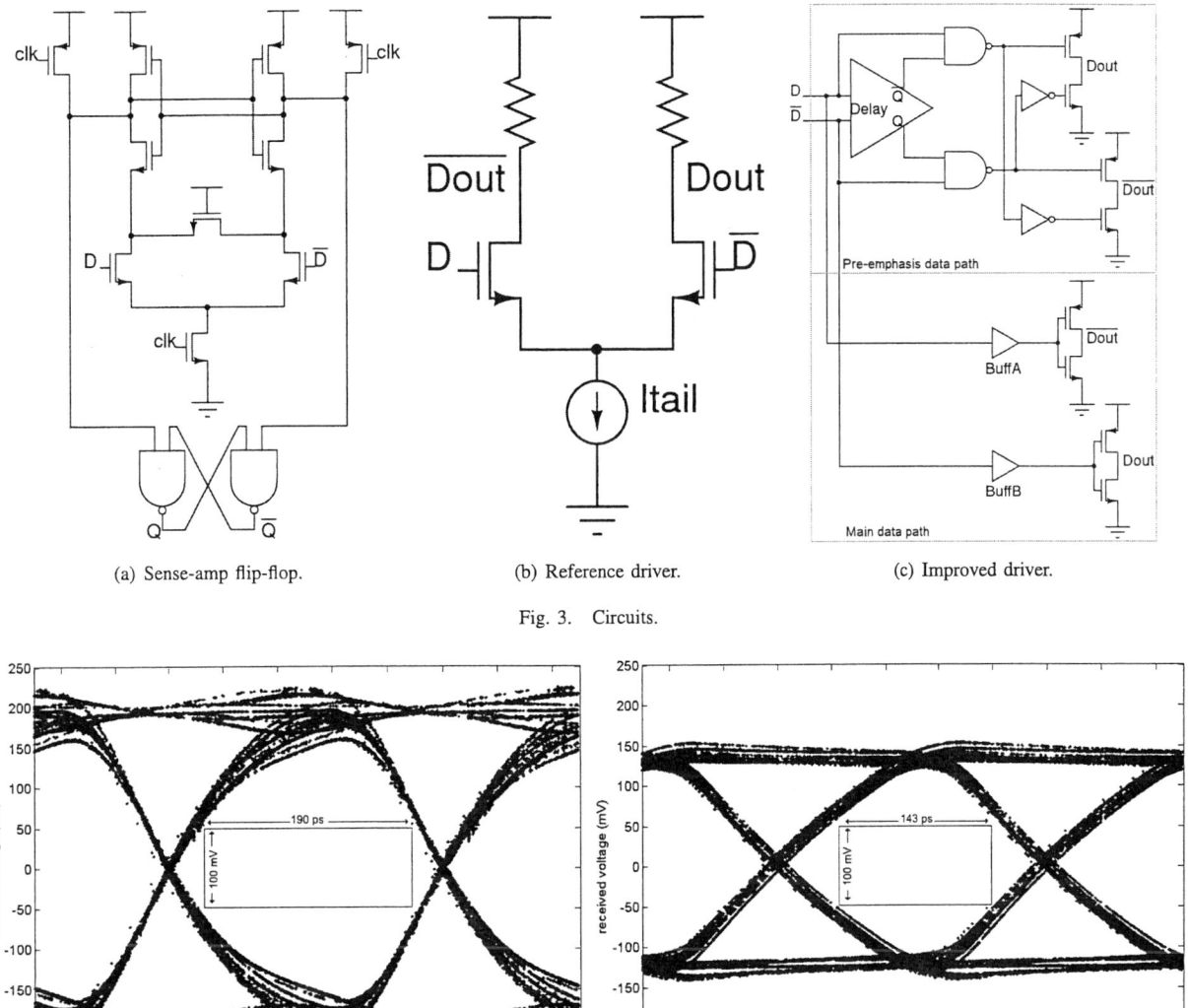

(a) Sense-amp flip-flop.   (b) Reference driver.   (c) Improved driver.

Fig. 3. Circuits.

(a) Reference driver   (b) Improved driver

Fig. 4. Driver eye diagrams at receiver input.

Eye diagrams were produced from the differential input to the receiver, and the receiver characteristics were used to compute the bit error rates. Fig. 4(a) shows the eye diagram for the reference driver on the 50 mm FR-4 interconnect, and Fig. 4(b) shows the eye diagram for the improved driver on the 50 mm silicon carrier interconnect. The eye diagrams show the eye masks in the center of the eyes. The masks show the minimum 100 mV swing required for the receiver to latch the bit by the length of time the eye is over the 100 mV. From the masks it is evident that both the interconnects and drivers hold the eyes open much longer than the required 100 ps; however, the reference channel and driver provide a larger margin for clock skew because the interconnect is much less lossy compared to the silicon carrier channel and driver.

The following metrics were measured on the eye diagrams: the 'one' and 'zero' voltage levels, and the one sigma noise level around the two logic voltage levels. Based on these metrics the signal-to-noise ratio (SNR) [9], eye opening factor [9], and bit error rate can be computed.

The bit error rate is computed by assuming a Gaussian distribution of the noise and computing the likelihood of a bit error occurring for each voltage level; it is assumed that the probability of transmitting a logical one or zero is the same. Based on

978-1-4244-4447-2/09 $25.00 © 2009 IEEE

TABLE I
DRIVER PERFORMANCE FOR 5 CM DIFFERENTIAL INTERCONNECT OPERATING AT 4 GBPS.

	Reference driver	Improved driver
'one' voltage level (μ_1)	178 mV	134 mV
'one' noise (σ_1)	22.9 mV	15.1 mV
High noise margin (NM$_1$)	128 mV	84 mV
'zero' voltage level (μ_0)	-179 mV	-127 mV
'zero' noise (σ_0)	22.8 mV	13.8 mV
Low noise margin (NM$_0$)	129 mV	77 mV
Bit error rate	2.08×10^{-15} errors/bit	3.39×10^{-15} errors/bit
Signal-to-noise ratio	17.85 dB	19.12 dB
Eye opening factor	0.87	0.89
Power	13.57 mW	3.32 mW
Energy per bit	3.39 mW/Gbps (pJ/bit)	0.85 mW/Gbps (pJ/bit)
Area (by gate widths)	260 μm	133.52 μm

these assumptions the bit error rate can be computed using Eqn. 1.

$$\text{Bit Error Rate} = \frac{1}{2} \left(\text{erfc}\left(\frac{\mu_1 - 50 \text{ mV}}{\sigma_1} \right) + \text{erfc}\left(\frac{|\mu_0 + 50 \text{ mV}|}{\sigma_0} \right) \right) \tag{1}$$

These measurements and computations are summarized in Table I to illustrate how both drivers performed over a 5 cm differential channel operating at 4 Gbps.

Both drivers achieved a bit error rate on the order of 10^{-15} errors per bit, but the improved driver on the silicon carrier channel utilized 1/4 of the energy per bit used by the reference driver on the FR-4 channel. Approximately 50% of the energy savings comes from the lower swing on the improved driver as is evident from the eye diagrams in Fig. 4. This reduced swing is possible since there are no input or output capacitive pads on the silicon carrier channel, which cause the signal to bounce at the receiver and allow the use of pre-emphasis to overcome the lossy behavior of the channel instead.

V. CONCLUSION

In this paper, a low power driver for a differential silicon carrier channel for chip-to-chip communications was presented, simulated, and compared to a reference driver operating under similar conditions for current technologies. A differential silicon carrier channel was simulated and characterized; S-parameters for the channel were extracted for use with circuit simulations.

Simulation results indicate that the improved driver for the silicon carrier channel is capable of achieving a bit error rate of 3.39×10^{-15} errors per bit while operating at 0.85 mW/Gbps, whereas the reference driver achieved a bit error rate of 2.08×10^{-15} errors per bit while consuming 3.39 mW/Gbps. The results indicate that there is a large power advantage to using the improved driver on the silicon carrier channel over the reference driver on the normal FR-4 channel. Utilizing pre-emphasis on the silicon carrier channel allows for a lower signal swing, resulting in a lower power driver despite having the added circuitry required for the pre-emphasis.

REFERENCES

[1] R. Tummala and V. Madisetti, "System on chip or system on package?" *Design & Test of Computers, IEEE*, vol. 16, no. 2, pp. 48–56, Apr-Jun 1999.

[2] S. Lim, "Physical design for 3d system on package," *Design & Test of Computers, IEEE*, vol. 22, no. 6, pp. 532–539, Nov.-Dec. 2005.

[3] "International technology roadmap for semiconductors," 2007. [Online]. Available: http://www.itrs.net/Links/2007ITRS/Home2007.htm

[4] J. U. Knickerbocker, P. S. Andry, L. P. Buchwalter, A. Deutsch, R. R. Horton, R. A. Jenkins, H. Kwark, Y, G. McVicker, C. S. Patel, R. J. Polastre, C. Schuster, A. Sharma, S. M. Sri-Jayantha, C. W. Surovic, B. C. Tsang, C. K. Webb, S. R. McKnight, S. R. Spongis, and B. Dang, "Development of next-generation system-on-package (SOP) technology based on silicon carriers with fine-pitch chip interconnection," *IBM Journal of Research and Development*, vol. 49, no. 4/5, pp. 725–753, July/September 2005.

[5] J. M. Rabaey, A. Chandrakasan, and B. Nikolić, *Digital Integrated Circuits: A Design Perspective*, 2nd ed. Prentice Hall, 2003.

[6] P. Gadfort, "Low power interconnect circuits using silicon carriers," Master's thesis, North Carolina State University, 2009.

[7] *FreePDK45*, North Carolina State University, March 2008. [Online]. Available: http://www.eda.ncsu.edu/wiki/FreePDK45:Contents

[8] L. L. Zhang, "Driver pre-emphasis signaling for on-chip global interconnects," Ph.D. dissertation, North Carolina State University, 2005.

[9] *40 Gb/s and Return-to-Zero Measurements Using the Agilent 86100A Infiniium DCA*, Agilent Technologies, 2001.

978-1-4244-4447-2/09 $25.00 © 2009 IEEE

Energy-Efficient Performance Budgeting in FEC-Based High-Speed I/O Links

Rajan Lakshmi Narasimha and Naresh Shanbhag
Coordinated Science Laboratory/ECE Department
University of Illinois at Urbana-Champaign
E-mail:[lakshmi,shanbhag]@illinois.edu

Abstract

In this paper, we look at how the introduction of Forward Error Correction (FEC) impacts system design in a high-speed I/O link. We present examples where coding gain maps to improvements in transmit swing, ADC precision, jitter tolerance and comparator offset tolerance.

1 Introduction

High-speed serial links operating at multi-Gb/s data rates today suffer from inter-symbol interference (ISI) caused by the band-limiting traces that carry the data. These links operate under stringent specifications - few tens of Gb/s data rates, power efficiencies of the order of $10 - 30$mW/Gb/s and a BER target of 10^{-15} and lower. State-of-the-art links employ transmit pre-emphasis (PE) for pre-cursor equalization and receive decision feedback equalization (DFE) to cancel the post-cursor ISI. In [1], for a fixed process technology node (130nm), a four-fold increase in power is predicted when the data-rate is increased from $5 - 12$Gb/s to 25Gb/s and higher, assuming an equalization based transceiver. As energy scales linearly with process technology, it is predicted that a technology node of 32nm is needed in order to meet the power budgets at these speeds. Higher size constellations such as 4-PAM help bring down the bandwidth requirement, but are limited by the peak-SNR constraint imposed by a given technology [2]. The need to optimize system performance and power has motivated recent work such as [3], where simultaneous system and circuit design space exploration is carried out to determine the optimal architecture and allocation of resources in a given system. [4,5] present results of optimizing the sum of transmit driver and receive amplifier power for links where the clock generation and distribution power can be amortized across several lanes. Fig. 1 depicts a high speed link as it

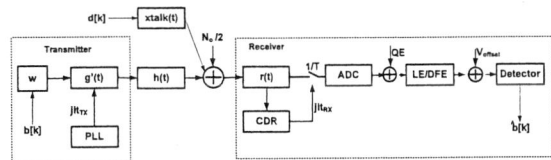

Figure 1: Prototypical high-speed communication link.

is implemented today, and also points to the various impairments leading to bit errors. At the heart of the reliability problem in state of the art links is the channel, which causes inter-symbol interference and crosstalk. Complexity constraints imply that residual ISI and cross-talk are present, resulting in reduced eye opening at the decision device. Timing jitter, quantization noise, resister thermal noise, and comparator offsets further add compound the reliability problem.

In this paper, we carry forward the work presented in [6] and examine how FEC with 2-PAM and 4-PAM modulations impacts the design budgets in a high speed serial link. In particular, we examine how the component-level budgets for transmitter, receiver amplifier, ADC, clock generation and recovery units and comparator (decision device) can be relaxed, by expending FEC power. Some of these trade-offs lead to lower system power, while the others lead to simpler design and better yields. The goal of this work, is to motivate through examples of component-level trade-offs, the need to migrate to FEC based links and larger constellations, for power savings.

978-1-4244-4447-2/09 $25.00 © 2009 IEEE

2 Link Performance Model

In this section, we describe the system model adopted to evaluate the performance trade-offs discussed in the following sections. The shaping filter $g(t)$, channel $h(t)$ and receiver front-end $r(t)$ in Fig. 1 are abstracted into an equivalent discrete-time channel at baud-rate $(p(m) = g(t) \star h(t) \star r(t)_{t=mT}$, where $\star$ denotes convolution) with additive colored Gaussian noise, $\nu(m)$. The term $\nu(m)$ represents the sum of voltage noise contributions from the receiver front-end input referred noise, and the jitter noise mapped to an equivalent voltage noise. The mapping is done using the approach illustrated in [7] to obtain the correlations of the voltage noise due to jitter at the sampler. The equalizer taps are used to compute the residual ISI taps and the resulting ISI noise distribution at the slicer. The ADC quantization noise is modeled as a uniform distribution, and the distribution corresponding to each of the equalizer input samples is convolved to obtain a resultant quantization noise distribution at the slicer. The cumulative distribution of quantization, jitter and thermal noise is convolved with the ISI distribution to obtain the total noise distribution at the slicer. The slicer sensitivity is modeled using a voltage offset V_{offset}, by shifting the signal at the slicer input by V_{offset}. The model just described is used to compute preFEC-BER, BER_{pre}. In the design phase, where to goal is to get a quick hold of the component budgets, the postFEC-BER, BER_{post}, is computed from BER_{pre} assuming that the effect of correlated errors can be handled by interleaving, without a power penalty [6]. For example, a DFE produces statistically significant burst errors of burst length equal to the DFE length (L_d). By implementing an interleaving factor of L_d (typically 4-6, can be achieved by the parallel encoding-decoding scheme proposed in [6]), one can guarantee burst correction, while employing random error correction capability to relax the specifications on impairments such as jitter, thermal noise and quantization.

3 Forward Error Correction: Advantages

The fundamental noise sources in a high-speed link are residual ISI (and crosstalk), timing noise such as transmit and receive jitter and circuit noise such as thermal noise due to the termination resistor, front-end amplifier and comparator circuits. We categorize the circuit induced noise sources into (a) those that can be handled by an increase in transmit swing and (b) those that can be alleviated by providing amplification in the receive chain. The termination resistor noise is an example of the former (denoted by $N_o/2$ in Fig. 1), while comparator input referred noise, static offset and metastability (accounted for by V_{offset}) are examples of the latter. Section 3.1 addresses the issue of improving performance in the presence of receive-amplification insensitive noise sources, while Section 3.4 deals with the performance improvements obtained in the presence random noise sources occurring downstream in the receiver chain by increasing the amplification factor. Section 3.2 presents reduction in ADC precision requirements, while Section 3.3 shows how coding gain is mapped to improved tolerance to uncorrelated jitter.

3.1 Reduced transmitter swing

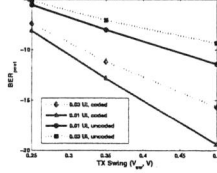 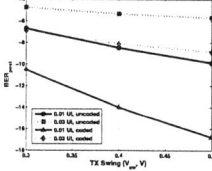 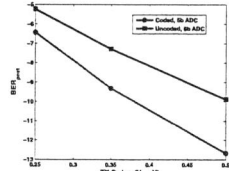 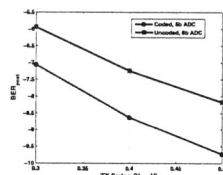

Figure 2: Coding gain provided by a (255,247,3) code leveraged to reduce: a) single-ended transmitter output swing with 2-PAM and b) 4-PAM modulation c) ADC precision requirement, with 2-PAM and d) 4-PAM modulation.

High speed links are peak-power constrained; this arises from the V_{dd} of the process node used in the design. The peak power constraint is a bottleneck in implementing larger constellations, as the minimum symbol distance decreases with constellation size for a fixed peak power. Timing jitter is a significant impairment, and the voltage noise induced by timing jitter increases with signal energy. This limits the extent to which performance can be improved by increasing transmit swing. To capture these aspects, 10Gb/s data transmission across a 20" FR4 trace (25 dB attenuation at Nyquist) was considered. Fig. 2(a) and Fig. 2(b) illustrate BER sensitivity to transmit swing for a $N_o/2$ = $4(mV)^2/GHz$ noise PSD and two different values of jitter, 0.01 UI and 0.03 UI. The effects of quantization are not considered in this analysis, as this is applicable to non-ADC based links as well. Receive equalization is carried out using a 4 feed-forward, 6 feedback tap equalizer (comparable in complexity to state-of-the-art). The figures illustrate that the performance sensitivity to V_{sw} decreases (lower slope) as we go from 0.01 UI to 0.03 UI jitter. We also note that for a given jitter value (0.01 UI, say), 2-PAM performance is more sensitive than that of 4-PAM owing to the dependence of jitter noise on transmit energy. For the same timing jitter, the induced voltage noise jitter relative to signal minimum distance is higher for a larger constellation, at a given peak swing. From Fig. 2(a), we infer that,

978-1-4244-4447-2/09 $25.00 © 2009 IEEE   42

1. FEC relaxes the transmit swing requirements.

2. The power savings are higher for a link dominated by jitter, owing to lesser sensitivity of BER to transmit swing increment. The FEC reduces swing requirement by 0.3 V ppd for the 0.01 UI jitter case as compared to 0.4 V ppd for the 0.03 UI jitter case.

3. The relaxed swing enables the use of a higher constellation such as 4-PAM. For example, at $V_{sw} = 1Vppd$, 2-PAM achieves $BER = 10^{-12}$, while 4-PAM achieves $BER = 10^{-10}$, not meeting the performance target. If a reduction in link operating rate is desired, for power savings, FEC enables 4-PAM to achieve the target BER at the same transmit swing level.

3.2 Reduced ADC precision requirement

Recently, several works on high speed ADCs [8, 9] have appeared, with the latter achieving 6 bits at 25GS/s and 1.2W power dissipation. While DSP based links offer flexibility and robustness, they have not been the favored design approach owing to the high power overhead of the ADC. Fig. 2(c) shows that the precision requirement for an ADC-based IO link can be relaxed by employing FEC. The performance of an FEC based link using a (255, 247, 3) BCH code and a 5 bit ADC is superior to an uncoded link using a 6b ADC. A similar observation is made for 4-PAM (Fig. 2(d)), where the 6b uncoded design does not achieve $BER = 10^{-12}$, while the coded system achieves it at lower precision. In order to evaluate the power trade-offs involved systematically, BCH codes with $n <= 511$ were considered and the minimum precision required to achieve $BER < 10^{-12}$, for a channel operating at $V_{sw} = 1Vppd$ was determined. Table 1(a) presents these results. As noted earlier, for coded 2-PAM modulation, a (255,247,1) code brings down the

M-PAM	(n,k,t)	CR	ENOB (bits)	M-PAM	(n,k,t)	TX jitter % UI, rms
2	Uncoded	1	>6	2	Uncoded	0.01
2	(255,247,1)	0.97	5	2	(255,247,1)	0.04
2	(511,502,1)	0.98	5	2	(511,502,1)	0.04
2	(511,493,2)	0.97	4.5	2	(511,493,2)	0.06
4	Uncoded	1	N/A	4	Uncoded	<0.01
4	(255,247,1)	0.97	5.5	4	(127,113,2)	0.02
4	(511,502,1)	0.98	5.5	4	(255,247,1)	0.02
4	(511,493,2)	0.97	4.8	4	(255,223,4)	0.04
4	(511,484,3)	0.95	5.5	4	(511,502,1)	0.02
				4	(511,448,7)	0.05

Table 1: Coding gain leveraged to a) reduce ADC precision and b) improve TX jitter permissible at $BER_{post} = 10^{-12}$

precision requirement from > 6 bits, to 5 bits. Using the stronger (511,493,2) code, at the same code rate reduces the required ENOB to 4.5 bits. Codes with rates lower than 0.95 are significantly affected by the increased ISI penalty and do not meet the performance target with 2-PAM. With 4-PAM, a larger subset of codes (down to the (511,438,8) code with CR = 0.86) meet the target (Table 1(a) only lists down to CR = 0.95, for brevity). However, an optimum in terms of ENOB reduction is achieved by the (511,493,2) code. Moving to a stronger code at this block length only increases line rate without any significant savings in precision, and hence is not justified.

3.3 Improved jitter tolerance

Residual ISI and jitter are known to be the two most significant impairments in high-speed links. A high speed link transmitter consists of a PLL that generates a clock to synchronize the data stream. Jitter in this clock arises from device noise, reference clock and power supply noise. It has been shown in [7] that high frequency transmitter jitter modulates the energy of the transmitted signal, increasing the effective jitter induced voltage noise at the comparator. The jitter induced noise increases proportional to signal amplitude, hence this cannot be handled by increasing SNR. FEC is a particularly effective strategy to handle this impairment. In order to evaluate the improvement in the transmitter jitter specification systematically, the codes considered in Section 3.2 were evaluated. A transmit swing of $1Vppd$ and thermal noise of $4(mV)^2/GHz$ rms were assumed. The admissible jitter is listed as a function of codeword length in Table 1(b), for a $BER_{post} <= 10^{-12}$.

The base reference performance for the uncoded link at 0.01 UI jitter are inferred from Fig. 2(a) and Fig. 2(b) for 2-PAM and 4-PAM respectively. While 2-PAM achieves the BER target, 4-PAM falls short. With 2-PAM modulation, a (511,493,2) code shows 6 times better tolerance to jitter. The effect of ISI penalty starts increasing in significance as we move to lower rate codes. With 4-PAM, jitter tolerance improves till CR = 0.88, for any codeword length (at any codeword length, only the weakest and strongest codes meeting performance are listed). ISI penalty starts assuming significance at lower rates. This increased jitter tolerance confirms that jitter is an impairment that can be handled by employing FEC. This result motivates the need for circuit designers to translate the improved jitter specifications into reduced power solutions.

3.4 Improving Comparator Offset Tolerance

In conventional design of IO links, comparator impairments (circuit impairments, in general) such as static offset, input-referred supply noise and metastability are modeled by adding an offset (V_{offset}, Fig. 1) to the comparator input signal. Figures 3(a) and 3(b) show BER at the slicer (BER_{pre}) as a function of voltage offset at the slicer

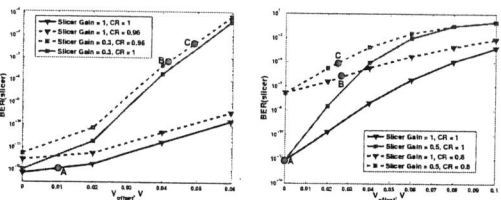

Figure 3: BER versus voltage offset a) 2-PAM, and b) 4-PAM.

for 2-PAM and 4-PAM modulations. 'Slicer gain' is a parameter that captures the effective amplification factor in the receive chain up-to the slicer. Higher the gain, higher the receiver sensitivity and higher the receiver power. In serial links where continuous time equalization is employed, and the clocking power is amortized across many lanes, the total link power is dominated by the driver power and receiver amplifier power. The results of this section show that requirements on the receive amplification (as reflected in the 'slicer gain') can be significantly reduced in an FEC based link. A slicer gain of one implies that the voltage swing input to the slicer is maintained at V_{dd} peak to peak differential. At lower slicer gain, more circuit error events occur (e.g, metastability) leading to worse BER. This way, receiver amplifier power (a function of its gain) can be traded off with error rate. The BER_{pre} for the link is also characterized as a function of code rate (CR), keeping the equalizer complexity fixed. Points A, B and C in Fig. 3(a) and Fig. 3(b) correspond to $BER = 10^{-12}$, with 2-PAM and 4-PAM modulations. Point A represents an uncoded link, while points B and C represent coded links with $CR = 0.96$ and $CR = 0.8$, for 2-PAM and 4-PAM. A higher code rate for PAM2 is chosen to reduce the ISI penalty. We see that an FEC assisted link, operating at $CR = 0.96$ offers 30-40 mV excess tolerance to voltage offset (points B and C compared to point A) at one third the amplification factor of the uncoded link for 2-PAM modulation. Similarly, with 4-PAM modulation, a 25 mV excess offset tolerance at one half the receive amplification is illustrated (comparing point A with C in Figure 3(b)).

References

[1] V. Stojanovic, "Channel-limited high-speed links: modeling, analysis and design," Ph.D. dissertation, Stanford University, U.S.A, 2004.

[2] J. T. Stonick, G.-Y. Wei, J. L. Sonntag, and D. K. Weinlader, "An adaptive PAM-4 5 Gbps backplane transceiver in 0.25 micron CMOS," *IEEE Journal of Solid State Circuits*, vol. 38, no. 3, pp. 436–443, 2003.

[3] R. Sredojevic and V. Stojanovic, "Optimization-based Framework for Simultaneous Circuit-and-System Design-Space Exploration: A High-Speed Link Example," in *International Conference on Computer-Aided Design*, 2008.

[4] B. Casper *et al.*, "Future Microprocessor Interfaces: Analysis, Design and Optimization," in *IEEE Custom Integrated Circuits Conference*, 2007, pp. 479–486.

[5] G. Balamurugan, B. Casper, J. Jaussi, M. Mansuri, F. O. Mahony, and J. Kennedy, "Modeling and Analysis of High-Speed I/O links," 2009.

[6] R. Narasimha and N. R. Shanbhag, "Forward error correction for high-speed I/O," in *Asilomar Conference on Signals, Systems and Computers*, 2008, pp. 1513–1517.

[7] V. Stojanovic, A. Amirkhany, and M. Horowitz, "Optimal Linear Precoding with Theoritical and Practical Data Rates in High Speed Serial-link Backplane Communication," in *IEEE International Conference on Communications*, 2004, pp. 2799–2806.

[8] M. Harwood *et al.*, "A 12.5 Gb/s SerDes in 65nm CMOS using a baud-rate ADC with digital RX equalization and clock recovery," in *IEEE International Solid-State Circuits Conference*, 2007.

[9] P. Schvan *et al.*, "A 24GS/s 6b ADC in 90nm CMOS," in *IEEE International Solid-State Circuits Conference*, 2008.

978-1-4244-4447-2/09 $25.00 © 2009 IEEE

DSP-based Multimode Signaling for FEXT Reduction in Multi-Gbps Links

Pavle Milosevic, José E. Schutt-Ainé, and Naresh R. Shanbhag

Department of Electrical and Computer Engineering, University of Illinois at Urbana-Champaign

1406 W Green St, Urbana, IL 61801 USA

phone: (217)244-6949, (217)244-7279, (217)244-0041; fax: (217)333-5962

email: pavle@emlab.uiuc.edu, jschutt@emlab.uiuc.edu, shanbhag@illinois.edu

Abstract

In order to alleviate the problem of far-end crosstalk induced jitter, signaling techniques based on exploiting the orthogonal property of fundamental transmission modes of multiline system have been proposed, but so far there have been no reports on practical transceiver realization for systems other than for the ones with completely degenerate channels. In this paper, a DSP-based implementation of a multimode transceiver for a bundle of chip-to-chip microstrip lines is investigated in terms of its jitter suppression performance versus design complexity and power.

I. Introduction

In today's high-speed chip-to-chip communication, off-chip interconnect bandwidth is increasingly becoming the system bottleneck. In order to keep up with the data throughput needed, high edge rates are being used for signaling and the interconnects are being routed at an increased density. However, this also increases the electromagnetic coupling between the lines, resulting in augmented levels of both near- (NEXT) and far-end crosstalk (FEXT). In particular, for most parallel-terminated microstrip interconnects, FEXT is the dominant noise source, causing crosstalk-induced jitter (CIJ). This in turn limits the system performance and enforces the data rates to be well below the Shannon limit of the channel capacity [1]. Various approaches have been suggested in order to mitigate the effects of FEXT, including channel coding [1], [2], active cancelation [3], adaptive equalization [4], and crosstalk-aware channel design [5], but at the expense of area overhead, power or design complexity.

A promising signaling scheme for dense interconnects called *multimode signaling* was suggested by [6], which takes advantage of the multiconductor transmission line theory [7] to encode the parallel signals as the linear combination of fundamental transmission modes. Due to mode orthogonality, the signals are decoupled; such signaling is theoretically free of crosstalk, and therefore could allow the data transfer at channel capacity. It has been demonstrated [8] that this method allows line spacing similar to conventional differential signaling for the same performance, with the density of a single line per signal. However, the issue of practical realization of the encoder/decoder system for a general channel remains open. For a completely degenerate multiconductor transmission line system (e.g. stripline), an implementation using analog circuits was suggested in [9]. The complexity of such a system for non-degenerate channels (e.g. microstrip) would grow quadratically with the number of signals [10], forcing the bus to be divided into bundles, thus limiting the usability of the multimode signaling technique. In this paper, we analyze the feasibility of implementing the multimode encoder in the digital domain, in order to take advantage of the power and throughput benefits provided by the scaling down of CMOS technology.

II. Example Channel and Encoder/Decoder

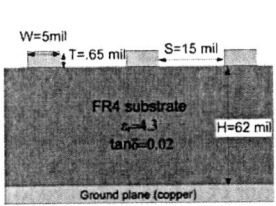

Figure 1. Cross-section of the multiconductor channel

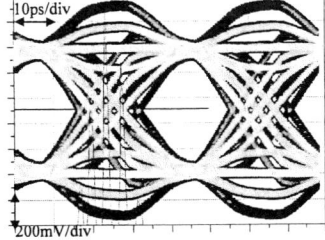

Figure 2. Eye diagram and jitter histogram using uncoded single-ended signaling

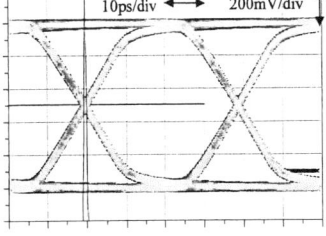

Figure 3. Eye diagram and jitter histogram with multimode signaling

To analyze the tradeoffs associated with the digital encoder design, an example multiconductor transmission line system has been set up in a form of three strongly coupled microstrip lines on FR4 substrate; the physical dimensions are as shown in Fig. 1. Line length was chosen to be 2 inches at the data rate of 12.8Gb/s, in order to ensure FEXT was the dominant noise source (as opposed to the inter-symbol interference (ISI), channel dispersion and attenuation). Matrices of impedance and admittance per unit

978-1-4244-4447-2/09 $25.00 © 2009 IEEE   45

length, $\mathbf{Z}$ and $\mathbf{Y}$, were first extracted using a quasi-static 2-D EM solver. Next, assuming that the channel loss is low compared to the reactive effects, encoder $\mathbf{T}$ and decoder $\mathbf{S}$ matrices were obtained following the well-known method [11] for diagonalizing $\mathbf{ZY}$ and $\mathbf{YZ}$ using frequency independent associated eigenvectors, thus achieving the total decoupling of the telegrapher's equations for the system. Note that, for this experimental setup, voltage signaling was chosen, however the same principles would still hold in the case of current-based signaling, which may be more suited to particular applications. From the characteristic impedance matrix $\mathbf{Z_C}$ of the system, an appropriate resistive matching network was constructed to parallel-terminate the lines at the far end and minimize reflections, thus reducing ISI and FEXT[1].

To explore the potential gains that could be obtained by multimode signaling, the encoder and decoder were initially implemented as ideal. The lines were first excited by uncoded single-ended PBRS sequences of length 2^{10}-1, with rise and fall times of 45 ps. The eye diagram and jitter histogram of the received signal on the central line are plotted in Fig. 2, clearly showing a five-modal distribution with peak-to-peak jitter J_{pp} of 37 ps. Next, the same input sequences were applied to the system with the encoder/decoder in place, resulting in the eye diagram and jitter histogram as plotted in Fig. 3. As expected, the CIJ is significantly reduced, with residual J_{pp} of 4.5 ps attributed to the effect of frequency-independent terminations [11].

III. DSP-based Transmitter Requirements

This section will discuss some of the issues connected with a DSP-based implementation of a multimode transmitter. The digital implementation of the encoder was proposed in [12], however there are practical issues with either of the two suggested methods. If the signals to be sent from the transmitter are in analog form, the digital processing block would have to perform sampling and processing at a Nyquist rate in order to preserve the signal shape. In case of a high-speed transmitter with a multi-Gb/s data rate, the digital block would have to run at a prohibitively high frequency, both from speed and power standpoint. However, in case of binary NRZ signaling this is also unnecessary, because the information to be transmitted is in essence digital, therefore we can assume that the digital block has at its inputs the digital bits (for example, obtained by a 1-bit slicer).

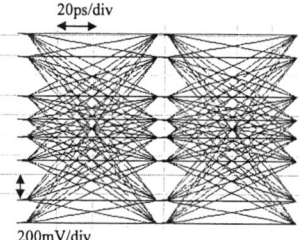

Figure 4. Transition diagram for ideal multimode signaling at line near-end

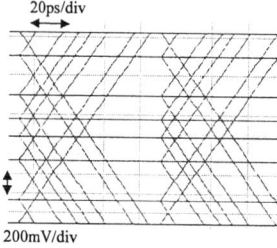

Figure 5. Transition diagram for an ideal slew-rate limited DAC

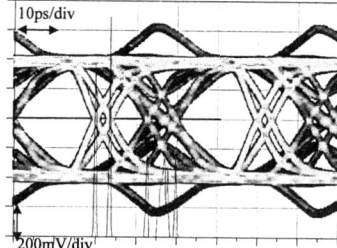

Figure 6. The effect of incorrectly generated analog transitions on multimode signaling

In case of digital information to be encoded, a simple DSP block performing matrix multiplication is not sufficient, due to the inherently analog nature of the multimode signaling. Let us revisit the ideal scenario from Section II. Assuming that the input word consists of three signals with equal rise and fall times t_r (in this example 75 ps), the linear combination of the three signals will also transition from the previous state in time t_r, regardless of the actual calculated values, as shown on the transition diagram in Fig. 4. On the other hand, the digital encoder performing matrix multiplication sends the new modal voltage values to the D/A converters at each bit interval. Modeling the D/A converter and line driver as being limited only by the slew rate of the driver, it is clear that the transition time on each line will depend on the previous and new voltage values, as shown in Fig. 5. In the modal space, this causes the decoded signals to exhibit different rise/fall times, or to change the slope during transition, all resulting in deviations from the orthogonal transmission modes and CIJ which is increased by a factor of 2 even compared to the uncoded case, as shown in Fig. 6.

In order to control the transition time and preserve the modal content of the coded signals, one option is to digitally control the slew rate of output drivers, by adjusting the biasing current depending on the previous and current transmitted bits. The physical realization of this can vary depending on the driver layout; one possible implementation would be to use a combinational network to multiplex the appropriate

---

[1] For this theoretic exploration, a fully matched resistive network was implemented. Depending on the system complexity and the amount of reflections that can be tolerated, a simpler M-terminating network or a parallel termination to ground may also be chosen.

978-1-4244-4447-2/09 $25.00 © 2009 IEEE

current for each output symbol to be sent. The system diagram is shown in Fig. 7. Due to multi-Gb/s data rate, it's expected that deserialization of the data stream before data processing will be necessary.

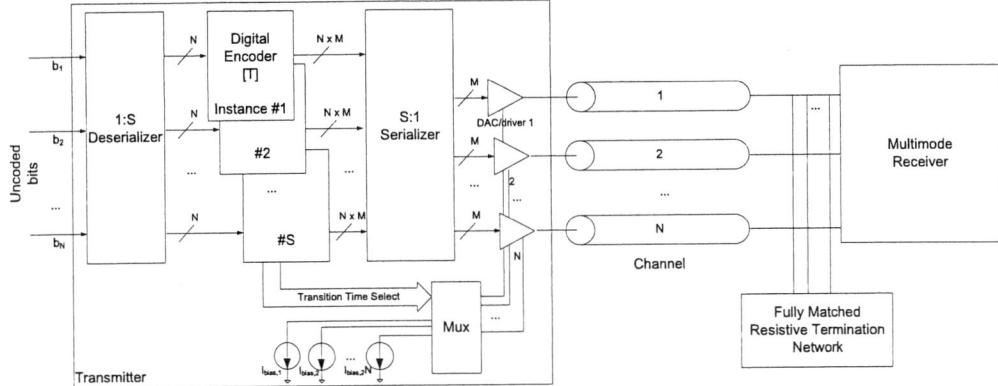

Figure 7. Block diagram of the proposed DSP-based transmitter system

In general, for a N-line system there exist N fundamental modes of propagation, and each line is driven by a linear combination of N input signals. In the case of digital input signals, there are 2^N values possible as the inputs to each D/A converter (refer to Fig. 4 for the case $N=3$). However, the exact values are in general nonuniformly spaced, and furthermore are functions of the physical configuration of the channel (since the encoder matrix T is derived from the Z and Y parameters p.u.l.). Therefore, instead of designing a N-bit DAC with nonuniform output levels for each line of the channel (and to investigate the finite resolution of adders and multipliers in the DSP block as well), it is assumed that the output DAC's are M-bit with uniform output levels, where $M \geq N$.

In order to determine the required resolution M of the DSP block, a series of simulations in Agilent's ADS/Ptolemy mixed-signal environment was next performed with different values of M, observing the jitter of the received decoded signals. As expected, choosing $M=N=3$ resulted in only a moderate jitter reduction to $J_{pp}=29$ ps (22% improvement compared to the uncoded signal transmission), due to the coefficients and generated voltages deviating from their ideal values, as shown in Fig. 8. Increasing the value of M helps better approximate the ideal transition levels, with $M=5$ improving CIJ by 70% compared to the uncoded case, as shown in Fig. 9. A plot of the RMS and peak-to-peak jitter values reveals the exponential improvement to both metrics with increasing resolution, with the ideal values being approached rapidly, as shown in Fig. 10. However, increasing the resolution adds to encoder area, design complexity and ultimately transmitter power, so a compromise must be made depending on the overall specifications of the system being designed.

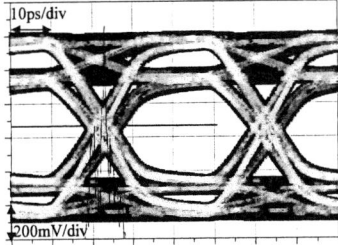

Figure 8. Eye diagram and jitter histogram of the decoded received signal for the case $N=3$, $M=3$

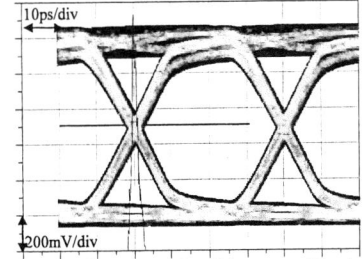

Figure 9. Eye diagram and jitter histogram of the decoded received signal for the case $N=3$, $M=5$

IV. Power Budgeting of the Digital Block

To explore the feasibility of the digital transmitter, the power of the DSP system (including the serializer/deserializer, but excluding ADC/DAC) was estimated for various combinations of number of lines N and DSP resolution M. For the assumed data rate of 12.8 Gb/s per line in 0.18 μm CMOS technology, serialization factor S of 16 was assumed. With no pipelining, the digital core runs at 800 Mb/s. In each DSP block, there are N additions of M-bit numbers for each of N channels. The main critical path delay is dominated by 1-bit adders. To achieve the required delay of 1.25 ns, the unit 1-bit full adder is first

978-1-4244-4447-2/09 $25.00 © 2009 IEEE   47

scaled so that its delay does not exceed $1.25/(N \cdot M)$ ns, and its power P_{IFA} was extrapolated accordingly. Assuming the multiplier (1-bit AND gate) power P_{1AND} to be approximately ¼ of P_{IFA}, and with total serializer/deserializer power estimated at P_{serdes}=100mW for every bit, the total DSP power is then:

$$P_{DSP}=S \cdot (P_{mult}+P_{add})+N \cdot P_{serdes},$$

$$\text{where:} \quad P_{mult}=N^2 \cdot M \cdot P_{1AND}, \quad P_{add}=N \cdot log(N) \cdot M \cdot P_{IFA}$$

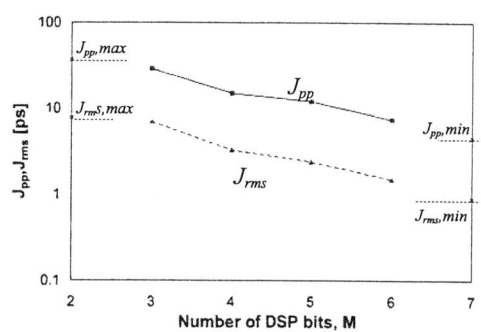

Figure 10. Jitter as a function of DSP precision M; uncoded (maximum) and ideal (minimum) limits also shown

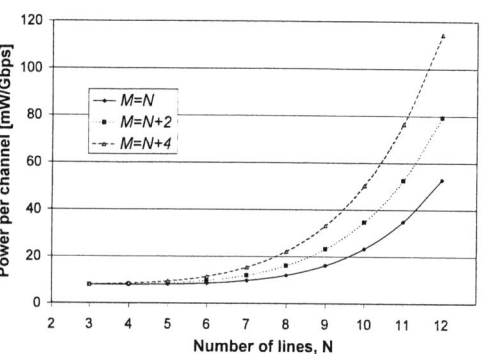

Figure 11. Estimated power of the DSP Tx block as a function of number of lines N and DSP precision M (normalized to 1Gb/s)

The power consumption of the transmitter's digital core per Gb/s is shown in Fig. 11, for various combinations of the number of lines N and DSP precision M. As expected, for low values of N the dominant factor is the serializer/deserializer power, while with increasing N the power grows quadratically with the number of lines, eventually limiting the bundle size. Increasing the number of DSP bits also adds to the power, but not as drastically, and its effect could be minimized with custom DAC levels for each line. A DSP-based receiver is estimated to have comparable levels of performance and complexity.

V. Conclusion

Based on an example three-line microstrip conductor channel exhibiting strong coupling, the tradeoffs inherent in a DSP-based transmitter for multimode signaling have been investigated. Depending on the target level of jitter, the transmitter power for bundle sizes of up to 10 lines is comparable to the 25mW/Gbps achieved by XTC equalization [4], with a greater level of FEXT cancelation and with ½ of the line density, but with no pre-emphasis and therefore shorter line length range. A fairer comparison with existing FEXT equalization solutions and a more universal encoding system could be achieved by shaping the transmitted waveform to mitigate ISI as well. Due to its mainly digital nature, this implementation is expected to take advantage of the power and performance benefits that the advanced submicron technology nodes provide.

Acknowledgment

The authors would like to thank Rami Abdallah, Rajan Narasimha and Chhay Kong for providing them with the metrics for power budgeting estimation.

References

[1] K.Farzan, D.A.Johns, "Coding Schemes for Chip-to-Chip Interconnect Applications," IEEE Trans.VLSI Syst., vol.14, no.4, pp. 393-406, April 2006.

[2] D.Oh et al, "Pseudo-Differential Vector Signaling for Noise Reduction in Single-Ended Signaling Systems," DesignCon 2009, Santa Clara, CA, Feb. 2009

[3] J.F.Buckwalter, A.Hajimiri, "Cancellation of Crosstalk-Induced Jitter," IEEE JSSC, vol.41, no.3, pp. 621-632, March 2006

[4] KJ.Sham et al, "FEXT Crosstalk Cancellation for High-Speed Serial Link Design," Proc. IEEE CICC, September 10-13, San Jose, CA, pp. 405-408

[5] KH.Lee et al, "A Serpentine Guard Trace to Reduce the Far-End Crosstalk Voltage and the Crosstalk Induced Timing Jitter of Parallel Microstrip Lines," IEEE Trans. Adv. Packaging, vol. 31, no. 4, pp. 809-817, November 2008

[6] T.H.Nguyen and T.R.Scott, "Propagation over multiple parallel transmission lines via modes," IBM Technical Disclosure Bulletin, vol. 32, no. 11, pp. 1–6, April 1990

[7] C.R.Paul, *Analysis of Multiconductor Transmission Lines*, 2nd ed., Wiley-Interscience, 2008.

[8] Y.Choi et al, "Analysis of inter-bundle crosstalk in multimode signaling for high-density interconnects," Proc.IEEE ECTC, May 27-30 2008, Lake Buena Vista, FL, pp. 664-668

[9] F.Broydé and E.Clavelier, "A Simple Method for Transmission with Reduced Crosstalk and Echo," Proc.IEEE ICECS, December 10-13 2006, Nice, France, pp.684-687

[10] D.C.Mansur, "Eigen-mode encoding of signals in a data group," U.S. Patent No. 6,226,330, Filed: July 16, 1998.

[11] F. Broydé, E. Clavelier, "A New Method for the Reduction of Crosstalk and Echo in Multiconductor Interconnections", IEEE Trans. Circuits Syst. I, vol. 52, No. 2, pp. 405-416, Feb. 2005.

[12] F. Broydé, E. Clavelier, "Digital method and device for transmission with reduced crosstalk", US Patent App. 10/547,083, Filed: February 18, 2004

Clock Jitter Modeling in Statistical Link Simulation

Dan Oh and Sam Chang

Rambus Inc.
4440 El Camino Real, Los Altos, CA 94022
Tel: 650-947-5363, Fax: 650-947-5001, doh@rambus.com

ABSTRACT

Statistical link analysis has gained significant importance as high-speed interconnect designs require accurate bit error rate prediction with device jitter and noise. Currently available statistical analysis techniques focus on modeling of data channels and the impact of a clock channel is often ignored or primitively approximated using a simple receiver sampling distribution. Thus, it ignores any jitter tracking between data and clock signals. In this paper, a general formulation is presented to model the common jitter source between data and clock signals in order to capture any jitter tracking between them. To demonstrate the usage of the proposed formulation, we have derived various models for commonly used clocking architectures, such as the forwarded clocking scheme in XDR™, DDR, and GDDR systems and the common source RefClk architecture used in PCIe channels. The formulation is verified numerically by using an internally developed CAD tool.

I. INTRODUCTION

As I/O interface speed continues to increase, accurate characterization of link performance is crucial in high-speed digital systems. The majority of link architectures, such as various chip-to-chip interconnects including memory and parallel bus channels, and off-chip network interfaces including HDMI cables and SERDES network, are running at the data rate of 5 Gbps and beyond. As the data rate continues to increase, device jitter contributes a significant portion of the overall link timing budget. Accurate channel simulation requires not only modeling of deterministic jitters, such as inter-symbol interference (ISI), and crosstalk, but modeling of transmitter and receiver device jitter components and their interaction with a passive channel [1]. High-speed channel design often requires a bit error rate (BER) of 10^{-12} or lower and traditional SPICE simulation is too slow in predicting such low BER numbers.

Recently, a statistical simulation approach has been used to predict the link BER performance [2]-[6]. So far, the method has been focused on modeling of the data channel since the method has been extensively used for SERDES applications where a clock-data recovery circuit (CDR) is used instead of an explicit clock signal[*]. Other synchronous I/O interfaces need some form of a clock signal, and the selection of the clocking architecture strongly affects the I/O interface design and system performance. Hence, accurate jitter model for the clock net is crucial in determining the link performance. In the previous study [2], [4], clock jitter for both transmitter and receiver sides is modeled as the receiver sampling distribution. Although the sampling distribution is ideal for modeling the receiver jitter, it cannot accurately represent the transmitter jitter, especially high-frequency jitter [5]. This limitation is mitigated by using an equivalent voltage noise concept in [3]. Although the separate transmitter and receive jitter models are quite useful in modeling the clock jitter at the transmitter and receiver sides, it fails to model any jitter component which is common to both transmitter and receiver sides. To overcome this limitation, the equivalent voltage noise for the common jitter source is generated through time-domain simulation in [6]. A simple dual-Dirac model is used to represent the equivalent voltage noise. In addition to the large transient simulation time required for generating the equivalent model, the accuracy of this model is limited by the dual-Dirac model which represents unbound Gaussian random jitter with two deterministic jitter peaks. The generalization of this apporach to other noise types is difficult as their distributions are unknown a priori.

In this paper, the statistical formulation described in [3] and [5] is generalized to include the common source clock jitter. The resulting model accurately accounts for any jitter tracking between the transmitted data and clock signals. The method extends the equivalent voltage noise concept to common jitter source for transmitter and receiver. In Section II, we present the general formulation for the statistical simulation framework. The explicit formulae are derived for the independent and common clock jitter cases. Any clock jitter can be represented by the combination of these two expressions. In Section III, a clock forwarding architecture and common clock referencing scheme are considered to demonstrate our formulation to real-world applications. The numerical verification of our formulation is shown in Section IV.

II. GENERAL STATISTICAL FORMULATION WITH CLOCK JITTER

Assuming a linear time invariance (LTI) system, the discrete samples of the output signal y_m can be approximated using Taylor series expansion as follows [5]:

---

[*] In [3], a Markov chain is used to model the dithering of CDR due to jitter in the transmitted signal. This model is further verified using a hardware measurement in [5].

$$y_m \cong y_0 + y_{ISI} + n^{TX} + n^{RX} \tag{1}$$

where y_0 is the ideal received signal without ISI, y_{ISI} is the amount of ISI at the current sampled location. n^{TX} and n^{RX} represent the equivalent voltage noise for the transmitter and receiver timing jitter, ε^{TX} and ε^{RX}, respectively. The expressions of n^{TX} and n^{RX} are given by

$$n^{TX} = \sum_k (b_{k-1} - b_k)\varepsilon_k^{TX} h_{m-k} = \vec{b}^T \mathbf{H}^{TX} \vec{\varepsilon}^{TX} \tag{2}$$

$$n^{RX} = \varepsilon_m^{RX} \sum_k (b_k - b_{k-1}) h_{m-k} = \vec{b}^T \vec{H}^{RX} \varepsilon^{RX} \tag{3}$$

where b_k is the input symbol pattern of the transmitter and k is the symbol index. h_m is the data-rate sampled impulse response of the passive channel. Note that the transmitter jitter is colored (modulated) by the channel impulse response and the matrix notation of $\mathbf{H}^{TX}$ represents the convolution process of the input jitter with the impulse response. On the other hand, the receiver jitter is not colored by the channel and the corresponding equivalent voltage is a simple scaling of the input jitter. When the transmitter jitter spectrum is sufficiently low, ε^{TX} can be taken out of the summation and the resulting expression is identical to the receiver case indicating there is no jitter modulation caused by the channel.

The system bit error rate can be calculated from the probability distribution function (PDF) of y_m which is a function of the three random variables y_{ISI}, n^{TX}, and n^{RX}. In general, these random variables are not independent since they are correlated with the same input symbol pattern b_k. A brute force way to computing PDF is first computing the PDFs of all possible data patterns and then averaging the resulting PDFs. This approach requires immense computation time and intractable[*]. In this paper, we ignore any correlation between y_{ISI} and the equivalent voltage noise n^{TX} and n^{RX}.

A. Independent Clock Jitter Sources

When the transmitter and receiver jitter, ε^{TX} and ε^{RX}, are independent, the PDF of y_m can be calculated by convolving individual PDFs for y_{ISI}, n^{TX}, and n^{RX}. In general, there are multiple transmitter and receiver jitter components depending on the jitter types and sources. The PDF calculations for the random Gaussian and bound jitter types are presented in [5]. Fig. 1 shows a general clock jitter model for both independent and common jitter types. We use ε^{TX} and ε^{RX} to represent independent transmitter and receiver jitter components and ε^{Common} to represent the common jitter for both transmitter and receiver. ζ^{TX} and ζ^{RX} represent the jitter transfer functions (JTF) for the transmitter and receiver clock paths, respectively. In general, the jitter transfer functions for common and independent jitter components may be different but we use the same notation for simplicity. The jitter transfer function for passive linear circuit can be derived analytically [7]. For nonlinear circuits, the jitter transfer function can be obtained through circuit simulation. Now the equivalent voltage noise expressions for the clock path with the independent jitter is written by

$$n^{TX} = \vec{b}^T \mathbf{H}^{TX} \zeta^{TX} \vec{\varepsilon}^{TX} = \vec{b}^T \Psi^{TX} \vec{\varepsilon}^{TX} \tag{4}$$

$$n^{RX} = \vec{b}^T \vec{H}^{RX} \left(\vec{\zeta}^{RX}\right)^T \vec{\varepsilon}^{RX} = \vec{b}^T \Psi^{RX} \vec{\varepsilon}^{RX} \tag{5}$$

Hence, the clock jitter transfer function is modeled using the above expressions and the same methods described in [5] can be used to calculate PDFs.

B. Common Clock Jitter Source

The equivalent voltage noise is correlated when they share a common jitter source as shown in Fig. 1. Let us introduce a new parameter n^{Common} to represent the equivalent voltage noise due to a common jitter. Then, (1) is modified as

$$y_m \cong y_0 + y_{ISI} + n^{TX} + n^{RX} + n^{Common} \tag{6}$$

Now, the expression of n^{Common} is given by

$$n^{Common} = \vec{b}^T \begin{bmatrix} \mathbf{H}^{TX} & \vec{H}^{RX} \end{bmatrix} \begin{bmatrix} \zeta^{TX} \\ \left(\vec{\zeta}^{RX}\right)^T \end{bmatrix} \vec{\varepsilon}^{Common} = \vec{b}^T \Psi^{Common} \vec{\varepsilon}^{Common} \tag{7}$$

Once again, the methods described in [5] can be used to derive the PDF for the common jitter case.

---

[*] Accounting the correlation between y_{ISI} and the effective voltage noises, n^{TX} and n^{RX}, can be done efficiently for certain types of jitter such as periodic jitter and duty-cycle distortion (DCD).

III. APPLICATION TO COMMONLY USED CLOCKING TOPOLOGIES

In this section, we consider practical applications of the previous formulation to common clocking topologies. GDDR5 memory interface is considered as an example channel. In contrast to other high-speed I/O interfaces where symmetric clocking architecture is used for both read/write (TX/RX) directions, GDDR5 employs an asymmetric clocking scheme. During the write operation, the controller forwards a clock signal to DRAM along with data signals, whereas there is no returning clock signal for during the read operation. Fig. 2 shows the system-level model for the write case where the clock is forwarded to sample data. This forwarded clocking architecture is commonly used for other on-board high-speed I/O channels such as DDR, XDR™, and FlexIO™ systems. In Fig. 2, ε^{src} represents the input jitter which may come from the reference clock or PLL phase noise. υ^{cont} and υ^{DRAM} represent additional power supply noise components from the controller and DRAM sides. Now the expressions of the corresponding the equivalent voltage noise at the receiver are given by

$$n^{Src} = \vec{b}^T \begin{bmatrix} \mathbf{H}^{Data} & \vec{H}^{Data} \end{bmatrix} \begin{bmatrix} \zeta^{Cntl} \\ \left(\vec{H}^{Clk} * \vec{\zeta}^{DRAM} \right)^T \end{bmatrix} \vec{\varepsilon}^{Src}, \tag{8}$$

$$n^{Cntl} = \vec{b}^T \mathbf{H}^{Data} \boldsymbol{\eta}^{Cntl} \vec{\upsilon}^{Cntl}, \quad \text{and} \quad n^{DRAM} = \vec{b}^T \vec{H}^{Data} \left(\vec{\eta}^{DRAM} \right)^T \vec{\upsilon}^{DRAM} \tag{9}$$

where ζ^{Cntl} and ζ^{Src} represent the jitter transfer functions of the clock paths at the controller and DRAM side, whereas η^{Cntl} and η^{Src} represents the voltage-to-jitter transfer functions, accordingly.

The read case model is shown in Fig. 3, and the corresponding expressions are written by

$$n^{Src} = \vec{b} \begin{bmatrix} \mathbf{H}^{Data} & \vec{H}^{Data} \end{bmatrix} \begin{bmatrix} \mathbf{H}^{Clk} * \zeta^{DRAM} \\ \left(\vec{\zeta}^{Cntl} \right)^T \end{bmatrix} \vec{\varepsilon}^{Src}, \tag{10}$$

$$n^{DRAM} = \vec{b}^T \mathbf{H}^{Data} \boldsymbol{\eta}^{DRAM} \vec{\upsilon}^{DRAM}, \quad \text{and} \quad n^{Cntl} = \vec{b}^T \vec{H}^{Data} \left(\vec{\eta}^{Cntl} \right)^T \vec{\upsilon}^{Cntl} \tag{11}$$

The above equations can be easily modified to include the jitter transfer function for any DLL or PLL if presents. For instance, the jitter modeling expression for the common Refclk Rx architecture of a PCIe channel [8], shown in Fig. 4, can be expressed as

$$n^{Src} = \vec{b}^T \begin{bmatrix} \mathbf{H}^{Data} & \vec{H}^{Data} \end{bmatrix} \begin{bmatrix} \mathbf{H}^{Clk1} * \zeta^{TX\,PLL} \\ \left(\vec{H}^{Clk2} * \vec{\zeta}^{RX\,PLL} \right)^T \end{bmatrix} \vec{\varepsilon}^{Src} \tag{12}$$

IV. NUMERICAL VERIFICATION

The proposed formulation is implemented in LinkLab which is a Rambus in-house statistical simulation tool. A GDDR5 channel at the data rate of 5Gb/s is considered to validate our formulation. The simulation setup is shown in Fig. 5. For the first analysis, we simulated the write case with 5ps of random jitter with 3db cutoff frequency at 700MHz injected at the transmitter. Jitter is added to both data and clock channels. Fig. 6(a) shows the effective eye shown at the sampler using the common jitter model and Fig. 6(b) shows the effective eye using the independent jitter model. As expected, the independent jitter model predicts worse performance (~30% reduction in margin) due to inability to model the jitter tracking. For the second analysis, the read case with varying delay values for the DRAM clock path is considered. Fig. 7 shows the system margin at BER=10^{-20}. As the delay increases, the jitter is less correlated at the controller side and the margin is decreased accordingly.

V. CONCLUSION

A general formulation to account for jitter tracking between data and clock signals is presented for statistical link simulation. This formulation enables the system bit error rate to be calculated strictly within the statistical domain without using any time-domain modeling process. The explicit formulae for various clocking topologies are also presented. The derivation to other clocking topologies can be easily done by following the procedure described in the paper. Although we have omitted in this paper due to space limitation, the simulation data correlated well with hardware measurement.

ACKNOWLEDGEMENTS

The authors wish to thank J. Ren, Q. Lin, C. Madden, B. Leibowitz, Y. Frans, and C. Yuan for helpful discussion.

REFERENCES

[1] D. Oh, F. Lambrecht, J. Ren, S. Chang, B. Chia, C. Madden, and C. Yuan, "Prediction of system performance based on component jitter and noise budgets," *IEEE 16th Topical Meeting on Electrical Performance of Electronic Packaging, EPEP '07*, pp. 33-36, Oct. 2007.

[2] B. K. Casper, M. Haycok, and R. Mooney, "An accurate and efficient analysis method for multi-Gb/s chip-to-chip signaling schemes," *IEEE Symposium on VLSI Circuits*, pp. 54-57, June 2002.

[3] V. Stojanovic and M. Horowitz, "Modeling and analysis of high-speed links", *IEEE Custom Integrated Circuits Conference*, September 2003.

[4] A. Sanders, M. Resso, and J. D'Ambrosia, "Channel compliance testing utilizing novel statistical eye methodology," *DesignCon*, February 2004.

[5] D. Oh, F. Lambrecht, S. Chang, Q. Lin, J. Ren, C. Yuan, J. Zerbe, and V. Stojanovic, "Accurate system voltage and timing margin simulation in high-speed I/O system designs," *IEEE Trans. Adv. Packag.*, vol. 31, no. 4, pp. 722-730, Nov. 2008.

[6] G. Balamurugan, B. Casper, J. E. Jaussi, M. Mansuri, F. O'Mahony, and J. Kennedy, "Modeling and analysis of high-speed I/O links," *IEEE Trans. Adv. Packag.*, vol. 32, no. 2, pp. 237-247, May. 2009.

[7] S. Chang, D. Oh, and C. Madden, "Jitter modeling in statistical link simulation," *IEEE EMC.*, vol. 31, no. 4, pp. 722-730, Nov. 2008.

[8] *PCI Express Card Base Specification, Rev.* 2.0, PCI-SIG, pp. 243-250, Dec. 20, 2006.

Figure 1. A general clock model for independent and common jitter sources

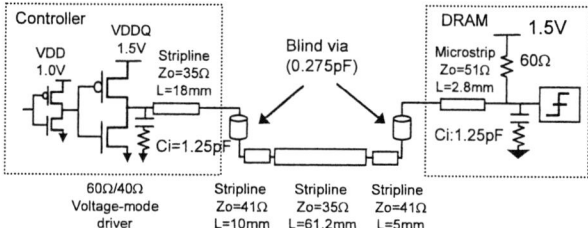

Figure 5. GDDR5 channel setup

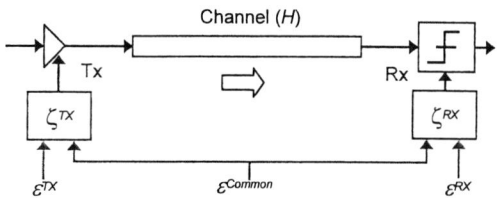

Figure 2. The write operation channel model for GDDR5 memory interface

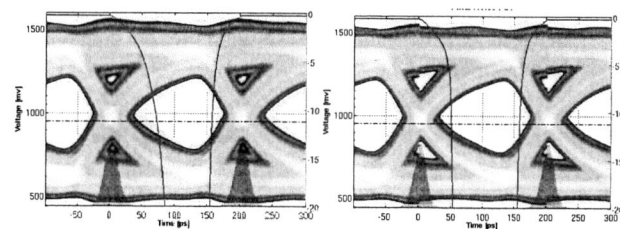

Figure 6. Eye diagrams for the write case with (a) the common jitter model and (b) the independent jitter model

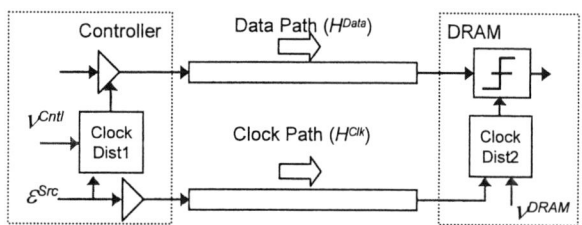

Figure 3. The read operation channel model for GDDR4 memory interface

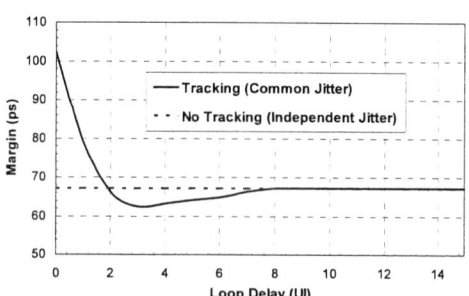

Figure 7. The read case channel margin vs. the DRAM clock path delay

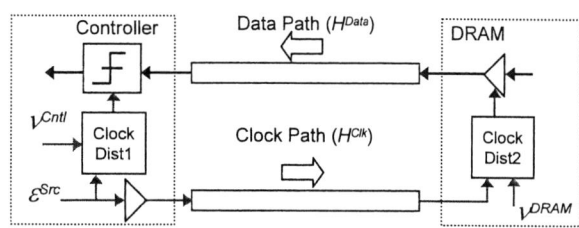

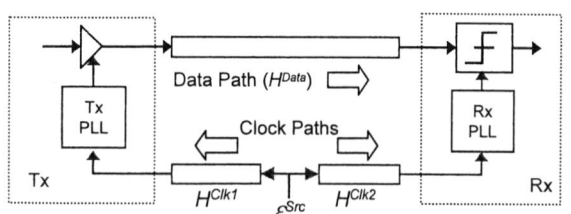

Figure 4. PCIe channel model

978-1-4244-4447-2/09 $25.00 © 2009 IEEE

Hybrid Equalizer Design for 12.5 Gbps Serial Data Transmission

Eakhwan Song, Jeonghyeon Cho, and Joungho Kim

Korea Advanced Institute of Science and Technology, Daejeon, South Korea
Phone) +82-42-879-9872, Fax) +82-42-869-8058, E-mail) sehnx@eeinfo.kaist.ac.kr;teralab@ee.kaist.ac.kr

Abstract – High-speed digital signals suffer from frequency-dependent losses on channels with degraded timing and voltage margin at the receiver. A hybrid equalization technique is proposed to compensate for the frequency-dependent loss. The proposed technique is achieved by combining a wide-band passive equalizer and a second-order high-pass active filter, and presents a noticeable improvement in timing jitter and eye-opening.

I. INTRODUCTION

In recent years, data rates of serial chip-to-chip digital I/O communication channels have increased considerably and are now over 10 Gbps. Due to these increasing data rates, high-speed digital signals suffer from frequency-dependent losses on Printed Circuit Boards (PCBs), packages, and at cables. Consequently, Inter-symbol Interference (ISI) problems result, and there is subsequent degradation of receiver timing jitter and voltage margin with declined system sensitivity [1]. In order to compensate for the frequency-dependent losses in a high-speed I/O channel, equalization methods have received significant attention as practical loss-compensation techniques to transmitter and receiver designs. In equalized serial links, high-speed signals that are significantly degraded from dispersive and lossy interconnects are recovered by equalizers, which pre-emphasize high-frequency components or de-emphasize low-frequency components of the signals, and consequently flatten the steep roll-off of the frequency response of the entire serial link [2].

Among various equalization methods, on-chip active equalization is a beneficial approach for applications that require active ac gain. However, on-chip active equalization inevitably sacrifices active power consumption and suffers from bandwidth limitations on the gain performance with current IC technology [3]. On the other hand, passive equalization techniques are preferred for applications with low power consumption or with a wide bandwidth requirement (Fig. 1-(a)). The passive equalizer can be integrated into a chip, package, or PCB, and each implementation has specific advantages and disadvantages. The on-chip passive equalizer enables wide-band equalization performance with the help of silicon technology advances, but unfortunately consumes a large chip area compared to other active circuitry [4]. Off-chip passive equalizers on the package or PCB can be implemented in the form of a lumped L-C filter, but they have limited bandwidth due to parasitics of the discrete components and the filter bonding materials [5]. To achieve a wide-bandwidth off-chip, we recently presented a sophisticated wide-band passive equalization technique using microwave effects on PCB: near-end crosstalk (NEXT) and reflections [6]. The passive equalization technique was successfully demonstrated for a serial data channel on a backplane PCB with a data rate greater than 10 Gbps; however, it still has a DC attenuation problem, which is an inherent drawback of passive equalizers.

In this paper, we propose a hybrid equalization technique to overcome the shortcomings posed by the active and passive implementations described above. The proposed technique is achieved by combining a wide-band passive equalizer using NEXT and reflections and an active equalizer on-chip, and it compromises the advantages of active and passive equalization techniques: active gain and wide-bandwidth, respectively (Fig 1-(b)). For the active equalizer part, a second-order equalizer was designed at the transistor level using a Samsung 0.18 um CMOS process. With the proposed hybrid equalization technique, significant improvement of eye-opening and timing jitter performance was successfully demonstrated for a 40-cm transmission line on PCB with a data rate of 12.5 Gbps.

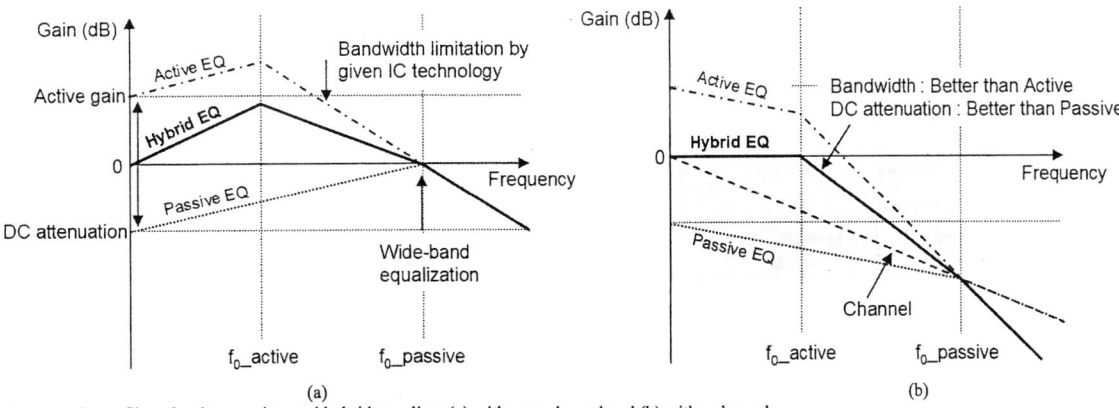

Fig. 1. Gain profiles of active, passive, and hybrid equalizer (a) without a channel and (b) with a channel.

978-1-4244-4447-2/09 $25.00 © 2009 IEEE

II. PASSIVE EQUALIZER DESCRIPTION

For the passive part of the proposed hybrid equalization technique, the recently introduced passive equalizer based on NEXT and reflections is used. The passive equalizer achieves wide-band equalization over 10 Gbps by designing wave propagation, reflection, and coupling effects on the coupled transmission line structure on a planar package and PCB [6]. Fig. 2.-(a) shows the design layout of the fabricated passive equalizer on a PCB with an area of 1.9 mm x 2.9 mm and a dielectric thickness of 0.6 mm, where $R_S = R_L = R_T = 50$ Ohms. To evaluate the fabricated equalizer, insertion loss measurements were performed for a 40-cm-long transmission line on PCB. The insertion loss of the transmission line (dotted line) is flattened by the passive equalizer from DC to 6.25 GHz by 6.2 dB, which results in the equalized insertion loss (solid line); while 6.25 GHz corresponds to the Nyquist frequency of a 12.5-Gbps data rate (Fig. 2-(b)). However, the DC attenuation of -6 dB limits the amplitude boundary of the equalized signal such that the eye-opening is unable to exceed the boundary. To improve the voltage margin at the receiver, active gain is still required.

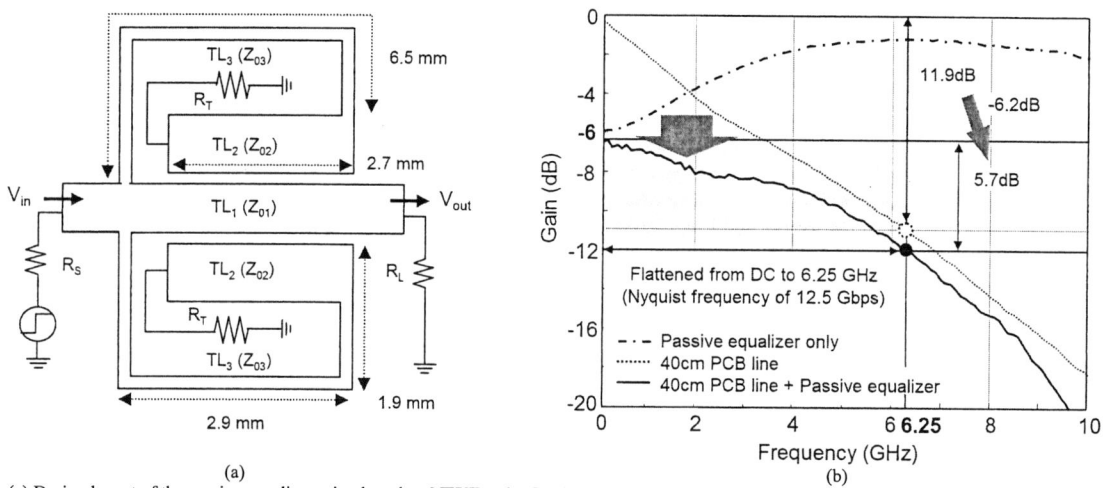

(a)   (b)

Fig. 2. (a) Design layout of the passive equalizer using based on NEXT and reflections on a PCB. (b) Measured insertion losses of the passive equalizer for a 40-cm-long transmission line on a backplane PCB.

III. ACTIVE EQUALIZER DESCRIPTION

In order to achieve active gain in the hybrid equalization technique, an active equalizer was designed to be combined with the passive equalizer. The active equalizer is a second-order high-pass filter composed of three different frequency-selective signal paths. Each path consists of pre-filters, a flat-band amplifier and a first-order differentiator, followed by a variable gain amplifier (VGA) (Fig. 3) [7]. Ideally, the transfer functions of the flat-band amplifier and the first-order differentiator are *1* and *s*, respectively, which results in the following total transfer function

$$H_{second-order}(s) = C_0 + C_1 \cdot s + C_2 \cdot s^2 \qquad (1)$$

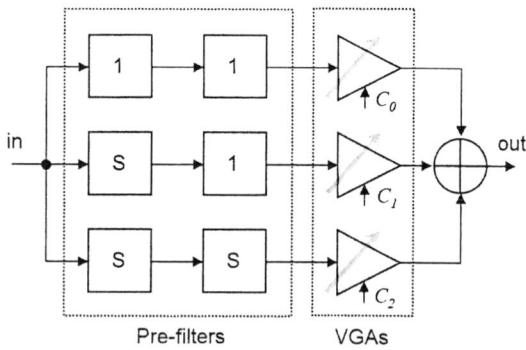

Fig. 3. Block diagram of the second-order active equalizer

where C_0, C_1, and C_2 are the variable gains of the three paths. The DC path (1-1) determines the DC and low frequency responses of the second-order equalizer. The first derivative path (S-1) and the second derivative path (S-S) correspond to compensation for the frequency-dependent loss of channel, which boosts high frequency components of signals. The filtered signals on each path are amplified by VGA depending on the variable gains; finally, they are linearly summed at the output, which results in the equalized signal. Fig. 4 shows the simulated step responses of each path with variable gain control. The total transfer function of the second-order equalizer is controlled by the combination of the variable gains. In this work, the active equalizer was designed using a Samsung 0.18 um process. While the active gain is achieved, the bandwidth is limited by the given IC technology.

978-1-4244-4447-2/09 $25.00 © 2009 IEEE   54

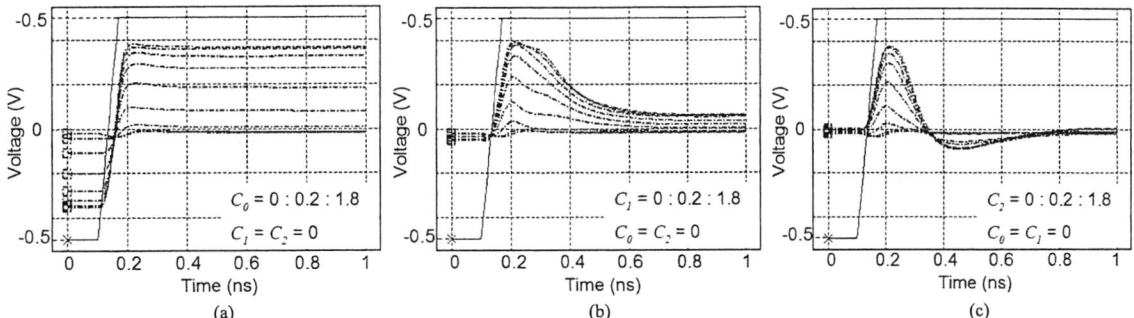

Fig. 4. The simulated step responses of (a) the DC path (b) the first derivative path (c) the second derivative path with variable gain control

IV. THE PROPOSED HYBRID EQUALIZER AND SIMULATION RESULTS

As shown in previous sections, the passive equalizer has the benefit of a wide bandwidth, but the signal amplitude is limited by the DC attenuation; however, the active equalizer provides active gain but suffers from limited bandwidth. The proposed technique, hybrid equalization, achieves both a wide band and an active gain by combining the passive and active equalizers. To validate the proposed technique, both frequency-domain and time-domain simulations were performed. For frequency-domain simulations, the gain profiles of the passive and active equalizers were extracted by measurement and SPICE simulation, respectively. In the SPICE simulation for the second-order active equalizer, the variable gains, C_0, C_1, and C_2, were set to 1.8, 0.7, and 0, respectively; this combination presents the best equalization performance when the passive equalizer is combined with the active equalizer. As shown in Fig. 5-(a), the measured insertion loss of the passive equalizer (dotted line) drifts up by the active equalizer gain (dash-dotted line) when the DC gain increased by 2.6 dB, which results in the hybrid equalizer gain (solid line). The bandwidth of the hybrid equalizer is improved by 2.65 GHz compared to that of the active equalizer. From the simulation results, it is demonstrated that the proposed hybrid equalization technique combines the advantages of the passive and active equalizers. Fig. 5-(b) depicts the gain profiles with and without hybrid equalization for a 40-cm-long transmission line on PCB. The insertion loss of the transmission line (dotted line) is successfully flattened by 5.3 dB using the proposed hybrid equalizer (dash-dotted line) within a frequency range from DC to 6.25 GHz, which results in the equalized insertion loss (solid line),;6.25

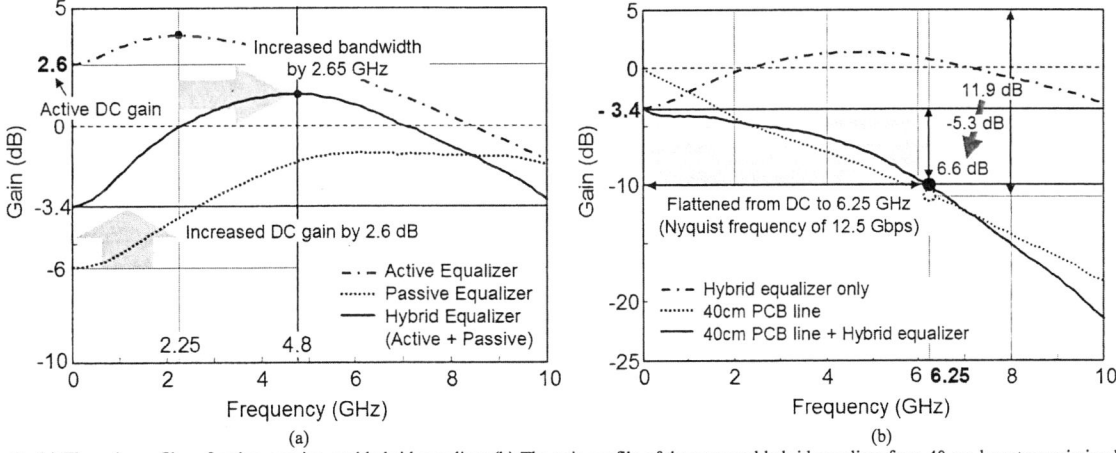

Fig. 5. (a) The gain profiles of active, passive, and hybrid equalizer (b) The gain profile of the proposed hybrid equalizer for a 40 cm-long transmission line on PCB

GHz corresponds to the Nyquist frequency of the target data rate of 12.5 Gbps.

In addition, eye-diagram simulations were conducted for the 40-cm-long transmission line with a data rate of 12.5 Gbps. The input signal was a 2^7-1 pseudo random bit sequence with an amplitude of 500 mV$_{pk-pk}$. As shown in Fig. 6, both the voltage and timing margin are considerably improved using the proposed hybrid equalization technique (Fig. 6-(d)) and achieve normalized eye-opening and pk-pk jitter of 33.1 % and 9 %, respectively; the improvements of them are 193 % and 80 % from the un-equalized eye-diagram (Fig. 6-(a)). Compared to the active equalization (Fig. 6-(b)) and the passive equalization (Fig. 6-(c)), the proposed hybrid equalization presents better performance in terms of eye-opening and pk-pk jitter (Table 1). These voltage and timing margin improvements are achieved by combining the advantages of active and passive equalizers, active gain and wide-bandwidth, respectively.

978-1-4244-4447-2/09 $25.00 © 2009 IEEE

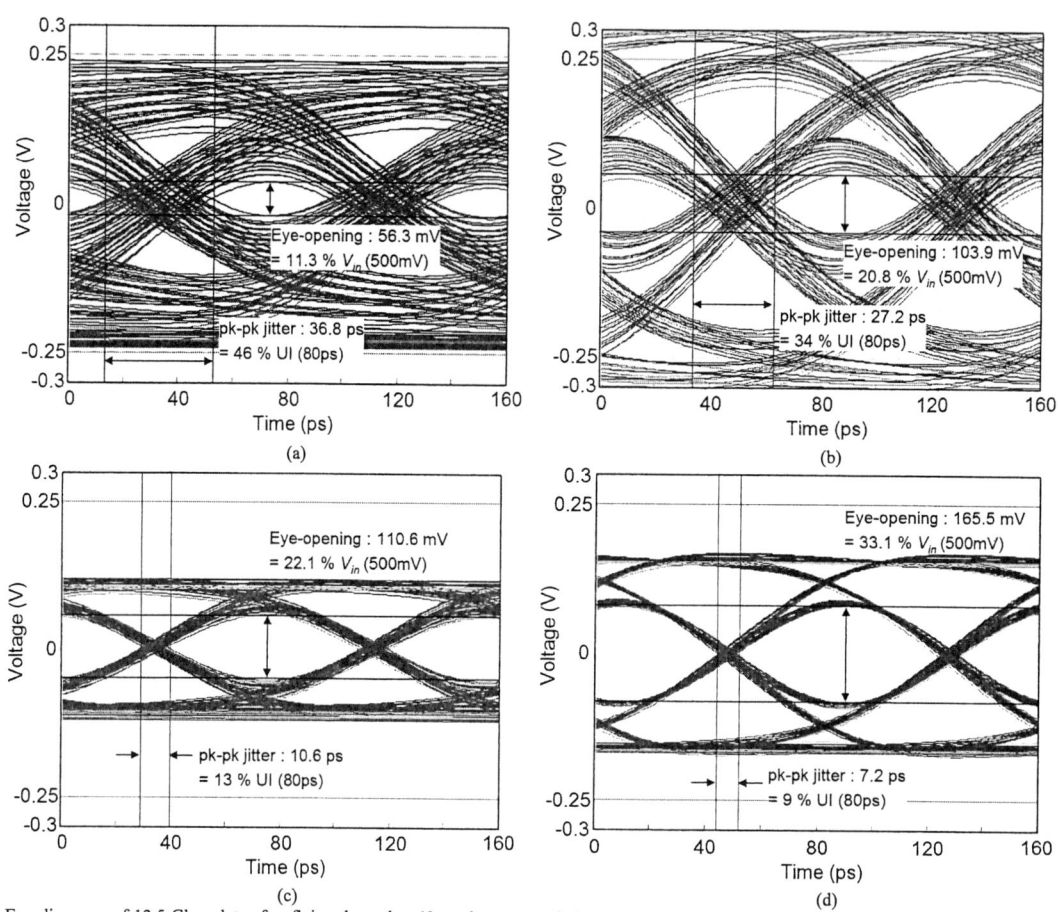

Fig. 6. Eye diagrams of 12.5 Gbps data after flying through a 40-cm-long transmission line on PCB (a) without equalization (b) with active equalizer (c) with passive equalizer (d) with the proposed hybrid equalizer

Table 1. Comparison of the eye-openings and the pk-pk jitters for various equalization methods

Metric	Without equalization	With active equalizer	With passive equalizer	With the proposed hybrid equalizer
Normalized eye-opening (V_{in})	11.3 %	20.8 %	22.1 %	33.1 %
Normalized pk-pk jitter (UI)	46 %	34 %	13 %	9 %

V. CONCLUSIONS

We have proposed a hybrid equalization technique by combining the passive equalizer based on NEXT and reflections and the second-order active equalizer. The advantages of active and passive equalization, active gain and wide bandwidth, were both achieved by the proposed technique. The hybrid equalization technique was validated for a data rate of 12.5 Gbps on a 40-cm-long transmission line on PCB, based on insertion loss and eye-diagram simulations. The proposed technique presents noticeable improvement of eye-opening and pk-pk jitter from the un-equalized one, 193% and 80%, respectively.

REFERENCES

[1] M. Horowitz, C.-K. K. Yang, S. Sidiropoulos, "High-Speed Electrical Signaling: Overview and Limitations," *IEEE Micro*, January 1998, pp. 12-24

[2] J. Liu and X. Lin, "Equalization in High-speed Communication Systems." *IEEE Circuits and Systems Magazine*, pp. 4–17, 2nd quarter, 2004

[3] J. Shin, Aygun, K.,"On-Package Continuous-Time Linear Equalizer using Embedded Passive Components, " *IEEE Symposium of Electrical Performance of Electronic Packaging*, October 2007, pp. 147-150

[4] Maxim Integrated Products, "Designing a Simple, Small, Wide-band and Low-power Equalizer for FR4 Copper Links," *Proceeding of DesignCon*, 2003

[5] R. M. Kurzrok, "A Review of Key Equalizer Specifications and What They Mean," *High Frequency Electronics Magazine*, 2004

[6] E. Song, J. Cho, W. Lee, M. Shin, and J. Kim, "A Wide-Band Passive Equalizer Design on PCB Based on Near-end Crosstalk and Reflections for 12.5Gbps Serial Data Transmission," *IEEE Microwave and Wireless Components Letters*, Vol. 18, No. 12, December 2008, pp. 794-796.ds

[7] H. Higashi et. al., "A 5-6.4-Gb/s 12-Channel Transceiver With Pre-Emphasis and Equalization," *IEEE Journal of Solid-State Circuits*, Vol. 40, No. 4, April 2005, pp. 978-985

978-1-4244-4447-2/09 $25.00 © 2009 IEEE

The Effects of Time Windowing on the Accuracy of the Short-Pulse Propagation Technique

Lionelle F. Wells[1], Alina Deutsch[2], Zhen Zhou[1], Kathleen L. Melde[1]

[1]Department of Electrical and Computer Engineering, Center for Electronic Packaging Research
The University of Arizona Tucson, Arizona 85721
Phone (520) 626-8439 or (520) 626-2538, email: {lwells,zhenz,melde}@ece.arizona.edu
[2]IBM T.J.Watson Research Center, 1101 Kitchawan Road, Yorktown Heights, N.Y. 10598
Phone (914) 945-2858, email: deutsch@us.ibm.com

Abstract – The accuracy of the short-pulse propagation (SPP) technique is evaluated for different time windowing methods. An algorithm written to replace the conventional program used for the signal processing step of the SPP technique allows control over the placement of the time window with respect to the pulse data. The values of the attenuation coefficient calculated using this algorithm are compared to VNA measurements to assess accuracy. Using the results from different time windowing techniques, a rule set for accurate time windowing of SPP data is developed.

I. Introduction

The SPP technique has been previously developed as a method to characterize the properties of transmission lines. In this technique, a short pulse obtained by differentiating the source of a sampling oscilloscope is propagated along two identical transmission line structures with lengths l_1 and l_2, where $l_1 > l_2$. The transmitted pulses are detected and digitized, and the data are input into a signal processing algorithm. The final step, referred to as the signal processing step, can introduce error and degrade the accuracy of the SPP technique if it is not performed properly. The signal processing step involves time windowing the pulse data to eliminate unwanted reflections from the probes, pads, vias, or cable connectors [1], performing a fast Fourier transform (FFT) on the time-windowed data, and calculating the complex propagation coefficient according to the following equation (1):

$$\Gamma(f) = \alpha(f) + j\beta(f) = \frac{1}{l_1 - l_2} \ln\left(\frac{A_1(f)}{A_2(f)}\right) + j\frac{\Phi_1(f) - \Phi_2(f)}{l_1 - l_2} \qquad (1)$$

where $\alpha(f)$ is the attenuation coefficient, $\beta(f)$ the phase coefficient, and $A_i(f)$ and $\Phi_i(f)$ are the amplitudes and phases of the transforms of the corresponding lines l_1 and l_2 [1].

Conventional signal processing algorithms used for SPP generate plots of the pulse data and then rely on the end user to visually select the extent of the time window. This subjective choice on the part of the end user may lead to inconsistent results and reduce the overall accuracy of SPP. The SPP technique has been previously applied to predict the effects of line parameter tolerances to provide guidelines to manufacturers [2] and in an application such as this it is essential that the technique be consistent and accurate. This paper presents a new algorithm for the signal processing step of the SPP technique that chooses the extent of the time window based on a validated rule set without the need for input from the end user.

Time windowing in SPP is best accomplished through the use of a rectangular window size because amplitude resolution is more important than frequency resolution [1]. Letting the vector w in the time domain which has the Fourier transform W represent the time window, the discrete Fourier transform of the windowed pulse is given by (2):

$$X_k = \sum_0^{N-1} x_n w_n e^{-j(2\pi/N)kn} = X * W \qquad (2)$$

where $*$ represents the convolution operator [3]. Considering the window to be a time-domain filter on the pulse data and taking into account the scaling property of the Fourier transform, given in the continuous case by (3):

$$\int_{t_1}^{t_2} X(at)e^{-j2\pi ft} dt = \frac{1}{|a|} X\left(\frac{f}{a}\right) \qquad (3)$$

where t_1 and t_2 are the edges of the time window and $t_1 < t_2$, it is evident that time windowing has the effect of smoothing the signal in the frequency domain.

978-1-4244-4447-2/09 $25.00 © 2009 IEEE

Assuming that a rectangular time window is used on the data, there are two fundamental choices that must be made either by the end user or an algorithm concerning how the time window is applied to the pulse. First, the relative size of the time window with respect to the pulse must be chosen. In the signal processing algorithm used, the relative size of the time window is defined by the percentage of total power underneath the pulse lost by application of that time window. Mathematically, the power lost due to time windowing is given by (4):

$$P_{lost} = \frac{\int_0^\infty | X_{tw}(t) |^2 \, dt}{\int_0^\infty | X(t) |^2 \, dt} \quad (4)$$

where $X(t)$ is the original pulse and $X_{tw}(t)$ is the part of the pulse lying inside the time window. To find the size of the time window such that the desired value of P_{lost} is reached, an iterative procedure is used. The algorithm first defines an initial time window that encompasses the entire pulse and then shrinks the window until the desired value of P_{lost} is reached.

In addition to the size of the time window, its temporal position with respect to the pulse is another important choice made by either the end user or the algorithm. For a time window of a given size that is smaller than the pulse, it can be positioned such that power is removed from the start of the pulse, the end of the pulse, or both. The relative accuracy of these three techniques is the focus of this paper. Section II covers the methods used to obtain the results, section III contains the results and discussion, and conclusions are given in section IV.

II. Analysis Methodology

The SPP signal processing algorithm was implemented in MATLAB and designed to be a substitute for other methods of implementing this step of the SPP technique. In addition to the pulse files, the algorithm has an input argument that specifies the percentage of power to be left outside of the time window. After reading in the data the algorithm performs time windowing on the data and iteratively reaches the desired power by removing it either from the start of the pulse, the end of the pulse, or symmetrically from both sides of the pulse. Time windowing is applied to the pulse associated with the longer transmission line first. Then, a time window of identical size is applied to the pulse on the shorter line while assuring that the same amount of power, if any, is left outside the window at the beginning of the pulse. The pulse on the longer line is used to define the time window because it has a longer duration than the pulse on the shorter line. This is due to the higher attenuation and larger phase misalignment on a longer length of transmission line. Additionally, the pulse associated with the longer line appears to have smaller reflections than the pulse on the shorter line due to the higher attenuation. Once the time windowing has been completed, the remaining data is zero-padded and cubic spline functions are used to interpolate between the zero-padded sections and the edges of the pulses. A FFT is then performed using the Nyquist condition to set the frequency scale, and the data for the attenuation and phase coefficients are written to an external file in a dB scale.

After the algorithm calculates the attenuation coefficients using the specified time window, a second program, also written in MATLAB, reads those values and also a separate file containing the attenuation coefficients obtained from VNA measurements on the same set of transmission lines. The two sets of data are compared for a frequency range from 1-19 GHz, and an average error is calculated over the entire frequency range.

III. Analysis and Discussion of Results

The purpose of this analysis is to determine how different temporal placements of the time window affect the accuracy of SPP. The pulses were measured on Nelco material for stripline with lengths of 2.794 cm and 10.414 cm. Fig. 1 shows the original pulse on the longer line that was analyzed. The pulse was time windowed with 1% of the power remaining outside of the window. The window was placed such that the power was removed from the beginning of the pulse, the end of the pulse, and from both sides of the pulse. The resulting time-windowed pulses are shown in Figs. 2, 3, and 4, respectively.

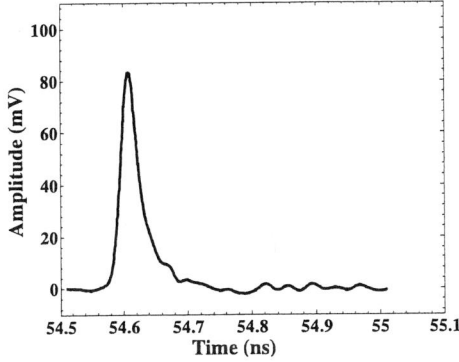

Fig. 1 Raw pulse for the longer line

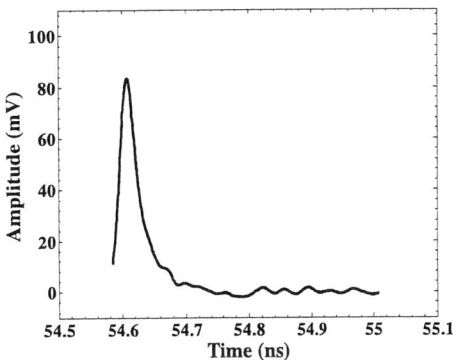

Fig. 2 Time windowed pulse with 1% of power discarded at start

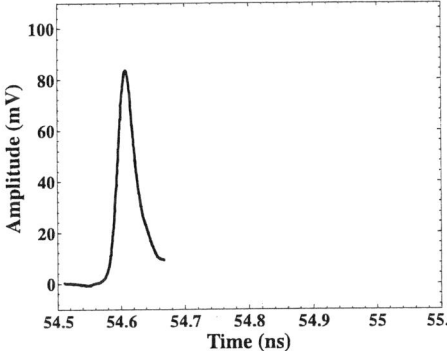

Fig. 3 Time windowed pulse with 1% of power discarded at end

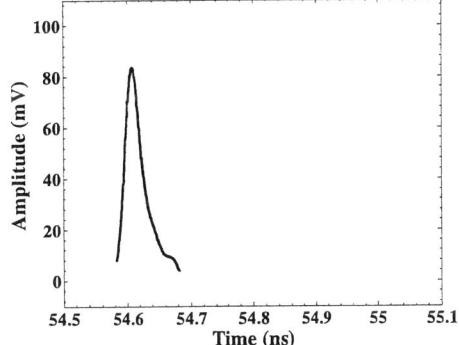

Fig. 4 Time windowed pulse with 1% total power discarded from both sides

For each of the three placements of the time window, the attenuation coefficients were calculated by using the MATLAB algorithm to accomplish the signal processing step of the SPP technique. The average error in the calculated attenuation coefficients with respect to the VNA-measured values was computed over a range of 1-19 GHz for each of the three time windowing cases. To calculate this error the values of the attenuation coefficients are first converted from units of *[dB/cm]* to units of *[1/cm]*, and then the following formula is used (5):

$$Error = \frac{1}{N} \sum_{n=1}^{N} \left| \frac{\alpha(f_n)_{SPP} - \alpha(f_n)_{VNA}}{\alpha(f_n)_{VNA}} \right| \qquad (5)$$

where N is the number of data points in the frequency range over which the average is taken.

Compared to the VNA measurements, the errors were 7.42% in the case where only the start of the pulse was discarded, 8.14% when only the end of the pulse was discarded, and 9.95% in the case where data were discarded from both sides of the pulse. These values show that the time windowing method in which only the start of the pulse is removed has the lowest average error with respect to the VNA measurements of the attenuation coefficient. To further investigate the difference in the attenuation coefficients at different frequencies, plots were created to compare the calculated SPP values to the VNA measurements. Fig. 5 shows the attenuation coefficients for the case where power was removed only from start of the pulse, Fig. 6 for the case where the power was removed only from the end of the pulse, and Fig. 7 for the case where power was removed from both sides of the pulse. These plots show that removing power only from the start of the pulse results in over-prediction of the attenuation coefficient values at high frequencies, while removing power only from the end of the pulse results in the over-prediction of the attenuation coefficients at low frequencies. Based on this data the proposed rule set for accurate time windowing will involve the removal of power only from the start of the pulse.

978-1-4244-4447-2/09 $25.00 © 2009 IEEE

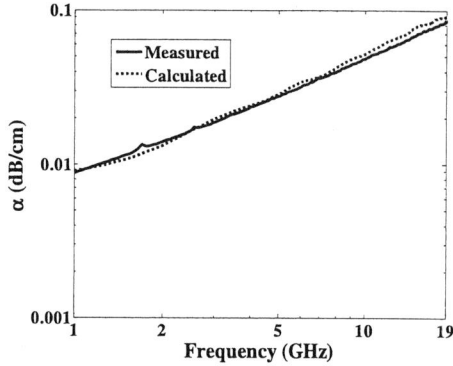

Fig. 5 Measured and calculated attenuation for 1% of power discarded at start of pulse

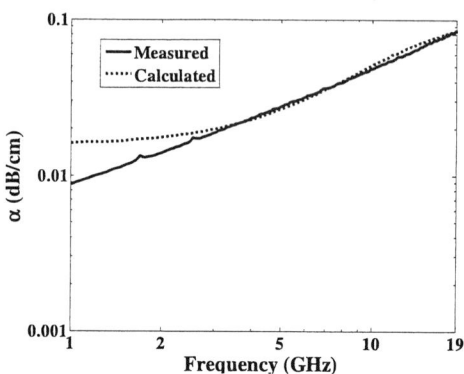

Fig. 6 Measured and calculated attenuation for 1% of power discarded at end of pulse

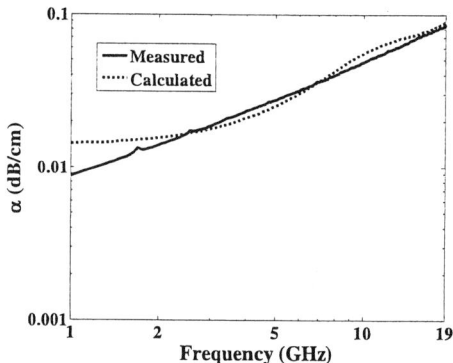

Fig. 7 Measured and calculated attenuation for 1% of total power discarded at both sides of pulse

IV. Conclusions

An algorithm was developed to assess the effect of different methods of time windowing on the accuracy of the SPP technique. The algorithm produces consistent and accurate results for the attenuation coefficient without the need for the end user to visually select a time window. The most accurate results are produced when the time window is placed in a position such that only the start of the pulse lies outside of it. The rule set for the signal processing step of the SPP technique developed here can be applied to improve the accuracy of the applications for which it is used.

Acknowledgements: This work is supported by an IBM Faculty Award.

References:
[1] Alina Deutsch, Thomas-Michael Winkel, Gerard V. Kopcsay, Christopher W. Surrovic, Barry J. Rubin, George A. Katopis, Bruce J. Chamberlin, and Roger S. Krabbenhoft, "Extraction of ε_r(f) and tanδ(f) for Printed Circuit Board Insulators Up to 30 GHz Using the Short-Pulse Propagation Technique," *IEEE Trans. Advanced Packaging,* vol. 28, No. 1, pp. 4-12, February 2005.
[2] Zhen Zhou, Alina Deutsch, Kathleen L. Melde, George A. Katropis, and Jason D. Morsey, "An analysis on measurement sensitivity of Short-Pulse Propagation technique using a virtual test bench," *Electrical Performance of Electronic Packaging, 2008 IEEE-EPEP*, pp. 213-216, October 2008.
[3] Pedro M. Ramos, Raul C. Martins, Sergio Rapuano, and Pasquale Daponte, "Frequency and Time—Frequency Domain Analysis Tools in Measurement," in *Data Modeling for Metrology and Testing in Measurement Science,* Franco Pavese and Alistair B. Forbes Ed. Birkhäuser: Boston, 2008, pp. 127-203.

On the performance of weighting schemes for passivity enforcement of delayed rational macromodels of long interconnects

A. Chinea, S. Grivet-Talocia, P. Triverio

Dip. Elettronica, Politecnico di Torino, C. Duca degli Abruzzi 24, 10129 Torino, Italy
Ph. +39 011 5644104, Fax +39 011 5644099 (e-mail stefano.grivet@polito.it)

Abstract: This paper focuses on robust black-box macromodeling of electrically long interconnects. These structures can be represented by very compact and efficient models that combine in their constitutive equations both closed-form delay operators and low-order rational coefficients. We present a simple perturbation approach for efficient passivity enforcement of these delayed rational macromodels, and we show how suitable norm weighting schemes may significantly improve in-band accuracy preservation.

1 Introduction

The main focus of this paper is the construction of highly efficient and guaranteed passive macromodels of long interconnects known from tabulated frequency responses. Availability of such macromodels may be a fundamental enabling factor for channel qualification via fast simulations including realistic driver and receiver models (i.e., including nonlinear and dynamic effects together with preemphasis and equalization). In fact, SPICE-based eye diagram simulation is still an overly demanding task, essentially due to the computational burden imposed by non-optimal interconnect models.

Rational macromodeling [1] has become a standard practice for SPICE-based modeling of lumped or electrically small structures, such as vias or connectors. In principle, the same technique can be applied directly to extract SPICE models of electrically long interconnects. However, the resulting model complexity may be excessive, since the delay operators related to wave propagation induce fast phase variations in the terminal responses, that, in turn, require many pole/residue pairs in their rational approximations. For this reason, explicit representations of such delay operators in the model structure, together with low-order rational coefficients, has been recently proposed and applied [2, 3, 4, 5], showing an important speedup (up to two orders of magnitude) in system-level simulations.

Passivity enforcement of these delay-based macromodels is still a partially open issue. A comprehensive theoretical framework for passivity check has been proposed in [6] with reference to the Method of Characteristic (MoC) model structure. Passivity enforcement based on perturbation within this framework has also been demonstrated in [7]. However, delayed rational macromodels have a more general structure than MoC models and call for a generalization. This paper extends the preliminary results of [8] by introducing a new norm weighting scheme based on a delay-preserving gramian formulation. This scheme allows perturbation of all model coefficients while optimizing accuracy of the final passive model. Numerical results show a significantly better performance of weighted schemes with respect to standard approaches.

2 Delayed-rational macromodeling

Our starting point is a set of tabulated frequency samples in scattering form

$$\widehat{\mathbf{H}}_k = \left[\widehat{H}^{i,j}(\mathrm{j}\omega_k) \right], \qquad i,j = 1,\dots,P, \quad k = 1,\dots,\bar{k} \tag{1}$$

where P denotes the number of interface ports of the interconnect and $\bar{k}$ is the number of samples. These may be available from direct measurement, numerical simulations and/or cascading of separate individual blocks, as in the case of a complex bus for chip-to-chip communication going through cards, boards, backplane and various discontinuities. Following the strategy of [4], each scattering response is processed separately in order to find a delayed rational model

$$H^{i,j}(\mathrm{j}\omega_k) \simeq \widehat{H}^{i,j}(\mathrm{j}\omega_k), \qquad H^{i,j}(s) = \sum_{m=0}^{M^{i,j}} Q_m^{i,j}(s) e^{-s\tau_m^{i,j}} + D^{i,j}, \qquad Q_m^{i,j}(s) = \sum_{n=1}^{N_m^{i,j}} \frac{R_{mn}^{i,j}}{s - p_{mn}^{i,j}}, \tag{2}$$

where $\tau_m^{i,j} \geq 0$ represents signal propagation delays, and $Q_m^{i,j}(s)$ are suitable rational coefficients representing other effects such as attenuation and dispersion. The first step is the identification of the dominant delays, here performed as in [4] by exploiting the time-frequency energy decomposition provided by the so-called Gabor transform. Once these delays are known, the Delayed Vector Fitting (DVF) or Delayed Sanathanan-Koerner (DSK) iterations [5] may be applied directly for the identification of the poles and residues for the rational coefficients in (2). A standard state-space realization process can finally be applied to the macromodel of each individual response (2), leading to a delayed state-space form

$$H^{i,j}(\mathrm{j}\omega) = \sum_{m=0}^{M^{i,j}} \mathbf{c}_m^{i,j} e^{-\mathrm{j}\omega\tau_m^{i,j}} \left(\mathrm{j}\omega\mathbf{I} - \mathbf{A}^{i,j}\right)^{-1} \mathbf{b}^{i,j} + D^{i,j}, \tag{3}$$

where $\mathbf{A}^{i,j}$ collects all the poles, column vector $\mathbf{b}^{i,j}$ provides single-input to state mapping, and vector $\mathbf{c}_m^{i,j}$ collects all residues in a single row. Without loss of generality, we assume a real-valued state-space realization.

3 Passivity enforcement

The delayed rational macromodel (3) is causal by construction (all delays are nonnegative) and strictly stable (all poles are constrained to have a negative real part by DVF or DSK, see [5]). Model passivity is not qualified yet, since all responses have been processed individually, whereas passivity conditions require analysis of all model responses at the same time [9]. Passivity is the fundamental property ensuring that the model cannot be the root cause for unstable behavior or lack of convergence during system-level transient simulations.

For scattering representations, passivity requires that the maximum singular value of the model transfer matrix does not exceed one at all frequencies [9]. Therefore, a passivity check is necessary to pinpoint the frequency bands where at least one of the singular values exceeds the unit threshold. Various techniques are available for this task, based either on adaptive sweep [7] or frequency-dependent Hamiltonian eigenvalue problems [6]. We assume that this information is available, and we show in the following how to obtain model passivity in case some violations are detected.

Let $\sigma > 1$ be a singular value of $\mathbf{H}(\mathrm{j}\bar{\omega})$ and $\mathbf{u}, \mathbf{v} \in \mathbb{C}^P$ the associated left and right singular vectors, respectively [1]. We want to perturb the vectors $\mathbf{c}_m^{i,j}$ by small amounts $\delta\mathbf{c}_m^{i,j}$ such that the perturbed singular value $\sigma + \delta\sigma \leq 1$. Application of standard results on singular value perturbation leads to the following first-order expansion

$$\delta\sigma = \Re\left\{\sum_{i,j=1}^{P} \sum_{m=0}^{M^{i,j}} u_i^* \delta\mathbf{c}_m^{i,j} \mathbf{z}_m^{i,j}\right\} \qquad \text{with} \qquad \mathbf{z}_m^{i,j} = e^{-\mathrm{j}\bar{\omega}\tau_m^{i,j}} \left(\mathrm{j}\bar{\omega}\mathbf{I} - \mathbf{A}^{i,j}\right)^{-1} \mathbf{b}^{i,j} v_j. \tag{4}$$

The passivity constraint on the singular value perturbation thus becomes

$$\sum_{i,j=1}^{P} \Re\left\{u_i^* (\mathbf{z}^{i,j})^T\right\} (\delta\mathbf{c}^{i,j})^T < 1 - \sigma, \tag{5}$$

where $\mathbf{z}^{i,j}$ and $\delta\mathbf{c}^{i,j}$ stack the partial contributions of individual delay terms $\mathbf{z}_m^{i,j}$ and $\delta\mathbf{c}_m^{i,j}$ in a single column and row, respectively. The constraint (5) is a linear inequality, which can be easily enforced using standard optimization engines.

A simplistic approach for enforcing global passivity solves the linearly constrained quadratic problem

$$\min \sum_{i,j=1}^{P} ||\delta\mathbf{c}^{i,j}||_2^2 \qquad \text{subject to (5)}, \tag{6}$$

where the standard $\mathcal{L}^2$ energy norm is used. A better approach adopts a more advanced cost function by defining a suitable weighting factor, described below. The following derivation is a generalization to the delayed rational case of the Gramian-based norm introduced in [10] within the scope of purely rational macromodels.

The cost function to be minimized in this second approach is the cumulative energy E of the perturbation induced by $\delta\mathbf{c}^{i,j}$ on the model impulse responses $h^{i,j}(t)$, defined as

$$E = \sum_{i,j=1}^{P} \int_0^{\infty} \left|\delta h^{i,j}(t)\right|^2 \mathrm{d}t, \qquad \text{with} \qquad \delta h^{i,j}(t) = \sum_{m=0}^{M^{i,j}} \delta\mathbf{c}_m^{i,j} \exp\left\{\mathbf{A}^{i,j}(t - \tau_m)\right\} \mathbf{b}^{i,j} u(t - \tau_m). \tag{7}$$

A combination of these two expressions leads to

$$E = \sum_{i,j=1}^{P} \sum_{m,n=0}^{M^{i,j}} \delta\mathbf{c}_m^{i,j} \mathbf{W}_{m,n}^{i,j} (\delta\mathbf{c}_n^{i,j})^T \tag{8}$$

---

[1]This is equivalent to $\mathbf{u}^H \mathbf{H}(\mathrm{j}\bar{\omega})\mathbf{v} = \sigma > 1$

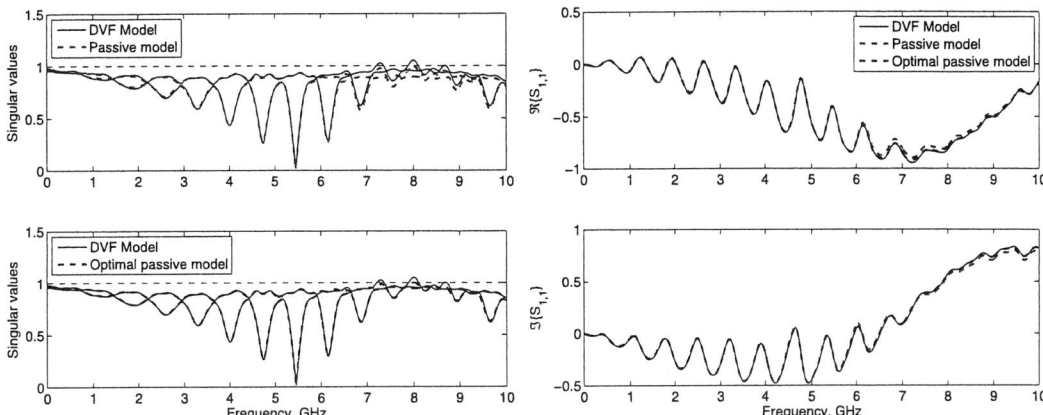

Figure 1: The original (non-passive) DVF model of a PCB link is compared to standard and optimal passive models. Left panels: singular values. Right panels: real and imaginary part of S_{11}.

with

$$\mathbf{W}_{m,n}^{i,j} = \int_{\tau_{m,n}}^{\infty} e^{\mathbf{A}^{i,j}(t-\tau_m)} \mathbf{b}^{i,j} (\mathbf{b}^{i,j})^T e^{(\mathbf{A}^{i,j})^T(t-\tau_n)} dt, \qquad \tau_{m,n} = \max\{\tau_m, \tau_n\}. \tag{9}$$

It turns out that $\mathbf{W}_{m,n}^{i,j}$ can be computed as the unique solution of the following Lyapunov equation

$$\mathbf{A}^{i,j} \mathbf{W}_{m,n}^{i,j} + \mathbf{W}_{m,n}^{i,j} (\mathbf{A}^{i,j})^T + \mathbf{Q}_{m,n}^{i,j} = 0 \tag{10}$$

where

$$\mathbf{Q}_{m,n}^{i,j} = \begin{cases} e^{\mathbf{A}^{i,j}(\tau_m-\tau_n)} \mathbf{b}^{i,j}(\mathbf{b}^{i,j})^T & \text{if} \quad \tau_m > \tau_n \\ \mathbf{b}^{i,j}(\mathbf{b}^{i,j})^T & \text{if} \quad \tau_m = \tau_n \\ \mathbf{b}^{i,j}(\mathbf{b}^{i,j})^T e^{(\mathbf{A}^{i,j})^T(\tau_n-\tau_m)} & \text{if} \quad \tau_m < \tau_n \end{cases} \tag{11}$$

Equation (8) can be further simplified by collecting submatrices $\mathbf{W}_{m,n}^{i,j}$ in a block-matrix $\mathbf{W}^{i,j}$, leading to

$$E = \sum_{i,j=1}^{P} \delta\mathbf{c}^{i,j} \mathbf{W}^{i,j} (\delta\mathbf{c}^{i,j})^T. \tag{12}$$

Finally, a coordinate change in the perturbation unknowns is performed as

$$\delta\tilde{\mathbf{c}}^{i,j} = \delta\mathbf{c}^{i,j}(\mathbf{K}^{i,j})^T, \qquad \text{where} \qquad \mathbf{W}^{i,j} = (\mathbf{K}^{i,j})^T \mathbf{K}^{i,j} \tag{13}$$

represents the Cholesky decomposition of $\mathbf{W}^{i,j}$. This allows the definition of the second proposed constrained optimization scheme for passivity enforcement

$$\min \sum_{i,j=1}^{P} ||\delta\tilde{\mathbf{c}}^{i,j}||_2^2 \qquad \text{subject to} \qquad \sum_{i,j=1}^{P} \Re\left\{ u_i^*(\mathbf{z}^{i,j})^T(\mathbf{K}^{i,j})^{-1} \right\} (\delta\tilde{\mathbf{c}}^{i,j})^T < 1 - \sigma. \tag{14}$$

4 Examples

We demonstrate the performance of proposed passivity enforcement scheme on two examples. First case is a delayed rational macromodel obtained by DVF from measured scattering parameters of a 10 cm PCB link (courtesy of IBM). A delayed rational model was obtained with DVF. As depicted in Fig. 1 (left panels), there is a localized passivity violation around 8 GHz. Both standard [8] and proposed passivity enforcement methods were applied. The resulting perturbation on the return loss is depicted in Fig. 1 (right panels), showing that both methods provide good results given the limited extent of the passivity violation. The accuracy of the proposed optimal scheme is slightly better.

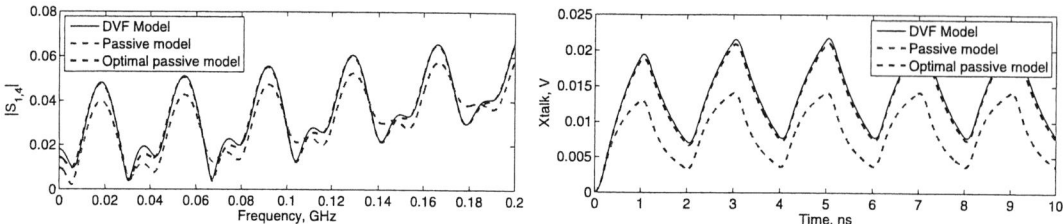

Figure 2: Shielded 12-port cable. The original (non-passive) DVF model is compared to standard and optimal passive models. Left panel: magnitude of scattering parameter $S_{1,4}$. Right panel: time-domain near-end crosstalk from transient SPICE simulation.

The second example is a shielded multiconductor cable of length 2.56 m, with 6 twisted inner signal conductors and an outer shielding layer made by several thin wires. A fully-coupled delayed rational model was obtained using DVF. This model resulted non-passive, with a maximum singular value $\sigma_{max} = 1.007$. Despite this small violation, standard passivity enforcement according to [8] led to some accuracy degradation in most sensitive crosstalk responses, see Fig. 2. Conversely, our proposed optimal scheme provided significantly better performance, as can be observed both in frequency domain (Fig. 2, left panel) and in time-domain (right panel). Time-domain results were obtained using SPICE by exciting one conductor with a 500 MHz clock signal and observing the near-end crosstalk on a 20 pF termination.

In summary, we have presented a passivity enforcement scheme that can be applied to delayed rational macromodels of long interconnects. This scheme provides a full generalization of existing methods for purely rational macromodels, including an optimality condition for the final accuracy on the passive macromodel. The resulting SPICE netlists can thus be safely used in transient Signal Integrity simulations, since potentially unstable behaviors are ruled out by the guaranteed model passivity.

References

[1] B. Gustavsen, A. Semlyen, "Rational approximation of frequency responses by vector fitting," *IEEE Trans. Power Delivery*, Vol. 14, N. 3, July 1999, pp. 1052–1061.

[2] S. Grivet-Talocia, "Delay-based macromodels for long interconnects via time-frequency decompositions," in *IEEE 15th Topical Meeting on Electrical Performance of Electronic Packaging, Scottsdale, Arizona*, pp. 199–202, Oct. 23–25, 2006.

[3] A. Charest, D. Saraswat, M. Nakhla, R. Achar, N. Soveiko, "Compact Macromodeling of High-Speed Circuits via Delayed Rational Functions," *IEEE Microwave and Wireless Components Letters* Vol. 17, No. 12, Dec. 2007, pp. 828–830.

[4] A. Chinea, P. Triverio, and S. Grivet-Talocia, "Compact macromodeling of electrically long interconnects," in *IEEE 17th Topical Meeting on Electrical Performance of Electronic Packaging (EPEP 2008), San Jose, CA*, pp. 199–202, October 27–29, 2008.

[5] A. Chinea, P. Triverio, S. Grivet-Talocia, "Delay-Based Macromodeling of Long Interconnects from Frequency-Domain Terminal Responses," *IEEE Transactions on Advanced Packaging*, 2009, in press.

[6] C. Chen, E. Gad, M. Nakhla, R. Achar, "Passivity Verification in Delay-Based Macromodels of Multiconductor Electrical Interconnects," *IEEE Transactions on Advanced Packaging*, Vol. 30, No. 2, May. 2007, pp. 246–256.

[7] A. Chinea, S. Grivet-Talocia, "Perturbation Schemes for Passivity Enforcement of Delay-Based Transmission Line Macromodels," *IEEE Transactions on Advanced Packaging*, vol.31, n.3, pp. 568-578, August 2008

[8] A. Charest, M. Nakhla, R Achar, C. Chen, "Passivity verification and enforcement of delayed rational function macromodels from networks characterized by tabulated data," in *IEEE Workshop on Signal Propagation on Interconnects, 2009. SPI'09*, pp. 1–4, May 12–15, 2009.

[9] P. Triverio, S. Grivet-Talocia, M. S. Nakhla, F. Canavero, R. Achar, "Stability, Causality, and Passivity in Electrical Interconnect Models," *IEEE Transactions on Advanced Packaging*, Vol. 30, No. 4, pp. 795–808, Nov. 2007.

[10] S. Grivet-Talocia, "Passivity enforcement via perturbation of Hamiltonian matrices," *IEEE Trans. CAS-I*, Vol. 51, No. 9, pp. 1755-1769, Sept. 2004

Passivity Verification and Enforcement of Delayed Rational Approximations from Scattering Parameter Based Tabulated Data

Andrew Charest, Michel Nakhla, and Ram Achar

Dept. of Electronics, Carleton University, Ottawa, ON, Canada, K1S-5B6

Ph: (613)520-5780, Fax: (613)520-5708, Email: {acharest, msn, achar}@doe.carleton.ca

ABSTRACT

Ensuring the passivity of macromodels obtained from tabulated data has become a critical issue for accurate signal integrity analysis. For delayed rational function based macromodels derived in the admittance domain, the issue of passivity has recently been addressed. In this paper, passivity verification and compensation algorithms for delayed rational function based macromodels derived in the scattering domain are presented. Numerical results validating the proposed algorithms are presented.

I. INTRODUCTION

Accurate signal integrity analysis and simulation of high-speed electronic modules has become one of the most critical components in modern digital design [1]. High-frequency effects such as signal distortion, crosstalk, attenuation, delays, and reflections can cause signal degradation, in some cases leading to system failure. Consequently, it is imperative that these effects are taken into consideration in the design and simulation phases of electronic production.

In order to accurately model the above high-frequency effects, high-speed networks are more prominently being characterized by tabulated data obtained from measurements or full-wave 2D or 3D electromagnetic simulators. For the purpose of transient analysis, the tabulated data is approximated using rational functions [2], or in the case of networks with long delays, delayed rational functions [3]–[5]. All of these approaches can be easily integrated in standard circuit simulators and are guaranteed to be stable by construction. However, transient analysis using these models may still produce spurious oscillations if the model is not passive.

For simple rational function based macromodels, there are several algorithms in the literature which verify and enforce passivity [6], [7]. Recently, for delayed rational function based macromodels obtained from admittance parameter based tabulated data, passivity verification and enforcement algorithms have been reported [8]. However, in the case of scattering parameter based delayed rational function macromodels there are no passivity verification and compensation algorithms currently available.

To address the above issue, this paper presents novel algorithms for passivity verification and compensation of delayed rational function macromodels that are obtained from scattering parameter based tabulated data. For passivity verification, a new algorithm based on a frequency dependent eigenvalue problem by restricting the search region along the imaginary axis. For nonpassive macromodels, a passivity compensation algorithm is developed which iteratively perturbs the residues of the nonpassive delayed rational macromodel until passivity is achieved. The necessary theoretical foundations as well as results validating the proposed algorithms are presented.

II. BACKGROUND AND PROBLEM REVIEW

Let the scattering equations of an L port network be in the form $b = Sa$, where a and b are the incident and reflected waves, respectively, and $S(s) = [S_{ik}(s)]$, $i, k \in \{1, 2, \ldots, L\}$ represents the scattering matrix. For macromodeling networks with long delays, S_{ik} is approximated with a delayed rational function in the form

$$S_{ik}(s) = Q_0^{i,k}(s) + \sum_{m=1}^{M} Q_m^{i,k}(s)e^{-s\tau_m}, \tag{1}$$

$$Q_m^{i,k}(s) = R_{m,0}^{i,k} + \sum_{n=1}^{N_m^{i,k}} \frac{R_{m,n}^{i,k}}{s - p_{m,n}^{i,k}}. \tag{2}$$

Stability of delayed rational functions can be ensured by forcing the poles to be in the left-half of the complex plane [2]. However, ensuring passivity is a challenging task. For $S(s)$ to be passive, it has to satisfy [9]

a) $S(s)$ is analytic for all values of s with $\mathfrak{Re}(s) > 0$,

b) $S(s^*) = S^*(s)$, where $(*)$ is the complex conjugate operator,

c) $S(s)$ is bounded real. That is,

$$z^{*T}\left[I - S^H(s)S(s)\right]z \geq 0, \tag{3}$$

for all complex values of s with $\Re\mathfrak{e}(s) > 0$ and for any arbitrary vector z. Here, the superscript 'H' represents the Hermitian conjugate operator.

III. Proposed Passivity Verification

In this section, the above passivity conditions, combined with the delayed rational formulation in (1), will be used to develop efficient techniques for passivity verification of delayed rational function based macromodels [10]. First, we define the following lemma which reduces the required search region to just the imaginary axis while verifying the passivity conditions a) to c):

Lemma 1: If $Q_m^{i,k}(s)$ is asymptotically stable for all $(i,k) \in \{1, 2, \ldots, L\}$, $m \in (0, 1, \ldots, M)$ and

$$\lim_{s \to \infty} Q_m^{i,k}(s) = 0, \quad \text{for } m \neq 0, \tag{4}$$

then $S(s)$ defined by (1) satisfies the above passivity conditions a) and b). In addition, the third condition is satisfied for all $\Re\mathfrak{e}(s) > 0$ if it is satisfied on the imaginary axis. That is, (3) can be replaced by the equivalent condition,

$$z^{*T}\left[I - S^H(j\omega)S(j\omega)\right]z \geq 0, \tag{5}$$

for $\omega \in \mathbb{R}$ and for any arbitrary vector z.

Next, the scattering parameter formulation in (1) can be converted to a set of delayed differential equations in the time-domain to obtain

$$\dot{x}(t) = \mathbf{A}x(t) + \mathbf{B}a(t),$$
$$b(t) = \sum_{m=0}^{M}\mathbf{C}_m x(t - \tau_m) + \sum_{m=0}^{M}\mathbf{D}_m a(t - \tau_m), \tag{6}$$

where x represents the state variables, $\mathbf{A} \in \mathfrak{R}^{P \times P}$, $\mathbf{B} \in \mathfrak{R}^{P \times L}$, $\mathbf{C}_m \in \mathfrak{R}^{L \times P}$, $\mathbf{D}_m \in \mathfrak{R}^{L \times L}$, $P = (N \times M) \times L$, and $N = \sum_{i,k,m} N_m^{i,k}$. Let $\mathbf{D} = \mathbf{D}_0$, then imposing the conditions of Lemma 1 on (6) and converting to the Laplace-domain, the scattering matrix becomes

$$S(s) = \left(\sum_{m=0}^{M}C_m e^{-s\tau_m}\right)(sI - A)^{-1}B + D, \quad \|D\|_2 < 1. \tag{7}$$

Using (5) and (7), Lemma 1, and the steps outlined in [11], the following theorem reduces passivity verification of delayed rational function based macromodels to solving a frequency dependent eigenvalue problem.

Theorem 1: A delayed rational function-based macromodel is passive if and only if there does not exist any pure imaginary eigenvalues for s that satisfy the following frequency-dependent generalized eigenvalue problem

$$s\zeta = H(s)\zeta, \tag{8}$$

where,

$$H(s) = \mathcal{V} + \sum_{m=0}^{M}\hat{W}_m e^{-s\tau_m} + \sum_{m=0}^{M}\bar{W}_m e^{s\tau_m} + \sum_{m,n=0}^{M}W_{mn}e^{s(\tau_m - \tau_n)}, \tag{9}$$

and,

$$\mathcal{V} = \begin{bmatrix} -A^T & 0 & 0 & 0 \\ BRB^T & A & 0 & BRB^T \\ BRB^T & 0 & A & BRB^T \\ 0 & 0 & 0 & -A^T \end{bmatrix}, \quad \hat{W}_m = \begin{bmatrix} 0 & 0 & 0 & 0 \\ 0 & BRD^T C_m & 0 & 0 \\ 0 & BRD^T C_m & 0 & 0 \\ 0 & 0 & 0 & 0 \end{bmatrix}, \quad R = \left(I - D^T D\right)^{-1}$$

$$\bar{W}_m = \begin{bmatrix} -C_m^T DRB^T & 0 & 0 & -C_m^T DRB^T \\ 0 & 0 & 0 & 0 \\ 0 & 0 & 0 & 0 \\ 0 & 0 & 0 & 0 \end{bmatrix}, \quad W_{mn} = \begin{bmatrix} 0 & -C_m^T DRD^T C_n & 0 & 0 \\ 0 & 0 & 0 & 0 \\ 0 & 0 & 0 & 0 \\ 0 & 0 & -C_m^T C_n & 0 \end{bmatrix}. \tag{10}$$

Next, the solution to the above frequency-dependent generalized eigenvalue problem can used to determine the points and regions of passivity violation [11], [12]. The proof of Lemma 1 and Theorem 1 are not given due to lack of space.

978-1-4244-4447-2/09 $25.00 © 2009 IEEE

IV. PASSIVITY COMPENSATION

For nonpassive macromodels, the following passivity enforcement algorithm is developed to iteratively perturb the residue matrices until the model is passive. For the purpose of illustration and without loss of generality, the necessary theory associated with the process is described below, with respect to perturbing a single residue matrix, C_m (defined in (7)).

Consider the Laplace-domain transfer function given by (7), where C_m is the residue matrix for any arbitrary $m \in (0, 1, \ldots, M)$. For the purpose of illustration and without loss of generality, assume that the nonpassive macromodel contains one region of passivity violation. Further, assume that the minimum eigenvalue of $I - S(j\omega)^H S(j\omega)$, λ_{min}, occurs at $\omega = \omega_\lambda$. The goal of passivity compensation is to perturb the scattering matrix, $S(j\omega)$, such that λ_{min} becomes positive. This can be achieved by perturbing the residue matrix, C_m, by ΔC_m to give $\hat{C}_m = C_m + \Delta C_m$.

To determine ΔC_m, let $\Delta \lambda$ be the required perturbation to make λ_{min} positive.i.e,

$$\Delta \lambda \geq -\lambda_{min}. \tag{11}$$

Then, first order perturbation theory can be used to relate $\Delta \lambda$ to $\Delta S(j\omega)$ in the form [6]

$$\Delta \lambda \approx \frac{\vartheta^H \left(-S^H(j\omega_\lambda)\Delta S(j\omega_\lambda) - \Delta S^H(j\omega_\lambda)S(j\omega_\lambda) \right) \xi}{\vartheta^H \xi}, \tag{12}$$

where ϑ and ξ are the left and right eigenvectors of $I - S(j\omega_\lambda)^H S(j\omega_\lambda)$, respectively, corresponding to the eigenvalue λ_{min}. It should be noted that the second order term, $\Delta S(j\omega_\lambda)^H \Delta S(j\omega_\lambda)$, has been neglected [6]. The remaining terms are described by

$$\begin{aligned}
S^H \Delta S &= (S_R^H + jS_I^H)\Delta C_m(F_R + jF_I)(\zeta_R + j\zeta_I) \\
&= S_R^H \Delta C_m F_R \zeta_R + jS_R^H \Delta C_m F_I \zeta_R + jS_R^H \Delta C_m F_R \zeta_I - S_R^H \Delta C_m F_I \zeta_I \\
&\quad + jS_I^H \Delta C_m F_R \zeta_R - S_I^H \Delta C_m F_I \zeta_R - S_I^H \Delta C_m F_R \zeta_I - jS_I^H \Delta C_m F_I \zeta_I,
\end{aligned} \tag{13}$$

and

$$\begin{aligned}
\Delta S^H S &= (F_R^H + F_I^H)\Delta C_m^H(\zeta_R^H + j\zeta_I^H)(S_R + jS_I) \\
&= F_R^H \Delta C_m^H \zeta_R^H S_R + jF_R^H \Delta C_m^H \zeta_I^H S_R + jF_R^H \Delta C_m^H \zeta_R^H S_I - F_R^H \Delta C_m^H \zeta_I^H S_I \\
&\quad + jF_I^H \Delta C_m^H \zeta_R^H S_R - F_I^H \Delta C_m^H \zeta_R^H S_I - F_I^H \Delta C_m^H \zeta_I^H S_R - jF_I^H \Delta C_m^H \zeta_I^H S_I.
\end{aligned} \tag{14}$$

Here, $F_R = \mathfrak{Re}((j\omega_\lambda I - A)^{-1}B)$, $F_I = \mathfrak{Im}((j\omega_\lambda I - A)^{-1}B)$, $\zeta_R = \mathfrak{Re}(e^{-j\omega_\lambda \tau_m})$, $\zeta_I = \mathfrak{Im}(e^{-j\omega_\lambda \tau_m})$, $S_R = \mathfrak{Re}(S(j\omega_\lambda))$, and $S_I = \mathfrak{Im}(S(j\omega_\lambda))$.

Substituting (12) into (11) and after some simple mathematical manipulations, we obtain

$$\Psi X \geq -\lambda_{min}, \tag{15}$$

where the matrix Ψ is a function of ϑ, ξ, and $S(s)$ at $s = j\omega$ and X is a column vector corresponding to the entries in the matrix ΔC_m.

Perturbing an arbitrary residue matrix using (15) will lead to an additional error in the macromodel response, $\Delta S(j\omega)$. In order to minimize the error introduced in the system response, the upper bound of the L_2-norm of the error is minimized. To illustrate this, we note that the L_2-norm of ΔS can be written as follows [13]

$$\begin{aligned}
\|\Delta S\|_2^2 &= \|e^{-j\omega\tau_m}\Delta C_m(j\omega I - A)^{-1}B\|_2^2 \\
&\leq \|e^{-j\omega\tau_m}\|_2^2 \|\Delta C_m(j\omega I - A)^{-1}B\|_2^2 \\
&= \|\Delta C_m(j\omega I - A)^{-1}B\|_2^2 \\
&= trace(\Delta C_m P \Delta C_m^T), \tag{16}
\end{aligned}$$

where P is the controllability Grammian obtained from solving the following Lyapunov equation,

$$AP + PA^H + BB^H = 0. \tag{17}$$

Therefore, to minimize the error added to the response from each perturbation, (15) is solved iteratively subject to the constraint in (16) for the entries in the matrix ΔC_m.

978-1-4244-4447-2/09 \$25.00 © 2009 IEEE

V. NUMERICAL RESULTS

In order to verify the proposed algorithm, it was tested on a two-port microwave network (a cascade of a microstrip-coaxial cable-microstrip) characterized by tabulated data. The data was approximated using the delayed vector fitting algorithm [5]. A total of 4 delay terms and 42 poles were required to fit the tabulated data. Applying the proposed passivity verification algorithm, the model was found to be nonpassive in the region $[0.9728, 1.4708]$ GHz. In order to verify the results from the proposed verification algorithm, a plot of $\|S(j\omega)\|_2$ is shown in Fig. 1(a). Fig. 1(a) shows a clear passivity violation ($\|S(j\omega)\|_2 > 1$) corresponding to the region of passivity violation found above.

Next, the proposed passivity compensation algorithm was applied to the macromodel. It is clear from Fig. 1(a) that the proposed compensation algorithm was successful. In addition, from Fig. 1(b) it is clear that a comparison of the scattering parameters from the original tabulated data, the nonpassive macromodel, and the compensated model are all in good agreement.

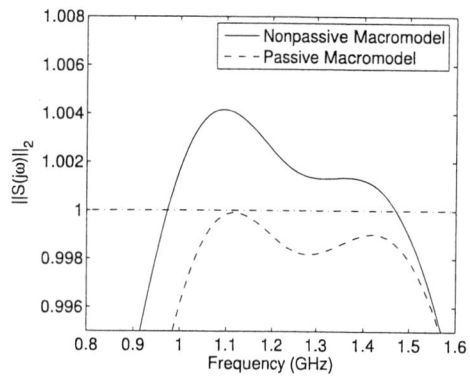

(a) Passivity verification of the macromodel.

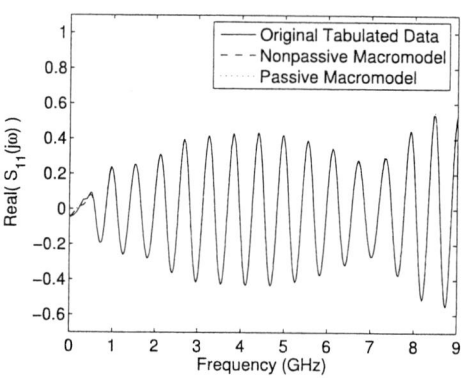

(b) Comparison of the accuracy of the scattering parameters.

Fig. 1. Numerical results for example.

VI. CONCLUSION

This paper presented novel algorithms for passivity verification and compensation of delayed rational function-based macromodels obtained from tabulated scattering parameter data. Necessary theoretical foundations for the proposed passivity verification algorithm have been developed. Passivity is verified using a frequency-dependent generalized eigenvalue problem by restricting the search region along the imaginary axis. If the model is found to be nonpassive, then passivity compensation is performed by iteratively perturbing the residues of the scattering matrix.

ACKNOWLEDGMENT

We would like to thank CST Gmbh, Germany, for providing the tabulated data used in the application example.

REFERENCES

[1] R. Achar and M. S. Nakhla, "Simulation of high-speed interconnects," *Proc. IEEE*, vol. 89, no. 5, pp. 693–728, May 2001.

[2] B. Gustavsen and A. Semlyen, "Rational approximation of frequency domain responses by vector fitting," *IEEE Trans. Power Del.*, July 1999.

[3] A. Charest, D. Saraswat, M. Nakhla, R. Achar, N. Soveiko, and I. Erdin, "Time domain delay extraction based macromodeling algorithm for long-delay networks," *IEEE Trans. Adv. Packag.*, accepted for future publication.

[4] A. Charest, D. Saraswat, M. Nakhla, R. Achar, and N. Soveiko, "Compact macromodeling of high-speed circuits via delayed rational functions," *IEEE Microw. Wireless Compon. Lett.*, vol. 17, no. 12, pp. 828–830, December 2007.

[5] A. Chinea, P. Triverio, and S. Grivet-Talocia, "Delay-based macromodeling of long interconnects from frequency-domain terminal responses," *IEEE Trans. Adv. Packag.*

[6] D. Saraswat, R. Achar, and M. S. Nakhla, "Fast passivity verification and enforcement via reciprocal systems for interconnects with large order macromodels," *IEEE Trans. VLSI Syst.*, vol. 15, no. 1, pp. 48–59, January 2007.

[7] S. Grivet-Talocia, "Passivity enforcement via perturbation of hamiltonian matrices," *IEEE Trans. Circuits Syst. I*, vol. 51, no. 9, September 2004.

[8] A. Charest, M. Nakhla, R. Achar, and C. Chen, "Passivity verification and enforcement of delayed rational function macromodels from networks characterized by tabulated data," *Proceedings of the IEEE Workshop on Signal Propagation on Interconnects*, 2009.

[9] E. Kuh and R. Rohrer, *Theory of Active Linear Networks*. San Francisco, CA:Holden-Day, 1967.

[10] A. Charest, M. Nakhla, and R. Achar, "Scattering domain passivity verification and enforcement of delayed rational functions," submitted to IEEE Microw. Wireless Compon. Lett.

[11] E. Gad, C. Chen, M. Nakhla, and R. Achar, "Passivity verification in delay-based macromodels of electrical interconnects," *IEEE Trans. Circuits Syst. I*, vol. 52, no. 10, pp. 2173–2187, October 2005.

[12] A. Chinea and S. Grivet-Talocia, "Perturbation schemes for passivity enforcement of delay-based transmission line macromodels," *IEEE Trans. Adv. Packag.*, vol. 31, no. 3, pp. 568–578, August 2008.

[13] C. Chen, D. Saraswat, R. Achar, E. Gad, M. Nakhla, and M. C. E. Yagoub, "Passivity compensation algorithm for method-of-characteristics based multiconductor transmission line interconnect macromodels," *IEEE Trans. Adv. Packag.*, To appear.

Order Estimation for Time-Domain Vector Fitting

Se-Jung Moon and A.C. Cangellaris*

Intel Corporation
Tel. 217-377-8479; E-mail: se-jung.moon@intel.com

*ECE Department, University of Illinois at Urbana-Champaign
Tel. 217-244-0975; Fax. 217-333-5962; E-mail: cangella@illinois.edu

Abstract. We derive a rule for estimating the order of the rational function fitting needed for accurate macromodel generation via the time-domain vector fitting. The proposed rule defines a relationship between the order, N, of the rational function approximation, the time duration, T, over which the response is desired and the frequency bandwidth W of the excitation.

Introduction

The proper choice of the order for the rational function approximation is important both in terms of achieving accuracy and in ensuring robustness of the fitting algorithm. While several techniques are available for predicting the rational function order in the context of frequency-domain vector fitting (VF), the topic has not been explored in the case of time-domain vector fitting (TDVF) [1]. For this reason, we propose the *WT* rule as an order estimation method for TDVF. The proposed rule is derived from the *2WT* theorem [2, 3]. We motivate the proposed rule in the context of the element-by-element fitting of the responses generated at each port of a passive network excited by a given band-limited signal [4].

Proposed Methodology

For simplicity and without loss of generality, a single-input single-output (SISO) passive system is considered because we are macromodeling the system network matrix element by element. TDVF constructs rational function approximation $H_{fit}(s)$ for the system transfer function using the input $x(t)$ and the excited output response $y(t)$. The excitation input $x(t)$ is set to be a Gaussian pulse with 60 dB bandwidth at the upper frequency limit of interest, W, and the output response $y(t)$, which are used as inputs to TDVF, are time-limited to $(T + t_0)$ as

$$\frac{|y(t)|}{\max(|y(t)|)} < \varepsilon \qquad \text{for } t > (T + t_0) \qquad (1)$$

where ε is a small enough value which is set to be 0.01 in this paper.

In the frequency domain, the input and output relation is presented as

$$Y(s) = H(s)X(s) \approx H_{fit}(s)X(s), \qquad (2)$$

where $H(s)$ is the transfer function, $X(s)$ and $Y(s)$ are, respectively, the Laplace transforms of $x(t)$ and $y(t)$. Because $X(s)$ is band-limited to $f = W$, $Y(s)$ is also band-limited to $f = W$, even if $H(s)$ is not. It is worthwhile to note that in this work, the time-limited and band-limited attributes of the response are defined based on the absolute value of a signal relative to its maximum value as presented in (1) [5,6], instead of being defined strictly according to [2,3].

From the *2WT* theorem [3], a signal $y(t)$ that is time-limited to $(T + t_0)$ and band-limited to W, has a Fourier series description as

$$y(t) = \sum_0^\infty a_n \cos 2\pi f_n t + \sum_1^\infty b_n \sin 2\pi f_n t,$$

$$f_n = \frac{n}{(T + t_0)}, \qquad (3)$$

which is valid for $0 < t < T + t_0$, and it can be rewritten as

$$y(t) = \sum_0^\infty \left(\frac{a_n}{2} + \frac{b_n}{2} \right) \exp(j2\pi f_n t) + \sum_0^\infty \left(\frac{a_n}{2} - \frac{b_n}{2} \right) \exp(-j2\pi f_n t) \qquad (4)$$

When the frequency interval of interest is $-W \le f \le W$, $y(t)$ has about $2W(T + t_0) + 1$ nonzero terms in its frequency spectrum.

The order of the rational function approximation is the number of poles, or equivalently, the dimension of the rational function space upon which the frequency data is projected. The projection is carried out in the context of VF through a least squares problem. Therefore, to solve the least squares problem, the order of the rational function approximation (which determines the number of columns of the matrix in the least squares problem) should be larger than or equal to the number of equations (which is the number of row of the matrix) which, in turn, is determined by the number of frequency data. In the fitting process used in VF, only positive frequencies are used [7]. Therefore, to solve the least squares problem, the number of frequency data, N_f, should be larger than the order, which leads to

$$N_y \le N_f, \qquad (5)$$

where N_y is the order of rational function fit for $Y(s)$. From the $2WT$ theorem, the time-limited signal $y(t)$ has $W(T + t_0)$ nonzero frequency terms in $0 \le f \le W$, and thus

$$N_y \le N_f \approx W(T + t_0) \qquad (6)$$

However, TDVF is designed to macromodel the transfer function $H(s)$ instead of $Y(s)$. Hence, the order for the fitting of $H(s)$ should be derived.

From (2), the order of $Y(s)$ can be presented in terms those for $X(s)$ and $H(s)$:

$$O(Y) \le O(H) + O(X), \qquad (7)$$

where $O(F)$ represents the order of a rational function representation of the function F. The validity of (6) can be simply justified as follows. Let $X(s)$ and $H(s)$ be represented in terms of rational functions as

$$X(s) = \sum_{m=1}^{N_x} \frac{R_{xm}}{s - p_{xm}} + D_x + sE_x$$

$$H(s) = \sum_{m=1}^{N} \frac{R_m}{s - p_m} + D + sE, \qquad (8)$$

where N_x and N are, respectively, the orders of the rational function fits for $X(s)$ and $H(s)$. The two rational function approximations are multiplied as in (2), to yield $Y(s)$ as follows.

$$Y(s) = \left(\sum_{m=1}^{N_x} \frac{R_{xm}}{s - p_{xm}} + D_x + sE_x \right) \left(\sum_{m=1}^{N} \frac{R_m}{s - p_m} + D + sE \right) \qquad (9)$$

$$Y(s) = \left(\frac{R_{x1}}{s - p_{x1}} + \frac{R_{x2}}{s - p_{x2}} + \cdots + D_x + sE_x \right) \left(\frac{R_1}{s - p_1} + \frac{R_2}{s - p_2} + \cdots + D + sE \right).$$

When the obtained expression for $Y(s)$ from (9) is decomposed into partial fractions, $Y(s)$ has at most all the poles of $X(s)$ and $H(s)$ in the form of a union, and the order of $Y(s)$ satisfies the relation in (7).

Upon further consideration, $X(s)$ is the spectrum of a delayed Gaussian function; hence, its magnitude is smooth while the phase exhibits variation due to the delay. N_x is determined to present the variation of phase incurred by the turn-on delay, t_0; hence, when the turn-on delay is subtracted from the time duration of $y(t)$ in (6), Equation (6) is modified for N,

$$N \le WT. \qquad (10)$$

Equation (10) informs the upper limit of the order, N for TDVF. The upper limit increases as the duration T and bandwidth W increase. Typically, N is chosen to be

$$N \approx WT. \qquad (11)$$

We call this result the WT rule.

Application

A numerical study concerns a SISO (single input/single output) interconnect circuit system depicted in Figure 1; the subcircuit, "sub_crt," involves seven transmission lines. All lines are lossy with the same p.u.l. resistance of 0.1 Ω/cm and zero p.u.l. conductance. The p.u.l. capacitance, C, p.u.l. inductance L, and length l of the lines are as follows. For T1, $C = 1$ pF/cm, $L = 0.6$ nH/ cm, and $l = 3$ cm. For T2, $C = 1$ pF/cm, $L = 1$ nH/ cm, and $l = 5$ cm. For T3, $C = 1.2$ pF/cm, $L = 0.6$ nH/ cm, and $l = 3$ cm. For T4, $C = 1$ pF/cm, $L = 0.6$ nH/ cm, and $l = 4$ cm. For T5, $C = 1.5$ pF/cm, $L = 1$ nH/ cm, and $l = 2$ cm [8].

The SISO interconnect system was analyzed using HSPICE to obtain the time-domain data required for the rational function fitting up to 5 GHz. The Gaussian pulse with a 60 dB bandwidth at 5 GHz ($W = 5 \times 10^9$), which was sampled every 5 ps, was assigned to V_{in} as an excitation. The turn-on delay, t_0, of the Gaussian pulse was 1.5 ns and $U_0 = \alpha\sqrt{\pi}$. The output voltage was recorded at the junction of the first and second subcircuits. The time response used for fitting was truncated at 8 ns ($T = 6.5$ ns). This was dictated by the criteria in (1) for a value of $\varepsilon = 0.01$. With T taken to be 6.5 ns and $W = 5 \times 10^9$, the order for TDVF was taken to be $WT = 34$ according to the WT rule and the result was plotted in Figure 2 from (a) to (c). As a counterexample, the order was taken to be larger than WT, 44 and the result was presented in Figure 2 from (d) to (f).

The plots in Figure 2 depict the accuracy achieved by the WT rule by comparing the two cases. The time-domain data from the generated rational fit were extrapolated via recursive convolution, and the values are presented as a dotted line in Figure 3.6(a) and (d). The reference time-domain data are obtained from HSPICE. Those for the frequency domain are from the Fourier transform of the transient response reached up to 40 ns. The reference values are plotted as dotted lines in the figure. Very good agreement is observed in both domains when the order is 34. The RMS error for the time-domain data is 4.19×10^{-6}, and for the frequency-domain data is 5.65×10^{-4}. However, when the order is 44, the frequency data of the macromodel does not show good agreement with reference value while the time-domain data correlate well with reference values. The RMS error for the time-domain data is 8.0614×10^{-6}, and for the frequency-domain data is 0.0094. The observed accuracy demonstrate the effectiveness of the WT rule for order estimation.

Conclusion

In summary, we introduced the WT rule as a robust means for estimating the order of the rational function approximation in the context of time-domain vector fitting. The proposed rule is shown to provide a reliable guideline for selecting the order to be used for accurate rational function approximation given the bandwidth, W, of the excitation and the duration, T, over which the response needs to be recorded in order to ensure that the output response has been subsided to negligible levels.

Acknowledgement

This work was supported in part by the Air Force Research for Scientific Research.

References

[1] S. Grivet-Talocia, F. G. Canavero, I. S. Stievano, and I. A. Maio, "Circuit extraction via time-domain vector fitting," in *IEEE Int. Symp. on EMC,* Aug. 2004, pp. 1005-1010.

[2] D. Slepian, "On bandwidth," *Proc. IEEE*, vol. 64, pp. 292-300, March 1976.

[3] D. Slepian, "Some comments on Fourier analysis, uncertainty and modeling," *SIAM Rev.*, vol. 25, pp. 379-393, July 1983.

[4] S. Moon and A. C. Cangellaris, "Passivity enforcement via quadratic programming for element-by-element rational function approximation of passive network matrices," in *Proc. IEEE 17th Topical Meeting on EPEP*, Oct. 2008, pp. 203-206.

[5] M. Yuan, A. De, T. Sarkar, J. Koh, and B. Jung, "Conditions for generation of stable and accurate hybrid TD-FD MoM solutions," *IEEE Trans. Microw. Theory Tech.*, vol. 54, pp. 2552-2563, June 2006.

[6] T. K. Sarkar, J. Koh, W. Lee, and M. Salazar-Palma, "Analysis of electromagnetic systems irradiated by ultra-short ultra-wideband pulse," *Meas. Sci. Tech.*, vol. 12, pp. 1757-1768, Nov. 2001.

[7] B. Gustavsen and A. Semlyen, "Rational approximation of frequency domain responses by vector fitting," *IEEE Trans. Power Del.*, vol. 14, pp. 1052-1061, July 1999.

[8] M. Celik and A. C. Cangellaris, "Simulation of dispersive multiconductor transmission lines by Padé via Lanczos process," *IEEE Trans. Microw. Theory Tech.*, vol. 44, pp. 2525-2535, Dec. 1996.

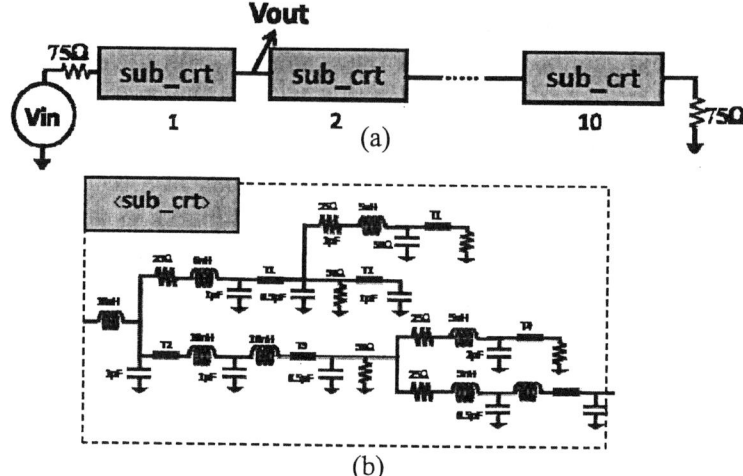

Figure 1 SISO interconnect system. (a) Whole circuit and (b) subcircuit.

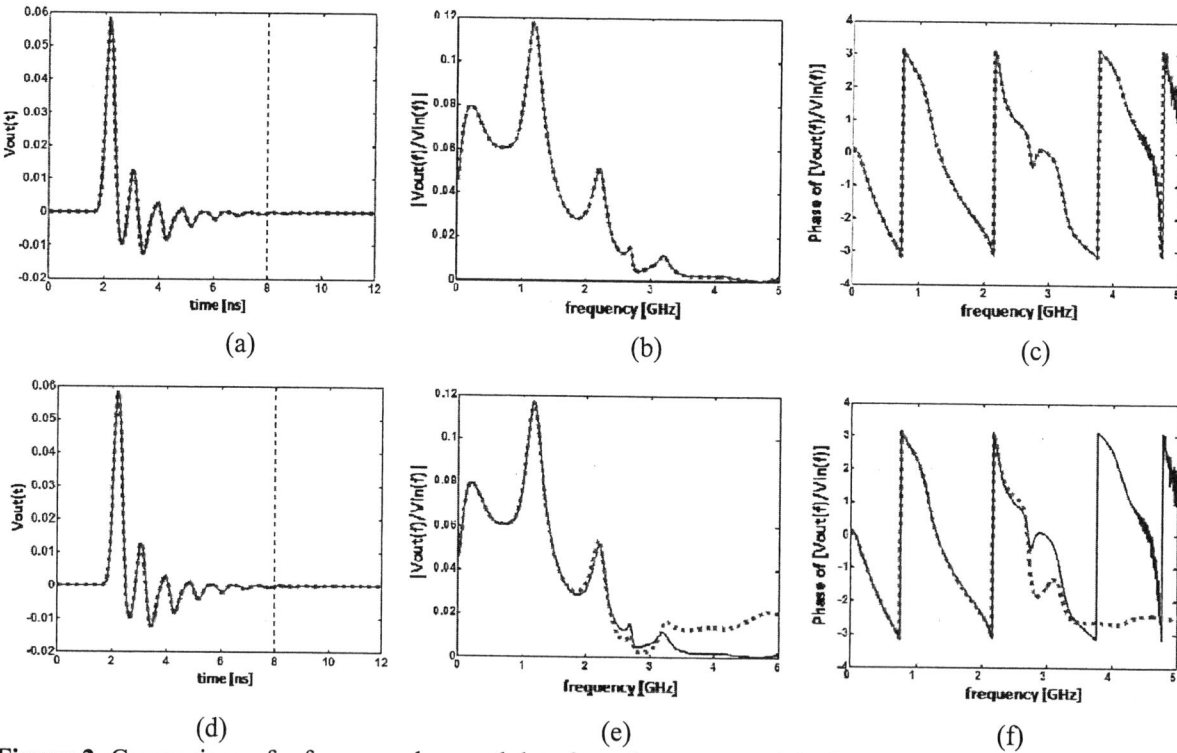

Figure 2 Comparison of reference values and data from the macromodel of TDVF; (a)-(c) for $N=34$, and (d)-(f) for $N=44$. (a)/(d) In the time domain (solid line: HSPICE reference solution; dots: TDVF interpolated/extrapolated response). (b)/(e) Magnitude and (c)/(f) phase in the frequency domain (solid line: frequency-domain data from Fourier transform; bold dots: frequency-domain data from the TDVF rational approximation).

Least Squares Convolution:
A Method to Improve the Fidelity of Convolution in Transient Circuit Simulation

Michael Tsuk
Ansoft LLC
67 South Bedford St.
Burlington, MA 01803
Phone: 781-229-8900 x351
Fax: 781-229-8624
Email: mtsuk@ansoft.com

Subramanian Lalgudi
Ansoft LLC
67 South Bedford St.
Burlington, MA 01803
Phone: 781-229-8900 x315
Fax: 781-229-8624
Email: slalgudi@ansoft.com

Abstract

The use of the Fast Fourier Transform to generate impulse responses from frequency-domain data is shown to have some pitfalls for transient circuit simulation. A new technique, Least Squares Convolution (LSC), resolves these.

I. INTRODUCTION

The use of frequency-domain data in transient circuit simulation has been increasing in recent years. The techniques used for this purpose are basically in two categories; convolution-based methods ([1] and the convolution-related references therein) and state-space (or recursive convolution) methods [2].

Of the two methods, convolution seems very straightforward, particularly in the case of uniform frequency distribution, including the zero-frequency (DC) point. In this case, the Fast Fourier Transform algorithm (FFT) [5] can be directly applied to convert the frequency-domain data into impulse responses. Convolution-based methods ([1] and references [7]-[13] therein) follow this approach.

However, there are some important subtleties involved in using the inverse FFT (iFFT) to generate high-quality impulse responses, which can be easily overlooked. An important drawback of iFFT-based approaches is that they expect data on a uniform frequency grid. Many times, data are not available this way. Of course, one can resort to interpolation (to get data on a uniform grid), but, unless special care is taken (see [6] and references therein), causality of the data can be violated.

Moreover, the impulse response (IR) computed through a direct iFFT of any finite bandwidth frequency response is non-causal [1], except in special cases. Some approaches enforce causality of IR with only finite bandwidth data [1], [3], [4]. However, such approaches can also introduce significant inaccuracy, primarily because of the omission of what happens to the frequency response outside the bandwidth. Some commercial circuit simulators overcome this inaccuracy issue by extrapolating the data outside the (given) bandwidth, before an iFFT. These simulators also explicitly make sure data remain causal not only in-band but also out-of-band, which makes the extrapolation more difficult.

The purpose of this paper is to describe a new algorithm, called Least Squares Convolution (LSC), that avoids falling into the traps engendered by these subtleties, and therefore leads to high-quality transient simulation with frequency domain data. Also, causality of the impulse responses are ensured without compromising accuracy much and without explicit extrapolation of the data. One of the important benefits of LSC is that it easily handles frequency-domain data that are not given on a uniform frequency grid.

The structure of this paper is as follows. First, we give an overview of the convolution algorithm and its application to transient circuit simulation. In the next section, we demonstrate the difficulties with the traditional use of the inverse FFT to generate impulse responses. In the next section, we describe LSC; in the following section, compare it with the traditional method on some representative examples. Finally, the paper ends with conclusions.

II. CONVOLUTION IN TRANSIENT CIRCUIT SIMULATION

The most common method for treating multi-port frequency-dependent data in circuit simulation is through the use of scattering parameters, more commonly known as S-parameters. Denoting $a(t)$ to be the forward wave into a port, $b(t)$ the reflected wave from the port, and $s(t)$ the impulse response, it is necessary to compute the convolution integral

$$b(t) = \int_{-\infty}^{t} s(t - t')a(t')dt', \qquad (1)$$

under the usual and necessary assumption of causality

$$s(t) = 0 \text{ for } t < 0. \tag{2}$$

To compute (1), it is often required to compute $s(t)$ for arbitrary t, which can be different from the times at which $s(t)$ is known. A straightforward approach is to interpolate at unavailable t's.

III. DIRECT APPLICATION OF INVERSE FFT

It seems straightforward to generate an impulse response from sampled data. But, as will be demonstrated, even in the simple cases, there are pitfalls to be avoided.

If we assume that the input data are on a uniform frequency grid, including the point at zero frequency (DC), then it is obvious that the iFFT is an appropriate tool for the job. If we have n frequency input points, including both positive and negative frequency data, on a grid with spacing of Δf, the iFFT will generate n real time data points, with a time spacing of $\Delta t = 1/(n\Delta f)$. [5]

The pitfall comes in figuring out how to take these n real data points and turn them into a function suitable for the convolution integral. The most natural way to do this is to linearly interpolate between the calculated data points. But it can be easily shown that this causes unwanted filtering of the transient signals.

Recall the assumptions that are required before one can turn a Fourier integral into a discrete Fourier transform (necessary so as to use the iFFT). We assume that the discrete samples in the frequency domain are, in fact, the entire information available about the input data. The fact that we have samples means that the time-domain waveform is infinitely periodic; the fact that the samples are limited in bandwidth implies that they are, in fact, one period of an infinitely periodic waveform in the frequency domain, leading to a series of discrete points in the time domain.

To restate: the FFT algorithm is a fast version of the discrete Fourier transform. The discrete Fourier transform takes a series of impulses in the time domain, one cycle of an infinitely periodic waveform, and takes it into a series of impulses in the frequency domain, one cycle of an infinitely periodic waveform.

If we take the values coming out of the iFFT and linearly interpolate them, that is equivalent to convolving the train of impulses with a triangular function:

$$w(t) = \begin{cases} 0 & t < -\Delta t \\ \frac{1}{\Delta t}(t + \Delta t) & -\Delta t < t < 0 \\ \frac{1}{\Delta t}(\Delta t - t) & 0 < t < \Delta t \\ 0 & \Delta t < t. \end{cases} \tag{3}$$

Since convolution in the time domain implies multiplication in the frequency domain, the effect of the linear interpolation is to multiply the original frequency data by the transform of the triangular function (3), which is given by

$$W(f) = \text{sinc}^2(f_n) = \frac{\sin^2(\pi f_n)}{(\pi f_n)^2}, \tag{4}$$

where

$$f_n = f/(2n\Delta_f). \tag{5}$$

At the end of the input frequency data, $f_n = 1/2$, and therefore, $W(f) = \text{sinc}^2(1/2) \approx 0.405285$. In other words, if the input to the S-parameter block is at the highest frequency of the data, the response will be down over 7.8 dB from the proper result.

This is shown in Figure 1. Here, a sinusoid of 10 GHz is input into a lossless transmission line model, on the one hand, and into the S-parameters for the same transmission line, on the other. The line has a characteristic impedance of 50 Ω, and an electrical length of 1ns; the S-parameter data of this line have a bandwidth of 10 GHz. The S-parameter simulation is done with convolution using a linear interpolation of the iFFT-calculated impulse response, as described above. As can be seen, the S-parameter results are smaller than the transmission line results, by exactly the ratio predicted by the formula above.

The simplest way to get around the problem of the filtering effect of linear interpolation is not to do the interpolation. One can just treat the impulse response generated by the iFFT as a train of impulses. However, this resurrects the artificial periodic behavior in the frequency domain, and also, a train of impulses is not the best-suited form in terms of ease of transient simulation.

In theory, extrapolation of the original frequency-domain data beyond the limit of the bandwidth will also resolve this problem, although a large amount of extrapolation is required to significantly reduce the filtering effect, and it is difficult to develop robust extrapolation algorithms that preserve causality and passivity.

978-1-4244-4447-2/09 $25.00 © 2009 IEEE   74

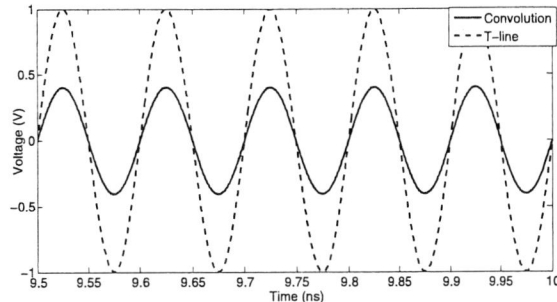

Fig. 1. Demonstrating the filtering effect of linear interpolation of iFFT output

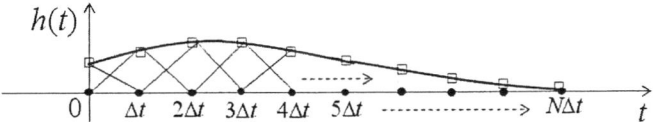

Fig. 2. Approximation of impulse response using triangular expansion functions.

IV. LEAST SQUARES CONVOLUTION

The idea used in this paper to circumvent these difficulties is quite simple. We assume *a priori* that the impulse response, $h(t)$, can be represented by a piecewise-linear function of time, i.e.,

$$h(t) \approx \sum_{i=1}^{N} a_n w(t - n\Delta t) + a_0 w(t) u(t), \tag{6}$$

where $w(t)$ is the triangular expansion function described in (3), $a_n \in \mathbb{R}$ is the unknown weight associated with $w(t - n\Delta t)$, and $u(t)$ is the Heaviside unit step function. The term after the summation in (6) is to ensure causality of $h(t)$. In (6), we choose $\Delta t = 1/(2f_{max})$, where f_{max} is the maximum frequency in the sampled data. In the frequency domain, Equation (6) is expressed as

$$H(\omega) \approx \sum_{n=1}^{N} a_n W(\omega) \exp(-j\omega n \Delta t) + a_0 W_{\frac{1}{2}}(\omega), \tag{7}$$

where $W_{\frac{1}{2}}(\omega)$ denotes the Fourier transform of the half triangular pulse, $w(t)u(t)$. We find the weights a_n's by equating (7) at each frequency of the sampled data and solving the resulting matrix system through a least-squares solve. This gives us the best possible fit of the spectral behavior of the piecewise-linear function to the given data. Once $h(t)$ is approximated this way, the integral (1) can be computed without the filtering effects of iFFT-based approaches.

There are many advantages to this method, Least Squares Convolution. The first, of course, is that it provides good fidelity to the input data, even up to the top edge of the input frequency band.

A second advantage of LSC is that the impulse response function is causal by construction. The traditional iFFT-based approach, because it generates a time-domain waveform that is periodic, has difficulty avoiding non-causal impulse responses.

A third advantage of LSC is that it can handle data on a non-uniform frequency grid, without the need for interpolation of the data. Interpolation, like extrapolation, can lead to causality and passivity problems.

The main drawback of LSC is that, by abandoning the FFT algorithm, the fantastic speed and memory benefits of that algorithm are also lost. The speed issue is not generally crucial; generating the impulse response, even with LSC, is a small portion of the transient simulation time. Memory can be a bigger issue, particularly if there are a large number of points in the frequency data. At present, decimation of the data is the only solution, although other potential improvements are under investigation.

V. RESULTS

Using LSC on the configuration of Figure 1 completely solves the filtering problem; the two curves, transmission line and LSC, exactly overlay.

A representative example consists of a series RLC circuit with a shunt resistor at each end of this series branch. The values of the components are as follows: in the series branch, $R = 1\Omega$, $L = 1$ nH, and $C = 4.5$ pF. The shunt resistors were 10 Ω each. The frequency response of this two-port circuit was captured in the form of a Touchstone file; the bandwidth of the data

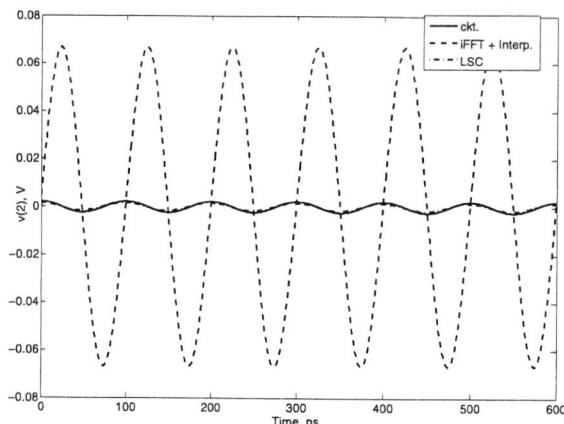

Fig. 3.   Comparing the two convolution methods on an LRC circuit

is 5 GHz, with a frequency step of 1 MHz. This two-port circuit is excited at one port by a sinusoid of frequency 10 MHz and terminated at the other end by a 50 Ω resistor. The objective is to compute the transient voltage at the second port.

One would expect that 5 GHz data are sufficient to simulate the transient response from a 10 MHz sinusoid. However, as can be seen in Figure 3, the iFFT-based convolution result is far off from both the circuit and LSC results, which overlay.

VI. CONCLUSIONS

In this paper, we have shown that the traditional method of generating impulse responses from frequency-domain data yields inaccuracies, particularly for excitation frequencies near the top of the bandwidth of the input data. This is due to the use of the inverse FFT data as knots in a piecewise-linear interpolation; since this is equivalent to convolving the train-of-impulses output of the inverse FFT with a triangle function, we get the filtering effect in the frequency domain from the transform of this triangle function. This filtering can be significant; on the order of 7.8 dB at the end of the band.

We have described a new algorithm, Least Squares Convolution, which by assuming a piecewise-linear function *a priori*, avoids this filtering effect. The results shown in this paper demonstrated the improvement in fidelity this new algorithm provides to transient simulation. The proposed method also ensures the causality of the impulse responses without compromising accuracy and without explicit extrapolation of the data.

REFERENCES

[1] S. N. Lalgudi, E. Engin, G. Casinovi, and M. Swaminathan, "Accurate transient simulation of interconnects characterized by band-limited data with propagation delay enforcement in a modified nodal analysis framework," *IEEE Transactions on Electromagnetic Compatibility*, vol. 50, no. 3, pp. 715–729, Aug. 2008.

[2] B. Gustavsen, and A. Semlyen, "Rational approximation of frequency domain responses by Vector Fitting," *IEEE Trans. on Power Delivery*, vol. 14, no. 3, July 1999.

[3] F. M. Tesche, "On the use of Hilbert transform for processing measure CW data," *IEEE Trans. Electromag. Compat.*, vol. 34, no. 3, pp 259–266, Aug. 1992.

[4] P. Perry and T. Brazil, "Forcing causality on *S*-parameter data using the Hilbert transform," *IEEE Microw. Guided Wave Lett.*, vol. 8, no. 11, pp. 378–380, Nov. 1998.

[5] W. H. Press, S. A. Teukolsky, W. T. Vetterling, and B. P. Flannery, *Numerical Recipes in C: The Art of Scientific Computing*. Cambridge: Cambridge Univ. Press, 1992.

[6] P. Triverio, and S. Grivet-Talocia, "Causality-constrained interpolation of tabulated frequency responses," *IEEE 15th Topical Meeting Elect. Perf. Electronic Packag.*, Scottsdale, AZ, Oct. 23-25, 2006, pp. 181-184.

978-1-4244-4447-2/09 $25.00 © 2009 IEEE

Application of Surrogate Modeling to Generate Compact and PVT-sensitive IBIS Models

Ting Zhu, Paul D Franzon

ECE, Box 7914, NCSU, Raleigh, NC27695, Phone- 919 515 7351, Fax-919 515 2285

{tzhu,paulf}@ncsu.edu

Abstract: A new proposal of applying surrogate-modeling in Input-output Buffer Information Specification (IBIS) is presented. It saves the IBIS data storage resource, extends the model utility to various process-voltage-temperature (PVT) simulations and eliminates the data interpolation deviations.

I. INTRODUCTION

Good macromodels of input/output circuits are essential for fast timing, signal-integrity, and power-integrity analysis in high-speed digital systems. Input/output buffer information specification (IBIS) [1] is a well-known standard of representing electrical behavior of I/O circuits. The core of IBIS consists of look-up-tables (LUTs) of I-V (current versus voltage) and rising/falling transition V-t (voltage versus time) at output ports. IBIS has been widely used. However, its LUT format meets challenges. First, thousands of LUTs are needed to model a large number of pins and options (like multiple choices of on-chip terminators) of the new devices. The increasing data size becomes a problem. Second, IBIS format only provides three types of corner data by individual LUTs within each model, and the data are insufficient to present the circuit behaviors under different process, voltage, and temperature (PVT) variations. Third, LUTs constrains to fixed, discrete data points only. Thus, in simulation, different interpolation methods of the data may cause uncertain deviations.

To handle those challenges, we propose a modified IBIS format based on surrogate-modeling (metamodeling), which aims to approximate circuit behavior as closely as possible while being computationally cheap to evaluate [2][3][4]. In our method, surrogate models of the PVT-sensitive I-V and V-t behavior are constructed and they replace the LUTs in classical IBIS models. The model extraction is similar to traditional IBIS extraction and only approximation models of the static behavior are required. Approximation models have ever been used in parametric macromodels for I/O devices [5][6]. In contrast, our method avoids the difficulty and complexity in approximating the whole dynamic behavior of the circuits, and requires neither well-designed system-identification input stimulus nor large numbers of training data.

By combining IBIS and surrogate-modeling, the new method exhibits several key benefits: (1) no LUTs are required and data storage resources are saved. (2) the utility of IBIS models is greatly extended to various combinations of PVT corners. Interpolation between corners is possible to support 'what-if' scenarios. (3) the uncertainty caused by different interpolation methods in simulations is eliminated. (4) fully based on traditional IBIS model, easy extraction and model-fitting.

This paper is organized as follows. In section II, we describe surrogate-modeling and its combination with IBIS model. Section III presents the modeling procedure based on an example driver circuit. Section IV presents the simulation results and the conclusion is addressed in Section V.

II. BASIS OF METHODOLOGY

A. Overview of Surrogate-modeling

The basic problem of surrogate-modeling is to find the best approximation function $\hat{f}$ for an unknown multivariate function f which values $f|_X = (f(x_1), ..., f(x_n))$ are known at limited sampling points $X = \{x_1, ..., x_n\}$.

$f = \hat{f} + \varepsilon$, ε is the approximation error and it's expected to be zero.

Once constructed, the surrogate model $\hat{f}$ can be reused instead of f in other stage of engineering. Surrogate models are particularly useful for design space exploration, sensitivity analysis, high dimensional visualization, and what-if analysis.

The modeling process involves steps of design of experiment (DOE), model construction and model validation. Reviews of the state-of-the-art are provided in [2][4]. Design of experiment decides how to select samples in the input variable space. Simulations will be executed for all the selected input variables. Widely used experiment designs include fractional factorial design, various Latin Hypercube Sampling (LHS) and Orthogonal Arrays (OA) etc. Model construction step chooses the surrogate model types and optimization of the model parameters. The most popular model types are Polynomial response surfaces, Kriging, Support Vector Machines (SVM) and Artificial Neural Networks (ANN). Algorithms such as Pattern

Search (PS), Genetic Algorithm (GA) and Simulated Annealing (SA) can optimize the model parameters. Model validation needs to reliably estimate the model accuracy. Split sample, cross-validation and bootstrapping are among the general validation methods. Absolute error functions such as Root Mean Square Error (RMSE) and Maximum Absolution Error (MAE), and relative error functions such as Root Relative Square Error (RRSE) can be used as measurement criteria.

Since one-fit-all method does not exist, strategies should be carefully chosen with consideration of the nature of the problems and computational complexity.

B. Combination of Surrogate-modeling and IBIS Model

Figure1 shows the basic elements of IBIS model, including tabulated pull-up and pull-down I-V characteristics, power/ground clamp I-V characteristics, rising and falling waveforms (V-t), buffer parasitic capacitance (C_comp) and package information [1]. Instead of using LUTs, we present I-V and V-t with surrogate models. Moreover, we extend the models to PVT-sensitive characteristics so that they can be simulated flexibly over different corner conditions.

Figure 2 shows the flow, which consists of circuit simulation in HSPICE, construction of surrogate model equation in SUrrogate Modeling (SUMO) toolbox [7] and implementation of the new IBIS format in Verilog-A. Methods of DOE, model type and validation can be specified before the flow starts. Data is obtained from transistor-level circuit simulation. The new IBIS model is implemented based on Verilog-A behavioral version IBIS model introduced by the IBIS open forum [8].

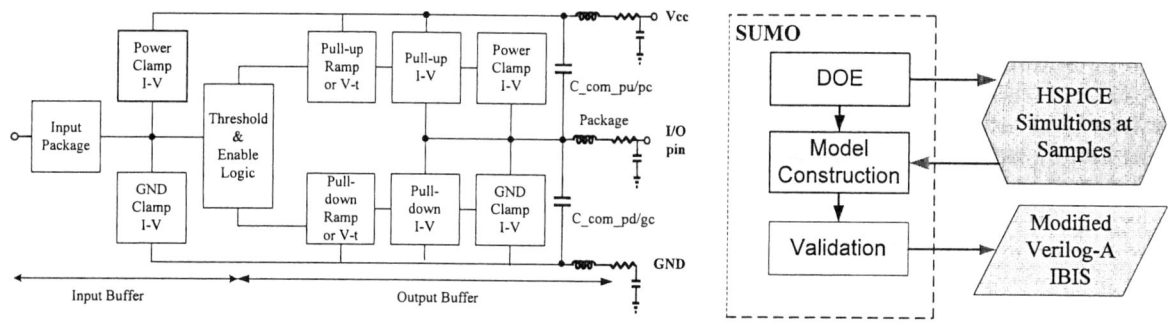

Figure 1. Basic elements in IBIS
Figure 2. Modeling Flow

Figure 3(a) shows the simulation setup for extracting I-V data, similar to the description in IBIS handbook [9]. Both the output port and power supply of the buffer are connected to voltage sources whose value are set by the sampling plan. Static pull-up output current I_{pu} and pull-down current I_{pd} are the modeling objectives, while V_{out}, V_{cc}, Temperature (T) and process factor $\Delta V_{th0}/V_{th0}$ (applied to both p-type and n-type MOSFETs) are the input variables. Latin Hypercube sampling generates the initial sampling plan. New samples are selected by using error-based adaptive sampling algorithm for each iteration. Rational function model is chosen since it is suitable for I/O behavior and can be easily implemented in Verilog-A. Other model types will be evaluated for the next step. Cross validation with Root-Relative-Square-Error (RRSE) assess the quality of the models.

Figure 3(b) shows the test setup for extracting V-t data, similar to [9] as well. The output port is connected to a specified loading resistor and termination voltage. Transient output voltage V_{out} of the rising waveform and the falling waveform are modeled with input variables time point t, V_{cc}, temperature (T) and process factor. Since the original design of SUMO tool is to tackle static input-output systems, to model the dynamic V-t behavior, a small fixed time interval is taken and a set of time series output values are recorded. The modeling strategies are set the same as I-V.

In the following section, the proposed modeling procedure is applied to an output driver circuit shown in Figure 4.

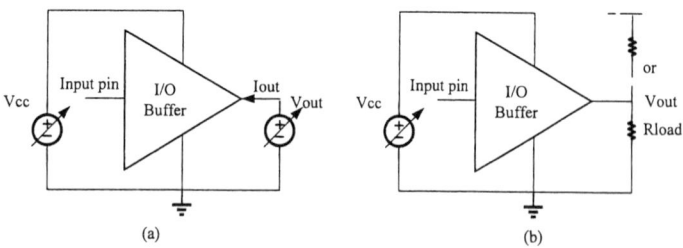

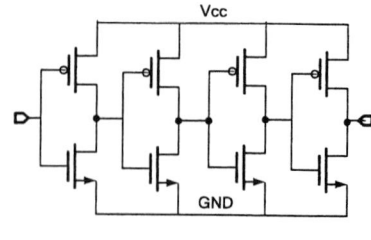

Figure 3 Test setup for (a) extraction of I-V (b) extraction of V-t

Figure 4 Driver circuit

978-1-4244-4447-2/09 $25.00 © 2009 IEEE

III. CONSTRUCTION OF THE MODELS

A simple 4-stage non-inverting driver circuit (Figure 4) was used in the experiment. We constructed models of the pull-up/pull-down I-V and the rising/falling V-t characteristics by the testing setup and the procedure discussed in Section II. The range of input variables for I-V extraction is defined in Table 1.

Table 1 Range of input variables in I-V extraction

Variable	V_{out}	V_{cc}	Temperature(T)	Process ($\Delta V_{th0}/V_{th0}$)
Range	-3.6-7.2V	3-3.6V	0-100℃	±10%

The rational function model of the pull-up current I_{pu} is shown in Figure 5. As the plots show, smooth model response was achieved. The slide plots for temperatures at 0℃, 50℃ and 100℃ provide an insight of the model surface. The complete rational model expression is shown in Equation-1, which can be further simplified without considering the elements with small coefficients. Modeling process of pull-down current I_{pd} was similar to I_{pu}. For V-t models, a small fixed time interval of 30ps was taken and a set of time series output voltages were sampled for each combinational PVT condition. The output data were then fitted in to rational function models as well.

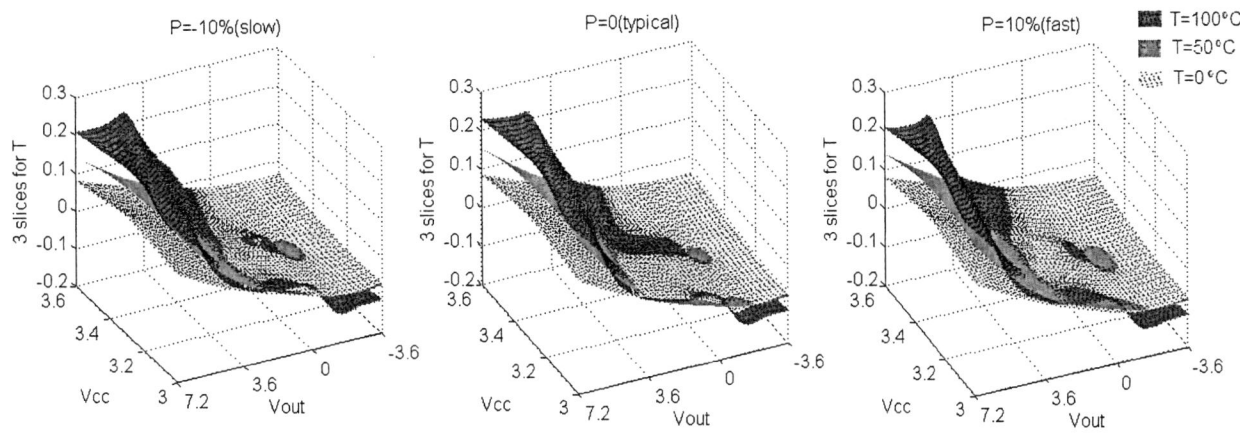

Figure 5. 4D rational function model of static pull-up output current I_{pu}. (T= temperature, P= Process= $\Delta V_{th}/V_{th}$).

Normalized inputs: Vo = (Vout - 1.8)/ 5.4, Vc = (Vcc - 3.3) / 0.3, T = (Temperature - 50)/ 50, P = Process= $\Delta V_{th}/V_{th}$

Ipu (Vo, Vc, T, P) = (-3.18e-2+5.79e-3*Vo-3.26e-3*Vc+2.68e-1*Vo^2-1.47e-3*P-1.14e-2*T+6.96e-4*Vo*Vc-2.28e-1*Vo^3-1.45e-3*Vo*P-2.13e-2*Vo*T
+2.607862e-2*Vo^2*Vc-3.82e-1*Vo^4-2.88e-4*Vc*P-8.16e-4*Vc*T+9.59e-3*Vo^2*P+5.73e-2*Vo^2*T+2.48e-5*P^2-8.56e-4*T*P+1.96e-2*T^2-1.32e-4
*Vo^3*Vc+7.76e-2*Vo^5-1.12e-5*Vo*Vc*P-1.89e-3*Vo*Vc*T+1.34e-3*Vo^3*P+4.52e-2*Vo^3*T+1.68e-4*Vo*P^2-6.14e-4*Vo*T*P+2.01e-2*Vo*T^2-3
.98e-2*Vo^4*Vc+2.19e-1*Vo^6+7.00e-4*Vo^2*Vc*P+4.18e-3*Vo^2*Vc*T-1.35e-2*Vo^4*P-5.47e-2*Vo^4*T-4.04e-5*Vc*P^2+1.11e-4*Vc*T*P+5.23e-4
*Vc*T^2-4.14e-4*Vo^2*P^2+6.54e-3*Vo^2*T*P-1.16e-1*Vo^2*T^2-1.89e-2*Vo^5*Vc+3.71e-2)/(1-2.11e*Vo-1.48*Vo^2+1.92e-2*P+3.59e-1*T+3.21*Vo^
3-2.21e-2*Vo*P+2.16e-4*Vo*T+2.81*Vo^4-5.64e-2*Vo^2*P-1.21*Vo^2*T-2.45e-3*P^2+2.46e-2*T*P-5.66e-1*T^2-8.33e-1*Vo^5+5.93e-2*Vo^3*P-1.06e-
1*Vo^3*T+2.505041e-3*Vo*P^2-4.42e-2*Vo*T*P+8.11e-1*Vo*T^2-1.38*Vo^6+2.06e-2*Vo^4*P3.80e-1*Vo^4*T+3.02e-3*Vo^2*P^2-1.82e-2*Vo^2*T*P
+5.73e-1*Vo^2*T^2-6.15e-1*Vo^7) ,   (Equation-1)

To implement the new model, we modified the Verilog-A behavioral version IBIS model [8] and used the new model equations instead of the data files for I-V and V-t behavior and the LUT functions $table_model( ).

IV. SIMULATION RESULTS

The new model was tested against the transistor-level (TL) model and classical IBIS model. The classical IBIS model was generated by s2ibis3 [10]. All the simulations were done in HSPICE 2009.03 SP1. Figure 6 shows the results during rising and falling transients at three corner conditions (Fast, Typical, Slow) defined in IBIS. As the results show, the new IBIS model achieves good accuracy in simulations. Moreover, the new method reduces model data size by 85.4%. A summary of the comparison is shown in Table 2.

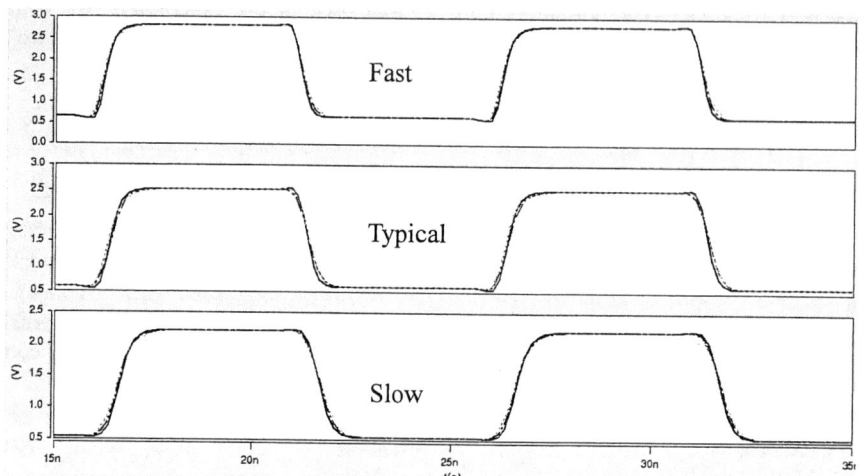

Figure 6. Transient simulation results of transistor-level model (continuous line), classic model (dashed line) and the proposed model-based IBIS model (dashed dot line).

Table 2 Comparison of the models size and accuracy with reference to transistor-level model SPICE simulations for the test case.

IBIS	Model Data Size	RMS Voltage Error	Max Time error
LUT-based	41 kbytes	5.9%	38ps
New Model-based	6 kbytes	4.6%	26ps

V. CONCLUSION

Classical IBIS model meets challenges as its look-up-table (LUT) format. A new method based on surrogate-modeling is proposed. In this paper, surrogate-modeling of I-V and V-t characteristics are discussed. Latin Hypercube sampling, rational function model and cross-validation are used in the flow. For demonstration, new model of a driver circuit was constructed and implemented in Verilog-A. Accuracy and flexibility of the model was validated by simulation results. In addition, the model size was greatly reduced compared to classical IBIS model.

VI. REFERENCES

[1] IO Buffer Information Specification [Online]. Available: http:// www.eigroup.org/ibis/ibis.htm
[2] G. G.Wang and S. Shan, "Review of metamodeling techniques in support of engineering design optimization,"Journal of Mechanical Design, vol. 129, no. 4, pp. 370–380, 2007
[3] N. V. Queipo, R. T. Haftka, W. Shyy, , etc., "Surrogate-based analysis and optimization," Progress in Aerospace Sciences, vol. 41, pp. 1–28, 2005.
[4] T.W. Simpson, J.D. Peplinski, P.N. Koch, J.K. Allen, " Metamodels for Computer-based Engineering Design: Survey and Recommendations", Springer-Engineering with Computer, Vol.17, pp.129-150, 2001
[5] I.S. Stievano, I.A. Maio, F. G. Canavero, "Parametric macromodels of differential drivers and receivers", IEEE Transactions on Advanced Packaging. 28(2), pp.189-196, May 2005
[6] B.Mutnury, M. Swaminathan, and J.P. Libous. Macromodeling of nonlinear digital I/O drivers. Advanced Packaging, IEEE Transactions on Advanced Packaging, 29(1), pp.102–113, Feb. 2006
[7] D. Gorissen, L. De Tommasi, K. Crombecq, T. Dhaene, "Sequential modeling of a low noise amplifier with neural networks and active learning", Springer - Neural Computing & Applications, Vol. 18, Nr. 5, pp. 485-494, June 2009.
[8] M.LaBonte,A.Muranyi "IBIS Advanced Technology Modeling Task Group Work-achievement: Verilog-A element library HSPICE test" [online]. Available: http://www.eda.org/pub/ibis/macromodel_wip/
[9] IBIS Modeling Cookbook [Online]. Available: http://www.vhdl.org/pub/ibis/cookbook/cookbook-v4.pdf
[10] A. Varma, A. Glaser, S. Lipa, M. Steer, P. Franzon,"The Development of a Macro-modeling Tool to Develop IBIS Models," 12'h Tropical Meeting on Electrical Performance of Electronic Packaging (EPEP 2003), Princeton, New Jersey, pp. 277-280, October, 2003.

Generalized Leapfrog Scheme for Large-Scale Circuit Simulation

Tadatoshi Sekine

Graduate School of Science and Technology, Shizuoka University

3-5-1 Johoku, Naka-ku, Hamamatsu-shi, 432-8561 Japan

Phone: +81-53-478-1237, Fax: +81-53-478-1269

Email: sekine@tzasai7.sys.eng.shizuoka.ac.jp

Hideki Asai

Dept. of Systems Eng., Shizuoka University

3-5-1 Johoku, Naka-ku, Hamamatsu-shi, 432-8561 Japan

Phone: +81-53-478-1237, Fax: +81-53-478-1269

Email: hideasai@sys.eng.shizuoka.ac.jp

Abstract

This paper describes a generalized leapfrog scheme for transient analyses of large-scale circuits. First, the leapfrog scheme developed in the basic latency insertion method (LIM) is described and the characteristics of the method and the scheme are discussed. Then, we propose the generalization technique of the leapfrog scheme which can be applied to the transient analysis of an ill-constructed circuit unsuitable for the basic LIM. Numerical results show that our proposed approach is applicable and efficient for the fast simulation of large-scale networks.

I. INTRODUCTION

Various effects dependent on the high-frequency characteristics of the high-speed signals are induced on the high-density electronic circuits. These effects cause the unexpected behaviors of the chips and packages on a printed circuit board and the errors caused by them seriously affect the signal and power integrity of the network. Thus, it becomes important to verify the electronic circuit behaviors including these effects. In the recent circuit design flow, a net-list for the circuit simulation of such verification is mainly provided by an extractor, which extracts circuit element parameters from the structure and the property of the object to be analyzed. Typically, whenever the extractor is used, the net-list tends to include an enormous number of parasitic elements in order to verify the exact behaviors of the chips and packages. Moreover, it may also include a number of coupling elements such as mutual inductance and branch capacitance because the components in the high-density circuits are very close to each other; they are connected each other electromagnetically, even though they are not connected physically or electrically. These facts cause the large amount of simulation time for a SPICE-like simulator based on the matrix solver. Therefore, fast simulation techniques different from SPICE-like ones are strongly demanded.

The latency insertion method (LIM) in [1] has proven to be accurate and extremely fast for the simulation of large networks but suffers from two limitations. The first is that the method requires a circuit to be composed of the branch topology and the node topology both of which must include the reactive elements. The second problem is that since the method is completely explicit, the time step size is restricted by a stability criterion. To overcome these limitations, several enhanced techniques have been proposed [2]–[10]. However, it is still difficult for the basic- and the enhanced-LIM to analyze the ill-constructed circuit such as the circuit without the reactive elements and the tightly coupled elements. The extracted networks may have those undesirable topologies if some kinds of commercial extraction tools such as Q3D Extractor (Ansoft) are used for the extraction. In order to circumvent the limitations of the basic LIM and deal with the ill-constructed circuit, we propose the generalized technique based on the leapfrog scheme for the simulation of large-scale networks.

The remainder of the paper is organized as follows. In II, the basic LIM algorithm is reviewed and the characteristics of the method, especially those related to the leapfrog scheme in the algorithm, are discussed. In III, we propose the generalized formulation based on the leapfrog scheme by using the block formulation approach and show the ways how to deal with the ill-constructed topologies to the basic LIM. Section IV shows some numerical results and conclusions are in V.

II. LEAPFROG SCHEME IN BASIC LIM

In the basic LIM algorithm [1], the circuit to be analyzed is required to be composed of a certain type of the topologies, namely, the branch topology and the node topology as shown in Fig. 1 (a) and (b). Although each branch in the network does not have to contain a resistor $R_{a,b}$ or a voltage source $E_{a,b}$, it must include an inductor $L_{a,b}$. Similarly, each node in the network does not have to contain a conductive path G_a or a current source H_a but must provide a capacitive path C_a to the ground. If there exists a branch without the inductor or a node without the grounded capacitor, a relatively small inductor or shunt capacitor is inserted into the corresponding branch or node to generate the latency, respectively. By this manipulation, the circuit completely consists of the branch topology and the node topology shown in Fig. 1 (a) and (b). Therefore, the all equations derived by applying the Kirchhoff's voltage law (KVL) to the branch topologies and the Kirchhoff's current law (KCL) to the node topologies contain derivative terms associated with the reactive elements $L_{a,b}$ and C_a as follows:

$$v_a - v_b = L_{a,b}\frac{di_{a,b}}{dt} + R_{a,b}i_{a,b} - E_{a,b}, \tag{1}$$

$$-\sum_{k=1}^{M_a} i_{a,k} = C_a\frac{dv_a}{dt} + G_a v_a - H_a, \tag{2}$$

978-1-4244-4447-2/09 $25.00 © 2009 IEEE

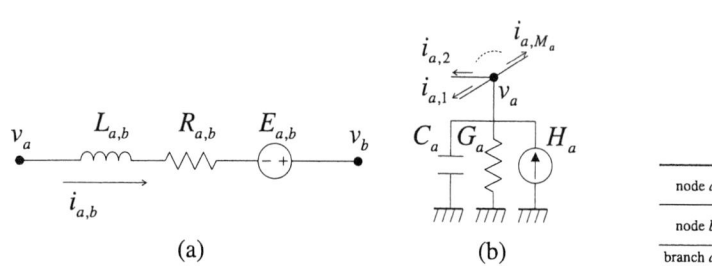

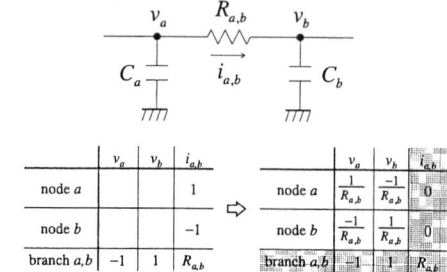

Fig. 1. Required circuit topologies for basic LIM. (a) Branch topology. (b) Node topology.

Fig. 2. Transformation of branch without inductive element. Dark gray parts are not used for stamps.

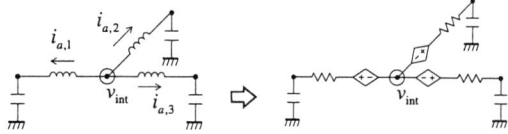

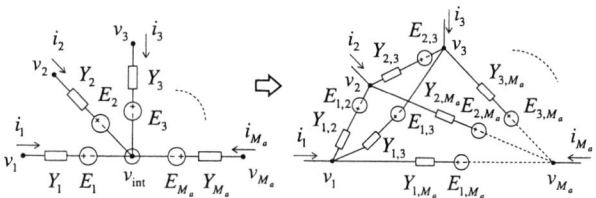

Fig. 3. Transformation of node without grounded path. First, inductor is transformed into its companion model in time domain. Then, Generalized Y-Δ transformation is applied to remove the intermediate node voltage v_{int}.

Fig. 4. Generalized Y-Δ transformation with voltage sources.

where $i_{a,b}$ is the current flowing through the branch between the nodes a and b, v_a and v_b are the voltages at the nodes and M_a is the number of the branches connected to the node a. Then, a finite difference method is applied to (1) and (2) so that the time points of the branch currents and the node voltages are collocated in half time steps, and the updating formulas of the branch current and the node voltage are derived as

$$i_{a,b}^{n+1} = \frac{L_{a,b} - \Delta t R_{a,b}}{L_{a,b}} i_{a,b}^n + \frac{\Delta t}{L_{a,b}} \left(v_a^{n+\frac{1}{2}} - v_b^{n+\frac{1}{2}} + E_{a,b}^{n+\frac{1}{2}} \right), \tag{3}$$

$$v_a^{n+\frac{1}{2}} = \frac{C_a}{C_a + \Delta t G_a} v_a^{n-\frac{1}{2}} + \frac{\Delta t}{C_a + \Delta t G_a} \left(-\sum_{k=1}^{M_a} i_{a,k}^n + H_a^n \right), \tag{4}$$

where n is a time step and Δt is a time step size. All of the branch currents and the node voltages in the network are calculated using (3) and (4) alternately as time progresses. The difference method and the updating process generated in the above basic LIM algorithm is called the "leapfrog scheme" from the viewpoint of the fact that the current and the voltage variables are defined at the different time points where only the current or the voltage exists. By this formulation, if a fixed time step size is used, the calculation amount at each time step results in $\mathcal{O}(N_b + N_n)$, where N_b and N_n are the number of the branches and the nodes in the network, respectively. Therefore, the CPU time of the basic LIM is linearly-increasing with the circuit size, and thereby it can reduce the calculation costs of the transient simulation significantly compared to the SPICE-like algorithm.

To clarify the basic LIM algorithm from another perspective, we should realize two principal facts of the method again. One is applying the leapfrog scheme to update the currents and the voltages in the time domain and the other is inserting the artificial latency into the branch and the node with no reactive elements. Additionally, the most important property of the basic LIM is its linear numerical complexity. These facts are the reasons why the *latency insertion approach* has been introduced in order to generate a scalar version of the leapfrog scheme. In other words, inserting the reactive latency is in fact one of the theoretical techniques to guarantee the linear numerical complexity for updating the variables. Therefore, a primal motivation of the basic LIM seems to be the application of the explicit leapfrog scheme to a transient analysis of large networks. As described above, one of the limitations of the latency insertion approach is the strict restriction of the time step size. The commonly-observed feature on the different stability conditions described in [11] and [7] is that the time step size is restricted by the smallest product of the inductor connected to a node and the grounded capacitor at the node. This condition indicates that if an extremely small reactive latency exists inherently in the network or is inserted artificially, the time step size of the whole system becomes extremely small to guarantee the stability of the method. It is clear that the small time step size induces increasing of the number of the total time steps. If there is the topology which does not have any reactive element, the methods based on the latency insertion approach suffer from the difficulty in achieving a good balance between the acceptable accuracy and the number of the total time steps. As a result, although the latency insertion approach is able to guarantee the linear computational complexity per time step, it may increase the number of the total time steps of the transient analysis significantly due to the extremely small value of, and especially lack of, the latency in networks.

978-1-4244-4447-2/09 $25.00 © 2009 IEEE

III. Generalized Leapfrog Scheme

The *block formulation approach* is one of the different techniques from the latency insertion approach. This approach has been developed to deal with the coupled elements such as the branch capacitance in the on-chip power distribution networks (PDNs) [5], [6] and the mutual inductance in the tightly coupled transmission lines [10]. In this section, we expand the approach to deal with more general topologies, especially the circuit without the reactive elements and the tightly coupled elements.

The vector-matrix version of the KVL and KCL equations are written as

$$\mathbf{A}^{\mathsf{T}} \cdot \mathbf{v} = \mathbf{L} \cdot \frac{d}{dt}\mathbf{i} + \mathbf{R} \cdot \mathbf{i} - \mathbf{e}, \tag{5}$$

$$-\mathbf{A} \cdot \mathbf{i} = \mathbf{C} \cdot \frac{d}{dt}\mathbf{v} + \mathbf{G} \cdot \mathbf{v} - \mathbf{h}, \tag{6}$$

where $\mathbf{i}$ and $\mathbf{v}$ are the vectors of the current and the voltage variables in whole circuit, $\mathbf{e}$ and $\mathbf{h}$ are the voltage and the current source vector and $\mathbf{A}$ is the incidence matrix. These equations are derived by using the RLCG-MNA method [12] in which the associated circuit equation is written as

$$\begin{bmatrix} \mathbf{C} & \mathbf{0} \\ \mathbf{0} & \mathbf{L} \end{bmatrix} \cdot \frac{d}{dt} \begin{bmatrix} \mathbf{v} \\ \mathbf{i} \end{bmatrix} + \begin{bmatrix} \mathbf{G} & \mathbf{A} \\ -\mathbf{A}^{\mathsf{T}} & \mathbf{R} \end{bmatrix} \begin{bmatrix} \mathbf{v} \\ \mathbf{i} \end{bmatrix} = \begin{bmatrix} \mathbf{h} \\ \mathbf{e} \end{bmatrix}. \tag{7}$$

where the matrices and the vectors are same as those in (5) and (6). It is clear that performing the block matrix operations for (7) is equivalent to calculating two equations (5) and (6). Note that if there is no coupling elements including mutual inductance, branch capacitance and some controlled sources in the circuit and the latency insertion approach is adopted, each matrix in (5) and (6) is a diagonal matrix. Therefore, the explicit leapfrog scheme can be obtained by applying the finite difference method in the basic LIM. However, if the coupling elements exist, the matrices have the dense parts at intervals around the diagonal elements. In the block formulation approach, the variables corresponding to each dense part are solved simultaneously, namely, the implicit leapfrog scheme is generated. In this case, although the dense parts are solved implicitly by using a direct method such as the LU decomposition method, those parts can be separated from each other and also divided from the non-dense parts where the coupling element does not exist. Actually, the implicit leapfrog scheme is more efficient because the matrix solver is applied only to each dense part separately, unlike the conventional direct methods which solve the whole circuit in a lump. Additionally, such an implicit leapfrog scheme can still use the different time step sizes for the different blocks by using the methods in [2] and [8] which are based on the partitioning of the network and the multirate behavior of the network.

The coupling elements are induced into the network if the equivalent transformations are performed for the ill-constructed topologies to the basic LIM. For example, the branch without the inductive element is shown in Fig. 2. In the basic LIM, since the resistance $R_{a,b}$ is a component of the branch topology, its value is stamped into the resistance matrix $\mathbf{R}$ associated with the current variable. However, the resistance can be regarded as a conductive pass between the nodes, therefore, its value is stamped into the conductance matrix $\mathbf{G}$ associated with the voltage variable. The current variable $i_{a,b}$ is no longer defined at the resistor as shown in Fig. 2. This is in fact equivalent to the partial Gaussian elimination, by which the current variable corresponding to the branch without the reactive elements is eliminated. By this transformation, the stamped part makes a dense block, which is solved by a direct method. Another ill-constructed topology is the node without the capacitive path to the ground shown in Fig. 3. In this case, the transformation by using the companion model of an inductor is applied as shown in Fig. 3, followed by the application of the generalized Y-Δ transformation to remove such an intermediate node. The value of the transformed elements shown in Fig. 4 are defined as

$$Y_{p,q} = \frac{Y_p Y_q}{\sum_{k=1}^{M_a} Y_k}, \quad E_{p,q} = E_p - E_q, \quad (1 \le p < q \le M_a). \tag{8}$$

The coupled admittances are induced between the nodes adjacent to the intermediate node in exchange for removing the intermediate node without the grounded capacitance. By these transformations, the ill-constructed topologies are removed and the generalized leapfrog scheme based on the block formulation approach can be applied. The block formulation approach is not restricted by the stability condition in the explicit leapfrog scheme because the extremely small latency is not introduced.

IV. Numerical Results

In this section, we apply the generalized leapfrog scheme to an example circuit to estimate the availability of the scheme. The example circuit is shown in Fig. 5. Analysing this type of network is the worst case analysis of the basic LIM because the circuit includes the branch without the inductance and the node without the grounded capacitance. Additionally, the mutual inductances as well as the branch capacitances exist to make tightly coupled connections among the branches and the nodes. The parameters of the grounded and the branch capacitance are $C_{\mathrm{g}} = 1.0$ pF and $C_{\mathrm{m}} = 1.0$ fF, respectively, the self inductance is $L = 2.0$ nH and the mutual inductance is defined as the coupling coefficient $K = 0.1$. The values of the all resistances are

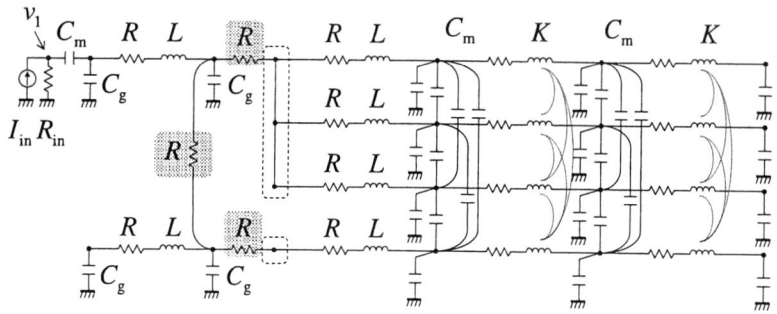
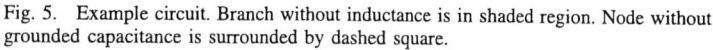

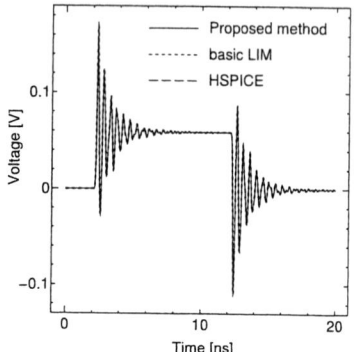

Fig. 5. Example circuit. Branch without inductance is in shaded region. Node without grounded capacitance is surrounded by dashed square.

Fig. 6. Waveform results from v_1.

assigned as $R = 0.1\ \Omega$ except for the resistance $R_{in} = 50\ \Omega$. We performed the transient analyses using the proposed method, the basic LIM and HSPICE and observed the voltage waveforms of v_1 in Fig. 5. In the basic LIM case, the branch capacitance and the mutual inductance are dealt with by the method proposed in [3]. The simulation interval is from 0 ns to 20 ns.

The waveform results obtained by each method are shown in Fig. 6. It is confirmed that there is a good agreement between the waveforms from three methods. In this case, the time step size of the proposed method is 1 ps while that of the basic LIM is restricted to 0.477 ps. Thus, the numbers of the total time steps are 20,000 for the proposed method and 41,928 for the basic LIM. As a result, the CPU time of the proposed method is 2.73 seconds and that of the basic LIM is 5.11 seconds, and the CPU costs include the factorization step before the transient analyses. These results indicate that our proposed method is more effective to the ill-constructed circuit. We have already shown the block-LIM is available for the simulation of the large-scale networks including tightly coupled components [10]. It is expected that the proposed method, which is not restricted by the network topologies, is more useful and practical.

V. CONCLUSIONS

In this paper, a generalized leapfrog scheme for transient analyses of large-scale circuits has been described. First, the leapfrog scheme developed in the basic latency insertion method (LIM) was described and the characteristics of the method and the scheme were discussed. Then, we proposed the generalization technique of the leapfrog scheme which can be applied to the transient analysis of an ill-constructed circuit to the basic LIM. Some numerical results show that our proposed approach is more suitable for the fast simulation of large-scale networks.

ACKNOWLEDGMENT

This work is partially supported by NEDO, and the authors would like to thank NEDO for the support in making this work possible.

REFERENCES

[1] J. E. Schutt-Ainé, "Latency insertion method (LIM) for the fast transient simulation of large networks," *IEEE Trans. Circuits Syst. I*, vol. 48, pp. 81–89, Jan. 2001.

[2] R. Gao and J. E. Schutt-Ainé, "Improved latency insertion method for simulation of large networks with low latency," in *Proc. IEEE EPEP 2002*, Monterey, CA, Oct. 2002, pp. 37–41.

[3] Z. Deng and J. E. Schutt-Ainé, "LIM-SPICE for the analysis of power distribution networks," in *Proc. IEEE SPI 2005*, Garmisch-Partenkirchen, Germany, May 2005, pp. 17–20.

[4] ——, "Turbo-spice with latency insertion method (LIM)," in *Proc. IEEE EPEP 2005*, Austin, TX, Oct. 2005, pp. 329–332.

[5] J. Choi, M. Swaminathan, N. Do, and R. Master, "Modeling of power supply noise in large chips using the circuit-based finite-difference time-domain method," *IEEE Trans. Electromagn. Compat.*, vol. 47, pp. 424–439, Aug. 2005.

[6] S. N. Lalgudi, Y. Kretchmer, and M. Swaminathan, "Simulation of switching noise in on-chip power distribution networks of FPGAs," in *Proc. IEEE EPEP 2005*, Austin, TX, Oct. 2005, pp. 319–322.

[7] S. N. Lalgudi, M. Swaminathan, and Y. Kretchmer, "On-chip power-grid simulation using latency insertion method," *IEEE Trans. Circuits Syst. I*, vol. 55, pp. 914–931, Apr. 2008.

[8] H. Asai and N. Tsuboi, "Multi-rate latency insertion method with RLCG-MNA formulation for fast transient simulation of large-scale interconnect and plane networks," in *Proc. IEEE ECTC 2007*, Reno, NV, May 2007, pp. 1667–1672.

[9] T. Sekine and H. Asai, "CMOS circuit simulation using latency insertion method," in *Proc. IEEE EPEP 2008*, San Jose, CA, Oct. 2008, pp. 55–58.

[10] ——, "Block latency insertion method (block-LIM) for fast transient simulation of tightly coupled transmission lines," in *Proc. IEEE EMC Symposium 2009*, Austin, TX, Aug. to be published.

[11] Z. Deng and J. E. Schutt-Ainé, "Stability analysis of latency insertion method (LIM)," in *Proc. IEEE EPEP 2004*, Portland, OR, Oct. 2004, pp. 167–170.

[12] Y. Tanji, T. Watanabe, H. Kubota, and H. Asai, "Large scale rlc circuit analysis using RLCG-MNA formulation," in *Proc. DATE'06*, Munich, Germany, Mar. 2006, pp. 45–46.

Modeling of IC power supply and I/O ports from measurements

I. S. Stievano[1], L. Rigazio[1], I. A. Maio[1], A. Girardi[2], R. Izzi[2], F. Vitale[2], T. Lessio[2]

[1] Politecnico di Torino, Italy

[2] Numonyx, R&D CAD Group, Italy

(e-mail igor.stievano@polito.it)

Abstract: This paper addresses the generation of behavioral models of digital ICs for signal and power integrity simulations. The proposed models are obtained by external port measurements and by the combined application of specialized state-of-the-art modeling techniques. The proposed approach is demonstrated on the I/O buffers and the core power supply ports of a commercial 90nm flash memory.

1 Introduction

Nowadays, the modeling of the power supply and I/O ports of ICs is of paramount importance for the simulation of many advanced electronic applications. The modeling from measured transient responses, in particular, is a key resource to handle ICs whose suppliers provide low-order or partial models only. Besides, when power supply ports are involved, the modeling from measured responses can be the best option to cope with the complexity of the problem.

Parametric modeling of I/O ports from measured transient responses has been addressed in [1] and the modeling of power supply ports is addressed in [2, 3, 4]. In this paper, we demonstrate those methods by developing models for both I/O and power supply ports for an high-speed IC memory, that is a 512Mb NOR Flash memory in 90nm technology produced by Numonyx. As shown in the schematic of Fig. 1, the test device has 16 high-speed I/O buffers (DQn terminals) for data communication and separate - weakly coupled - supply ports for the core logic (VDD-VSS terminals) and the high-speed I/O buffers (VDDQ-VSSQ terminals). This device is a medium complexity IC for System-in-Package (SiP) application and is therefore a good test case for the modeling approach being addressed.

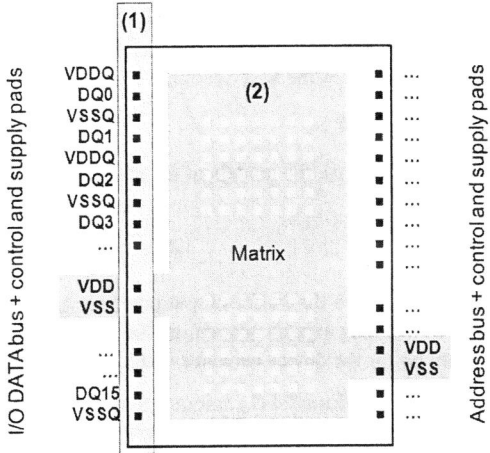

Figure 1: Simplified structure of the example IC highlighting the data bus (*i.e.,* the I/O buffers DQn in the scheme) and the different sub-networks VDDQ-VSSQ (1) and VDD-VSS (2) composing the power delivery network of a digital memory.

The test board that has been developed to carry out the transient measurements used in the modeling process is also described. The test board allows accurate measurements of the IC supply currents as well as of the voltage and currents of the I/O ports at the same time. Besides, the developed board can be easily adapted to test different ICs.

2 Model structures

This paper focuses on the modeling of the I/O ports (DQn) and of the core power supply port (VDD-VSS). The structure of the models and the data required for the estimation of the model parameters are shortly outlined in this Section.

2.1 VDD-VSS core power supply

According to [2, 3, 4], the models for the core power supply of ICs are defined by simplified - physically inspired - circuit equivalents that attempt to describe the different blocks involved in the power delivery network of the digital IC. A common assumption in these approaches is the description of the power delivery network of the IC by means of a Norton equivalent where the short-circuit current generator accounts for the internal switching activity of the device and the equivalent impedance accounts for the passive interconnect structure (see Fig. 2). It is ought to remark that this assumption holds when the physical dimension of the silicon die and the frequency bandwidth of interest are compatible with lumped modeling and the effects of possible nonlinearities in the supply port behavior are negligible. When these conditions are met, this simplification is the best solution to estimate the model parameters from external measurements.

The estimation of the Norton model of the core power supply port, however, requires care in collecting, interpretating and processing the measured data. The estimation of the equivalent impedance has been proven to be best achieved via on-chip two-port scattering parameter measurements as suggested in [4]. In contrast to measurements based on fixtures including package and external interconnects, this solution generates responses free from resonant effects, leading to improved accuracy levels. Once the port impedance at die-level is known, the actual impedance at the IC port is obtained by taking into account the package effects via electromagnetic modeling.

For the test IC of this paper, the measurement and estimation of the impedance of the model is addressed in [4]. Here we concentrate on the estimation of the equivalent current source via the measurement of the switching activity. This measurement must be carried out with the device mounted on a board and operating in nominal conditions, as it is explained in the next Section.

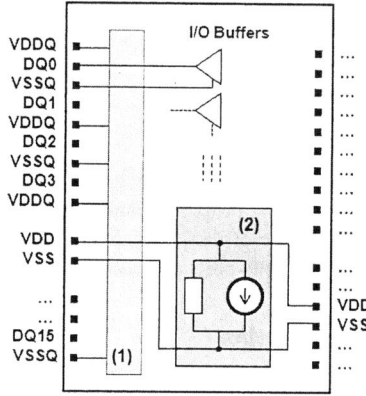

Figure 2: IC structure with highlighted the different blocks representing the different behavioral models considered in this study.

2.2 I/O buffers

According to [1, 6], the models of the output buffers DQn exploit the following two-piece parametric relation:

$$i(t) = w_H(t)i_H(v(t), d/dt) + w_L(t)i_L(v(t), d/dt) \qquad (1)$$

where i and v are the buffer output port voltage and current variables, with associated reference directions, w_H and w_L are switching signals accounting for the device state transitions and i_H and i_L are nonlinear parametric relations accounting for the device behavior in the fixed high and low logic states, respectively.

The estimation of model (1) amounts to computing the parameters of submodels i_H and i_L and the weighting signals w_H and w_L from suitable port voltage $v(t)$ and current $i(t)$ responses. These estimation responses must be recorded while the buffer drives different loads and performs state switchings. Model parameters are then computed by minimizing suitable error functions between the model and the measured responses, which are used as references to be fitted.

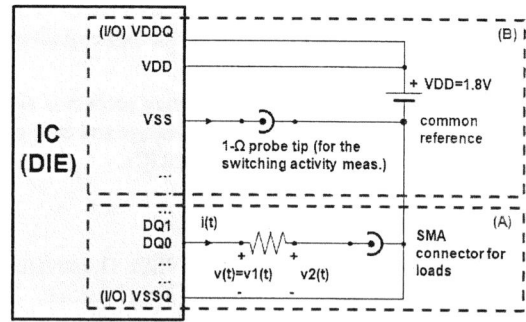

Figure 3: Schematic of the test setup for the measurement of the core switching activity (box A) and of the buffer port voltage and currents waveforms (box B).

Figure 4: Measurement board for recording the core switchig activity and the buffer waveforms for the example IC. The key elements of the setup can be recognized.

978-1-4244-4447-2/09 $25.00 © 2009 IEEE

3 Test board

As outlined in the previous Section, the transient measurement required for the modeling study of this paper, are the switching activity at the core power supply port and the voltage and currents waveform at the buffer ports recorded during state switching and for different loads. The ideal schematic of a setup for the measurement of these quantities is shown in Fig. 3. In this setup, the switching current of the core power supply port is measured by means of a 1-Ω current probe that has been suitably designed and series connected to the VSS pad. This method and the probe design follows the guidelines of [5].

The buffer voltage waveform is obtained by direct measurement, whereas the current waveform is obtained by means of the series resistor of Fig. 3. The connector on the buffer output is used to plug-in the different loads needed to generate the estimation waveforms [1].

In order to implement this setup, a test board composed of a general purpose control circuitry for the operation of the device under test and of a measurement board holding the IC under test and the measurement fixture has been developed. The measurement board is connected to the control board via a pair of 40-pin QTE connectors, and can be replaced to test different ICs. A close-up of the measurement board for the example IC of this paper is shown in Fig. 4.

In all the measurements carried out within this activity, a LeCroy WavePro 7300A scope (3GHz bandwidth, 10GS/s) and two single-ended passive voltage probes P6158 (3GHz bandwidth, 1 kΩ, 1.5 pF, 20x attenuation) have been used to collect the transient waveforms. To reduce the effects of the measurement noise, the memory buffers have been forced to produce a periodic bit pattern and the averaging feature of the scope has been set (16 waveforms are considered for the average).

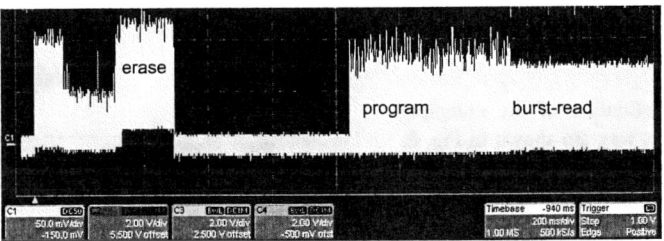

Figure 5: Measured power supply current obtained while the memory is driven to perform the three basic cycles (*i.e.*, the erase, programn and burst-read programs).

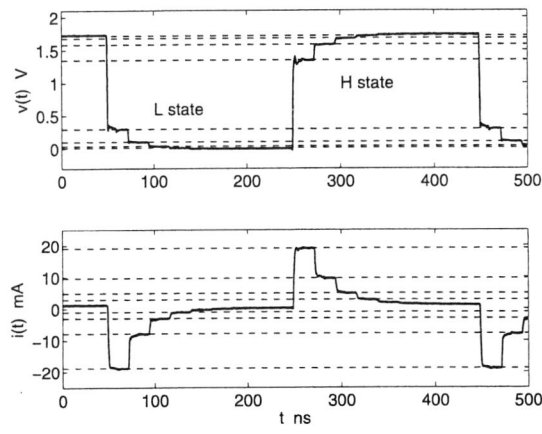

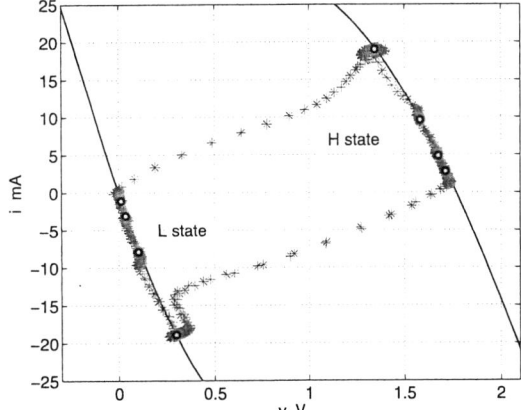

Figure 6: Responses obtained by driving the buffer DQ0 in Fig. 3 to produce a periodic '010' bit pattern on suitable transmission line load. The dashed lines highlight the steps of the voltage and current responses that allows for the extraction of the static parts of submodles i_H and i_L.

Figure 7: Static characteristics of the DQ0 buffer. Solid lines: transistor-level response; gray stars: region explored by the $v(t), i(t)$ samples of the transient responses of Fig. 6; Circles: static points extracted from the curves of Fig. 6.

978-1-4244-4447-2/09 $25.00 © 2009 IEEE

4 Results

This Section illustrates measurements and models developed in this study.

Figure 5 shows an example of the switching current recorded as explained in the previous Section. The different patterns corresponding to the basic erase, program and read cycles can be clearly distinguished. Supplemental study based on numerical simulations allowing for the package effects proved that the value of the signal-to-noise ratio of this measurement is good for frequencies up to 500 MHz. Besides, the relation between the impedance of the supply network of our test board and the Norton model impedance guarantees that the measured switching current is close to the equivalent current source of Fig. 2. The cross-validation of the complete model obtained in this way is currently under development.

The model of the output buffers is based on i_H and i_L submodels (see eq. 1) defined as sum of a static and a dynamic part. The static parts is extracted form the measured responses of Fig. 6, that are obtained by driving the buffer to produce a periodic '010' sequence on a transmission line load. The $\{i, v\}$ pairs corresponding to the flat parts of the waveforms of Fig. 6 are shown in Fig. 7, which proves the excellent accuracy of the static behavior extracted by this approach. The remaining model parameters, including the dynamic parts of submodels and the weighting signals are estimated as suggested in [1] from the edges of the waveforms of Fig. 6.

The waveforms of a validation of the complete buffer model built in this way are shown in Fig. 8. This Figure compares the measured and predicted response of the modeled buffer for a load different from those involved in the parameter estimation process. The load consists of the shunt connection of an open ended coaxial cable ($Z_0 = 50\,\Omega$, $L = 2.6\,m$) with a coaxial cable ($Z_0 = 50\,\Omega$,

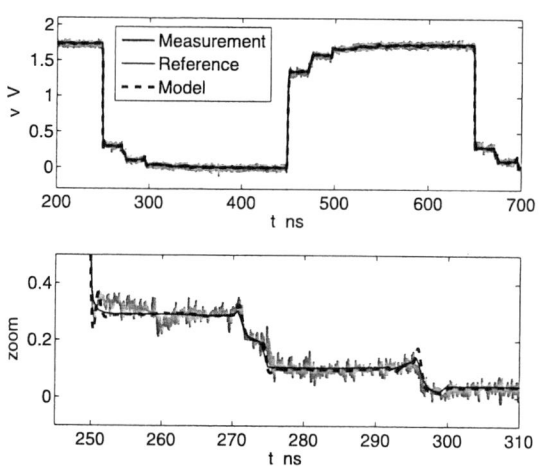

Figure 8: Port voltage and current responses of the DQ0 buffer for the validation test considered in the study (see text for details).

$L = 2\,m$) terminated by a $C = 15\,pF$. Both the response of the estimated model and the response of the high-order transistor level model of the buffer provided by the foundry are shown in the Figure.

The very good agreement among the curves of Fig. 8 confirms the strengths of the proposed methodology in generating accurate models from measured transient responses. Such models can be easily obtained by the proposed measurement setup and can effectively replace the hardly available and less efficient transistor-level models of ICs.

Acknowledgement. The research leading to these results has received funding from the European Community's Seventh Framework Programme FP7-ICT-2007-1 under the MOCHA (MOdeling and CHAracterization for SiP - Signal and Power Integrity Analysis) grant n. 216732.

References

[1] I.S. Stievano, I.A. Maio, F.G. Canavero, "Behavioral models of IC output buffers from on-the-fly measurements," IEEE Transactions on Instrumentation and Measurement, vol. 57, No. 4, pp. 850–855, 2008.

[2] "Integrated Circuits Electrical Model (ICEM)", International Electro-technical Commission (IEC) 61967, March 2001.

[3] C. Labussiere-Dorgan, S. Bendhia, E. Sicard et Al., "Modeling the electromagnetic emission of a microcontroller using a single model,"IEEE Transactions on EMC, Vol. 50, No. 1, Feb. 2008.

[4] I.S. Stievano et Al., "Characterization and modeling of the power delivery networks of memory chips," accepted for presentation at the *13-th IEEE Workshop on Signal Propagation on Interconnects, Strasbourg, F, May. 12–15, 2009*.

[5] "International Electro-technical Commission, IEC 61967 Part 4: Measurement of conducted emission - 1 Ω/150 Ω direct coupling method," 2006.

[6] I.S. Stievano, I.A. Maio, F.G. Canavero, "Mπlog, Macromodeling via Parametric Identification of Logic Gates," IEEE Transactions on Advanced Packaging, Vol. 27, No. 1, pp. 15–23, Feb. 2004.

Solver for Current Source Type Driver(s) and Interconnect with Linear or Nonlinear Loads

Albert E. Ruehli and Jerry Hayes*,
Emeritus, IBM T. J. Watson Research Center, Yorktown Heights, NY 10598, USA
*IBM Austin Research Laboratory. Austin, TX 78703

Abstract

Current source type models are widely used for noise and timing analysis of on-chip drivers and interconnects. A time domain solution using Spice may be too time consuming. Here, we consider small solver for nonlinear driver(s) with interconnects and linear or nonlinear loads.

I. INTRODUCTION

Applying a Spice transient circuit solver for on chip timing and noise analysis is too time consuming. Today, current source type macromodels are used for Static Timing Analysis (STA) and noise analysis of on-chip drivers with interconnects. An input to output driver model is needed for noise analysis. In this paper, we use a standard IVV type current source model for the driver circuit. The load on some ports of the circuit can be linear or nonlinear. Hence, several different circuit topologies may be of interest, especially for noise solutions where a nonlinear driver output impedance must also be modeled for passive drivers. For efficiency reasons, we pursue two different partitioned approaches to solve the problem which may involve a nonlinear circuit loaded with linear interconnects which are also loaded with linear or nonlinear loads. The approach chosen depends on the configuration. We call the first approach Nonlinear-Linear *NL-L* where a nonlinear driver is connected to a linear linear interconnect model which in turn is connected to linear loads only. The second case we call *NL-NL*, where both the driver(s) and the loads (receiver(s)) can be nonlinear. The fast solution approach allows the appropriate trade-off required between accuracy and speed.

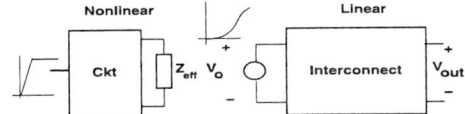 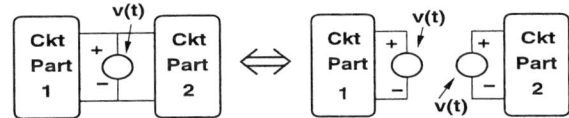

Fig. 1.   Left: Model split into separate driver and load   Right: Partitioning of a circuit with voltage source theorem

II. NONLINEAR DRIVER WITH LINEAR LOAD *NL-L*

First we consider a nonlinear driver connected to a linear interconnect with linear loads shown in Fig. 1,L, which we call the *NL-L* case. We can solve the nonlinear circuit and the linear load as separate units. Our approach resembles the interesting effective capacitor C_{eff} approach [1], where interconnect models are taken into account with an iterative approximation and the driver is characterized for multiple load C_{eff} capacitances. In contrast, we use a *CS* driver that is compatible with today's IVV models with a single static nonlinear characterization. The interconnect circuit part may be small or include a large portion of a 2D or a 3D model [2] with linear loads. Our *NL-L* approach gains its efficiency from the use of the voltage substitution theorem. A general rule for partitioning is that the partition circuit impedances should be represented with the input impedance, admittance characteristics, etc., as close as possible to those of the original circuit. As is shown in Fig. 1,L one part is a current source driver model loaded with the input impedance of the interconnect part. We call this load impedance Z_{eff} to relate to the earlier C_{eff} load work [1]. The current source output I of the driver leads $V_o = Z_{eff} I$. Hence, we us an impedance Z_{eff} model rather than an admittance while the voltage source driven interconnect part is in the admittance form $I = Y V_o$. The partitioning of the circuit is accomplished with the following:

Theorem 2.1 (Voltage source substitution): A circuit is divided into port connected parts. Voltage sources V are placed at the common boundary ports for the partitioned circuit without altering the solution if we assign the voltages $v(t)$ of the sources to correspond to the solution of the original circuit. Figure 1,R shows the case of a single source.

By applying the substitution theorem, the linear interconnect part of any size can be solved separately from the nonlinear solution. Reduced Order Macromodeling (ROM), a conventional Spice type circuit solution, or an interconnect solver [2] can be applied to the linear part. It is clear from Fig. 1,L that the driver circuit is loaded with the appropriate input impedance Z_{eff} of the interconnect circuit so that we can compute $v_o(t)$. The most simple Z_{eff} model consists of a simple capacitance like the C_{eff} model. A more complex macromodel which includes inductances is shown in Fig. 2,L, for which we can compute the input impedance as $Z_{eff} = (s^2 + sR_1/L_1 + 1/(L_1 C_2))/ \left[sC_1(s^2 + s R_1/L_1 + (C_1 + C_2)/(L_1 C_1 C_2)) \right]$

Details of this model are given in [3]. We used the *VectorFit* approach [4] to obtain a rational model for a given number of poles. While the circuit interpretation is desirable, it is not necessary, especially for systems which require higher order expansions of the form for $M > 3$,

$$Z_{eff} = d + \sum_{m=1}^{M_r} \frac{r_m}{s - p_m} + \sum_{m=1}^{M_c} \left[\frac{r_{m,r} + j\, r_{m,i}}{s - p_{m,r} - j\, p_{m,i}} + \frac{r_{m,r} - j\, r_{m,i}}{s - p_{m,r} + j\, p_{m,i}} \right] \tag{1}$$

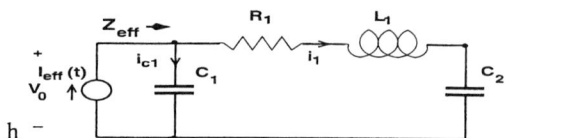 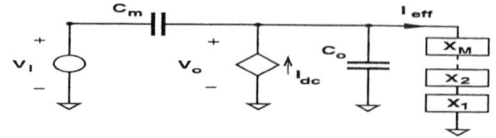

Fig. 2. Left: CLRC inteconnect macromodel for Z_{eff} input impedance   Right: Circuit model for driver loaded with Z_{eff} macromodel

Fig. 3. Left: Example of interconnect macromodel with three ports   Right: Reduced order macromodel with time domain solver

where M_r are the real poles and M_c are the complex pole pairs. Additional capacitances are added due to the connected driver type circuit as is shown in Fig. 2,R. The constant device capacitances C_m, C_o may be absorbed into the interconnect model. To provide a solution for more complex interconnects, we give matrix stamps in the usual MNA formulation. To solve the driver with the load we augment the MNA approach with state stamps as shown in Fig. 2,R. The form of Z_{eff} is a series connection of impedances, where the m-th term is of the form $V_m = Z_m I_m$ in (1). The constant d can be interpreted as a series resistance, while we can write the equivalent differential equation for the real poles and residues, or $dx_m/dt - p_m x_m = r_m i(t)$ where the current $i(t) = i_{eff}(t)$, is the same for all the terms in series. We note that the state variable x_m is equivalent to a voltage.

Hence, we can augment the MNA circuit matrix by using a stamp which corresponds to the differential equation with the BE or BD2 discretization as

Ro\Col.	$x_m^{(p)}$	$i_{eff}^{(p)}$	RHS
m	$k_p - p_m$	$-r_m$	$k_{p-1} x_m^{(p-1)} - k_{p-2} x_m^{(p-2)}$
aux (v)	1	0	0

which in our case, $i^{(p)}$ is $i_{eff}^{(p)}$. Note that we use (p) to indicate the present time step. Complex poles are best treated as pole pairs with the poles $p_{m,r} \pm p_{m,i}$ and the residues $r_{m,r} \pm r_{m,i}$ leading to a differential equation [5]. Here, the solution is derived by assuming a complex pair of solutions $x_{m,r} \pm x_{m,i}$ leading to the stamp

Row\Col.	$x_{m,r}^{(p)}$	$x_{m,i}^{(p)}$	$i_{eff}^{(p)}$	RHS
m1	$k_p - p_{m,r}$	$p_{m,i}$	$-r_{m,r}$	RHS_1
m2	$-p_{m,i}$	$k_p - p_{m,r}$	$-r_{m,i}$	RHS_2.
aux (v)	2	0	0	0

where $RHS_1 = k_{p-1} x_{m,r}^{(p-1)} - k_{p-2} x_{m,r}^{(p-2)}$ and $RHS_2 = k_{p-1} x_{m,i}^{(p-1)} - k_{p-2} x_{m,i}^{(p-2)}$. Finally, the stamps connect the impedance part of the MNA to the state matrix shown in Fig. 2,R. As is the case for an inductance stamp, the current through all the elements and the voltage rows are shared. It is clear that the circuit matrix shares a single column for the current i_{eff} and a single row for all the voltages $aux(v)$.

III. NONLINEAR DRIVER NONLINEAR LOAD *NL-NL*

The source substitution theorem cannot be applied directly for nonlinear loads. So, we pursue another macromodeling approach to reduce compute time. For the more general case, terminations may consist of linear or nonlinear loads, driver input circuits or multiple driver outputs. We use a reduced order macromodel to represents the port admittances in residue/pole form. Figure 3,L gives an example for three connection ports $I = YV$ where we labeled them $1, 2, 3$ for convenience. In these matrices, each element is represented by the residue/pole model of the form

$$y_{k\ell} = g_{k\ell} + \sum_{m=1}^{M_r} \frac{r_{k\ell}^m}{s - p_m} + \sum_{m=1}^{M_c} \left[\frac{r_{k\ell}^{(m,r)} + j\, r_{k\ell}^{(m,i)}}{s - p_{m,r} - j\, p_{m,i}} + \frac{r_{k\ell}^{(m,r)} - j\, r_{k\ell}^{(m,i)}}{s - p_{m,r} + j\, p_{m,i}} \right] \tag{2}$$

where the same set of poles is used for all admittances. The residues and poles are computed by using a reduced order modeling approach or the *VectorFit* technique.

The second part of the *NL-NL* approach involves the stamping of the residue/pole model into a fast time domain MNA circuit solver. We re-write the system using (2) in the form

$$\begin{bmatrix} I_1 \\ I_2 \\ I_3 \end{bmatrix} = \begin{bmatrix} g_{11} & g_{12} & g_{13} \\ g_{21} & g_{22} & g_{23} \\ g_{31} & g_{23} & g_{33} \end{bmatrix} \begin{bmatrix} V_1 \\ V_2 \\ V_3 \end{bmatrix} + \frac{1}{s - p_1} \begin{bmatrix} r_{11}^1 & r_{12}^1 & r_{13}^1 \\ r_{21}^1 & r_{22}^1 & r_{23}^1 \\ r_{31}^1 & r_{23}^1 & r_{33}^1 \end{bmatrix} \begin{bmatrix} V_1 \\ V_2 \\ V_3 \end{bmatrix} + \frac{1}{s - p_2} \begin{bmatrix} r_{11}^2 & r_{12}^2 & r_{13}^2 \\ r_{21}^2 & r_{22}^2 & r_{23}^2 \\ r_{31}^2 & r_{32}^2 & r_{33}^2 \end{bmatrix} \begin{bmatrix} V_1 \\ V_2 \\ V_3 \end{bmatrix} + \cdots$$

978-1-4244-4447-2/09 $25.00 © 2009 IEEE

where we only considered the real poles for space reasons. The first term can be interpreted as a resisistive circuit while for the $m-th$ pole can be written in matrix form as $\boldsymbol{I_m} = \frac{1}{s-p_m}\boldsymbol{R_m V_\ell}$ with the contributions to the currents as

$$\begin{bmatrix} I_{k,1} \\ I_{k,2} \\ I_{k,3} \end{bmatrix} = \boldsymbol{G\,V_\ell} + \begin{bmatrix} I_{1,1} \\ I_{1,2} \\ I_{1,3} \end{bmatrix} + \cdots + \begin{bmatrix} I_{m,1} \\ I_{m,2} \\ I_{m,3} \end{bmatrix} + \cdots \tag{3}$$

Each of the individual currents in (3) must be entered into the MNA matrix of the nonlinear solver. Hence, it is clear that poles can be included by the addition of residue/pole stamps. The next step is to represent the currents in the time domain. This is accomplished by writing the equation as $\frac{d\boldsymbol{I_m}}{dt} - p_m\boldsymbol{I_m} = \boldsymbol{R_m V_\ell}$ and by using BD2 for the numerical integration. The combined currents are of the form $(k_p - p_m)\boldsymbol{I}_m^{(p)} - \boldsymbol{R_m V}_\ell^{(p)} = k_{p-1}\boldsymbol{I}_m^{(p-1)} + k_{p-2}\boldsymbol{I}_m^{(p-2)}$

Finally, we are ready to develop the matrix stamps so that we can add the ROM to the fast MNA circuit solver. We recognize that the stamp has to involve all the voltages corresponding to the port voltages for the interconnect model. However, for space reasons we only include two of the three ports in the matrix stamp from the example in Fig. 3,R. Each pole m is added to circuit matrix as

Row\Col.	$I_{m,1}^{(p)}$	$I_{m,2}^{(p)}$	$v_1^{(p)}$	$v_2^{(p)}$	RHS
v_1	$k_p - p_m$	0	$-r_{12}^m$	$-r_{13}^m$	RHS_1
v_2	0	$k_p - p_m$	$-r_{21}^m$	$-r_{22}^m$	RHS_2
I_1	1	0	0	0	0
I_2	0	1	0	0	0

where $RHS_1 = k_{p-1}I_{m,1}^{(p-1)} + k_{p-2}I_{m,1}^{(p-2)}$, $RHS_2 = k_{p-1}I_{m,2}^{(p-1)} + k_{p-2}I_{m,2}^{(p-2)}$. From (3), it is evident that adding more poles to the model is rather simple. A schematic example for this nonlinear solver, which includes the above linear interconnect model, is shown in Fig. 2,R.

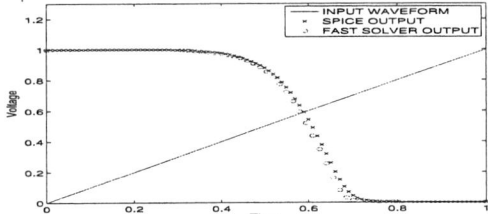 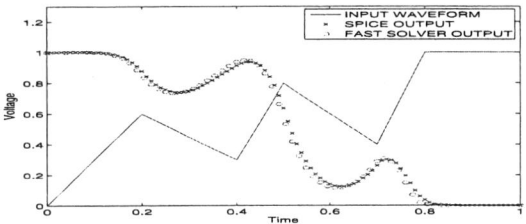

Fig. 4. Left: Comparison for falling output waveform.   Right: Comparison for input waveform with noise

IV. NONLINEAR DRIVER MODEL

Today's Spice type general purpose circuit solvers are very efficient for a broad range of circuits. However, we need to further speed-up the solution. The known waveform voltages, model currents and time scales are used for the normalization of the circuit variables. The logical function of the circuit determines the initial and final conditions of the input and output voltages and the *dc* solution. Less accuracy is required than for Spice allowing for larger time steps. Also, simpler, less time consuming algorithms can be used. Time steps can be omitted for input waveforms which start with a time delay. Also, an initial time step can be computed based on the input waveform. The time steps can be increased or decreased without an elaborate algorithms. We can terminate the time solution as soon as the output waveform has reached a steady state. The special purpose solver only includes a few Z_{eff} stamps. Hence, the solver does not need to include many nodes even for an elaborate linear load circuit and we can model interconnect macromodels with very few poles and residues. This is true for both the *NL-L* and the *NL-NL* interconnect macromodels. Next, we elaborate on the strategy outlined above for a current source driver loaded with Z_{eff} in Fig. 2,R. To trade accuracy for speed in the nonlinear solution, we use the explicit updating of the source current at the present time step p, $i^{(p)} = I_{dc}(v_i^{(p)}, v_o^{(p-1)})$. This requires that the voltage changes at each time step are small enough. We can reduce the time step if the computed voltage step is too large, since we know the maximum amplitude of v_o which determines the voltage norm used. A different approach can be used for the evaluation of the Miller capacitance current which is given by $i_m = C_m^{(p-1)}\frac{dv_o}{dt} - C_m^{(p-1)}\frac{dv_i}{dt} = i_{mo}(t) - i_{mi}(t)$ The derivative of v_i is known at each time step. Further, the second term of the equation, which is a function of the derivative of v_o, results in additional capacitance C_m in parallel to the output capacitance C_o. We use an MNA matrix which also includes stamps for the loads. The best integration methods for this purpose are the implicit Back Euler (BE) and the BD2 preferable for inductive circuits. We represent the BE and the BD2 methods in the following form $\dot{x} = k_p x^{(p)} + k_{p-1} x^{(p-1)} + k_{p-2} x^{(p-2)}$ where for BE the coefficients are $k_p = -k_{p-1} = \frac{1}{h_p}$ and $k_{p-2} = 0$. The coefficients for BD2 are given by $k_p = \frac{h_p + h_{p+}}{h_p h_{p+}}$, $k_{p-1} = -\frac{h_p + h_{p-1}}{h_{p-1} h_p}$ and $k_{p-2} = \frac{h_p}{h_{p-1} h_{p+}}$ where h_p is the present time step, h_{p-1} is the previous time step and where we define $h_{p+} = h_p + h_{p-1}$.

We employ variable time steps for the widely varying signal and noise waveforms. We use the methods listed above for step size control. To start initial time stepping we take advantage of the fixed topology. For Fig. 2,R, the circuit equation for the V_0 node is given by

$$i_{mo} - i_{mi} - i_{dc}(v_i, v_o) + C_o(v_o)\frac{dv_o}{dt} + i_{eff} = 0 \qquad (4)$$

where the first two currents are due to the Miller capacitance and the third current is due to the IVV current source. Hence, we can step forward in time without solving the entire circuit while monitoring that the changes in the currents stays within a given norm. This approach can be effective in saving steps as is apparent from the resultant waveforms in Fig. 4. During the time domain solution, the size of the step h_p can be controlled with a simpler step control mechanism. The algorithm is based on concepts similar that in a Spice type solver. We use an inexpensive second order Lagrange based predictor or estimator or a first order Lagrange extrapolation given by $x_p^E = h_p/h_{p-1} x^{(p-2)} + (t_p - t_{p-2})/h_{p-1} x^{(p-1)}$. At each time point, we can alter the present time step h_p depending on the evaluation of the local error given by $E_p = |v_o^{(p)} - v_o^E|/2 < \epsilon_r v_i^{max} + \epsilon_a$ where we compare the estimated solution with the integrated solution for the output voltage $v_o^{(p)}$ for the integration method, where v_o^E is the estimated output voltage using the above prediction formula, where ϵ_r is the relative error, and ϵ_a is an absolute error. For the switching waveforms, we normalize the relative error with the input amplitude v_i^{max}. The time step h_p is increased if the error E_p is strongly underestimated or, it is decreased if the error estimate is strongly exceeded or else is left unchanged. Unfortunately, otherwise A-stable integration schemes are unstable during even conservative increases in the step size. The termination time for v_o is unknown at the start. It may be much shorter or longer than the end time of v_i. The end of the input switching and the end time of the input waveform file may be different. To terminate the solution, we make sure that a minimum number of time steps have been exceeded to avoid the unwanted termination. Then, we again test if the changes in v_o and device current are small to check both the static and dynamic behavior are close to a steady state so that the time solution can be terminated. Clearly, a key factor for the speed-up is given by the fact that we use only poles which are necessary for the Z_{eff} model. It is shown below that the relatively simple model in Fig. 2,L, is sufficient to model the interconnect. Hence, the two part partitioning approach keeps the nonlinear driver circuit small, which reduces the compute time. The circuit matrix can be implemented with the MNA matrix stamps given in Section II. Further, the interconnect circuit, *Part2*, can be analyzed in a pole/residue form in the time domain using the approach in Section III. This again leads to a reduced MNA time domain solver based on the low order pole/residue based model.

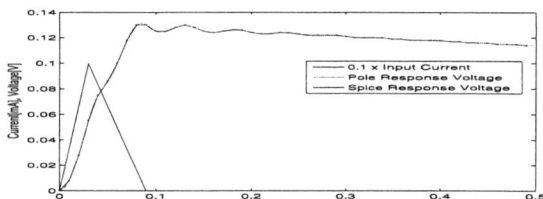 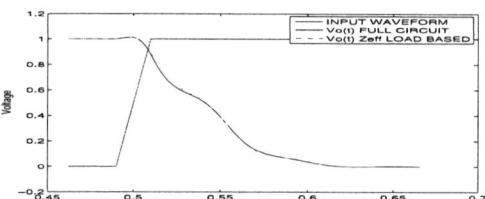

Fig. 5.   Left: Circuit using voltage substitution theorem vs. full Spice solution   Right: Intermediate voltage $v_o(t)$ in comparison to full load

V. RESULTS

For noise analysis, we need to be able to handle arbitrary input signal waveforms. Hence, we give two examples, one for a conventional ramp input and another noise waveform with several slope changes. We compare the results with a Spice solver using the same circuit and IVV model. Figure 4,L gives an example of the response for a rising input signal. Time-voltage points are omitted wherever possible by the end point removal and the variable time step algorithms. The right side figure shows a comparison for the multi-slope noise waveforms. The single slope example used 37 points with 3 sec run time in Matlab and the noise waveform took 58 points and 3.6 sec in the experimental code. For a speed comparison, we also run Spice and a small C++ implementation. The Matlab program speed was dominated by special purpose functions used for the IVV evaluation. Spice was about ten times faster and the C++ program was almost 10 times faster than Spice. Another important result is to show that the Z_{eff} model is accurate. For this purpose, we start from a more challenging load circuit with eight elements and three inductor. For our comparison, we reduced it to a $CLRC$ model shown in Fig. 2,L using *VectorFit*. Details are given in [3]. As a test we used the rapidly changing input current shown in Fig. 5,L. From the output waveforms, we see that the agreement between the three pole impedance model using BD2 and Spice is excellent in spite of the oscillations in the waveforms. As a further test, the responses for the voltage substitution theorem Z_{eff} model are compared with a complete Spice analysis. Figures 5,R show the very good waveform fidelity both at v_o in Fig. 1,L and also at the output v_{out}. This shows the ability to model inductive loads which is key for fast changing circuits with inductances.

REFERENCES

[1] J. Qian, S. Pullela, and L. T. Pileggi. Modeling the effective capacitance for the RC interconnect of CMOS gates. *IEEE Transactions on Computer-Aided Design*, 13:1526–1535, December 1994.
[2] A. Grivet-Talocia, I. S. Stievano, and F. G. Canavero. Hybridization of FDTD and device behavioral-modeling techniques. *IEEE Transactions on Electromagnetic Compatibility*, 45:31–42, February 2003.
[3] A. Ruehli and J. Hayes. Nonlinar solver with linear interconnect load. In *Digest of Electr. Perf. Electronic Packaging*, volume 12, pages 195–198, San Jose, CA, October 2008.
[4] B. Gustavsen and A. Semlyen. Rational approximation of frequency domain responses by vector fitting. *IEEE Transactions on Power Apparatus and Systems*, 14:1052–1061, July 1999.
[5] R. Achar and M. Nakhla. Simlation of higb-speed interconnects. *Proceedings of the IEEE*, 89:693–728, May 2001.

Challenges and Solutions for Next Generation Main Memory Systems

Joong-Ho Kim, Dan Oh, Ravi Kollipara, John Wilson, Scott Best, Thomas Giovannini,
Ian Shaeffer, Michael Ching, and Chuck Yuan

Rambus Inc.
4440 El Camino Real, Los Altos, CA 94022
Tel: 650-947-5524, Fax: 650-947-5001, jhkim@rambus.com

ABSTRACT

Today's high performance computing memory systems mainly consist of with DDR3 DRAMs offering 800Mb/s to 1600Mb/s data rates. Extending the performance of these main memory systems beyond the current data rate is quite challengeable as the signal integrity issues with physical channel remains relatively constant compared to the device performance which improves as process advances. This paper presents three key technologies which help the current memory architecture to reach the data rates of 1600~3200Mb/s without sacrificing memory capacity, increasing power consumption, or switching to more advanced differential signaling. These key features include FlexPhase™ timing adjustment to eliminate trace length matching, dynamic point-to-point signaling to increase memory capacity at high data rates, and near ground signaling to reduce IO signaling power. This paper demonstrates the benefits of these features from signal and power integrity point of view.

I. CHALLENGES OF MAIN MEMORY SYSTEM DESIGNS

Driven by recent multi-core computing, virtualization and processor integration trends, the data rate of a next generation main memory system for in high-end computer and workstation applications is expected to reach 3200Mb/s. In order to minimize system cost and provide backward compatibility, single-ended signaling based on Stub-Series Terminated Logic (SSTL) is the logic choice for next generation main memory I/O interfaces. Unfortunately, as the target data rate for memory interface continues to increase, signal integrity issues such as crosstalk, simultaneous switching output (SSO) noise, and reference voltage (VREF) noise, have become major limiting factors for potential data rates in high-speed memory interfaces [1].

In addition to these conventional signal integrity issues, the main memory applications pose additional challenges due to the signal noise generated by multiple memory module loadings. In these multi-drop topologies, the major factor that determines the maximum speed of the memory bus is the worst-case loading characteristics in which all connectors are fully populated with memory modules. As a consequence, the number of modules that can be supported in a multi-drop architecture decreases with increasing bus speed. Fig. 1 shows the number of device drops per data pin versus data rate. As illustrated in this figure, the high data rate prohibits any multi-drops beyond 1333 or 1600Mb/s. This limitation has the effect of reducing the total system memory capacity. As such, alternative methods for achieving high capacity are desirable.

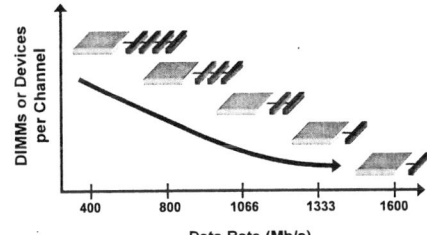

Figure 1. Main memory DIMM trend [2].

This paper introduces three key technologies to enhance the current DDR3 technology to beyond 1600Mb/s [3]. The FlexPhase™ technology removes significant timing errors extending the potential data rate. The dynamic point-to-point technology enables multiple memory modules to be implemented as point-to-point signaling to increase the memory capacity. Finally, the near ground signaling technology reduces the total I/O interface power without sacrificing the signal quality. To demonstrate the effectiveness of the proposed technologies, we consider a DDR3 system setup for a dual-rank expansion system, as shown in Fig. 2. To predict the maximum channel performance, we used the

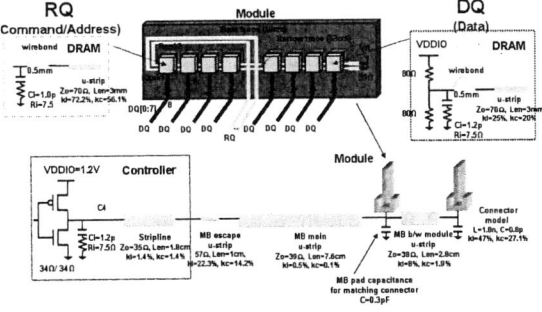

Figure 2. DDR3 system setup.

optimized passive channel: the motherboard impedance is optimized with compensating filters in order to minimize ISI impact. We used a simulation methodology described in [4] to perform simulation including analysis SSO noise.

II. IMPROVING SYSTEM PERFORMANCE USING FLEXPHASE™ TIMING ADJUSTMENT

As a data rate increases, timing error becomes more important and it is crucial to match timing parameters between different pins. This is especially true for DDR3 systems as the single-ended signaling bus is wider than the differential bus as it operates at lower speed. FlexPhase™ provides controller timing adjustment circuitry to calibrate pin-to-pin timing variation. Since

978-1-4244-4447-2/09 $25.00 © 2009 IEEE

placing timing adjustment circuitry in DRAM is more expensive than placing it in the controller, the timing adjustments for both read and write cases are performed in the controller. During the write operation, the data is sent with different delays in order to arrive at the DRAM at the same time. During the read operation, DRAM sends data without any delay adjustment and controller takes care of any timing mismatch by sampling the data at different sampling times.

Using FlexPhase™, a physical layout design for trace routing can be greatly simplified since trace length matching is no longer required. Fig. 3 shows the typical motherboard design with matched trace length. It is important to note that due to limited routing area, traces contain many serpentine sections and they are tightly coupled. Two motherboard designs with and without trace matching are considered to study the crosstalk impact. The simulation eyes are shown in Figs. 4 and 5. As noted in [1], the crosstalk is one of the major bottlenecks for single-ended signaling systems such as DDR3, and the FlexPhase™ technology can greatly reduce this crosstalk due to relaxed routing constraints.

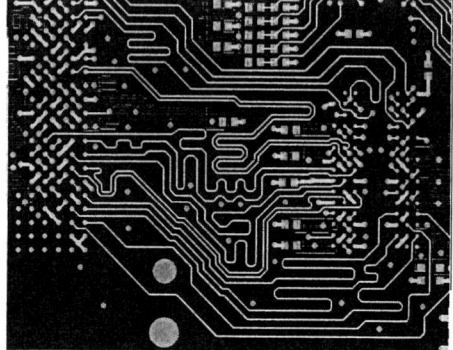

Figure 3. Routing example using trace length match.

Additional benefits of FlexPhase™ include the removal of any pin-to-pin timing variation due to process variation, on-chip timing skew ability for debugging purpose such as generating bathtub curves, and real-time sampling location adjustment for an optimum eye center. The timing adjustment can be performed during the system initialization stage and periodically updated during the DRAM refresh periods.

III. EXTENDING CAPACITY USING DYNAMIC POINT-TO-POINT (DPP) TOPOLOGY

Dynamic point-to-point (DPP) is a signaling topology for increasing memory capacity at high data rates. It enables high-performance computer systems to add modules without sacrificing memory system performance while maintaining backwards-compatibility with existing signaling technologies such as SSTL, PODL, DRSL, etc. DPP uses a technique that dynamically configures the memory channel so that all data connections are "point-to-point". Fig. 6 shows typical two-DIMM multi-drop and DPP systems, and Fig. 7 illustrates logical representations of DPP architecture examples for supporting a two dual-rank module system at 3200Mb/s. With DPP-enabled systems, the half of the memory bus is routed to each of the memory channel's DIMM sockets. DPP can be applied to even a larger number of modules, such as 4 modules, but we focus on the two-module case for illustration.

When a single active memory module is used in a dual-rank expansion system, the 2nd module socket is populated with a low-cost passive module called a continuity module. The continuity module is used to maintain the point-to-point connectivity for half of the signal traces between the active module and the memory controller, as illustrated in Fig. 7(a). The memory capacity of the system is defined by the bit capacity of the active rank. When the 2nd active module rank is added to the configuration, as shown in Fig. 7(b), the 2nd module provides the data for the second half of the data lanes to the memory controller. The 2nd active module rank replaces the continuity module. With the addition of the 2nd module rank, the active modules are reconfigured to each have twice the number of addressable locations and one-half the number of data bits as compared to the non-configured modules. The trace connection between the 1st and 2nd active modules is not used in this configuration. The total memory capacity of the system has been doubled, without compromising the high-speed, point-to-point signaling topology.

The address/command (RQ) bus in DPP is still fly-by signaling for supporting two dual-rank modules (16 DRAM devices per module). The worst case loading of each RQ bus (RQ_A and RQ_B) is the same as it would be for a full module, with the addition of two connector crossings to support the inter-module routing. In the single-module case, both RQ buses are still utilized, each addressing a portion of the module in place. As shown in the Fig. 7, RQ_A bus is first fly-by routed from the 1st active module to the continuity module (or 2nd active module) and is terminated on the continuity module (or 2nd active module). RQ_B is similarly routed from the continuity module to the first active module.

Fig. 8 shows DQ channel performance comparison for both standard multi-drop topology and DPP topologies. Each eye diagram shows the worst case configuration for a system topology supporting two dual-rank modules. For the multi-drop topology, the worst case signal integrity is observed in a system with both modules loaded. In contrast, the worst case signaling in the DPP configuration is when only one module is loaded and the signals must cross a continuity module, adding extra connector crossings, i.e. additional connector crosstalk, as shown in Fig. 6(b). With two modules loaded in a DPP configuration, the signaling improves since the electrical path is shortened and the extra connector crossing is eliminated. As can be seen in Fig. 8, the signaling eye of the multi-drop topology does not allow enough voltage and timing margin to successfully transmit data at 3200Mb/s, even with the memory channel optimized to reduce signal crosstalk and impedance mismatches, as illustrated in Section II. However, both the timing and voltage margin increases significantly for the DPP configuration vs. the multi-drop. With DPP, a two dual-rank module configuration can be supported at a data rate of 3200Mb/s.

978-1-4244-4447-2/09 $25.00 © 2009 IEEE

IV. REDUCING POWER CONSUMPTION USING NEAR GROUND SIGNALING (NGS)

The total memory system power consumption is dominated by three major components: system clocking power, IO signaling power, and DRAM core access power. In this section, to reduce IO signaling power, we introduce a new signaling scheme named near ground signaling (NGS). The proposed signaling scheme enables high data rates at significantly reduced IO signaling power and design complexity while maintaining good signal integrity. The system topology for NGS is built up as shown in Fig. 9. NGS uses push/pull (PU/PD) output drivers made up with NMOS over NMOS, and terminates to a ground rail by on-die-termination (ODT). IO driver circuits are powered from an internally regulated supply (500mV DC) to provide a low IO signal swing. IO signals swing from the ground rail (low) to $0.5V \times Z^{PU}_{DRV} / Z^{ODT}_{TERM}$ (high), where $Z^{PU}_{DRV} = 34\Omega$ and $Z^{ODT}_{TERM} = 40\Omega$.

Compared to the SSTL currently used for the DDR system assuming a 1.2V supply rail, NGS can save ~60% of IO driver power consumption. To save further power consumption, NGS can apply a data bus inversion (DBI) coding technique, which is commonly used for the GDDR system [5]. Since steady-state current is only generated during the transmission of a 'high' value, the data bit pattern is inverted whenever more than half the transmitted data bits would be 'high' at any given time. With the DBI encoding, NGS can obtain not only provide ~80% power savings, but also ~40% of SSO noise reduction from the worst case.

From the signal/power integrity point of view, there are two additional advantages in NGS. Since the device input capacitance due to ESD protection can be reduced by ~10% under the low IO signal swing system, better signal integrity is obtained. This termination scheme can also help reduce simultaneous switching output noise (SSO), as the ground rail is typically the lowest-impedance plane in a memory channel. From the power delivery system point of view, NGS can overcome the design constraint that the supply rail between the controller and DRAM should remain the same. Since NGS is signaling at near ground levels, the controller and DRAM power supplies can be selected independently. The NGS signaling concept can be also applied to Pseudo Open Drain Logic (PODL), which is used for graphic memory systems (GDDR3/4/5). However, it would require the same power supply voltage for controller and DRAM.

Fig. 10(a) shows NGS signaling performance under the worst case configuration of the DPP topology. The eye response trend is very similar to the full swing case shown in Fig. 8. NGS provides the same timing margin performance only with a smaller IO signal swing. To examine the impact of SSO noise on the voltage and timing margin, the channel model shown in Fig. 9 was simulated with ideal and non-ideal power supply models [4]. Fig. 10 shows the additional margin loss caused by the worst SSO noise effect while the channel simulation with an ideal power supply reflects crosstalk and ISI effects only. It is interesting to note that the eye opening of the open-terminated rank 1 is more significantly degraded by SSO than that of the 40Ω-terminated Rank 2. Fig. 10(b) shows SSTL signaling performance. All simulation set-ups remained the same except that ODT is connected between the power and ground rails. The eye diagram becomes slightly worse by SSO as the power rail is designed with higher impedance. In overall, SSO noise level is proportional to the IO signal swing, and its relative impact on the system margin remains the same.

IV. CONCLUSION

This paper presented three key technologies for a next generation main memory system that address the anticipated demand for higher data rate, lower power dissipation, and higher capacity. FlexPhase™ provides a timing calibration which compensates any device or trace length mismatches allowing traces to be routed without much crosstalk. Dynamic point-to-point signaling topology maintains the ideal point-to-point signaling for high data rates even with multiple memory modules. Using the continuity memory module, it also provides a user upgradable memory system. Finally, near ground signaling further reduces the overall system power consumption by using a low swing driver. The ground referencing signaling also allows that distinct power supply voltages can be used for controller and DRAM devices.

REFERENCES

[1] D. Oh, W. Kim, J.-H. Kim, J. Wilson, R. Schmitt, C. Yuan, L. Luo, J. Kizer, J. Eble, and F. Ware, "Study of signal and power integrity challenges in high-speed memory I/O designs using single-ended signaling schemes", *DesignCon*, Santa Clara, Feb. 2008.

[2] Intel Corporation, "Intel® Xeon® Processor 500 Series (Nehalem) DDR3 DIMM System-level Validation Results", http://www.intel.com/technology/memory/ddr/valid/ddr3_UDIMM_RDIMM_results.htm.

[3] M. Ching, "Beyond DDR3: advancing the main memory roadmap", *MemCon09*, Jun., 2009.

[4] J.-H. Kim, W. Kim, D. Oh, R. Schmitt, J. Feng, C. Yuan, L. Luo, and J. Wilson, "Performance impact of simultaneous switching output noise on graphic memory systems," *IEEE EPEP '07*, pp. 197-200, Oct. 2007.

[5] J. Ihm, *et al.*, "An 80nm 4Gb/s/pin 32bit 512Mb GDDR4 graphics DRAM with low-power and low-noise data-bus inversion", *ISSCC Dig. Tech.* papers, pp. 492-493, Feb., 2007.

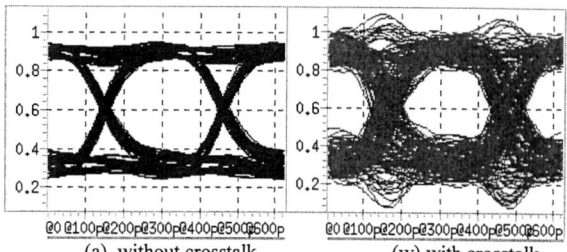

(a) without crosstalk   (w) with crosstalk

Figure 4. Eye diagram with trace matching in 1 DIMM system

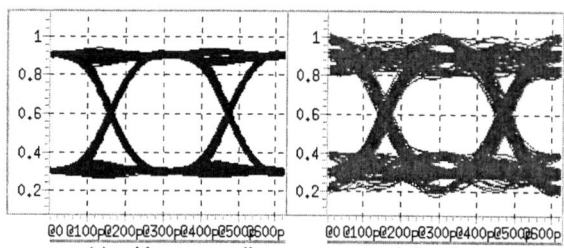

(a) without crosstalk   (w) with crosstalk

Figure 5. Eye diagram without trace matching in 1 DIMM system (FlexPhase™).

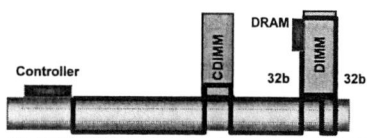

(a) Typical 2-DIMM multi-drop system

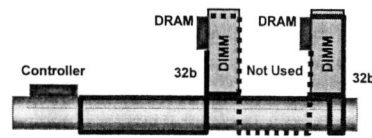

(b) DPP with CDIMM and one active DIMM

(c) DPP with two active DIMMs

Figure 6. 2-DIMM multi-drop and DPP systems.

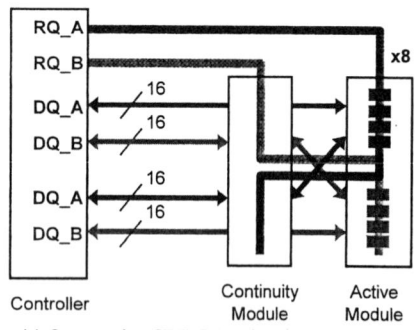

(a) One passive CDIMM and one active DIMM

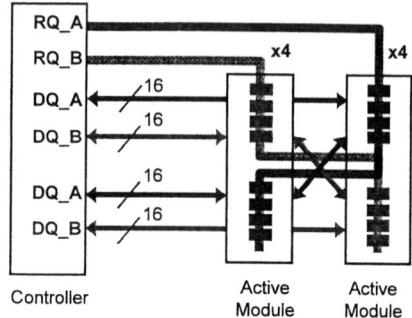

(b) Two active DIMMs

Figure 7. 2-DIMM DPP Configuration for 3200Mb/s.

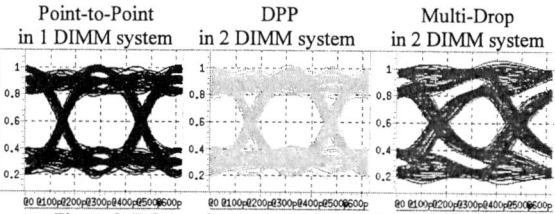

Figure 8. DQ eye diagram comparison for 3200Mb/s.

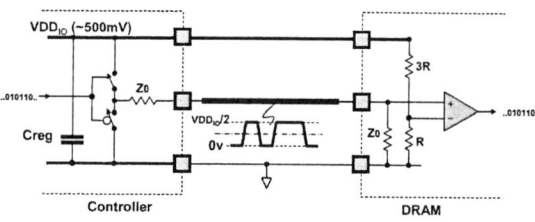

Figure 9. Near ground signaling (NGS).

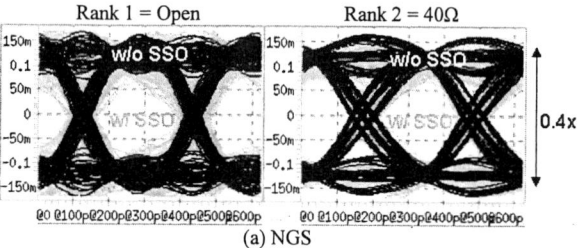

(a) NGS

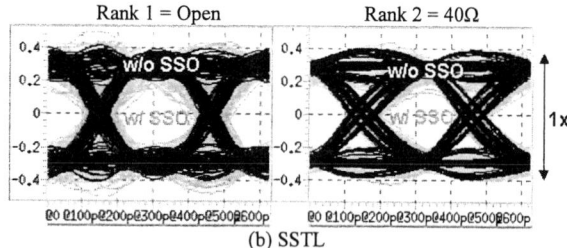

(b) SSTL

Figure 10. Signaling performance and SSO impact at 3200Mb/s.

Active Circuit to Through Silicon Via (TSV) Noise Coupling

[1]Jonghyun Cho, [1]Jongjoo Shim, [1]Eakhwan Song, [1]Jun So Pak, [2]Junho Lee, [2]Hyungdong Lee, [2]Kunwoo Park and [1]Joungho Kim

[1]Terahertz Interconnection and Package Laboratory, Division of Electrical Engineering,
School of Electrical Engineering and Computer Science, KAIST, Daejeon, South Korea
[1]Phone:+82-42-879-9870, Fax:82-42-869-8058, E-mail : jonghyun@eeinfo.kaist.ac.kr
[2]Advanced Design Team, Hynix Semiconductor Inc. Icheon Kyoungki-do, South Korea
[2]E-mail : Junho.lee@hynix.com, Hyungdong.lee@hynix.com, kunwoo.park@hynix.com

Abstract – **In this paper, we propose a coupling model between through silicon via (TSV) and substrate based on a 3-Dimensional transmission line matrix (3D-TLM), which utilizes equivalent lumped circuit model of silicon substrate and TSV. The proposed model is verified by S-parameter simulations using a 3D field solver and analyzed with various structural parameters: TSV diameter, distance between TSV and noise source, and silicon substrate height. Based on the model, timing jitter degradation on phase locked loop (PLL) caused by substrate noise coupling is investigated. A shielding technique using a guard ring structure is applied to suppress the coupling noise.**

I. INTRODUCTION

Currently, conventional 2-dimensional system in package (2D-SiP) must be modified to achieve low cost, small area and high performance. In 2D-SiP, a few chips are located on a package substrate with long wire bonding has been used. Consequently, though chip speed has increased, the conventional packaging technology is not available for high-speed digital signals [1]. In recent years, 3D technology has been exploited in order to resolve the interconnection problem. The through silicon via (TSV) process has become one of the most significant breakthrough in 3D technology that achieves a significantly shorter interconnection compared to the conventional interconnects using wire-bonds [2].

Fig.1 shows a TSV structure fabricated by a via-first TSV process. For DC isolation between the TSV and the highly conductive silicon substrate, an SiO₂ insulation layer is formed surround the TSV structure. The insulation layer has a thickness of a few hundredths of a nanometer, which results characteristic high capacitance between the TSV and the silicon substrate. Therefore, high frequency noise in silicon substrate can be coupled to the TSV through the large capacitance [3]. The coupled noise propagates through the TSV and corrupts the circuits connected to the TSV which degrades circuit performance as illustrated in Fig.1. In a traditional 2D system, several noise isolation techniques have been introduced: split power planes, the deep-nwell process, and the guard ring structure. The techniques have been proven for 2D noise isolation, but they are problematic if the TSV structure is connected for 3D integration.

In this paper, we propose a model of the noise coupling effect between TSV and the substrate by combining an equivalent circuit model for TSV and a 3D-TLM model for silicon substrate. The proposed model is verified by coupling simulation using a 3D field solver. Based on the model, we investigated the coupling noise from the silicon substrate to TSV and applied a shielding technique using a guard ring structure to the TSV for the noise isolation. The benefit of the isolation is evident in the jitter performance of a practical PLL circuit.

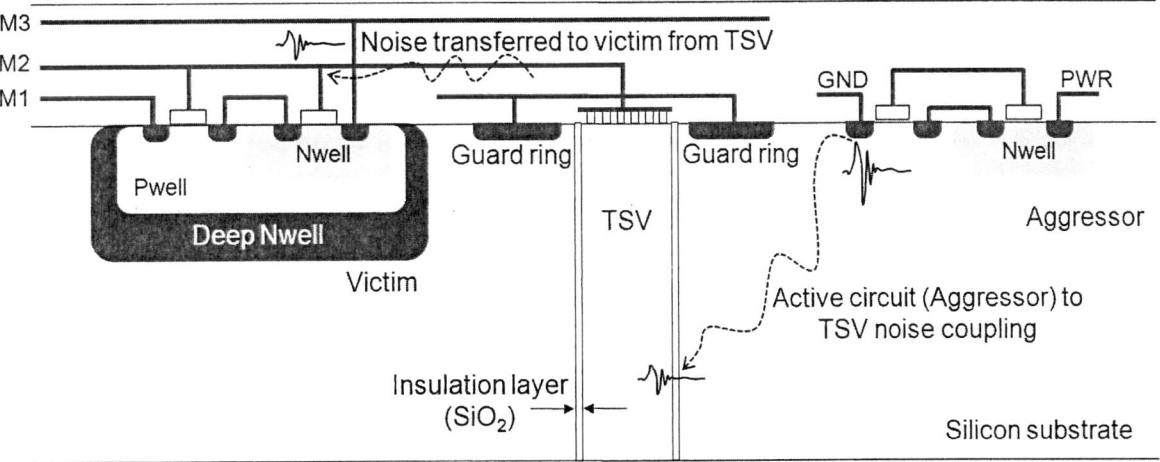

Fig.1. Active circuit to TSV coupling through the silicon substrate

II. THE PROPOSED MODEL FOR SILICON SUBSTRATE TO TSV NOISE COUPLING

3D-TLM is a modeling method that divides the entire system into many 3D unit cells. This modeling method enables each cell to be modeled as lumped elements and connects all the unit cells in order to estimate the entire system performance. Each unit cell should be smaller than the

978-1-4244-4447-2/09 $25.00 © 2009 IEEE

wave length of interest divided by 20 for lumped modeling. Furthermore each unit cell must be modeled accurately based on numerical or experimental equations. To model the coupling, we need silicon substrate and TSV equivalent circuit models.

For TSV, we adopted an equivalent circuit model of GSG type TSV [4]. From this circuit model, we only used the lumped circuit model of TSV. A TSV unit cell model is illustrated in Fig.2-(a). The R, L and C parameter values of TSV in Fig.2-(a) can be calculated by equations (1a), (1b) and (1c) [5], where r_{TSV}, h_{TSV}, t_{oxide} and p_{TSV} are the radius of TSV, the height of substrate unit cell, thickness of SiO$_2$, and pitch between ground and signal TSV's, respectively. For the substrate, we used the 3D-TLM unit cell model illustrated in Fig.2-(b), which is composed of R and C networks [6] that can be calculated using equations (2a)-(2f), where w_{SUB}, l_{SUB} and h_{SUB} are the width, length and height of the substrate unit cell, respectively.

$$R_{TSV} = \frac{h_{TSV}}{2\pi \sigma_{TSV} r_{TSV}^2} \quad (1a) \qquad\qquad C_{TSV} = \frac{\pi \varepsilon_{OX} h_{TSV}}{2\ln\left(\frac{r_{TSV}+t_{oxide}}{r_{TSV}}\right)} \quad (1b)$$

$$L_{TSV} = \frac{\mu_0 h_{TSV}}{4\pi}\left[\left\{\ln\left(\frac{h_{TSV}}{r_{TSV}}+\sqrt{\left(\frac{h_{TSV}}{r_{TSV}}\right)^2+1}\right)+\frac{r_{TSV}}{h_{TSV}}-\sqrt{\left(\frac{r_{TSV}}{h_{TSV}}\right)^2+1}\right\}-\left\{\ln\left(\frac{h_{TSV}}{p_{TSV}}+\sqrt{\left(\frac{h_{TSV}}{p_{TSV}}\right)^2+1}\right)+\frac{p_{TSV}}{h_{TSV}}-\sqrt{\left(\frac{p_{TSV}}{h_{TSV}}\right)^2+1}\right\}\right] \quad (1c)$$

$$R_{SUB\_1} = \frac{1}{2\rho}\frac{h_{SUB}}{w_{SUB} l_{SUB}} \quad (2a) \qquad C_{SUB\_1} = 2\varepsilon\frac{w_{SUB} l_{SUB}}{h_{SUB}} \quad (2b) \qquad R_{SUB\_2} = \frac{1}{2\rho}\frac{w_{SUB}}{h_{SUB} l_{SUB}} \quad (2c)$$

$$C_{SUB\_2} = 2\varepsilon\frac{h_{SUB} l_{SUB}}{w_{SUB}} \quad (2d) \qquad R_{SUB\_3} = \frac{1}{2\rho}\frac{l_{SUB}}{w_{SUB} h_{SUB}} \quad (2e) \qquad C_{SUB\_3} = 2\varepsilon\frac{w_{SUB} h_{SUB}}{l_{SUB}} \quad (2f)$$

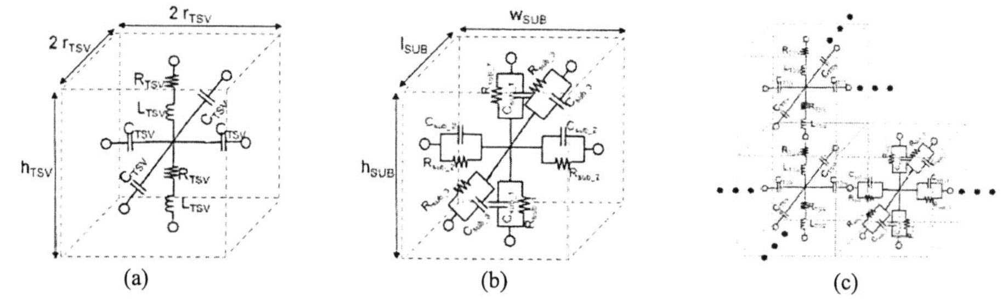

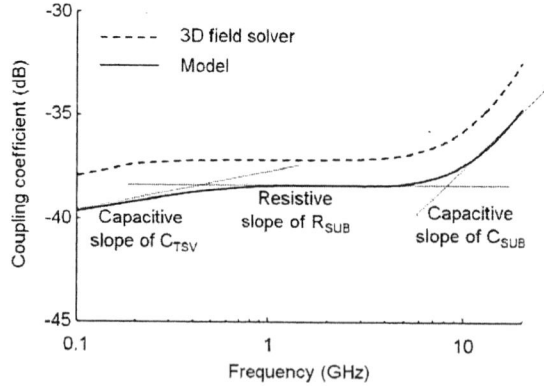

Fig.2. Unit cell of the equivalent circuit model for (a) TSV, (b) silicon substrate, and (c) a combination of TSV and silicon substrate

As shown in Fig.3, the coupling coefficients from the proposed model and the 3D field solver have good correlation. A difference between coupling coefficient of model and 3D field solver results is caused by the lack of unit cell number. The proposed model and 3D field solver simulation was performed on the environment of 100 μm substrate height, 30 μm TSV diameter, and 120 μm distance between noise source and TSV. In low frequency range that is less than 1 GHz, the capacitance of SiO$_2$ surrounding TSV (C_{TSV}) determines the amount of coupling. The coupling coefficient increases as frequency increases. In the mid-frequency range from 1 GHz to 10 GHz, the coupling coefficient remains almost unchanged, which is evidence that the substrate resistor (R_{SUB}) mainly contributes to coupling. In high frequency range over 10 GHz, the coupling coefficient increases as frequency increases. In this range, the substrate capacitor of substrate unit cell (C_{SUB}) mainly determines the amount of coupling.

Fig.3. Coupling coefficient between substrate noise source and TSV

III. Degradation of Phase Locked Loop (PLL) Performance of the Silicon Substrate to TSV Noise Coupling

Coupling through the silicon substrate has been an issue in traditional 2D systems, especially analog and digital mixed-mode systems. A number of models for the coupling effect and the isolation techniques have been introduced [6], [7]. Among many solutions, deep-nwell is the most powerful method to prevent coupling through the substrate. With deep-nwell, substrate coupling level was lowered by -70dB, up to 20GHz [7]. Therefore, the deep-nwell is generally used when designing a noise-sensitive circuit, e.g., PLL.

However, a conventional deep-nwell structure does not provide a sufficient isolation level in 3D-IC. Fig.1 illustrates the coupling issues in 3D-IC. The circuit itself can be isolated by the deep-nwell method, but power, ground, and I/O signals are transferred by TSVs in 3D-IC; this TSV cannot be isolated using the deep-nwell method. The coupled noise from the substrate to TSV degrades circuit performance. For PLL circuit in 3D-IC, TSV is used as an external route for the reference clock, which is the most important signal for PLL operation. Based on the proposed model for substrate noise coupling to TSV, we can estimate the substrate coupling noise. Fig.4 illustrates the coupled noise with various design parameters: substrate height, coupling distance, and TSV diameter. We modeled the TSV input driver as a 100 Ω resistor, the TSV output driver as a 200 fF capacitor and substrate noise source with a 1.6-GHz and peak-to-peak 200 mV clock. The coupled noise is highly dependent on the substrate thickness and the distance between the TSV and noise source, while the TSV diameter has no significant effect on the coupled noise. However, the coupling noise is almost constant when the substrate height is greater than 70 μm.

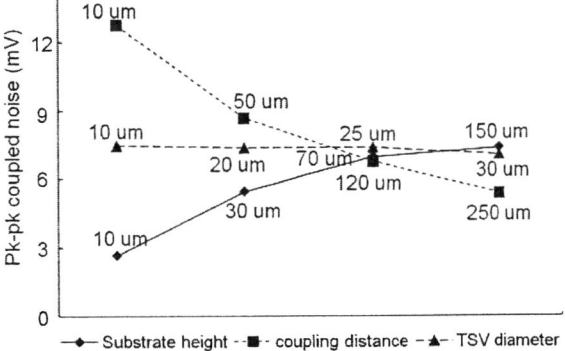

Fig.4. The peak-to-peak voltage of coupled noise depending on the parameters of substrate and TSV.

We designed a 2.4 GHz operating PLL using a TSV structure as an external route for the reference clock signal with a 150-MHz frequency and 1.8 V amplitude. The reference clock is corrupted by the substrate noise coupled from the noise source described above. PLL jitter simulation including this external clock and substrate noise is performed with HSPICE, and the results are illustrated in Fig.5. The peak-to-peak jitter is increased from 7.2 ps to 9.4 ps, 30.5% increased the peak-to-peak jitter from intrinsic status, due to coupled noise from the silicon substrate to the TSV.

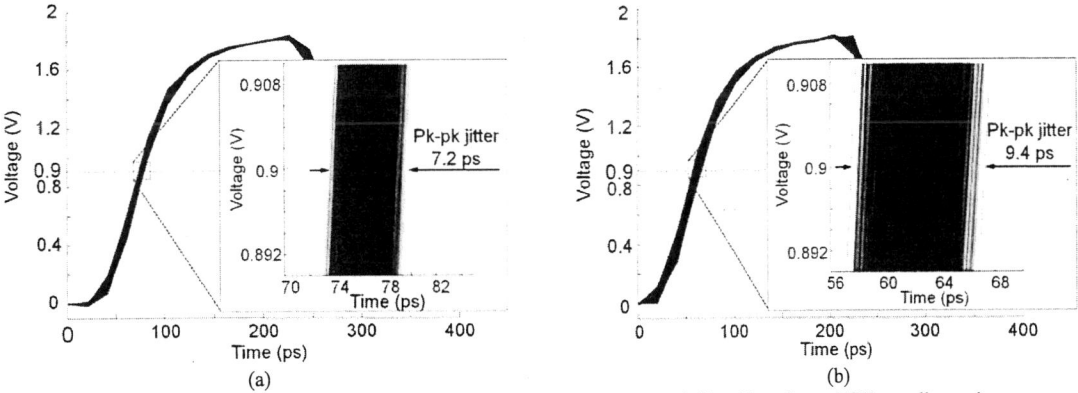

Fig.5. Simulated peak-to-peak jitters of PLL (a) without substrate noise, and (b) with active to TSV coupling noise

IV. A Guard Ring Structure to Suppress Silicon-Substrate-to-TSV Coupling Noise

In the previous chapter, we showed that silicon-substrate-to-TSV coupling degraded PLL jitter performance. For better performance in a 3D-IC environment, substrate noise suppression is needed. A guard ring is the most popular solution to the substrate coupling problem. In order to apply a guard ring structure to 3D-IC, guard ring structure surrounding victim TSV is used. The coupling effect between the substrate and the TSV is investigated by varying the distances between the TSV and guard ring. The simulation environment is the same as the one of Fig.3 except one thing that we used guard ring structure having 10 μm width and 10 - 30 μm distance from TSV. The results are illustrated in Fig.6

For a 2D system, guard rings with a smaller distance to the victim circuit achieve better isolation performance [7]. However, simulation shows the opposite result for 3D-IC. Guard rings with a greater distance to the victim TSV achieve better isolation performance. This means that the guard ring near TSV is less beneficial because the height of the TSV is a few hundred times larger than the thickness of guard ring. If the guard ring is located near the TSV, coupling to the lower part of the TSV cannot be blocked because of the great height of the TSV and the thin guard ring. However if the guard ring is located far from the TSV, which means it is a shorter from the noise source, the coupling noise to the lower part of TSV can be blocked because the noise source and guard ring have the same thickness.

Fig.7 shows the peak-to-peak jitters on PLL induced by the substrate-to-TSV coupling noise. The peak-to-peak jitter is decreased from 9.4 ps to 7.8 ps, with a guard ring structure that is 10 μm wide and 30 μm away from the TSV. The guard ring structure led to a 17.0 % of jitter improvement. The guard ring near TSV caused some substrate noise reduction. For greater isolation, guard ring should be located farther from the victim TSV, and this means that a large area is needed for TSV isolation.

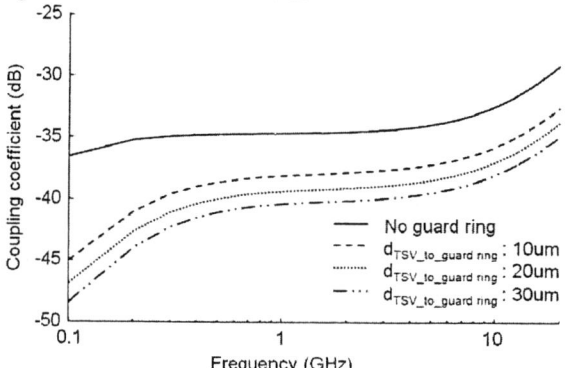

Fig.6. Active to TSV coupling depending on the TSV to guard ring distance

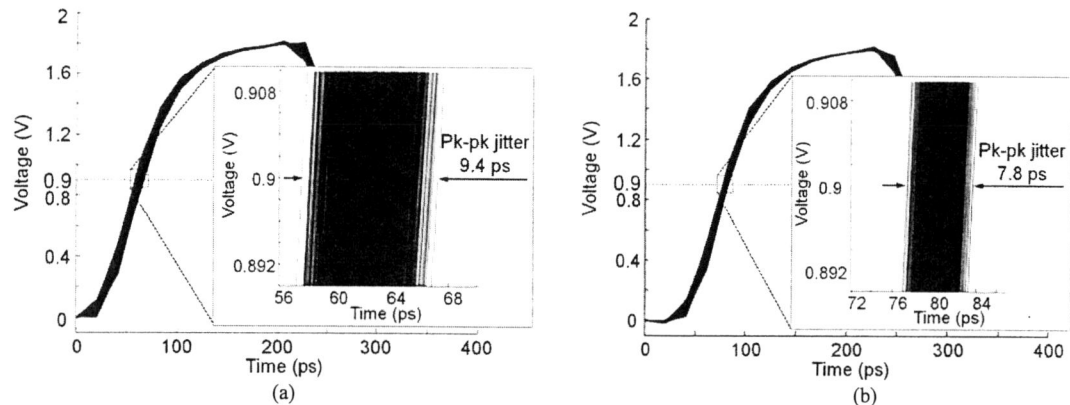

Fig.7. Simulated peak-to-peak jitters of PLL (a) without a guard ring structure, and (b) with a guard ring structure

V. CONCLUSION

The TSV is an essential structure for a 3D integration system, but has low immunity to substrate noise coupling, which results in degradation of the RF and analog circuit performance. In this paper, we proposed a coupling model between the TSV and substrate by combining an equivalent circuit model for TSV with a 3D-TLM model for silicon substrate. The proposed model was verified with S-parameter simulation using a 3D field solver. Based on the model, we investigated the effect of the substrate coupling noise on the jitter performance of a practical PLL circuit, and applied a shielding technique using a guard ring structure for the noise isolation. With the noise isolation technique, the jitter performance of the PLL was improved by 17 %.

REFERENCES

[1] J. Pak, C. Ryu, J. Kim, "Electrical Characterization of Through Silicon Via (TSV) depending on Structural and Material Parameters based on 3D Full Wave Simulation", *Electromagnetic Electronic Materials and Packaging, 2007.Conf.*, May.2008 pp.351-354

[2] J. U. Knickerbocker, P. S. Andry, B. Dang, R. R. Horton, M. J. Interrante, C. S. Patel, R. J. Polastre, K. Sakuma, R. Sirdeshmukh, E. J. Sprogis, S. M. Sri-Jayantha, A. M. Stephens, A. W. Topol, C. K. Tsang, B.C. Webb, and S. L. Wright, "Three dimensional silicon integration", IBM J. Res. & Dev. 52, 553-569,(2008).

[3] M. Rousseau, O. Rozeau, G. Cibrario, G. Le Carval, M. Jaud, P. Leduc,A. Farcy, and A. Marty, "Through-silicon via based 3D IC technology: Electrostatic simulations for design methodology," ,*JMAPS Device Packaging Conf.2008*, 2008.

[4] C. Ryu, J. Lee, K. Lee, T. Oh, J. Kim, "High Frequency Electrical Model of Through Wafer Via for 3-D Stacked Chip Packaging", *Electronics System-integration Technology Conf.*,Sept.2006,vol 1,pp.215-220.

[5] F. B. J. Leferink, "Inductance calculations: methods and equations," in Proc. *IEEE Int. Symp. Electromag. Compat.*, Atlanta, GA, 1995, pp.16–22.

[6] Afzali-Kusha. A, Nagata. M, Verghese. N.K, Allstot. D.J, "Substrate Noise Coupling in SoC Design : Modeling, Avoidance, and Validation", *Proceedings of the IEEE.*,Dec.2006,vol 94,pp.2109-2138

[7] A. Helmy and M. Ismail, "A Design Guide for Reducing Substrate Noise Coupling in RF Applications", *IEEE Circuits and Devices Magazine*, Oct.2006

978-1-4244-4447-2/09 $25.00 © 2009 IEEE

Experimental Characterization of Metal Fill Placement and Size Impact on Spiral Inductors

Vikas S. Shilimkar, Steven G. Gaskill and Andreas Weisshaar

School of Electrical Engineering and Computer Science,
Oregon State University, Corvallis, OR 97331, USA.
{shilimvi, gaskill, andreas} @eecs.oregonstate.edu

Abstract—**This paper experimentally studies the parasitic impact of metal fill on a spiral inductor fabricated in a 180nm Bi-CMOS process. Metal fill design aspects such as placement (in-plane vs. off-plane) and size are considered. We demonstrate larger impact on the quality factor (Q) and self-resonance frequency for off-plane metal fill compared to in-plane metal fill. A 70% decrease in fill size with unchanged metal density gives a 13% improvement in measured Q.**

Index Terms—**CMP, metal fill, spiral inductors, transmission lines**

I. INTRODUCTION

ADVANCED IC manufacturing processes employ Chemical Mechanical Polishing (CMP) to planarize metal and dielectric layers [1]. CMP alone provides good local uniformity, but does not guarantee global uniformity. However, CMP in conjunction with a metal density rule improves global uniformity. The metal density rule requires adding dummy metal fill structures in areas of low metal density [2]. The additional metal fill results in parasitic capacitance and added loss in passive components and cause or worsen mismatches in critical circuit blocks.

The metal density design rules for RF/analog-mixed signal (AMS) systems processes are not as aggressive as for the smaller process nodes for digital designs. In system-on-chip (SoC) designs, the RF/AMS blocks and voltage regulators are designed on the same die as digital blocks. Also, AMS designs are trending towards smaller process nodes. Hence, it is important to consider the impact of the metal fills on RF/analog passive components, and, in particular, on spiral inductors. Most of the research done in this area focuses on analyzing the impact on the inductance (L), quality factor (Q) and self-resonance frequency (f_r) of the device [3-5]. Tsuchiya et al. [3] show the effects of metal fills and structures to reduce the impact of in-plane metal fills, i.e., on fills added in the same layer as the spiral inductor coil. The proposed patterned floating dummy fill (PFD) technique, which is similar to patterned ground shields (PGS), reduces the impact on Q. Pastore et al. [4] analyzed the impact of metal fill on spiral inductor performance by varying the buffer distance and metal fill density, and misaligning the metal fills. Tiemeijer et al. [5] provide a compact model for spiral inductors including metal fill and suggest smallest fill to be best for reducing eddy losses.

In this paper, we experimentally investigate the impact of placement and size of metal fills on spiral inductors. We first consider a section of transmission line and demonstrate the impact of metal fill on distributed capacitance and resistance as function of frequency. We then extend the work to a measurement-based investigation of spiral inductors. We compare the significance of in-plane and off-plane metal fill parasitics. We also study the impact of metal fill size in terms of additional capacitance and loss in the transmission line and spiral inductor and experimentally demonstrate reduced impact of smaller metal fills, as suggested in [5].

II. TRANSMISSION LINE TEST CASE

Before analyzing the complex spiral inductor structure, we study a section of the structure in terms of a transmission line by fullwave simulation with the commercial simulator HFSS [6]. Figure 1(a) shows a single turn spiral inductor, and Figures 1(b) and 1(c) show a zoomed-in section of the coil. Figure 1(b) shows the arrangement of metal fills in the same layer as that of the line (M6). The fills are separated by buffer distance, B (2μm in this case). Figure 1(c) shows the arrangement for off-plane metal fills. The inter-layer dielectric thickness between the line (M6) and off-plane metal fill (M5) is 2μm. In both cases the vertical and horizontal separation between the line and metal fill are kept the same. Four cases are compared with the no metal fills case: (i) in-plane nominal size (9μm x 9μm), (ii) off-plane nominal size (9μm x 9μm), (iii) in-plane small size (3μm x

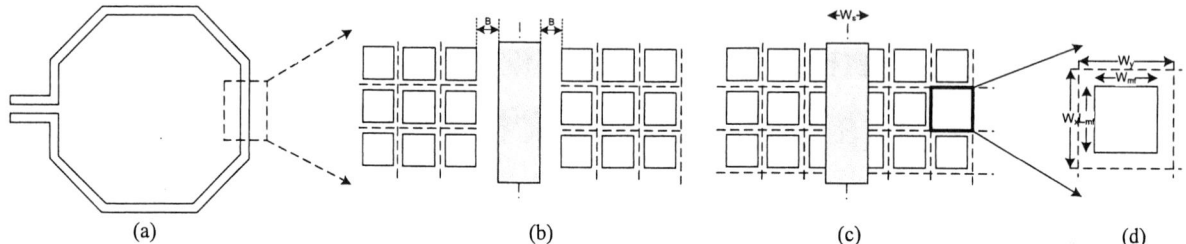

Fig. 1. HFSS simulation setup (a) one turn octagonal spiral inductor; (b) in-plane metal fills, B: buffer distance; (c) off-plane metal fills, W_s: width of the line (5μm); (d) individual metal fill cell, W_{mf}/L_{mf}: width/length of the metal fill, W_x/W_y: x and y lengths of one unit cell

3μm), (iv) off-plane small size (3μm x 3μm). In all cases, the metal fill density is kept constant at 50%. Figure 2 shows the fullwave simulation results obtained with HFFS [6], separated into the capacitive and magnetic impact. It can be seen from the

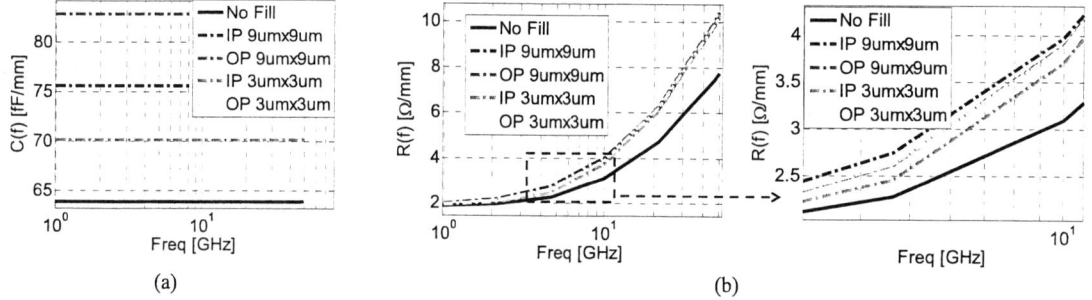

Fig. 2. HFSS simulations results to demonstrate the effects of metal fill placement and size (a) per unit length capacitance and (b) resistance. IP: in-plane; OP: off-plane

results that the off-plane fills have a larger impact on the capacitance as well as resistance. This is because the off-plane metal fills are directly underneath the line and cover a larger electric flux area as compared to the in-plane fills. In the case of smaller fills, the eddy current loss and parasitic capacitance are reduced. The same results were found for wider lines.

III. SPIRAL INDUCTOR FABRICATION

The findings for the canonical transmission line structure are extended to the more complex spiral inductor geometry. Several inductor test structures have been fabricated in a Jazz 0.18μm process [7] to investigate the impact of in-plane metal fill vs. off-plane metal fills and metal fill size. Table I summarizes the electrical and physical design specifications of the device fabricated.

TABLE I
SPIRAL INDUCTOR AND METAL FILL DESIGN

Symbol	Quantity	Specification
L_{DC}	DC inductance	2nH
Freq.	Operating range	1-30GHz
R_o	Outer dimension	150μm x 150μm
R_i	Inner opening	56μm
W	Coil width	6μm
S	Coil separation	2.1μm
N	Number of turns	3.5
W_{mf}	Metal fill size	5μm, 2.74μm
D	Metal fill density	30%
KOZ	Keep out zone	20μm

The coil is placed on the top metal layer (M6), ground connection for all the test pads is on the bottom layer (M1), and metal fill is placed on M6-M2, as shown in Figure 3d. The spiral inductor device includes a patterned ground shield (PGS). A minimum 20μm keep out zone (KOZ) is maintained between the outer edge of the spiral coil and the metal fill. Five different spiral inductor test cases are considered: (i) no metal fill (Figure 3(a)), (ii) in-plane metal fill (Figure 3(b)), (iii) off-plane metal fill (Figure 3(c)), (iv) in- and off-plane metal fill, and (v) smaller size metal fill.

The device measurements were de-embedded by removing the probe pad shunt admittance and the series impedance of the interconnect connecting the probe pads and device terminal. We used two de-embedding structures: open and through. The probe parasitics can be de-embedded from the measured results by (1) assuming the through structure to be symmetric.

$$\left[Z_{dut} \right] = \left[Y_{meas} - Y_{open} \right]^{-1} - \frac{1}{2}\left(\frac{-1}{Y_{thru(1,2)}} \right) \tag{1}$$

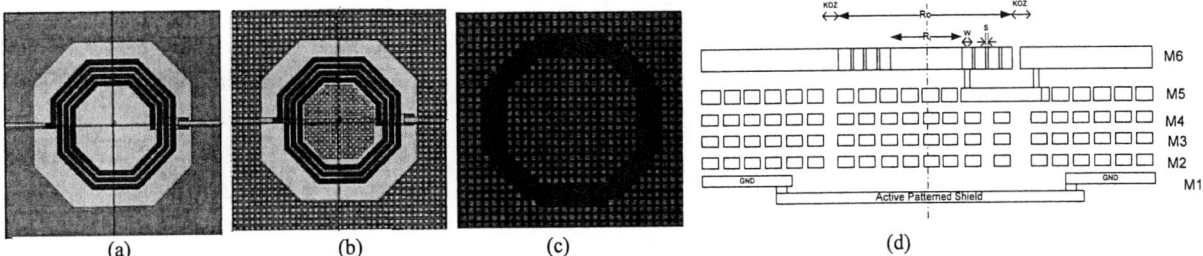

(a)   (b)   (c)   (d)

Fig. 3. Spiral inductor geometry with (a) no fills, (b) in-plane fills, (c) off-plane metal fills. (d) Cross section of the structure with in-plane and off-plane fills W: coil width, S: coil separation, R_i: inner opening, R_o: outer dimension, KOZ: keep out zone.

IV. SPIRAL INDUCTOR π MODEL

Figure 4 shows a simple π model, which is used to separate the electric and magnetic impact of metal fills on the spiral inductor. L_s is the inductance of the coil, R_s is the series resistance, which accounts for DC and high frequency eddy current losses, C_s represents the inter-winding capacitance, and C_p and G_p are shunt oxide capacitance and conductance, respectively. As metal fill is inserted, C_p increases reflecting the electric impact, and R_s increases reflecting the magnetic loss impact. The single-ended port 1 L and Q characteristics are calculated from the measured Y-parameters, as shown in (2) and (3).

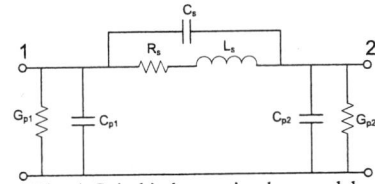

Fig. 4. Spiral inductor simple π model

To discriminate the electric and magnetic effects, we calculate Rs and Cp

$$L = \frac{\text{Im}\{1/Y_{11}\}}{\omega} \qquad (2)$$

$$Q = -\frac{\text{Im}\{Y_{11}\}}{\text{Re}\{Y_{11}\}} \qquad (3)$$

for each of the test case. From the π-model, it is straight forward to calculate the shunt elements, Cp and Gp as shown in (4) and (5),

$$C_{p1} = \frac{\text{Im}\{Y_{11} + Y_{12}\}}{\omega} \qquad (4)$$

$$G_{p1} = \text{Re}\{Y_{11} + Y_{12}\} \qquad (5)$$

The series Rs and Ls parameters are affected by the skin and proximity effects at higher frequencies. These effects lead to an increase in R_s and a decrease in L_s with rising frequency. To first order, examining the real part of $-1/Y_{12}$ indicates that the inter-winding capacitance (C_s) does not show a significant effect until higher frequencies. Hence, C_s can be ignored to first order and R_s and L_s are directly obtained from $-1/Y_{12}$ as

$$R_s = \text{Re}\left\{\frac{-1}{Y_{12}}\right\} \qquad (6)$$

$$L_s = \frac{\text{Im}\{-1/Y_{12}\}}{\omega} \qquad (7)$$

V. MEASUREMENT RESULTS

The two-port VNA measurements were de-embedded and the π model elements were extracted using (1)-(7) as described in the previous sections. Figure 5 shows a comparison of the impact of in-plane vs. off-plane metal fills. Figures 5(a) shows the

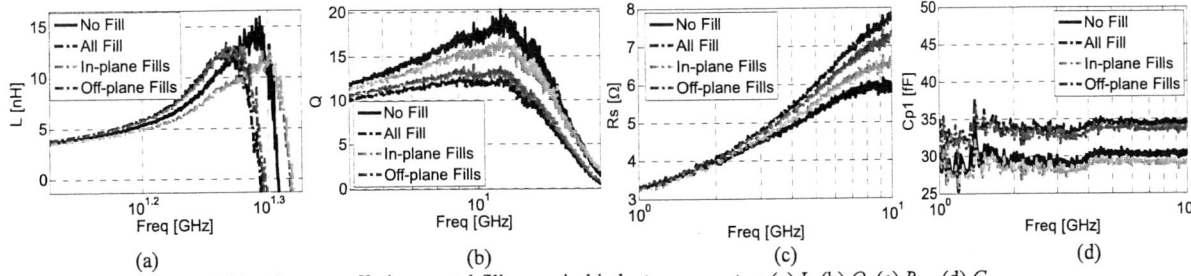

(a)   (b)   (c)   (d)

Fig. 5. Effect of metal fill in-plane vs. off-plane metal fills on spiral inductor parameters (a) L, (b) Q, (c) R_s, (d) C_{p1}.

impact on inductance. It can be seen that the peak L frequency and self-resonance frequency change based on the metal fill placement. The self-resonance frequency reduces from 20.3 GHz to 20.0 GHz with off-plane metal fill and increases to 21.0 GHz with in-plane metal fill, as expected due to the 11.6% increase and 3.6% decrease in capacitance C_{p1}, respectively (see

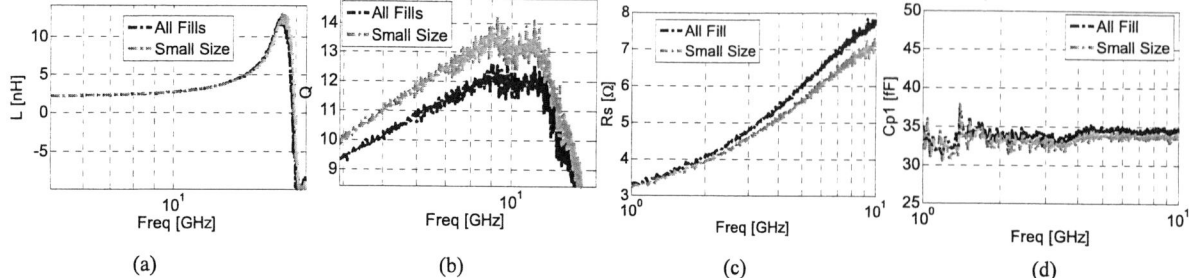

(a) (b) (c) (d)

Fig. 6. Effect of metal fill on parasitic parameters R_s and C_{pl}: comparison of off-plane vs. in-plane (a) R_s, (b) C_p; performance improvement using smaller fill size (c) R_s, (d) C_{pl}.

Figure 5(d)). The reduction in Q is larger in the case of off-plane metal fill as compared to in-plane fill due to larger parasitic resistance, R_s (see Figure 5(c)) as well as larger parasitic capacitance, C_{pl}. The larger parasitic effect could be explained as follows: (i) The vicinity to the coil: the off-plane fills on M5 are just 2µm in vertical direction from the coil, (ii) overlap area: as can be seen in Figure 3(d), the off-plane fills cover more area under the coil, (iii) larger number: the off-plane fills are placed on M5-M2, and hence, are greater in number as compared to the in-plane fills. Figure 6 shows the effect of metal fill size on spiral inductor performance. As explained in Section II, the smaller fills reduce the parasitic impact. Here, instead of 5µm x 5µm fill cells, 2.74µm x 2.74µm fills (30% fill area) are used while keeping the metal density the same. Although the self-resonance and peak L frequencies do not change much because of the smaller reduction in capacitance (from 13.9% to 11.3%), the smaller fills improve the peak Q from 11.8 to 13.4. This is mainly because of the reduction in resistance from 32.9% to 22.1% (shown in Figure 6(c)).

TABLE II
SUMMARY OF METAL FILL EFFECT ON SPIRAL INDUCTOR PARAMETERS

Test Case	Q_{peak}	f_r	f_{Qmax}	ΔC_{pl}	ΔR_s
No Fill	18.9	20.3 GHz	11.3 GHz	0%	0%
In-plane	16.3	21.0 GHz	11.4 GHz	-3.6%	13.6%
Off-plane	13.4	20.0 GHz	11.5 GHz	11.6%	24.5%
All fills	11.8	19.6 GHz	11.5 GHz	13.9%	32.9%
Small fills	13.4	19.9 GHz	11.5 GHz	11.3%	22.1%

Table II summarizes the results at 10 GHz. The small metal fill improves the series resistance more than the shunt capacitance. As can be seen from the summary table, ΔR_s changes from 32.9% to 22.1% compared to ΔC_p changing from 13.9% to 11.3%.

VI. CONCLUSION

We have experimentally investigated the impact of metal fill on spiral inductors in terms of quality factor and self-resonance frequency. The electric and magnetic parasitic effects were separated using a simple π model of the spiral inductor. We have demonstrated that the impact of off-plane fills on the spiral inductor characteristics is larger compared to in-plane metal fills. Hence, inductor designs as well as mitigation techniques to reduce the metal fill impact should foremost consider the impact of off-plane metal fill. A simple mitigation technique suggested by other researchers was demonstrated experimentally. A 70% reduction in fill size while keeping the metal density constant improved Q by about 13%. It was also noticed that the smaller metal fill size reduces magnetic loss more than parasitic capacitance, as expected.

ACKNOWLEDGMENT

This research was supported in part by the Center for Design of Analog-Digital Integrated Circuits (CDADIC). In addition, A. Weisshaar acknowledges the support through the Independent Research/Development (IR/D) Program while working at the National Science Foundation (NSF). The findings do not necessarily reflect the views of the NSF.

REFERENCES

[1] M. Fury, *"Emerging developments in CMP for semiconductor planarization,"* Solid State Technol., vol. 38, no. 5, pp 47-54, April 1995.

[2] B. Stine et al., *"The Physical and Electrical Effects of Metal-Fill Patterning Practices for Oxide Chemical-Mechanical Polishing Processes,"* IEEE Trans. Electron Devices, vol. 45, no. 3, pp 665-679, March 1998

[3] A. Tsuchiya, and H. Onodera, *"Patterned Floating Dummy Fill for On Chip Spiral Inductor Considering the Effect of Dummy Fill,"* IEEE Trans. Microwave Theory And Techniques, Vol. 56, No. 12, Dec. 2008.

[4] C. Pastore, F. Gianesello, D. Gloria, E. Serret, and P. Benech, *"Innovative and Complete Dummy Filling Strategy for RF Inductors Integrated in an Advanced Copper BEOL,"* Proceedings of the 3rd European Microwave Integrated Circuits Conference, October 2008, Amsterdam, The Netherlands.

[5] L.F. Tiemeijer, R. J. Havens, Y. Bouttement, and H. J. Pranger, *"Physics-Based Wideband Predictive Compact Model for Inductors With High Amounts of Dummy Metal Fill"*, IEEE Trans. Microwave Theory and Techniques, vol. 54, no. 8, pp. 3378-3386, Aug. 2006.

[6] Ansys Corporation, Pittsburg, MA.

[7] Jazz Semiconductors, Newport Beach, CA, USA.

Power Integrity Optimization of 3D Chips Stacked Through TSVs

Waqar Ahmad, Li-Rong Zheng, Roshan Weerasekera, Qiang Chen, Awet Yemane Weldezion, Hannu Tenhunen

Department of Electronics, Computer and Software Systems,
KTH School of Information and Communication Technologies,
Forum 120, 164 40 Kista, Sweden.
Email: {ahmadw, lirong, roshan, qiangch, aywe, hannu }@kth.se

Ph: +4687904259, Fax: +4687511793

Abstract: On-chip power distribution network model for simultaneous switching of 3D ICs stacked through TSVs to choose TSV pattern, maximum number of chips in a stack and location of the decoupling capacitor for early design trade-offs.

Keywords: Power integrity, Power distribution network, Peripheral TSVs.

Introduction:

On-Chip frequency, current density and power density increases whereas physical size of logic cells and interconnects reduces. Logic cell density follows Moor's law whereas supply voltage decreases with each technology node. Stacking different chips through TSVs improves power supply integrity by reducing interconnect length, size of package and interconnect losses. Section 1 describes previous work in 3D power distribution. Section 2 shows sample physical model of three chips, supply grid interconnection of two chips and TSVs interconnecting three chips. Section 3 shows equivalent electrical model of TSV and LRC model after adding TSVs between two chips. Section 4 shows how electrical model is converted to equivalent mathematical model by extending [1]. Section 5 gives general details including physical dimensions of an example model and simulation set up. The results in fig. 3 show comparison between proposed model and equivalent spice model. Section 6 gives some of the conclusions based on simulation results and finally the references in section 7. The model is extension of [1] which has been proved to be faster and accurate by author for 2D chips through comparison with equivalent spice model. The model proposed here is applicable to any number of chips stacked through any number of TSV pairs and is easily extendable to power distribution network for specific applications such as 3D Network on Chip (NOC) architecture.

1. Previous work:

So far there is no general model for on-chip power distribution in 3D ICs on circuit level design. The reference [2] incorporated on-chip voltage regulator module and effect of on-chip inductance whereas [3] addresses only design challenges and power delivery issues in 3D chips. The reference [4] gives impact of 3D integration on power supply noise. The reference [1] is for core switching noise but easily extendable to 3D power distribution by adding TSVs to interconnect global supply grid of one chip to the other.

2. Physical Model:

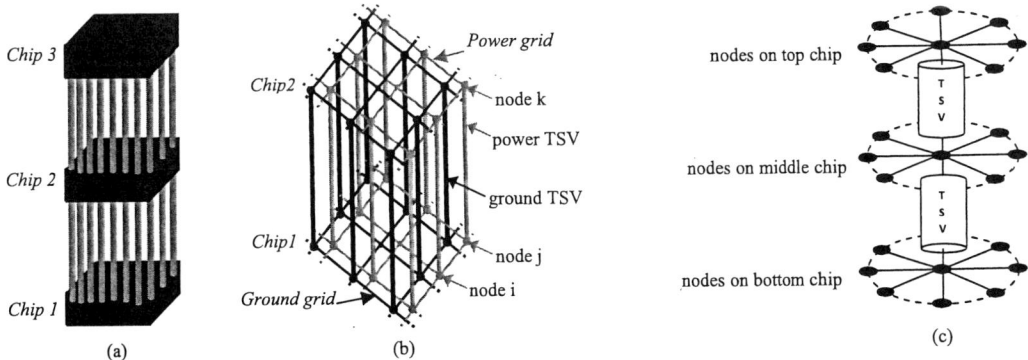

Fig. 1. *(a) Three chips in 3D stack interconnected through TSVs along periphery of the chips (b) Sample 3X3 supply grid of orthogonal layers connecting two chips through peripheral TSVs at crossing points, where i,j are two arbitrary nodes on global supply layer of chip 1 and k is arbitrary node on global supply layer of chip 2 connected to node j through TSV. Inner crossing points are connected through vias. (c) TSV interconnecting three chips where each TSV includes bump and pad pair at each side which is not shown here for the sake of simplicity.*

The author would like to acknowledge European Union research funding under grant FP7-ICT-215030
(ELITE) of the 7th framework program for the support of this work.

3. Electrical Model:

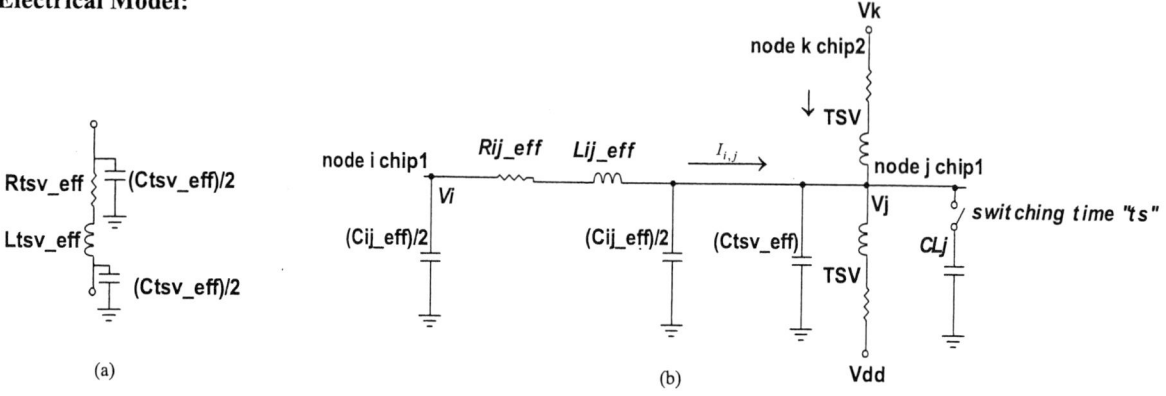

Fig. 2. *(a) Effective Pi-Model of TSV pair. (b) Effective RLC Model of node j on chip1 connected to node i on same chip and node k on chip2 where Vdd is supply voltage and CL$_j$ is switching load connected to node j with switching time t$_s$.*

4. Mathematical Model:

Adding the effective Pi-Model of TSV in equation (4) of [1] and simplifying for minimum switching voltage V_j at an arbitrary node j in 3D stack gives:

$$V_j^{min} = \frac{1}{\lambda_j}\left\{\sum_{\substack{i=1\\i\neq j}}^{n} X_{i,j}V_i^{min} + \sum_{k=n+1}^{2n} X_{k,j}V_k^{min} + \frac{1}{2}\left(\sum_{\substack{i=1\\i\neq j}}^{n} C_{i,j} + \sum_{k=n+1}^{2n} C_{k,j}\right)V_{dd}\right\} \rightarrow (1)$$

where

$$\lambda_j = \sum_{\substack{i=1\\i\neq j}}^{n} X_{i,j} + \sum_{k=n+1}^{2n} X_{k,j} + \frac{1}{2}\left(\sum_{\substack{i=1\\i\neq j}}^{n} C_{i,j} + \sum_{k=n+1}^{2n} C_{k,j}\right) + C_{L_j}$$

$$X_{i,j} = \frac{t_s^2}{(6L_{i,j} + 3R_{i,j}t_s)}, X_{k,j} = \frac{t_s^2}{(6L_{k,j} + 3R_{k,j}t_s)}$$

nxn matrix form of (1) where a given chip has 1 to n nodes and other chip connected to it through TSVs has n+1 to 2n nodes:

$$\begin{bmatrix} -1 & \frac{X_{1,2}}{\lambda_1} & \frac{X_{1,3}}{\lambda_1} & \cdots & \frac{X_{1,m}}{\lambda_1} & \cdots & \frac{X_{1,n}}{\lambda_1} \\ \frac{X_{2,1}}{\lambda_2} & -1 & \frac{X_{2,3}}{\lambda_2} & \cdots & \frac{X_{2,m}}{\lambda_2} & \cdots & \frac{X_{2,n}}{\lambda_2} \\ \cdot & & & & & & \cdot \\ \frac{X_{m,1}}{\lambda_m} & \frac{X_{m,2}}{\lambda_m} & \frac{X_{m,3}}{\lambda_m} & -1 & & & \frac{X_{m,n}}{\lambda_m} \\ \cdot & & & & & & \cdot \\ \frac{X_{n,1}}{\lambda_n} & \frac{X_{n,2}}{\lambda_n} & \frac{X_{n,3}}{\lambda_n} & \cdots & \frac{X_{n,m}}{\lambda_n} & \cdots\frac{X_{n,n-1}}{\lambda_n} & -1 \end{bmatrix}\begin{bmatrix} V_1^{min} \\ V_2^{min} \\ \cdot \\ V_m^{min} \\ \cdot \\ V_n^{min} \end{bmatrix} + \begin{bmatrix} \frac{X_{1,n+1}}{\lambda_1} & \frac{X_{1,n+2}}{\lambda_1} & \frac{X_{1,n+3}}{\lambda_1} & \cdots & \frac{X_{1,n+m}}{\lambda_1} & \cdots & \frac{X_{1,2n}}{\lambda_1} \\ \frac{X_{2,n+1}}{\lambda_2} & \frac{X_{2,n+2}}{\lambda_2} & \frac{X_{2,n+3}}{\lambda_2} & \cdots & \frac{X_{2,n+m}}{\lambda_2} & \cdots & \frac{X_{2,2n}}{\lambda_2} \\ \cdot & & & & & & \cdot \\ \frac{X_{m,n+1}}{\lambda_m} & \frac{X_{m,n+2}}{\lambda_m} & & & \frac{X_{m,n+m}}{\lambda_m} & & \frac{X_{m,2n}}{\lambda_m} \\ \cdot & & & & & & \cdot \\ \frac{X_{n,n+1}}{\lambda_n} & \frac{X_{n,n+2}}{\lambda_n} & & & \frac{X_{n,n+m}}{\lambda_n} & & \frac{X_{n,2n-1}}{\lambda_n}\frac{X_{n,2n}}{\lambda_n} \end{bmatrix}\begin{bmatrix} V_{n+1}^{min} \\ V_{n+2}^{min} \\ \cdot \\ V_{n+m}^{min} \\ \cdot \\ V_{2n}^{min} \end{bmatrix} = \begin{bmatrix} -\frac{1}{2\lambda_1}\left(\sum_{i=2}^{n}C_{i,1} + \sum_{k=n+1}^{2n}C_{k,1}\right)V_{dd} \\ -\frac{1}{2\lambda_2}\left(\sum_{\substack{i=1}}^{n}C_{i,2} + \sum_{k=n+1}^{2n}C_{k,2}\right)V_{dd} \\ \cdot \\ -\frac{1}{2\lambda_m}\left(\sum_{\substack{i=1\\i\neq m}}^{n}C_{i,m} + \sum_{k=n+1}^{2n}C_{k,m}\right)V_{dd} \\ \cdot \\ -\frac{1}{2\lambda_n}\left(\sum_{i=1}^{n-1}C_{i,n} + \sum_{k=n+1}^{2n}C_{k,n}\right)V_{dd} \end{bmatrix}$$

$$\begin{bmatrix} \phi_{ff} & \phi_{fp} \\ \phi_{pf} & \phi_{pp} \end{bmatrix}\begin{bmatrix} V_f \\ V_p \end{bmatrix} = \begin{bmatrix} b_f \\ b_p \end{bmatrix}, \quad \phi_{ff_1} = \begin{bmatrix} \phi_{ff_1} & + & \phi_{ff_2} \end{bmatrix}, \quad \phi_{fp} = \begin{bmatrix} \phi_{fp_1} & + & \phi_{fp_2} \end{bmatrix}, \quad \phi_{pf} = \begin{bmatrix} \phi_{pf_1} & + & \phi_{pf_2} \end{bmatrix}$$

Where

$$\phi_{ff_1} = \begin{bmatrix} -1 & \dfrac{X_{1,2}}{\lambda_1} & \cdot & \cdot & \cdot & \dfrac{X_{1,m}}{\lambda_1} \\[2mm] \dfrac{X_{2,1}}{\lambda_2} & -1 & \cdot & \cdot & \cdot & \dfrac{X_{2,m}}{\lambda_2} \\[2mm] \cdot & \cdot & \cdot & & & \cdot \\ \cdot & \cdot & & \cdot & & \cdot \\ \cdot & \cdot & & & \cdot & \cdot \\[2mm] \dfrac{X_{m,1}}{\lambda_m} & \dfrac{X_{m,2}}{\lambda_m} & \cdot & \cdot & \dfrac{X_{m,m-1}}{\lambda_m} & -1 \end{bmatrix}, \phi_{ff_2} = \begin{bmatrix} \dfrac{X_{1,n+1}}{\lambda_1} & \dfrac{X_{1,n+2}}{\lambda_1} & \cdot & \cdot & \cdot & \dfrac{X_{1,n+m}}{\lambda_1} \\[2mm] \dfrac{X_{2,n+1}}{\lambda_2} & \dfrac{X_{2,n+2}}{\lambda_2} & \cdot & \cdot & \cdot & \dfrac{X_{2,n+m}}{\lambda_2} \\[2mm] \vdots & & \cdot & & & \cdot \\ & & & \cdot & & \\ & & & & \cdot & \\[2mm] \dfrac{X_{m,n+1}}{\lambda_m} & \dfrac{X_{m,n+2}}{\lambda_m} & \cdot & \cdot & \cdot & \dfrac{X_{m,n+m}}{\lambda_m} \end{bmatrix}, V_f = \begin{bmatrix} V_1 \\ V_2 \\ \cdot \\ \cdot \\ \cdot \\ V_m \end{bmatrix}, V_p = \begin{bmatrix} V_{m+1} \\ V_{m+2} \\ \cdot \\ \cdot \\ \cdot \\ V_n \end{bmatrix}$$

$$\phi_{fp_1} = \begin{bmatrix} \dfrac{X_{1,m+1}}{\lambda_1} & \dfrac{X_{1,m+2}}{\lambda_1} & \cdot & \cdot & \cdot & \dfrac{X_{1,n}}{\lambda_1} \\[2mm] \dfrac{X_{2,m+1}}{\lambda_2} & \dfrac{X_{2,m+2}}{\lambda_2} & \cdot & \cdot & \cdot & \dfrac{X_{2,n}}{\lambda_2} \\[2mm] \cdot & \cdot & \cdot & & & \cdot \\ \cdot & \cdot & & \cdot & & \cdot \\ \cdot & \cdot & & & \cdot & \cdot \\[2mm] \dfrac{X_{m,m+1}}{\lambda_m} & \dfrac{X_{m,m+2}}{\lambda_m} & \cdot & \cdot & \cdot & \dfrac{X_{m,n}}{\lambda_m} \end{bmatrix}, \phi_{fp_2} = \begin{bmatrix} \dfrac{X_{1,n+m+1}}{\lambda_1} & \dfrac{X_{1,n+m+2}}{\lambda_1} & \cdot & \cdot & \cdot & \dfrac{X_{1,2n}}{\lambda_1} \\[2mm] \dfrac{X_{2,n+m+1}}{\lambda_2} & \dfrac{X_{2,n+m+2}}{\lambda_2} & \cdot & \cdot & \cdot & \dfrac{X_{2,2n}}{\lambda_2} \\[2mm] \cdot & \cdot & \cdot & & & \cdot \\ \cdot & \cdot & & \cdot & & \cdot \\ \cdot & \cdot & & & \cdot & \cdot \\[2mm] \dfrac{X_{m,n+m+1}}{\lambda_m} & \dfrac{X_{m,n+m+2}}{\lambda_m} & \cdot & \cdot & \cdot & \dfrac{X_{m,2n}}{\lambda_m} \end{bmatrix}, b_f = \begin{bmatrix} -\dfrac{1}{2\lambda_1}\left(\sum_{i=2}^{n} C_{i,1} + \sum_{k=n+1}^{2n} C_{k,1} \right) V_{dd} \\[4mm] -\dfrac{1}{2\lambda_2}\left(\sum_{\substack{i=1 \\ i\neq 2}}^{n} C_{i,2} + \sum_{k=n+1}^{2n} C_{k,2} \right) V_{dd} \\[4mm] \cdot \\ \cdot \\[2mm] -\dfrac{1}{2\lambda_m}\left(\sum_{\substack{i=1 \\ i\neq m}}^{n} C_{i,m} + \sum_{k=n+1}^{2n-1} C_{k,m} \right) V_{dd} \end{bmatrix}$$

5. Detail of parameters and simulation set up:

Fig. 1(b) shows the model based on orthogonal paired power and ground grids. The drop across TSV reduces from bottom to top in a stack. The system of matrices shows each node on a chip connected to n neighboring nodes on same chip through supply line segment and corresponding node on other chip through TSV. Fig. 1(a) and 1(b) show peripheral TSVs in example model. $R_{i,j}$, $R_{k,j}$, $L_{i,j}$, $L_{k,j}$, $C_{i,j}$ and $C_{k,j}$ are wire segment resistance, inductance and capacitance in fig. 2(b). Each segment includes respective ground path resistance, inductance and capacitance. Segment ij between any two neighboring nodes has same length for uniform square grid but it can vary for non-uniform grid. Segment kj represents a TSV but it can be a metallic contact. Logic load is standard cell (containing fifty thousand 2-input NAND gates). We followed [8], [6] and ITRS,2008 for line load and switching time "t_s". C_{Lj} includes wire capacitance, symbiotic bypass capacitance and input capacitance of logic load assigned to each node. We followed [7] for line width, pitch, spacing and supply grid area relationship using 50% of metal area for supply grid. Line *(width=35µm, length=1mm , thickness=1.2µm)*, grid*(pitch=100µm,spacing=15µm,length=width=1mm)* and TSV*(diameter =5µm, height =50µm, S_{iO2} barrier thickness around TSV=1µm and pitch=100µm)*. Metal used for both grid and TSVs is Copper and dielectric is S_{iO2}. *LRC* values for supply lines and TSVs are found using Ansoft Q$_{3D}$ Extractor. Symbiotic bypass capacitance is determined following [6] and [8] based on 20% switching activity. *LRC* values of solder bump and pad have been added to TSVs for peripheral nodes. Example model has been solved through Matlab after adding effective Pi-Model of TSV(shown by fig. 2(a)) in [1]. In example model each chip has hundred nodes (n=100) out of which thirty six node on periphery have known voltage V_p and rest of the sixty four nodes have unknown voltage V_f. The supply voltage Vdd is equal to 1V. In first step the unknown node voltages (matrix V_f) at chip1 are calculated by adding TSV connections between chip1 and chip2 and using known voltages as 1V (matrix V_p). In second step drop across TSV is deducted from 1V to find known voltages (matrix V_p) for chip 2. TSVs are connected from chip2 to chip1 and chip3 and unknown potentials (matrix Vf) for chip 2 are found. Same steps are repeated for chip 3 and in this way we can add further chips in the stack one by one. The elements of matrix system have different values accordingly if pattern of TSVs is changed.

6. Conclusions:

The results in fig. 3 show that minimum switching voltage has no regular pattern both for proposed model and equivalent spice model however the trend of decrease in value is towards the centre of the chips as compared to periphery as shown by small patches on surface plots. The decrease in switching voltage from bottom to top of the stack is due to drop across TSVs whereas the decrease in switching voltage from periphery to center of the chips is due to drop across supply lines. The results of proposed model are almost compliant with spice model with small difference at some nodes. Values of the matrix system depend on physical dimensions of tracks and TSVs as well as the pattern of TSVs and type of supply grid. The model is applicable to any number of chips stacked through any number of TSVs following any TSV pattern using uniform or non-uniform grid. It is also dependant on type of logic load and how it is distributed on each chip. Minimum switching voltage distribution can have different pattern if load is not uniformly distributed or grid is not uniform or TSV pattern is different. Minimum switching voltage spots can be identified on each chip through this model which helps to locate and determine the value of on-chip decoupling capacitor. The supply network is expanded gradually by adding more chips in the stack. There should be a limit of maximum number of chips in a stack for a particular supply network design. Suitable number of chips in a stack can be chosen using this model depending on what is minimum switching voltage limit for on-chip logic load. Designers are not fully aware of the actual behavior of switching load unless the design is fabricated and tested. The model is quite useful for early design trade offs for power distribution network design.

978-1-4244-4447-2/09 $25.00 © 2009 IEEE

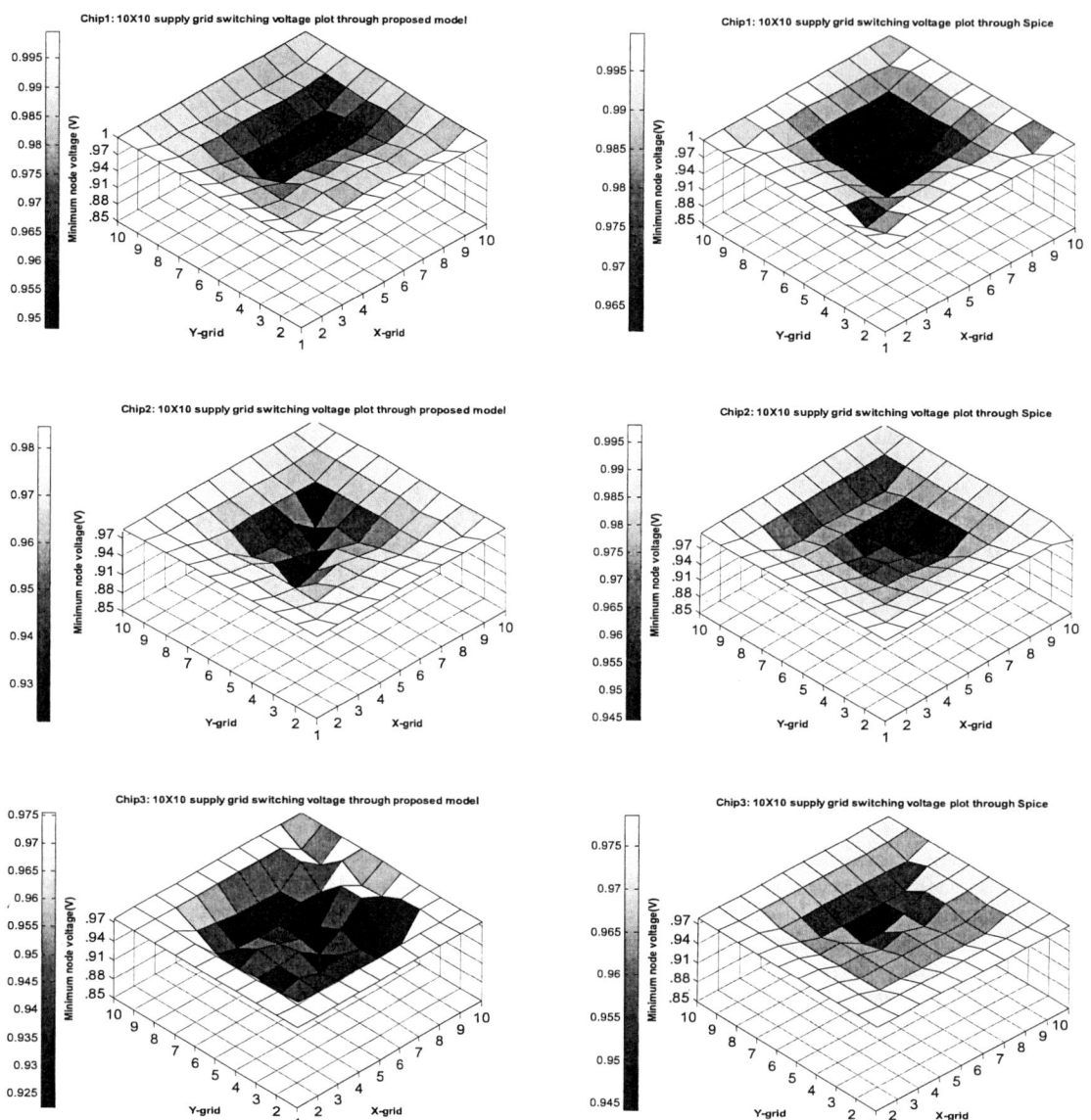

Fig. 3. *Proposed Model vs. Spice Model minimum switching voltage distribution within three similar chips of 3D stack. Chip1 is bottom chip and chip3 is top chip. Each chip has 10X10 supply grid i.e. there are 10 power/ground pairs in x-direction and 10 in y-direction. The grid lines are orthogonal to each other. Each point of intersection in xy-plane shows node of on-chip supply grid. Each chip has 100 switching nodes.*

7. References:

[1] Li-rong Zheng and Hannu Tenhunen, "Fast Modeling of Core Switching Noise on Distributed LRC Power Grid in ULSI Circuits," IEEE Transaction on Advanced Packaging, Vol.24, No.3, August 2001.

[2] Shayan A, et al. "3D Power Distribution Network Co-design for Nanoscale Stacked Silicon ICs," IEEE Electrical Performance of Electronic Packaging and Systems, Oct. 2008, San Diego, A.

[3] Pulkit Jain, Tae-Hyoung Kim, John Keane, and Chris H. Kim," A Multi-Story Power Delivery Technique for 3D Integrated Circuits," International Symposium on Low Power Electronics and Design, Aug. 2008, Bangalore.

[4] Gang Huang, M. Bakir, A.Naeemi, H. Chen, and J. D. Meindl, "Power Delivery for 3D Chip Stacks: Physical Modeling and Design Implication," IEEE Electrical Performance of Electronic Packaging and Systems, Oct. 2007, Atlanta.

[5] R. Weerasekera, System interconnection design trade-offs in three dimensional integrated circuits, PhD dissertation: Royal Institute of Technology (KTH) Sweden, Nov. 2008.

[6] Dennis Sylvester, William Jiang and Kurt Keutzer, BACPAC (Berkeley Advanced Chip Performance Calculator), http://www.eecs.umich.edu/~dennis/bacpac (Accessed on 15th of June, 2009).

[7] Andrey V. Mezhiba and Eby G. Friedman, Power Distribution Networks in High Speed Integrated Circuits, Kluwer USA: Press 2004, pp. 197-204.

[8] William J.Dally and John W. Poulton, Digital System Engineering, Cambridge UK: Cambridge Univ. Press 1998, pp. 237-242.

978-1-4244-4447-2/09 $25.00 © 2009 IEEE

Extraction of Equivalent Inductance in Package-PCB Hierarchical Power Distribution Network

*Jingook Kim, **Jaemin Kim, *Liehui Ren, *Jun Fan, **Joungho Kim, and *James L. Drewniak

*EMC Laboratory, Missouri University of Science and Technology (formerly University of Missouri-Rolla),
4000 Enterprise Dr., Rolla, MO 65401, USA
Homepage: http://emclab.mst.edu, Tel: +1-573-341-4139, Fax: +1-573-341-4532,
E-mail: kimjing@mst.edu, youvine@gmail.com
** Teralab., EECS, KAIST, Guseong-dong, Yuseong-gu, Daejeon, KOREA

ABSTRACT

A method to calculate equivalent inductances of multiple vias in arrays has been proposed. The method has been applied to the modeling of a hierarchical power bus and validated by measurements. It is accurate at the frequencies lower than the cavity resonant frequencies.

I. INTRODUCTION

Highly integrated circuits (ICs) can operate at a clock frequency of more than several giga-Hertz and consume a current of up to hundreds of milli-Amperes. The large switching current drawn by an IC is usually provided through a hierarchical power distribution network (PDN) consisting of chip-, package-, and PCB-level PDN components, as shown in Figure 1. This switching current gives rise to significant ripples in the supply voltage, mainly resulting from the parasitic inductances in the PDN. Therefore, quantifying the parasitic inductances common in a PDN is critical to evaluate its performance in terms of power integrity. Power distribution network has been widely investigated. Many approaches and methods have been developed to calculate via parasitic inductances and to determine the PDN impedance at a specific location, including full-wave methods, the cavity resonator model, and the transmission-line grid method (TLM) [1-3]. Also, a closed-form expression for via self inductance in a rectangular power bus has been derived [4]. However, in this paper, the focus is put on the parasitic inductances associated with multiple vias in arrays. It first proposes a rigorous method to calculate an equivalent inductance matrix for via arrays, based on the cavity resonator model. Then the equivalent inductances of multiple vias are further calculated through matrix simplification by grouping corresponding inductance terms together. Furthermore, the equivalent inductances can be used to build a lumped-circuit model for the entire hierarchical PDN. Lumped-circuit models suitable for SPICE implementation are much more preferable for fast and simple prediction of system performance at the design stage. The proposed circuit model cannot capture the distributed cavity resonances in the planar structure; however, it is acceptable in the view of power integrity because the distributed resonances are usually at the high frequencies near giga-Hertz where on-chip decoupling capacitors dominate.

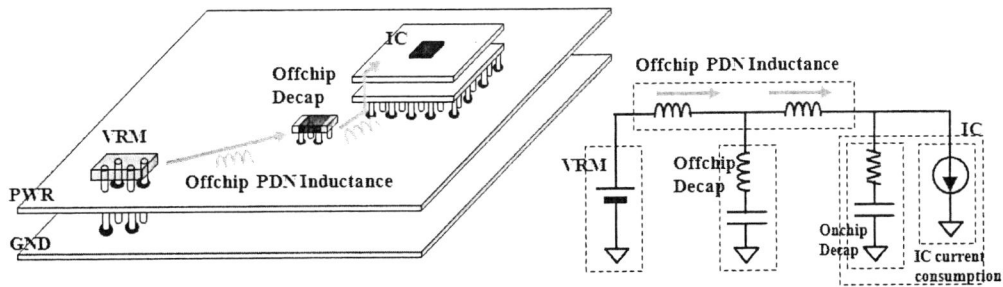

Fig 1. A hierarchical IC-Packag-PCB Power Distribution Network with Decoupling Capacitors

II. Equivalent Inductance Calculation for Multiple Vias in Arrays

(a) Via Inductance Formulation from Cavity Resonator Model

Common multilayer PDN structures can be decomposed into blocks so that each block contains only one pair of parallel planes. This "divide and conquer" approach can be applied because the interactions between the blocks can be neglected in most practical structures of interest. For each block, the impedance of its parallel plane pair can be obtained using the cavity resonator model [2]. For simplicity, a rectangularly-shaped parallel plane pair is used below as an example. For irregularly-shaped plane pair, the segmentation technique can be used, combined with the cavity resonator model, to calculate its impedance. In the impedance expression (1) of a rectangular plane pair, each individual term in the series summation represents an individual resonant mode of the cavity. The first term with the mode numbers m, $n = 0, 0$ is the impedance of the low-frequency inter-plane capacitance of the parallel plane pair. The sum of the rest of the terms represents an inductive term. The inductive term shows the inductive property in the frequency range from dc to the lumped model limit before the first cavity resonant frequency [5]. Therefore, the inductance expression is obtained as in (2). As frequency increases, the inductance value increases sharply near the resonant frequency, although it changes little at lower frequencies. In most practical cases, the dc inductance values of (2) can be used since the inductance value is almost constant in the frequency range lower enough than the first cavity resonant frequency. For multiple vias as shown in Figure 2, Eqn. (2) can be used to obtain both the self and mutual inductance values by adjusting the i and j values.

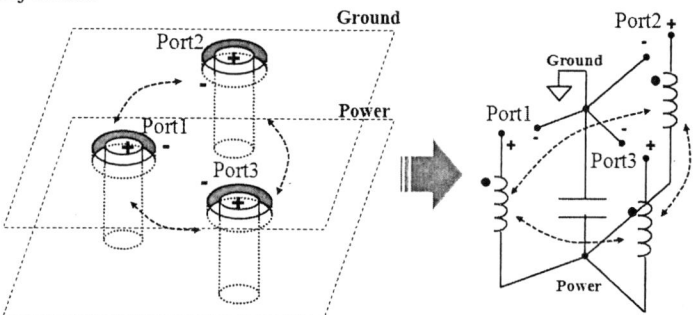

Fig 2. Physics-based L-C circuit model of via ports in parallel planes. The inductance values are calculated based on the cavity resonator model.

$$Z_{ij}(\omega) = -\frac{j\omega\mu d}{ab}\frac{1}{k^2} + j\omega\left[\frac{\mu d}{ab}\sum_{m=0}\sum_{n=0}\frac{\varepsilon_m^2\varepsilon_n^2}{k_{mn}^2-k^2}f(x_i,y_i,x_j,y_j)\Big|_{m=n\neq 0}\right]$$

$$= \frac{1}{j\omega\frac{\varepsilon ab}{d}} + j\omega L_{ij}(\omega) = \frac{1}{j\omega C} + j\omega L_{ij}(\omega) \qquad (1)$$

,where

$$f(x_i,y_i,x_j,y_j) = \cos\left(\frac{m\pi x_i}{a}\right)\sin c\left(\frac{m\pi t_{xi}}{2a}\right)\cos\left(\frac{n\pi y_i}{b}\right)\sin c\left(\frac{n\pi t_{yi}}{2b}\right)\cos\left(\frac{m\pi x_j}{a}\right)\sin c\left(\frac{m\pi t_{xj}}{2a}\right)\cos\left(\frac{n\pi y_j}{b}\right)\sin c\left(\frac{n\pi t_{yj}}{2b}\right)$$

$$L_{ij} = \frac{\mu d}{ab}\sum_{m=0}\sum_{n=0}\frac{\varepsilon_m^2\varepsilon_n^2}{k_{mn}^2}f(x_i,y_i,x_j,y_j)\Big|_{(m,n)\neq(0,0)} \qquad (2)$$

(b) Equivalent Inductances of Via Arrays

The total equivalent inductances of multiple vias in arrays can be calculated using the inductance formula (2). Figure 3 (a) illustrates multiple vias consisting of two different groups. Some vias are placed closely to each other, and connected in parallel constructing Group A, and others are also closely connected in parallel constructing Group B. If Group A is assumed as IC power pin vias and Group B as decoupling capacitor vias, the via currents in each group are along the same direction. Then the equivalent inductances are the equivalent self inductance of each group and the equivalent mutual one in between. The corresponding lumped circuit models are represented as in Figure 3 (b). If the numbers of vias in Groups A and B are m and n, respectively, the relationship between via voltages and currents can be expressed as in (3) from the inductance formula (2). If the voltages across the vias closely located to each other are assumed to be the same, the $(m+n)\times(m+n)$ matrix $\mathbf{L}$ can be reduced to a 2×2 matrix in (5), where L_{e_11}, L_{e_22}, and L_{e_12} (L_{e_21}) denote the equivalent self inductances of Group A, Group B, and the equivalent mutual inductance between Groups A and B, respectively.

$$[V] = j\omega[L][I], \qquad (3)$$

where $V = \begin{bmatrix} V_P & \cdots & V_P & | & V_S & \cdots & V_S \end{bmatrix}^T$ and $I = \begin{bmatrix} I_{P1} & \cdots & I_{Pm} & | & I_{S1} & \cdots & I_{Sn} \end{bmatrix}^T$. Then,

$$j\omega \begin{bmatrix} I_P \\ I_S \end{bmatrix} = \begin{bmatrix} \sum\limits_{i=1}^{m}\sum\limits_{j=1}^{m} B_{ij} & \sum\limits_{i=1}^{m}\sum\limits_{j=m+1}^{m+n} B_{ij} \\ \sum\limits_{i=m+1}^{m+n}\sum\limits_{j=1}^{m} B_{ij} & \sum\limits_{i=m+1}^{m+n}\sum\limits_{j=m+1}^{m+n} B_{ij} \end{bmatrix} \begin{bmatrix} V_P \\ V_S \end{bmatrix} , \text{ where } [B]=[L]^{-1}, \ I_P = \sum_{i=1}^{m} I_{Pi}, \ I_S = \sum_{i=1}^{n} I_{Si}, \quad (4)$$

$$\text{and, } \begin{bmatrix} V_P \\ V_S \end{bmatrix} = j\omega \begin{bmatrix} L_{e\_11} & L_{e\_12} \\ L_{e\_21} & L_{e\_22} \end{bmatrix} \begin{bmatrix} I_P \\ I_S \end{bmatrix}. \quad (5)$$

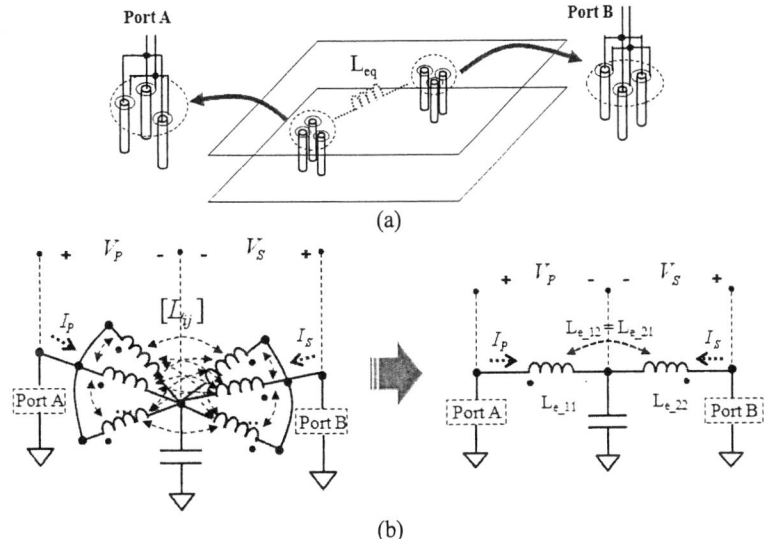

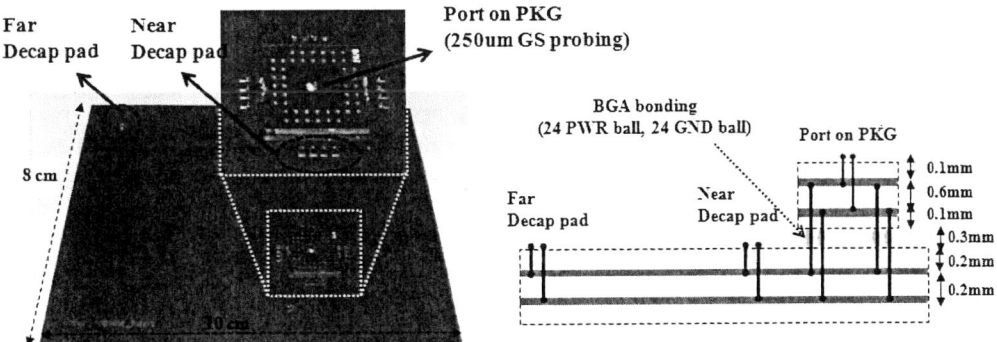

Fig 3. (a) Multiple vias consisting two different groups A and B; (b) Simplification to the equivalent inductances associated with the two via groups.

III. Application to the Modeling of a Package-PCB Hierarchical Power Distribution Network

Fig. 4. Geometry of a Package-PCB hierarchical PDN. Its impedance is measured at the port on the package.

A lumped circuit model of a multi-layer hierarchical PDN can be simply built by using (2) and merging inductances for multiple via groups. A test vehicle has been manufactured to validate the circuit model. The test vehicle consists of a board and a package, each of which has a single plane pair as shown in Figure 4. The power bus in the PCB and the package are connected to each other by 24 power and 24 ground balls. Decoupling capacitor pads are located at two different locations in the board. One is far away from the package, and the other is near the package. The input impedances were measured on the package with the far and the near decoupling capacitor pads shorted, respectively. Thus, the overall PDN can be considered as three cavities cascaded together as shown in Fig. 5 (a), where the interlayer structure between the top plane in the PCB and the bottom plane in the package is also modeled as a cavity [2]. Inductances of all the vias and balls are calculated from (2) and are reduced to the equivalent inductance matrix associated with the power and ground via groups for each of the three cavities, whose values are shown in Fig. 5 (b). It is found that the inductance matrix of the PCB changes according to the location of the shorting connection, where L_{e12} and L_{e21} change dominantly. Shorting at the near pad increases the mutual inductance. The impedance obtained from the lumped circuit model agrees well with measurements at the frequencies lower than the distributed cavity resonances.

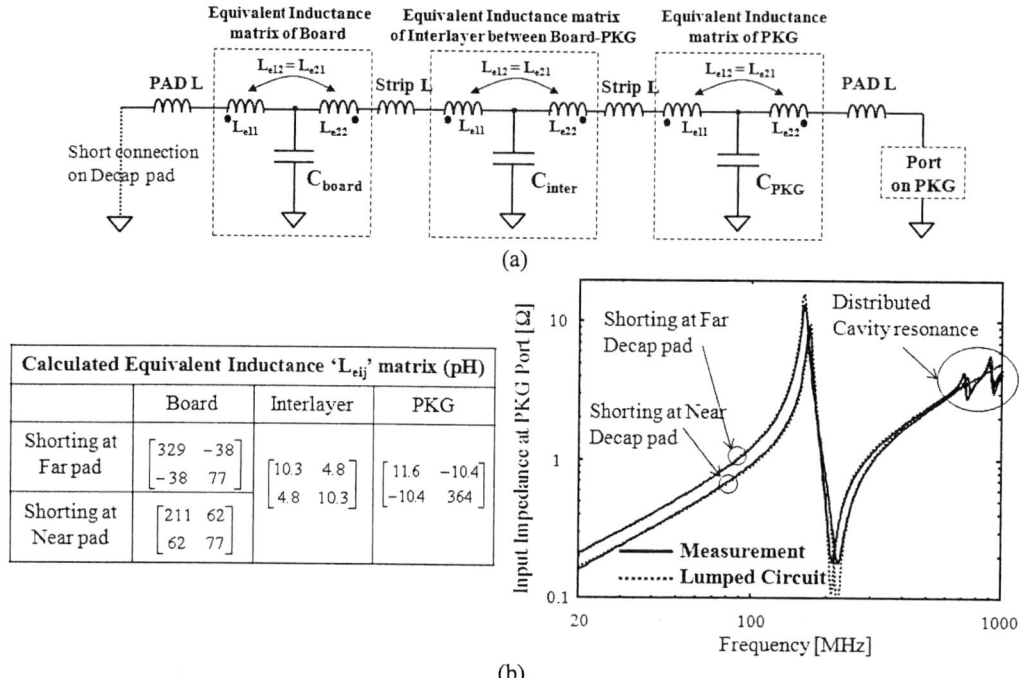

(a)

(b)

Fig. 5 (a) Equivalent L-C model of the hierarchical PDN shown in Fig. 4; (b) Comparison of modeled and measured results.

IV. CONCLUSION

The parasitic inductances existing in power distribution network can cause voltage ripples and are critical factors in power integrity design. In this paper, a method to calculate the equivalent inductances of multiple vias in arrays has been proposed using a via inductance formula based on the cavity resonator model. The method has been applied to the modeling of a package-PCB hierarchical PDN and validated by measurements. The lumped circuit model with the equivalent inductances provides not only engineering insights for BGA and decoupling designs, but also a useful tool to estimate the integrity of a power distribution network accurately and quickly. The lumped circuit model is only valid at the frequencies lower enough than the cavity resonant frequencies; however, it is still useful because on-chip decoupling capacitors usually dominate at higher frequencies.

V. REFERENCES

[1] X. D. Cai, G. L. Costache, R. Laroussi, and R. Crawhall, "Numerical extraction of partial inductance of package reference (power/ground) planes," in *Proc. IEEE Int. Symp. Electromagn. Compat.*, 1995, pp. 12–15.

[2] Jaemin Kim, et al., "Modeling and Measurement of Interlevel Electromagnetic Coupling and Fringing Effect in a Hierarchical Power Distribution Network Using Segmentation Method with Resonant Cavity Model" *IEEE Trans. Advanced Packaging*, vol. 31, pp. 544-557, August 2008.

[3] K. Lee and A. Barber, "Modeling and analysis of multichip module power supply planes," *IEEE Trans. Comp., Packag., Manufact. Technol. B*, vol. 18, pp. 628–639, Nov. 1995.

[4] J. Fan, W. Cui, J. L. Drewniak, T. P. Van Doren, and J. L. Knighten, "Estimating the Noise Mitigation Effect of Local Decoupling in Printed Circuit Boards," *IEEE Transactions on Advanced Packaging*, vol. 25, no. 2, pp. 154-165, May 2002.

[5] Liehui Ren, Jingook Kim, Gang Feng, Bruce Archambeault, James L. Knighten, James Drewniak, and Jun Fan, "Frequency-Dependent Via Inductances for Accurate Power Distribution Network Modeling", *IEEE Int. Symp. Electromagn. Compat.*, Aug. 2009

[6] A. E. Ruehli and A. C. Cangellaris, "Application of the partial element equivalent circuit (PEEC) method to realistic printed circuit board problem," in *Proc. IEEE Int. Symp. Electromagn. Compat.*, vol. 1, 1998, pp. 182–187.

978-1-4244-4447-2/09 $25.00 © 2009 IEEE

Effect of System Components on Electrical and Thermal Characteristics for Power Delivery Networks in 3D System Integration

Jianyong Xie[1], Daehyun Chung[1], Madhavan Swaminathan[1],
Michael Mcallister[2], Alina Deutsch[3], Lijun Jiang[3], Barry J Rubin[3]

[1]School of Electrical and Computer Engineering, Georgia Institute of Technology, Atlanta, GA 30332
Email: {jianyong.xie, d.chung}@gatech.edu, madhavan@ece.gatech.edu
[2]IBM Package Design, Development, and Electrical Services Group, Poughkeepsie, N.Y. 12601
Email: mcallism@us.ibm.com
[3]IBM T. J. Watson Research Center, Yorktown Heights, N.Y. 10598
Email: {deutsch, ljiang, brubin}@us.ibm.com

Abstract - In this paper, parameterized electrical-thermal co-analysis for power delivery networks (PDN) in 3D system integration is carried out. A 3D integrated system including glass-ceramic substrate, single and stacked dies, power delivery network, through-silicon vias (TSVs), controlled collapse chip connections (C4s), underfill material, and thermal interface material (TIM) is analyzed with several variable parameters. The analysis results show that temperature effects on DC IR drop can not be neglected. The TIM thermal conductivity, C4 density, stacking order of stacked dies, and voltage source location affect the final IR drop and hot spot temperature in the system.

I. INTRODUCTION

Through-silicon via (TSV) technology is a key enabler of high density 3D system integration. However, due to the increased heat flux density of stacked chips in 3D system integration, thermal problem is becoming more critical compared to 2D system [1]. Since electrical circuits will be affected by the thermal field due to their temperature-dependent characteristics, the thermal effect should be included in the electrical design procedure in order to validate the accurate electrical performance. On the other hand, the electrical elements are the sources to generate heat in the system, and these heat sources will affect the thermal field again. Thus, the electrical and thermal fields are coupled together in the system.

Recently, an electrical-thermal co-analysis method for power delivery networks in 3D integration system has been proposed by the authors and a 3D integrated system was analyzed using the method [2]. In the analysis, all the system components such as planes, package, stacked chips, TSVs, C4s and TIM were included. In the design of PDN for 3D system integration, there are many variable parameters which could affect the electrical and thermal performance such as TIM thermal conductivity, C4 density, stacking order of stacked dies, voltage source location, etc. Being aware of the effects of those variable parameters on IR drop and hot spot temperature of the PDN will be critical to designing the PDN which has thermal robustness and low IR drop.

In this paper, a complete parameterized electrical-thermal co-analysis is performed focusing on four variable parameters, that is, TIM thermal conductivity, C4 density, stacking order of stacked dies, and voltage source location. Electrical-thermal co-analysis with those system level parameters provides accurate characteristics of the PDN in 3D integrated system.

In section II, the electrical-thermal co-analysis methodology for PDN analysis is briefly explained. A 3D integrated system example and parameterized analysis results are shown in section III followed by conclusions in section IV.

II. ELECTRICAL-THERMAL CO-ANALYSIS METHODOLOGY

In this paper, the steady state electrical and thermal effects have been analyzed. In the steady state, the governing equations for voltage distribution and thermal distribution can be expressed as:

$$\nabla \cdot [\sigma(x,y,z,T)\nabla\phi(x,y,z)] = 0 \qquad (1)$$

$$\nabla \cdot [k(x,y,z,T)\nabla T(x,y,z)] = -P(x,y,z) \qquad (2)$$

where, $\sigma(x,y,z,T)$ and $k(x,y,z,T)$ represent the temperature-dependent electrical and thermal conductivity in PDN, $\phi(x,y,z)$ and $T(x,y,z)$ represent the voltage distribution and temperature distribution, respectively [2]. The electrical and thermal fields are coupled together due to the temperature-dependent electrical conductivity $\sigma(x,y,z,T)$ and the Joule heat generated by the power delivery networks expressed by

$$P(x,y,z) = \bar{J}(x,y,z) \cdot \bar{E}(x,y,z) = \bar{J}(x,y,z) \cdot (-\nabla\phi(x,y,z)) \qquad (3)$$

in which $\bar{J}(x,y,z)$ is the current density in PDN [3]. The relationship between electrical and thermal fields is shown in Fig. 1.

In order to capture the coupling effects between the electrical and thermal fields, the electrical-thermal co-analysis method for PDN in [2] is employed. In the iteration loop shown in Fig. 1, the DC IR drop simulation tool "*Rgen*" and thermal simulation tool "*ChipJoule*" which are IBM EIP Suite tools [4] are employed for steady state electrical and thermal analysis, respectively. In electrical analysis, the electrical conductivities of conductors in PDN including copper, tungsten for TSVs, Sn-0.7Cu for C4 balls are all considered as temperature-dependent parameters. As an example, the temperature-dependent electrical conductivities of copper and tungsten are shown in Fig. 2. In the thermal analysis, the material thermal conductivities are considered as constant for simplicity. The heat sinks are modeled as ideal heat sinks with constant room temperature of 25 °C. Only heat conduction boundary condition is applied in the thermal modeling. Initially, the system temperature including dies' temperature is assumed to be room temperature of 25 °C at the beginning of the co-analysis.

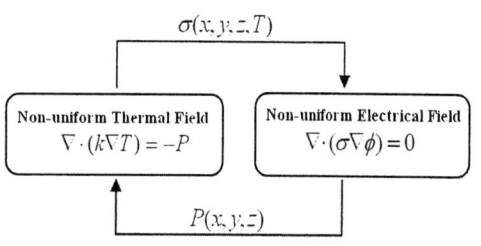

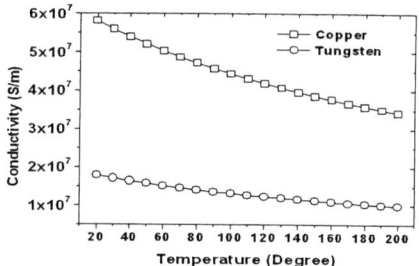

Figure 1. The relationship between electrical and thermal fields.

Figure 2. The temperature-dependent electrical conductivities of conductors.

III. SYSTEM EXAMPLE AND ANALYSIS RESULTS

An example of 3D integrated system is shown in Fig. 3(a). Two metal layers in glass-ceramic substrate are shunted together with multiple vias to reduce the electrical resistance of the PDN. A 2.5 V voltage source is applied at the location P1, the corner of the package. The stacked dies are stacked using TSVs and C4 interface. The TSV configuration for stacked dies is illustrated in Fig. 3(b), in which top die and bottom die use different power supply TSVs. The power consumptions for Die1, Die2, Die3, and Die4 are 75 W, 10 W, 40 W and 20 W, respectively. The uniform power maps are used for Die2, Die3 and Die4, while non-uniform power map is adopted for Die1 as shown in Fig. 3(c). The geometry and material parameters are summarized in Table I.

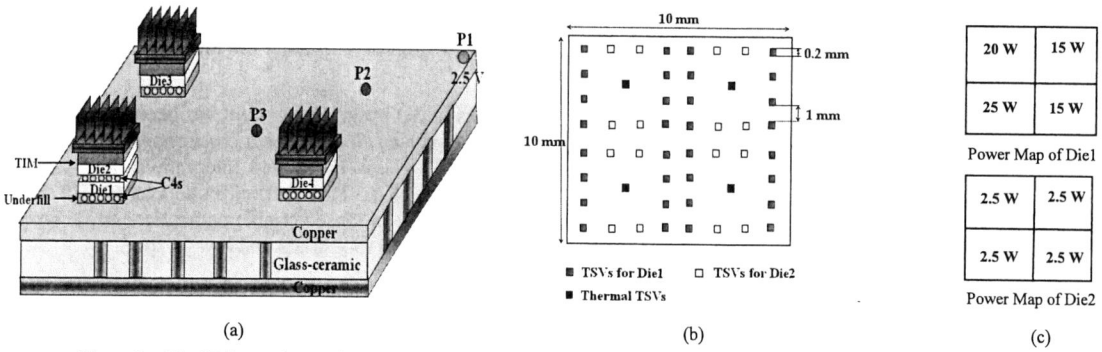

Figure 3. The 3D integration package. (a) Whole system view (b) TSV configuration for stacked dies (c) Power maps of Die1 and Die2.

A. TIM Thermal Conductivity

The above 3D integration package is analyzed with three different kinds of TIMs. The thermal conductivities of TIM1, TIM2 and TIM3 are 2, 5, 8 W/mK, respectively. The IR drop and temperature of Die1 with multiple electrical-thermal iterations are captured in Fig. 4. It shows that both the IR drop and temperature converge after 4 iterations. Fig. 4(a) shows that with TIM which has higher thermal conductivity, the Die1's temperature becomes lower. Die1's temperature is about 130 °C with TIM1 thermal conductivity of 2 W/mK, while its temperature is reduced to 77.3 °C with TIM2 thermal conductivity of 5 W/mK. Fig. 4(b) shows the effects of TIM and temperature difference on IR drop. With TIM1, the final voltage of Die1 is 2.455 V, while it is 2.459 V with TIM2. The corresponding IR drop is reduced from 45

TABLE I. PACKAGE GEOMETRY AND MATERIAL PARAMETERS

	Size (mm * mm)	Material Thickness (mm)	Thermal conductivity (W/mK)
Glass-ceramic	200 * 200	0.35	5
Copper	200 * 200	0.036	400
Die	10 * 10	0.5	110
Underfill	10 * 10	0.2	4.3
C4	0.2 * 0.2	0.2	60
TSV (Tungsten)	0.2 *0.2	0.5	174

978-1-4244-4447-2/09 $25.00 © 2009 IEEE

mV to 41 mV, about 9% decease when TIM1 is changed to TIM2. It demonstrates that temperature effect on IR drop in PDN can not be neglected and TIM will affect not only the temperature but also the IR drop.

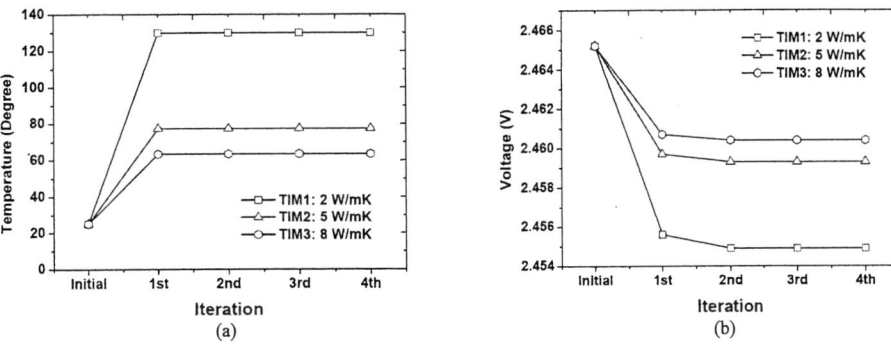

Figure 4.   The temperature and IR drop with various TIMs. (a) Temperature (b) IR drop.

B.   C4 Density

In the example of 3D package configuration, it is assumed that C4s occupy about 33% of the interface area. In order to study the relationship between C4 density and system performance, the 3D package with TIM2 and three different C4 densities are analyzed. The comparisons are provided in Fig. 5. It shows that with higher C4 density, the bottom Die1 has lower temperature. However, compared to the TIM, the effect of C4 density on temperature is much smaller. This is because the thermal conductivity of C4 is about 60 W/mK, which is much larger than 5 W/mK of TIM2 used in the analysis. Due to relatively smaller temperature change, the C4 density has much less effect on the voltage of Die1 compared to that of TIM as shown in Fig. 5(b).

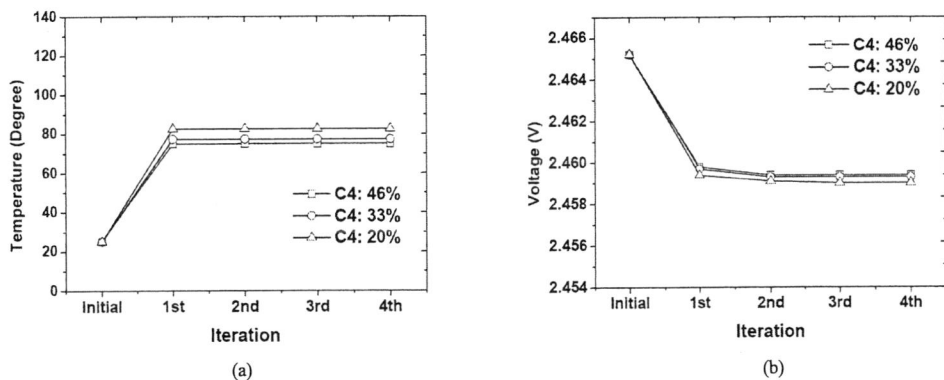

Figure 5.   The temperature and IR drop of Die1 with various C4 densities. (a) Temperature (b) IR drop.

C.   Stacking Order

In the package configuration in Fig. 3(a), it is assumed that memory chip (Die2) is stacked on top of CPU chip (Die1). In order to show the stacking order effect, the case of CPU stacked on top of the memory is also analyzed. The IR drop and temperature with TIM2 for those two cases are provided in Fig. 6. Fig. 6(a) illustrates when memory is stacked on top of CPU, the CPU's temperature is much larger than that of the memory and the difference is about $12\,°C$. Since the CPU consumes more power and has longer heat conduction path to the heat sink, it has much higher temperature than that of the memory. The highest temperature can be reduced by stacking a high power chip (CPU) on top of a low power chip (Memory). Fig. 6(b) shows that the highest temperature of CPU is reduced to $65.4\,°C$, and the temperature difference between the two chips is only $0.5\,°C$. From the temperature perspective, it shows that placing the high power consumption chip on top could reduce its hot spot temperature effectively. However, from the IR drop perspective, it is different. Fig. 6(a) and Fig. 6(b) demonstrate that CPU has more IR drop in CPU on top case than that of memory on top case. This is due to longer TSVs used for power supply of CPU when CPU stacks on top of the memory. Therefore, stacking order should be determined by considering both hot spot temperature and maximum IR drop allowed at the same time. There must be a trade off between them.

D.   Voltage Source Location

The 3D package example has three different voltage source locations P1, P2 and P3 (see Fig. 3(a)). Fig. 7(a) and Fig. 7(b) illustrate that voltage source location will affect the IR drop without changing the temperature. With voltage source location at P3 which is closest to Die1, the Die1's final voltage is 2.482 V compared to 2.459 V when applying voltage source at P1. The

978-1-4244-4447-2/09 $25.00 © 2009 IEEE   115

temperature of Die1 is not affected by the voltage source locations as shown in Fig. 7(a), since its temperature is dominated by the power consumption of the die itself. Therefore, it is important to locate the voltage source as close to the die as possible to reduce its IR drop.

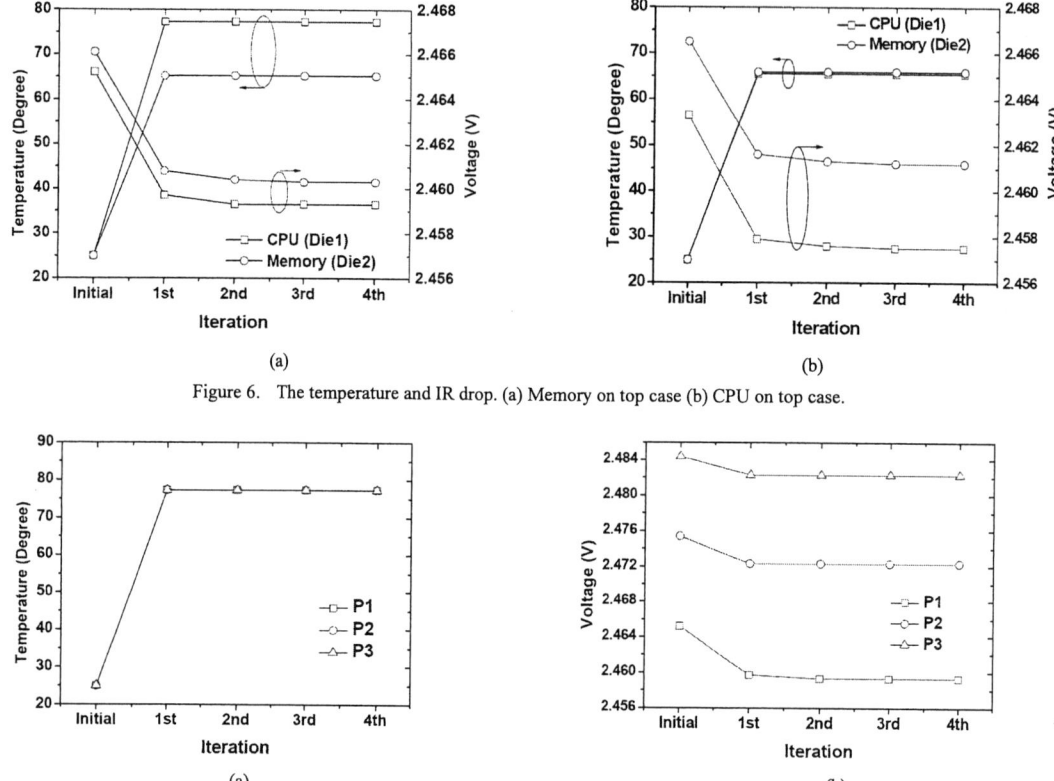

Figure 6.   The temperature and IR drop. (a) Memory on top case (b) CPU on top case.

Figure 7.   The temperature and IR drop of Die1 with TIM2 and three different voltage source locations. (a) Temperature (b) IR drop.

IV.   CONCLUSION

In this paper, electrical-thermal co-analysis for PDN of 3D integrated system has been performed with variable parameters analyzed. The co-analysis results demonstrate that the temperature effect on IR drop can not be neglected and it has to be taken into account in PDN design. The thermal conductivity of TIM affects both IR drop and temperature of dies, while the C4 density has much less effects than the TIM. Chip stacking order is also an important factor for determining hot spot temperature and IR drop. The stacking order should be determined by considering trade off between hot spot temperature and maximum IR drop. Finally, voltage source location is a factor which can reduce IR drop without affecting die's temperature.

ACKNOWLEDGMENT

This work was supported by IBM Corporation under the project in Georgia Institute of Technology.

REFERENCES

[1]   D. W. Gerlach, Y. K. Joshi, "Parametric thermal modeling of 3D stacked chip electronics with interleaved solid heat spreaders," *The Tenth Intersociety Conference on Thermal and Thermomechanical Phenomena in Electronics Systems*, pp. 1208-1212, Jun. 2006.

[2]   J. Xie, D. Chung, M. Swaminathan, M. Mcallister, A. Deutsch, L. Jiang, B. J Rubin, "Electrical-thermal co-analysis for power delivery networks in 3D system integration," *IEEE International Conference on 3D System Integration (3DIC)*, Sept. 2009.

[3]   B. Vermeersch, G. De Mey, "Device level electrothermal analysis of integrated resistors," *14th International Conference on Mixed Design of Integrated Circuits and Systems*, pp. 375-380, Jun. 2007.

[4]   http://www.alphaworks.ibm.com/tech/eip

[5]   S. Wunsche, C. Clauss, P. Schwarz, F. Winkler, "Electro-thermal circuit simulation using simulator coupling," *IEEE Trans. on VLSI*, vol. 5, no. 3, pp 277-282, Sept. 1997.

[6]   Y. Zhan, B. Goplen, and S. S. Sapatnekar, "Electrothermal analysis and optimization techniques for nanoscale integrated circuits," *Asia and South Pacific Conference on Design Automation*, pp. 219-222, Jan. 2006.

[7]   K. Fukahori, P. R. Gray, "Computer simulation of integrated circuits in the presence of electrothermal interaction," *IEEE Journal of Solid-State Circuits*, vol. 11, no. 6, pp. 834-846, Dec. 1976.

Electrical Modeling of Annular and Co-axial TSVs Considering MOS Capacitance Effects

Tapobrata Bandyopadhyay, Ritwik Chatterjee, Daehyun Chung, Madhavan Swaminathan and Rao Tummala

School of Electrical and Computer Engineering
Microsystems Packaging Research Center
Georgia Institute of Technology
Atlanta, GA 30332.
Email: tapobrata@gatech.edu
Tel.: (404)894-2028, Fax: (404)894-3842

Abstract - This paper presents analytical modeling and parametric study of the voltage dependent metal-oxide-semiconductor (MOS) capacitance of annular and co-axial TSVs. 3D electromagnetic (EM) simulations of TSVs are performed considering the depletion region. A low loss TSV structure is proposed utilizing the MOS capacitance effect.

I. INTRODUCTION

There have been several publications focusing on the electrical modeling and simulation of TSVs. However, in these efforts the semiconductor substrate has been treated as a lossy dielectric [1, 2]. Ignoring the semiconducting properties of the substrate and the resulting MOS capacitance introduces significant inaccuracies in the modeling of the capacitance in these structures. The voltage dependent MOS capacitance of TSVs was measured and reported by Chow et.al. [3]. The variable MOS capacitance of cylindrical TSVs was first studied analytically by the authors in [4].

This paper extends the analytical modeling to annular and co-axial TSV structures. A parametric study of via dimensions, substrate resistivity, liner thickness, liner material and, the TSV metal on the MOS capacitance is performed. 3D EM simulations are performed for annular and co-axial TSVs by modeling the depletion region. Results are presented to show the importance of the MOS capacitance effect in accurate electrical modeling of TSVs. A low-loss co-axial TSV structure is designed such that the Si between the core and shell is completely depleted to result in improved electrical performance.

II. ANALYTICAL MODELING

TSVs have a MOS structure as illustrated in Figure 1. The Si substrate is doped and is usually electrically grounded. Power TSVs follow the high frequency (HF) curve as shown in Figure 2 in inversion region whereas signal TSVs follow the deep depletion curve in the inversion region [4, 5].

The MOS capacitance analysis of an annular TSV comprises of two parts: (i) the inner surface and, (ii) the outer surface of the TSV. The following section analyzes the MOS effect in the inner surface of the TSV. The analysis is carried out for a p-type Si but a similar analysis is applicable for n-type Si substrates, as well.

Neglecting charge in the oxide liner and in the Si-SiO$_2$ interface, the gate voltage or the potential of the metal core is:

$$V_G = V_{FB} + \phi_s + \frac{Q_M}{C_{ox,i}} \qquad (1)$$

where, V_{FB} = Flatband voltage, Q_M = Charge on the metal and, $C_{ox,i}$ = Capacitance of the inner SiO$_2$ liner. The flatband energy diagram of a MOS structure is obtained when the energy band of the semiconductor is flat. In order to obtain the flatband condition, a flatband voltage (V_{FB}) has to be applied on the gate. The flatband voltage equals the work function difference between the gate metal (ϕ_M) and the semiconductor (ϕ_{Si}).

$$V_{FB} = \phi_M - \phi_{Si} = \phi_M - \chi - \frac{E_g}{2q} - V_t \ln \frac{N_a}{n_i} \qquad (2)$$

where, χ = Electron affinity of Si (4.05 V), E_g = Band gap energy of Si (1.12 eV at room temperature), q = Electronic charge (1.6022 × 10^{-19} coulombs), V_t = KT/q = 0.026 V (at 300 °K), K = Boltzmann's constant (1.3807 × 10^{-23} m^2 kg s^{-2} K^{-1}), T = Absolute temperature, N_a = Acceptor concentration and, n_i = Intrinsic carrier concentration of Si (1.18 x 10^{10} cm^{-3} at 300 °K).

The Poisson's equation in cylindrical co-ordinates is:
$$\frac{1}{r}\frac{d}{dr}\left(r\frac{d\phi}{dr}\right) = \frac{qN_a}{\varepsilon_s} \qquad (3)$$

where, ϕ = potential, ρ = charge density and, ε_s = Dielectric constant of Si. We have assumed that the electric charge distribution is axisymmetric and does not vary along the length of the TSV. We consider the full depletion approximation (FDA) which assumes that there are no mobile charge carriers (holes or electrons) in the depletion region. Thus, the potential and the electric field at the edge of the depletion region (r = r$_0$ in Figure 1) is zero.

978-1-4244-4447-2/09 $25.00 © 2009 IEEE

Integrating equation 3 from r_0 to r (where $r_0 < r < r_1$), $\quad r\dfrac{d\phi}{dr} = \dfrac{qN_a}{2\varepsilon_s}\left(r^2 - r_0^2\right)$ (4)

Integrating from r_0 to r_1, $\qquad \phi_s = \dfrac{qN_a}{2\varepsilon_s}\left(\dfrac{r_1^2 - r_0^2}{2} - r_0^2\ln\dfrac{r_1}{r_0}\right)$ (5)

where ϕ_s = Surface potential at the Si-SiO$_2$ interface. The charge on the metal, Q_M is equal to the charge in the depletion region due to the ionized acceptor atoms. $Q_M = qN_a\pi(r_1^2 - r_0^2)L$ ...(6) The oxide liner capacitance at the inner and outer surfaces are: $C_{ox,i} = 2\pi\varepsilon_{ox}L / \ln(r_2/r_1)$ ..(7); $C_{ox,o} = 2\pi\varepsilon_{ox}L / \ln(r_4/r_3)$ ..(8) where L is the TSV length. Using equations 2, 5, 6 & 7 in equation 1:

$$V_G = \phi_M - \chi - \frac{E_g}{2q} - V_t\ln\frac{N_a}{n_i} + \frac{qN_a}{2\varepsilon_s}\left(\frac{r_1^2 - r_0^2}{2} - r_0^2\ln\frac{r_1}{r_0}\right) + \frac{qN_a\left(r_1^2 - r_0^2\right)}{2\varepsilon_{ox}}\ln\frac{r_2}{r_1}$$ (9)

At threshold, the Si-SiO$_2$ surface potential, $\phi_s = 2\phi_F = 2V_t\ln(N_a/n_i)$. The radius of the inner depletion region edge at threshold (r_{0T}) is calculated by equating equation 5 to $2\Phi_F$. $\qquad \dfrac{qN_a}{2\varepsilon_s}\left(\dfrac{r_1^2 - r_{0T}^2}{2} - r_{0T}^2\ln\dfrac{r_1}{r_{0T}}\right) = 2\phi_F$ (10)

Using equations 9 and 10, the threshold voltage for the inner surface of the annular TSV,

$$V_{Ti} = \phi_M - \chi - \frac{E_g}{2q} + V_t\ln\frac{N_a}{n_i} + \frac{qN_a\left(r_1^2 - r_{0T}^2\right)}{2\varepsilon_{ox}}\ln\frac{r_2}{r_1}$$ (11)

A similar analysis can be performed for the outer surface of an annular TSV [4] to obtain the following expressions:

$$V_G = \phi_M - \chi - \frac{E_g}{2q} - V_t\ln\frac{N_a}{n_i} + \frac{qN_a}{2\varepsilon_s}\left(r_5^2\ln\frac{r_5}{r_4} - \frac{r_5^2 - r_4^2}{2}\right) + \frac{qN_a\left(r_5^2 - r_4^2\right)}{2\varepsilon_{ox}}\ln\frac{r_4}{r_3}$$ (12)

The radius of the outer depletion region edge at threshold (r_{5T}) is obtained from: $\dfrac{qN_a}{2\varepsilon_s}\left(r_{5T}^2\ln\dfrac{r_{5T}}{r_4} - \dfrac{r_{5T}^2 - r_4^2}{2}\right) = 2\phi_F$ (13)

The threshold voltage for the outer surface of the annular TSV, $V_{To} = \phi_M - \chi - \dfrac{E_g}{2q} + V_t\ln\dfrac{N_a}{n_i} + \dfrac{qN_a\left(r_{5T}^2 - r_4^2\right)}{2\varepsilon_{ox}}\ln\dfrac{r_4}{r_3}$ (14)

The depletion capacitance at the inner and outer surfaces are: $C_{di} = 2\pi\varepsilon_{ox}L / \ln(r_1/r_0)$ ...(15); $C_{do} = 2\pi\varepsilon_{ox}L / \ln(r_5/r_4)$ ...(16) The TSV capacitance is a parallel combination of the inner and outer surface capacitances: $C_{TSV} = [C_{ox,i}^{-1} + C_{di}^{-1}]^{-1} + [C_{ox,o}^{-1} + C_{do}^{-1}]^{-1}$ ...(17) For co-axial TSVs, 3 surfaces need to be analyzed: (i) outer surface of the co-axial core, (ii) inner surface of the co-axial shell and, (iii) outer surface of the co-axial shell. Thus, the co-axial TSVs are analyzed by using equations 8-10 and equations 11-13 as applicable.

III. PARAMETRIC STUDY

Parametric studies were performed on the annular as well as co-axial TSVs varying: 1) TSV diameter, 2) SiO$_2$ liner thickness, 3) TSV liner material, 4) Resistivity of the Si substrate and, 5) TSV metal. The results of the co-axial TSVs and annular TSVs showed similar trends. In this section, the parametric study results of the annular TSVs are presented.

This study used a base TSV model which had 0.1 μm thick SiO$_2$ dielectric liner around the Cu-filed TSV. The outer diameter of this TSV was 30 μm, the annular metal thickness was 5 μm, Si resistivity was 10 ohm-cm and the voltage on the TSV was 1.45 V. Various parameters of this base TSV model were varied to observe their effect on capacitance. Figures 3-7 show the results of the parametric studies. These figures also show the error in calculating the TSV capacitance if the MOS capacitance effects are ignored.

It is observed from Figure 3 that the TSV capacitance as well as the error increases with TSV diameter. This is because the depletion capacitance increases at a slower rate than the oxide capacitance with increase in TSV diameter. Increasing the liner thickness reduces the surface potential (at the Si-SiO$_2$ interface) thus creating a thinner depletion region. This, in turn, results in a larger depletion capacitance. However, the oxide capacitance decreases with liner thickness. This leads to a smaller error for TSVs with thick liners. The variable MOS capacitance effects are negligible for TSVs with liners thicker than 1 μm as shown in Figure 4.

A liner material with a larger dielectric constant increases the liner capacitance but has a negligible effect on the depletion capacitance and additionally, the depletion capacitance is significantly smaller than the liner capacitance as shown in Figure 5. Therefore, the TSV capacitance increases negligibly but the error increases significantly with increase in the dielectric constant of the liner. Reducing the doping level in the Si substrate increases its resistivity. For the same voltage on the TSV, a wider depletion region is created in a Si substrate with higher resistivity, thereby reducing the depletion capacitance. The oxide

978-1-4244-4447-2/09 $25.00 © 2009 IEEE

capacitance is unaffected by the Si resistivity. The TSV capacitance decreases and the error increases with increasing Si resistivity as shown in Figure 6. Filling the TSVs with a metal which has a higher work function increases the flatband and threshold voltages (equations 2, 10 & 13), increasing the capacitance for a TSV biased in the depletion region as shown in Figure 7. Al, Ni, Cu, W, Au and Ag which have work functions of 4.1, 4.55, 4.65, 4.67, 5 and 5.1 volts respectively were considered in this study.

Ignoring the MOS capacitance effects can lead to a significant error in estimating the TSV capacitance as seen from Figures 3-7 and would therefore result in significant inaccuracies in high speed signal analysis with TSVs.

IV. 3D EM SIMULATIONS

Annular and co-axial TSVs were simulated in CST Microwave Studio (CST-MWS) up to 10 GHz. The TSVs were simulated with ports on their top and bottom surfaces. The vias were 100 μm long, filled with copper with 0.1 μm thick SiO$_2$ liner. The TSVs were modeled in Si substrate with 10 ohm-cm resistivity and 1 μm thick layer of SiO$_2$ on its top and bottom surface.

Two models were simulated each for the annular and co-axial TSVs. The first model neglected the depletion region. The second model considered a depletion region around the TSV. The depletion region was modeled as a very high resistivity Si (because depletion regions are practically devoid of any mobile charge carriers). Figures 8 and 9 show the comparison of the insertion losses of these two models for annular TSV and co-axial TSV respectively. It is observed that neglecting the depletion region in 3D EM simulations can lead to a large error in modeling and analyzing TSVs in a 3D system.

Co-axial TSVs can be designed utilizing the low loss depleted Si region. In ICs, co-axial TSV can be designed such that the Si (between the core and shell) is completely depleted by the signal voltage on the TSV. In Si interposers, a negative bias (positive bias for n-type Si) can be applied on the Si substrate to completely deplete the Si between the core and shell. The result of such a design is showed in Figure 10. It compares two co-axial TSVs with identical dimensions – one with completely depleted Si and, the other with the Si only partially depleted. It is seen that completely depleting the Si reduces the signal loss by more than 0.2 dB. This effect is significant for signal paths through multiple TSVs (for e.g., in 3D stacks with multiple chips).

V. CONCLUSIONS

Modeling the MOS capacitance effects associated with TSVs is necessary for accurate electrical modeling of TSVs. The MOS capacitance was modeled analytically for annular and co-axial TSVs. The importance of modeling the depletion region and the resulting depletion capacitance were observed from the results of the parametric study and 3D EM simulation. Important design guidelines can be proposed for signal and power TSVs based on the parametric study results. Low-loss TSVs can be designed by utilizing the concept of depleting the Si to reduce the dielectric loss. The variable TSV capacitance can be utilized as a design parameter in 3D systems for novel applications.

REFERENCES

[1] C. Ryu, J. Lee, H. Lee, K. Lee, T. Oh, and J. Kim, "High Frequency Electrical Model of Through Wafer Via for 3-D Stacked Chip Packaging," 2006 Electronics Systemintegration Technology Conference, Dresden, Sept. 2006, Vol. 1, pp. 215-20.

[2] Dong Min Jang, Chunghyun Ryu, Kwang Yong Lee, Byeong Hoon Cho, Joungho Kim, Tae Sung Oh, Won Jong Lee, and Jin Yu, "Development and Evaluation of 3-D SiP with Vertically Interconnected Through Silicon Vias (TSV)," Proc. of ECTC 2007, pp. 847-52.

[3] E.M. Chow, V. Chandrasekaran, A. Partridge, T. Nishida, M. Sheplak, C.F. Quate, and T.W. Kenny, "Process Compatible Polysilicon-Based Electrical Through-Wafer Interconnects in Silicon Substrates," Journal of MEMS, Vol. 11, No. 6, Dec. 2002, pp. 631-40.

[4] T. Bandyopadhyay, R. Chatterjee, D. Chung, M. Swaminathan and R. Tummala, "Electrical Modeling of Through Silicon and Package Vias," 2009 IEEE International Conference on 3D System Integration (3D IC), San Francisco, California, USA.

[5] E. H. Nicollian and J.R. Brews, MOS Physics and Technology, Wiley (1982).

[6] Narain Arora, Mosfet Modeling for VLSI Simulation, World Scientific Publishing Co. Inc. (March 2007), pp. 150.

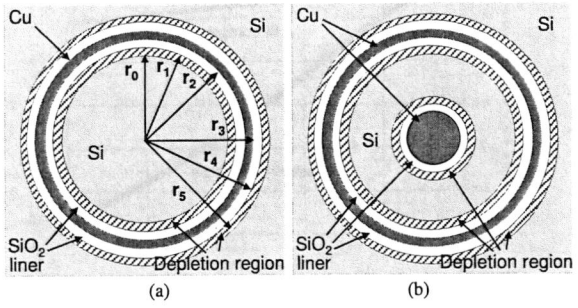

Figure 1. Schematic cross-section diagram of TSV with SiO$_2$ liner biased in the depletion region of operation. (a) Annular TSV. (b) Co-axial TSV.

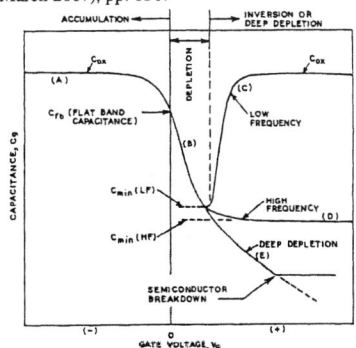

Figure 2. Typical Capacitance vs. Gate Voltage plot for planar MOS Capacitors on p-type Si [6].

978-1-4244-4447-2/09 $25.00 © 2009 IEEE

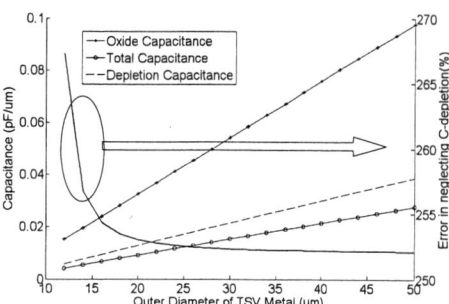

Figure 3. Capacitance variation of annular TSV as a function of its outer diameter. Left axis: Capacitance plots. Right axis: Error in estimating TSV capacitance if depletion capacitance is ignored (solid line).

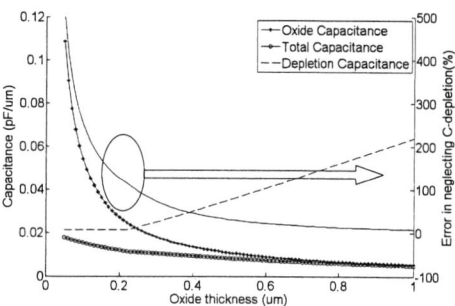

Figure 4. Capacitance variation of annular TSV as a function of the oxide liner thickness. Left axis: Capacitance plots. Right axis: Error in estimating TSV capacitance if depletion capacitance is ignored (solid line).

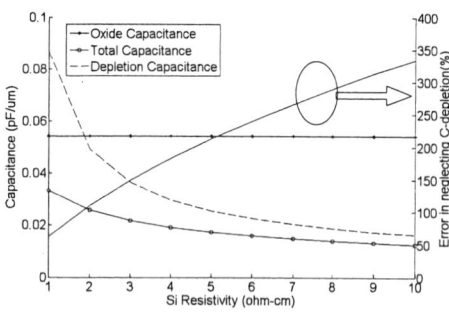

Figure 5. Capacitance variation of annular TSV as a function of the dielectric liner material. Left axis: Capacitance plots. Right axis: Error in estimating TSV capacitance if depletion capacitance is ignored (solid line).

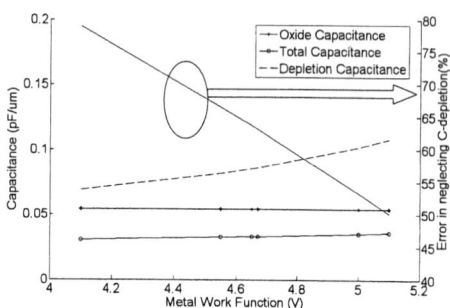

Figure 7. Capacitance variation of annular TSV as a function of the TSV metal work function. Left axis: Capacitance plots. Right axis: Error in estimating TSV capacitance if depletion capacitance is ignored (solid line).

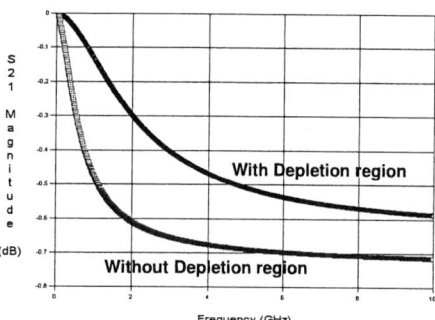

Figure 8. Insertion loss (S21) comparison of annular TSVs. The outer and inner diameter of the annular TSVs are 15 μm and 5 μm respectively while the via pitch is 30 μm.

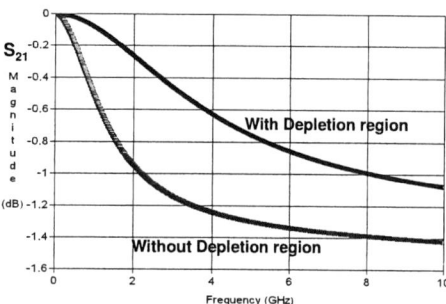

Figure 9. Insertion loss (S21) comparison of co-axial TSVs. The co-axial TSV has a core diameter of 15 μm. The inner and outer diameter of the TSV shell is 30 μm and 40 μm respectively.

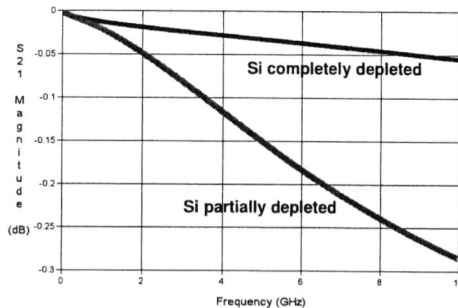

Figure 10. Insertion loss (S21) comparison of co-axial TSVs with the Si (between core & shell) either completely or partially depleted. The TSVs have 3 μm core diameter, 9 μm inner shell diameter and 15 μm outer shell diameter.

978-1-4244-4447-2/09 $25.00 © 2009 IEEE

Multi-Bit Fractional Equalization for Multi-Gb/s Inductively Coupled Connectors

Evan Erickson[1], John Wilson[2], Karthik Chandrasekar[3], Paul D. Franzon[1]

1. North Carolina State University, ECE Dept., MRC 429, 2410 Campus Shore Dr., Raleigh, NC 27695, Phone: 919-513-2007, Fax: 919-515-2285 { elericks, paulf } @ncsu.edu
2. Rambus Inc., 1512 E Franklin St # 200, Chapel Hill, NC 27514, jwilson@rambus.com
3. nVidia Corporation, 2701 San Tomas Expressway, Santa Clara, CA 95050, karthikc@nvidia.com

Abstract   **Multi-bit fractional equalization at the driver side allows for multi-Gb/s signaling across transformers which suffer from excessive ISI in inductively coupled connectors and backplanes. When an inductively coupled element is place in a transmission line or the gap between inductors in a transformer increases, the amplitude of the coupled pulse is decreased while the natural decay of the pulse is maintained. By removing the effects of the natural decaying tail of the pulse created by coupling an NRZ signal across a transformer, high-speed inductive coupling can be achieved over the larger transformers required by connectors and backplanes.**

I. INTRODUCTION

AC coupled interconnect (ACCI), using either capacitors or inductors, provides contactless signaling for chip-to-chip, chip-to-package, package-to-socket, and board-to-board interfaces. By removing the DC component, ACCI provides a zero insertion force connection that can withstand misalignment due to thermal expansion and doesn't suffer from the wear and tear of mechanically mated connectors [1]. For an inductively coupled interconnect, inductors are fabricated on opposing surfaces and brought into close proximity to form a transformer. Coupled inductors are not as sensitive to the gap spacing between elements as coupled capacitors and they provide a variety of geometric parameters than can be fine tuned, such as the width and spacing of the lines creating the inductor and the number of turns. Additionally, inductive coupling has an advantage over capacitive coupling when placed in a transmission line since the frequency-dependent impedance of a series capacitor can lead to significant return loss due to reflections. When a low-loss 1:1 turns ratio transformer is used, the impedance at the input to the transformer is very close to the impedance at the output, thus minimizing reflections [2-4].

Inductively coupled interconnects have previously shown the potential for multi-Gb/s signaling in 3D integrated circuits (ICs) and for chip-to-chip communications (level 1 interconnections) by utilizing small diameter inductors physically mated in close proximity to each other [1, 5-8]. For level 2 & 3 interconnections, connectors and sockets, multi-Gb/s inductively coupled interconnect has been demonstrated at sub-millimeter pitch (Fig. 1) [2-4]. For all three levels of interconnection, high-speed signaling has been demonstrated using inductors composed of 5 μm or less trace widths and spacings, creating transformers with sub 200 μm diameters and gap spacings of 5 μms or less. As the gap spacing between the inductors increases, the amount of coupled signal significantly decreases, requiring the use of larger transformers. Channels with lossy transmission line stubs on either or both ends also may require the use of larger transformers due to signal degradation from the stubs. As larger transformers with higher inductances are used for longer distance signaling over larger air gaps, the natural decay time of the transformer can lead to significant inter-symbol interference (ISI) when signaling at multi-Gb/s data rates. We examine the effects of ISI in high inductance transformers and propose a driver-side equalization scheme to facilitate high-speed signaling.

II. INDUCTIVELY COUPLED SIGNALING

Pulse signaling in a sub-millimeter pitch inductively coupled system is inherent due to the high-pass filter response of the transformer in the frequency domain that acts as a differentiator in the time domain (Fig. 2). Binary non-return-to-zero (NRZ) signals sent into the transformer (Fig. 3) are output as pulses with a magnitude directly related to the rise or fall time of the NRZ edge and a decay time dependant on the coupling of the transformer (Fig. 4).

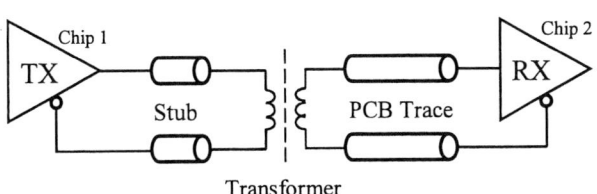

Fig. 1 Drawing of an inductively coupled ACCI connector with transmission line stubs and traces

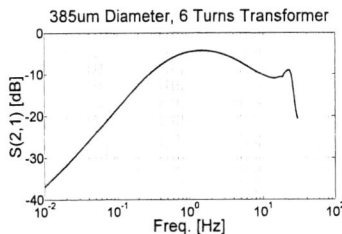

Fig. 2 Frequency domain (S21) for a 385 μm transformer

978-1-4244-4447-2/09 $25.00 © 2009 IEEE

As the inductance of the transformer increases due to an increase in the number of turns and/or the overall diameter, the magnitude of the coupled signal increases along with the natural decay time. This effect can be seen in the frequency and time domain as the outer diameter of a transformer with 15 μm width and spacing lines and a 25 μm via is varied from 145 μm (2 turns) to 385 μm (6 turns) (Fig. 5 & 6).

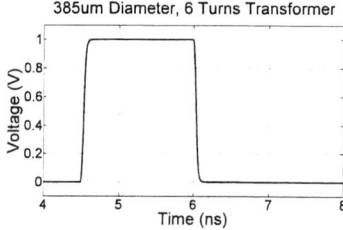

Fig. 3 NRZ input to a 385 μm transformer

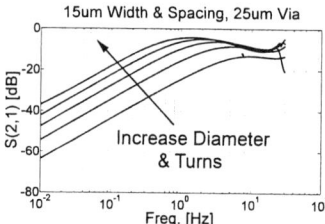

Fig. 4 Output from a 385 μm transformer

Fig. 5 Frequency domain s-parameters (S21) for a range of transformer sizes

When a transformer is placed in a lossy transmission line, such as in a backplane or a connector with stubs (Fig. 1), the transmission line acts as a low-pass filter in the frequency domain, while the transformer acts as a high-pass filter. The result in the time domain for the complete channel is a reduction in the peak magnitude of the pulse, while the slow decay time of the pulse is unchanged. Additionally, as the gap spacing between inductors is increased, the magnitude of the coupled signal is decreased, while the natural decay of the pulse remains the same (Fig. 7). Therefore larger diameter transformers with more turns are required to successfully transmit signals over larger gaps and longer transmission lines. This increase in transformer size leads directly to slower signaling data rates to avoid the ISI caused by the slow decay of the pulse's tail (Fig. 8). A transformer's size must be independent from its maximum signaling rate to enable multi-Gb/s signaling over large gaps and long transmission lines.

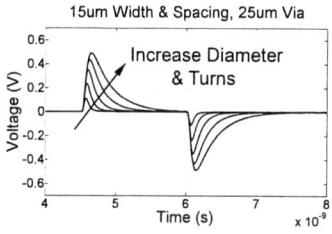

Fig. 6 Time domain output for a range of transformer sizes

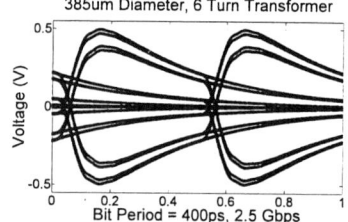

Fig. 7 Time domain comparison of gap spacing between inductors

Fig. 8 Eye diagram illustrating ISI due to the natural decay of the pulse tail

III. EQUALIZATION FOR INDUCTIVE COUPLING

A driver-side equalization methodology for overcoming the frequency-dependent losses associated with signaling over a transmission line was proposed by Dally & Poulton [9]. By applying the effects of a high-pass filter to the NRZ bit stream input on a transmission line, previously un-receivable bits (such as a lone '1' amongst a series of '0's) can be transmitted successfully. This was accomplished using a digital finite impulse response (FIR) filter to de-emphasize the magnitude of the non-transition bits in a data stream. Applying the same technique for inductive coupling is only slightly beneficial, however, since the high-pass response of the transformer results in the edges of input NRZ signals producing pulses while the DC content is removed. While equalizing at the bit level has little effect, fractionally-spaced equalization can be used to sub-divide a bit into multiple levels, potentially minimizing ISI [10, 11].

Multi-bit fractional equalization has the potential to drastically increase the maximum signaling speed of inductive coupling, especially when used in conjunction with a transmission line. An optimally equalized input signal is created by preserving the rising and falling edges of the NRZ data. The larger and sharper the edge input to a transformer, the greater the amplitude of the coupled pulse. The tail of the pulse can then by removed by de-emphasizing the DC component of the NRZ input. Too little de-emphasis fails to adequately remove the tail, while too much reduces the swing of the input signal, thereby coupling less signal. By dividing each input bit into fractions based on a clock, more precise equalization can be achieved. In order to equalize out longer tails and enhance the effects of equalization, multiple bits after a transition bit may have to equalized (Fig. 9). A simple FIR filter composed of multiple flip-flops enables the detection of transition bits and the bits immediately following a transition, while phases of a clock can be used to create different amplitude levels within a single bit and over multiple bits.

To model the transformers in this paper Ansoft's HFSS, an electromagnetic field solver which allows for simple manipulation of all transformer properties, was used to extract s-parameters [12]. Time domain simulations were completed using the extracted s-parameters with the RF Toolkit of MathWorks Matlab [13]. A custom Matlab based inductive channel model was developed to simulate the effects of various equalization schemes in the time domain and compute the optimal equalizer tap weights.

A multi-bit fractionally equalized input bit stream for a 385 μm (6 turns, 15 μm width & spacing) transformer is illustrated in Fig. 10 along with the non-equalized NRZ signal. The corresponding output signals (Fig. 11) show the reduction in the tail ISI achieved through equalization. A simple pulse receiver, without the need for complex logic or clocking circuitry, can then be used to recover the full-swing NRZ data. A secondary benefit of multi-bit fractional equalization can also be observed during long sequences of '1's or '0's, during which the amplitude of the equalized input signal can be drastically reduced. Depending on the channel, the optimally equalized driver output is closer to true pulse-signaling than NRZ signaling (Fig. 10). A significant amount of driver-side power can potentially be saved, while still allowing for the use of a simple low-power pulse receiver.

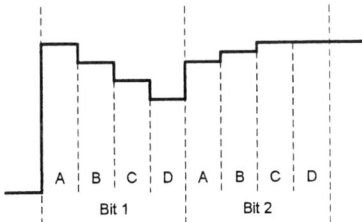

Fig. 9 2-bit, 4 phase fractional equalization. Enables different signal levels for the first two bits in a series and for each phase (A-D) within the bit

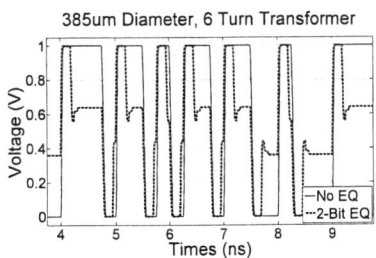

Fig. 10 NRZ and multi-bit fractionally equalized input to a 6-turn transformer at 4 Gbps

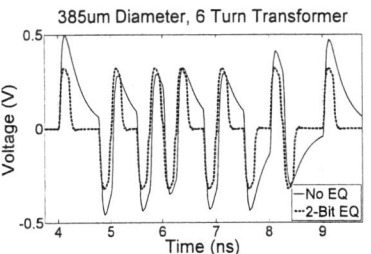

Fig. 11 Output from a 6-turn transformer with and without fractionally equalized input at 4 Gbps

The potential to decouple the physical parameters of a transformer from the rate at which data can be transmitted across it using equalization is readily apparent. For example, without any form of equalization the natural decay time of a 205 μm diameter, 15 μm width and spacing, 3 turn transformer (Fig. 12) makes it ideally suited to signal at 4 Gbps or less. If a larger amplitude signal is required at the receiver for error-free operation, a larger transformer, such as a 525 μm diameter, 25 μm width & spacing, 5 turn transformer can be used. Due to the increased natural decay time of the larger transformer, ISI at 4 Gbps makes transmission impossible (Fig. 13). However, by applying fractional equalization to the input data, successful transmission at 10+ Gbps can be achieved while providing the benefits of a larger, wider pulse inherent to larger transformers (Fig. 14). Minimizing the pitch of transformers for a specific channel still requires a tradeoff between coupling and area, but the maximum speed at which a particular transformer can signal is no longer a limiting factor.

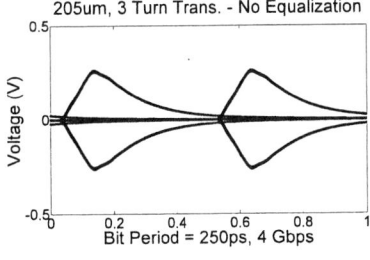

Fig. 12 An ideally sized transformer for 4 Gbps signaling without equalization

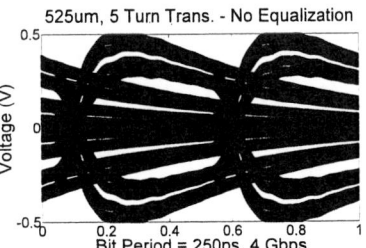

Fig. 13 4 Gbps signaling over a transformer with too slow a natural decay, leading to ISI

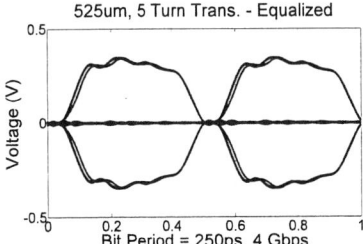

Fig. 14 Multi-bit fractionally equalized 4 Gbps signaling over the same transformer as Fig. 12

The simulated eye diagrams in Fig. 15 and 16 detail the improvement in eye opening for a 385 μm, 15 μm width and spacing, 6 turn transformer operating at 10 Gbps. Without equalization (Fig. 15) the eye is completely closed, while driver-side equalization is able to open the eye (Fig. 16). An overview of multi-bit fractional equalization shows that for 10 different transformers, each signaling at 4 different data rates, the benefit of equalization increases as transformer size and signaling data rate increases (Fig. 17). Transformers that were previously impossible to use for high-speed signaling can now be utilized.

IV. CONCLUSION

By applying multi-bit fractional equalization at the driver-side, ISI in inductively coupled channels can be drastically reduced, enabling high-speed, low-power signaling, while decoupling transformer design from the desired data rate. Without equalization, the physical design of a transformer limits the maximum achievable signaling rate due to the slow decay of the pulse created when an NRZ signal is sent across a transformer. When a transformer is placed in a transmission line, the low-pass filter response of the transmission line reduces the peak amplitude of the pulse while maintaining the slow decay of the pulse's tail. In order to successfully signal across such a channel, larger transformers with better coupling may be required. As coupling increases, the maximum ISI-free signaling rate decreases, limiting high-speed operation. A large range of transformer sizes can be used for a variety of signaling rates by equalizing at the driver-side, while providing low-power signaling with minimal ISI.

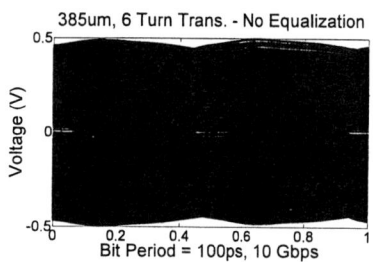

Fig. 15 Eye diagram for non-equalized 10 Gbps signaling across a 385μm transformer

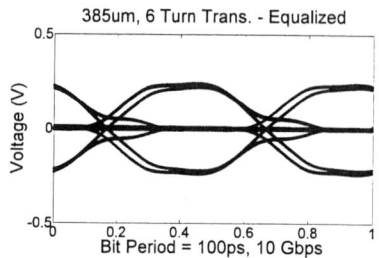

Fig. 16 Eye diagram for multi-bit fractionally equalized 10 Gbps signaling across a 385μm transformer

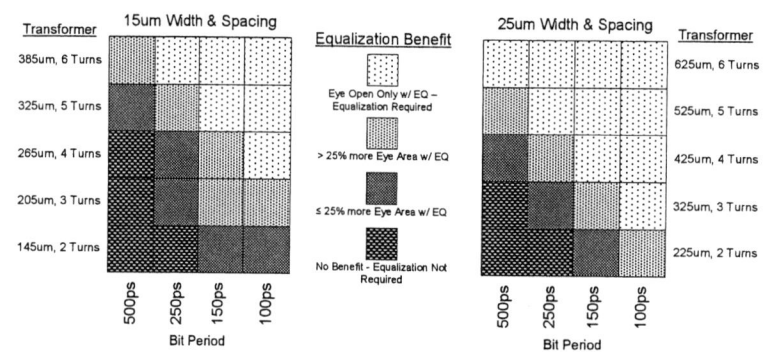

Fig. 17 The benefit of multi-bit fractional equalization for inductive coupling increases as the overall diameter of the transformer and the rate of signaling increases.

Acknowledgements

The authors wish to thank AFRL and NSF for support of this work.

References

[1]   S. Mick, J. Wilson, and P. Franzon, "4 Gbps high-density AC coupled interconnection," in *Proc. IEEE Custom Integrated Circuits Conf.*, 2002, pp. 133-140.

[2]   K. Chandrasekar, Z. Feng, J. Wilson, S. Mick, and P. Franzon, "Inductively Coupled Board to Board Connectors," in *Proc. Electronic Components and Tech. Conf.*, 2005, pp. 1109-1113.

[3]   K. Chandrasekar, J. Wilson, E. Erickson, Z. Feng, J. Xu, S. Mick, and P. Franzon, "Fine pitch inductively coupled connectors for multi-Gbps pulse signaling," in *IEEE Topic Meeting on Electrical Performance of Electronic Packaging*, 2006.

[4]   K. Chandrasekar, J. Wilson, E. Erickson, Z. Feng, J. Xu, S. Mick, and P. Franzon, "Inductively Coupled Connectors and Sockets for Multi-Gb/s Pulse Signaling," *IEEE Trans. Advanced Packaging*, vol. 31, pp. 749-758, 2008.

[5]   J. Xu, S. Mick, J. Wilson, L. Luo, K. Chandrasekar, E. Erickson, and P. D. Franzon, "AC coupled interconnect for dense 3-D ICs," in *IEEE Nucl. Sci. Symp. Record*, 2003, pp. 125-129 Vol.1.

[6]   D. Mizoguchi, Y. B. Yusof, N. Miura, T. Sakura, and T. Kuroda, "A 1.2Gb/s/pin wireless superconnect based on inductive inter-chip signaling (IIS)," in *IEEE Int'l Solid-State Circuits Conf., Dig. Tech. Papers*, 2004, pp. 142-517 Vol.1.

[7]   J. Xu, J. Wilson, S. Mick, L. Luo, and P. Franzon, "2.8 Gb/s inductively coupled interconnect for 3D ICs," in *Symp. on VLSI Circuits*, 2005, pp. 352-355.

[8]   N. Miura, D. Mizoguchi, M. Inoue, K. Niitsu, Y. Nakagawa, M. Tago, M. Fukaishi, T. Sakurai, and T. Kuroda, "A 1 Tb/s 3 W Inductive-Coupling Transceiver for 3D-Stacked Inter-Chip Clock and Data Link," *IEEE J. of Solid State Circuits*, vol. 42, pp. 111-122, 2007.

[9]   W. J. Dally and J. Poulton, "Transmitter equalization for 4-Gbps signaling," *IEEE Micro*, vol. 17, pp. 48-56, 1997.

[10]   G. Ungerboeck, "Fractional Tap-Spacing Equalizer and Consequences for Clock Recovery in Data Modems," *IEEE Trans on Communications*, vol. 24, pp. 856-864, 1976.

[11]   R. D. Gitlin and S. B. Weinstein, "Fractionally-Spaced Equalization: An Improved Digital Transversal Equalizer," *The Bell System Technical Journal*, vol. 60, pp. 275-296, 1981.

[12]   Ansoft HFSS - User's Manual.

[13]   MathWorks Matlab - User's Manual.

978-1-4244-4447-2/09 $25.00 © 2009 IEEE

Hybrid Substrate Integrated Waveguides Developed Using Flexible Substrates

Mohammad S. Mahani, Asanee Suntives, and Ramesh Abhari

McGill University, Dept. of Electrical and Computer Engineering, 3480 University St., Montreal, QC, H3A 2A7, Canada

Phone: (514) 398-1451, Fax: (514) 398-4470

Emails: {mohammad.shahidzadehmahani, asanee.suntives}@mail.mcgill.ca, ramesh.abhari@mcgill.ca.

Abstract: In this paper, hybrid substrate integrated waveguide (SIW) structures developed using multi-layer flexible laminates are investigated for application in off-the-board chip-to-chip communication systems. It is shown through full-wave simulations that the transmission characteristics of the TEM and TE_{10} channels of the hybrid SIW are virtually unaffected by the substrate bending. This new implementation platform is found to be suitable for high-speed applications due to its relatively low dielectric losses and offers new possibilities for deployment of the hybrid SIW interconnect.

I. Introduction

In modern high-speed electronic systems, serial links are widely utilized as the efficient method for multi-gigabit data transmission. However, the maximum attainable data rate is limited by the conductor and dielectric losses of the conventional electrical interconnects used in these architectures. Additionally, the parasitics due to the discontinuities in the junctions and routing paths as well as the crosstalk between the links further degrade the signal integrity. To improve the quality of the high-speed communication channels, off-the-board transmission links as shown in Fig. 1 have been proposed in order to bypass the routing discontinuities and the backplane [1]. In [2], low-loss flexible printed circuit boards with custom-made connectors were used in which more than triple the bandwidth of an FR-4 board was achieved demonstrating the viability of this alternative substrate material.

Recently, substrate integrated waveguides (SIWs) have become an attractive interconnect structure for high-speed and high-frequency applications due to their low-loss and excellent immunity to electromagnetic interference [3]–[6]. In order to make a more efficient use of the SIW's substrate volume, the hybrid configuration was proposed in [7]. In this manner, several stripline interconnects are embedded in an SIW and utilized as the additional signaling channels to ultimately increase the aggregate transmission rate. It was also shown that the coupling between the waveguide and single-ended stripline channels is negligible [7]. In [8] and [9], SIWs have been fabricated using flexible PCB substrate materials, which demonstrate the feasibility of adapting this new interconnect technology to other fabrication platforms.

In this paper, hybrid SIWs constructed using multi-layer flexible laminates are investigated. A comparative study of the transmission performance of the stripline and waveguide channels in a flexible hybrid SIW structure is presented using full-wave S-parameter simulations. This is followed by evaluations of single-ended and differential signaling as well as the effects of structural bending on the transmission characteristic.

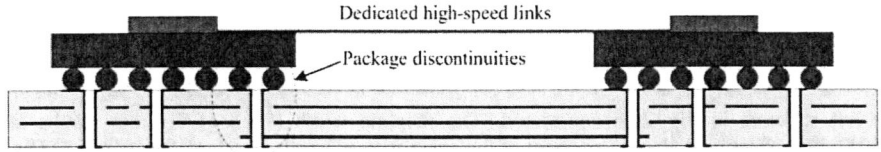

Fig. 1. Chip-to-chip communication system containing off-the-board high-speed links.

II. Flexible Hybrid SIW

The geometry of a hybrid SIW is shown in Fig. 2 in which a single stripline is embedded to create a dual-mode transmission channel. A number of striplines can be placed in the waveguide structure as long as $\frac{S_2}{h} \gg 1$ [7]. The right side of Fig. 2 depicts an SIW with two embedded striplines, which can be excited as a differential pair. Thus, the structures of the hybrid SIWs of Fig. 2 with single-ended and differential excitations are studied. For routing the embedded lines in and out of the waveguide, the side-entry transition introduced in [7] can be used. In addition, a coplanar microstrip transition can be

978-1-4244-4447-2/09 $25.00 © 2009 IEEE   125

employed for connecting the SIW to source and load [10]. These transitions are not considered in this paper in order to focus on the performance of the flexible hybrid structure.

The hybrid SIW requires a multi-layer fabrication process. Fig. 3 shows the construction of the flexible laminate considered in this study. Coverlays are not shown herein as the circuit performance is not affected by them due to the shielding presence of the copper layers. The flexible substrate is made of a polyimide composite (Dupont Pyralux AP double-sided adhesiveless laminate). Since the adhesive material is relatively lossy, this multi-layer construction employs adhesiveless laminates, and its fabrication process aims at minimizing the number of adhesive layers. The electrical and geometrical parameters of the substrate layers are presented in Table I.

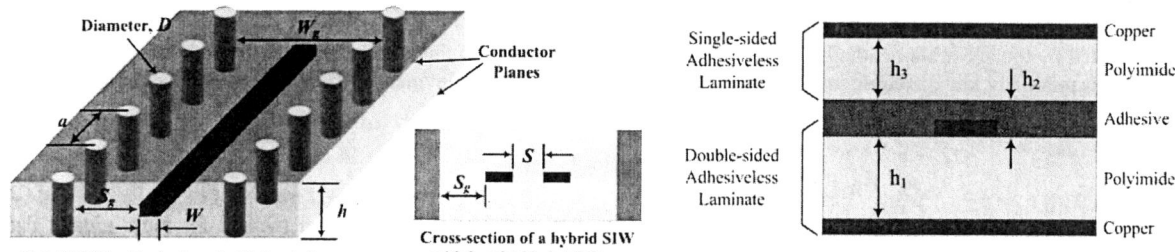

Fig. 2. Hybrid substrate integrated waveguides.

Fig. 3. Diagram of the multi-layer flexible laminate considered in this study.

Table I Electrical and Geometrical Parameters of the Flexible Laminate Construction

Copper		
	Conductivity	5.8×10^7 S/m
	Thickness	0.018 mm (0.5 oz/ft^2)
Pyralux AP Dielectric (Polyimide)		
	Dielectric Constant (ε_r)	3.3
	Dissipation Factor ($\tan\delta$)	0.003 at 1 GHz
	h_1	0.1016 mm (4 mils)
	h_3	0.0762 mm (3 mils)
Adhesive		
	Dielectric Constant (ε_r)	3.0
	Dissipation Factor ($\tan\delta$)	0.025
	h_2	0.0434 mm (1 mil + 18μm)

III. Transmission Characteristic of the Hybrid Channels

In this section, the performances of the different channels in a hybrid SIW structure are investigated using Ansoft HFSS [11]. First, the left structure in Fig. 2 is considered. The stripline width (W) of 0.1 mm results in a characteristic impedance of 50 Ω. The effective dielectric constant of the composite substrate is calculated to be 3.2. The SIW dimensions are chosen to be: W_g = 5 mm, a = 0.8 mm and D = 0.4 mm. Consequently, its TE_{10} cutoff frequency is 15.7 GHz. In addition, a differential stripline pair designed to have an odd-mode impedance (Z_{odd}) close to 100 Ω is considered in another hybrid SIW structure (the case on the right side of Fig. 2). The spacing between the striplines (S) is chosen to be 0.2 mm, which results in Z_{odd} of 90 Ω. The total length of each structure is 24 mm.

Fig. 4 shows a comparison between the simulated per-unit-length channel insertion losses. It can be observed that the SIW channel exhibits the lowest attenuation in its usable bandwidth, which starts a few GHz above the TE_{10} cutoff. For instance, the insertion loss of the waveguide is between 0.276–0.3 dB/cm from 19.86 to 30 GHz. The transmission losses of the single-ended (single strip) and differential striplines are 0.374–0.493 dB/cm and 0.5–0.688 dB/cm, respectively, over the same frequency band. It is expected that the coupling between the striplines will be strong due to the small spacing of the

978-1-4244-4447-2/09 $25.00 © 2009 IEEE

differential pair. Therefore, part of the electromagnetic fields that propagate along the differential lines is concentrated in the lossy adhesive layer around and above the copper traces. Hence, in general, the differential pair shows a higher attenuation than that of the single-ended line for this type of laminate construction over the entire simulated frequency range. Nonetheless, these levels of insertion loss seem to be comparable to those required for high-speed interconnects [2].

IV. The Effects of Bending

The laminate understudy is flexible and expected to be utilized under different deployment scenarios. In order to investigate the impact of bending the flexible laminate on signal integrity, 90° and 180° bends as shown in Fig. 5 are considered. In addition, a straight hybrid SIW having the same total length of 47 mm is analyzed as the benchmark. The simulated transmission coefficients of these 3 cases are shown in Fig. 6. It can be observed that there is virtually no change in the transmission performance of the SIW channel when the substrate is subjected to bending. Fig. 7 shows the results for the stripline channel in which the change in the insertion loss can be as high as 0.44 dB at 24 GHz. In addition, the couplings between the waveguide and TEM modes for the three cases are presented in Fig. 8. It can be seen that the coupling is generally well below -40 dB for all cases. Thus, it can be concluded that bending this type of laminate at 90° and 180° does not adversely affect the performance of the hybrid SIW.

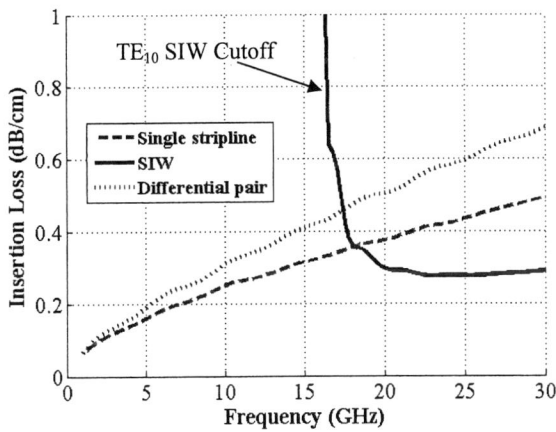

Fig. 4. Simulated hybrid SIW channel insertion losses.

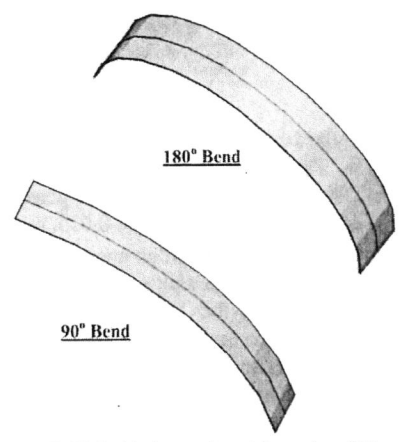

Fig. 5. Hybrid channels subjected to 90° and 180° bends.

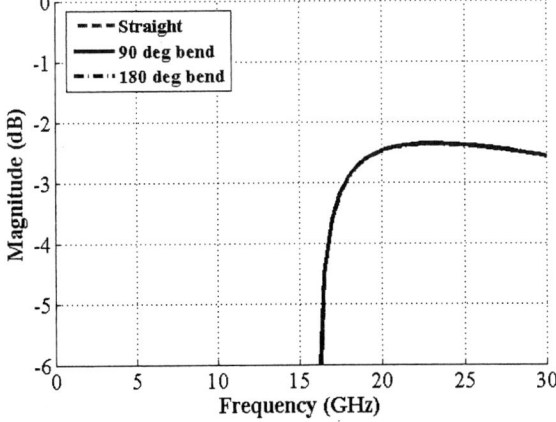

Fig. 6. Simulated transmission coefficients of the SIW channel for straight, 90° and 180° bends.

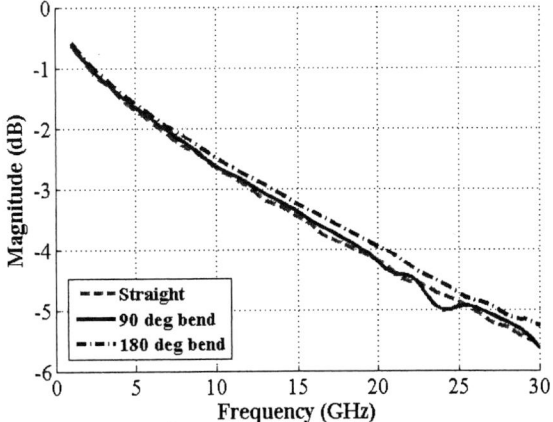

Fig. 7. Simulated transmission coefficients of the stripline channel for straight, 90° and 180° bends.

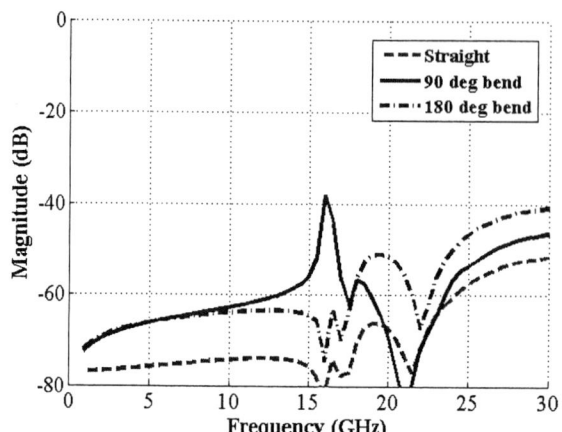

Fig. 8. Simulated mode coupling between SIW and stripline channels for straight, 90° and 180° bends.

VI. Conclusion

In this paper, hybrid substrate integrated waveguides (SIWs) are investigated for multi-layer flexible printed circuit board implementation. The data conduits of the hybrid structure are composed of single-ended or differential striplines and an SIW channel. The flexible laminate is constructed from a combination of low-loss polyimide composite and lossy adhesive material with loss tangents of 0.003 and 0.025, respectively. From full-wave simulations, the waveguide channel is found to exhibit the lowest insertion loss when operating beyond its TE_{10} cutoff frequency. The transmission loss in the differential pair is higher than that of the single-ended stripline interconnect. This can be attributed to the stronger concentration of electromagnetic fields in the lossy adhesive material. Nonetheless, these channels exhibit excellent insertion loss performance, which is the desirable feature for high-speed interconnects. The effect of bending the flexible laminate is also investigated. In comparison with the straight hybrid SIW, bending the substrate by 90° and 180° does not show significant degradation in the transmission and coupling performance of the hybrid channels.

References

[1] J. Fjelstad, K. Grundy, and G. Yasumura, "3D PCB architecture for next generation high speed interconnections," *Circuit World*, vol. 31, no. 4, pp. 25–33, 2005.

[2] H. Braunisch, J. E. Jaussi, and J. A. Mix, "Flex-circuit chip-to-chip interconnects," in *Proc. 15th IEEE Electrical Performance of Electronic Packaging (EPEP)*, Scottsdale, AZ, Oct. 23–25, 2006 , pp. 273–276.

[3] K. Wu, D. Deslandes, and Y. Cassivi, "The substrate integrated circuits – a new concept for high-frequency electronics and optoelectronics," in *Proc. 6th International Conference on Telecommunications in Modern Satellite, Cable, and Broadcast Services (TELSIKS)*, Nis, Yugoslavia, Oct. 1–3, 2003, pp. P-III-P-X.

[4] A. Suntives and R. Abhari, "Design and characterization of the EBG waveguide-based interconnects," *IEEE Trans. Adv. Packag.*, vol. 30, no. 2, pp. 163–170, May 2007.

[5] A. Suntives and R. Abhari, "Experimental evaluation of high-speed data transmission in a waveguide-based interconnect," in *Proc. 15th IEEE Electrical Performance of Electronic Packaging*, Scottsdale, AZ, Oct. 23–25, 2006, pp. 269–272.

[6] J. J. Simpson, A. Taflove, J. A. Mix, and H. Heck, "Substrate integrated waveguides optimized for ultrahigh-speed digital interconnects," *IEEE Trans. Microw. Theory Tech.*, vol. 54, no. 5, pp. 1983–1990, May 2006.

[7] A. Suntives and R. Abhari, "Dual-mode high-speed transmission using substrate integrated waveguide interconnects," in *Proc. 16th IEEE Electrical Performance of Electronic Packaging*, Atlanta, GA, Oct. 29–31, 2007, pp. 215–218.

[8] H. Zhu, W. Hong, L. Tian, F. He, and B. Liu, "Experimental investigations on substrate integrated flexible waveguide," *Asia-Pacific Microwave Conference*, Bangkok, Thailand, Dec. 11 – 14, 2007.

[9] H. Yousef, S. Cheng and H. Kratz, "79 GHz slot antennas based on substrate integrated waveguide (SIW) in a flexible printed circuit board," *IEEE Trans. Antennas Propag.*, vol. 57, no. 1, pp. 64 – 71, Jan. 2009.

[10] D. Deslandes and K. Wu, "Integrated microstrip and rectangular waveguide in planar form," *IEEE Microw. Wireless Comp. Lett.*, vol. 11, no. 2, pp. 68–70, Feb. 2001.

[11] High Frequency Structure Simulator (HFSS). Ver. 11.2, Ansoft Corporation, Pittsburgh, PA, 2009.

Frequency-Dependent Circuit Models of Carbon Nanotube Networks

Mahmoud A. EL Sabbagh and Samir M. El-Ghazaly

Dept. of Electrical Engineering, University of Arkansas, Fayetteville, AR, USA 72701

msabbagh@ieee.org; elghazal@uark.edu

Abstract

Frequency-dependent circuit models for carbon nanotube networks are developed. The procedure of extraction include: fabricating planar transmission-line structures for microwave test where CNT networks replace the metallic traces; extracting the transmission-line complex impedance and propagation constant; computing the transmission line parameters per unit length: resistance, inductance, capacitance, and conductance; finally deducing circuit models for CNT networks.

I. INTRODUCTION

Since the discovery of carbon nanotubes (CNTs) [1], the material has been investigated in many applications such as: mechanical resonator [2], nanotube transistors [3], antenna effect in arrays of aligned CNTs [4], [5], planar waveguide with saturable absorber based on CNTs [6], and interconnects [7]. Many other applications are mentioned in [8]. CNTs have superior mechanical, thermal, and electrical properties compared to copper which make them appealing for electronic packaging applications. There are theoretical studies about modeling of CNTs properties. For example, in [9], an RF circuit model is developed for single CNT transmission line placed parallel to ground plane. A model for single CNT working as dipole transmitting antenna is presented in [10]. However, the theoretical models were developed only for a single-walled metallic CNT. It is to be noted that as-produced CNT networks are a mixture of both metallic and semiconductor nanotubes depending on diameter and chirality. In recent work, electrical properties of CNT-based transmission lines are characterized [11] where fabricated transmission lines are of stripline configuration with different lengths and CNT networks replaced stripline inner conductor. It is found that the random nature of CNT networks leads to slight variation of extracted properties specifically at lower frequencies of the measured frequency range due to percolation theory [12]. Based on those characterizations, prototype of miniaturized resonator is developed and presented in [13]. Several other CNT-based RF components are possible to design, yet using current available commercial software in design procedure is quite impossible if not reaching a deadlock end. The main objective of this paper is to develop frequency-dependent circuit models of CNT networks based on microwave measurements of test structures. The procedure of extraction include: fabricating planar transmission-line structures for microwave test where CNT networks replace only the metallic traces not the ground planes; extracting the transmission line complex impedance and propagation constant; computing the transmission line parameters per unit length: resistance, inductance, capacitance, and conductance; finally separating the circuit parameters related to test structures from those pertaining to CNT networks. The circuit models are important CAD tool in design of RF/Microwave CNT-based components, which is considered to speed up initial design steps. Within the context of this work, two equivalent circuit models are proposed. Validity of each model is verified. The circuit models included in this paper correspond in general to randomly oriented CNT networks.

II. EXPERIMENTAL SETUP AND REALIZATION

Schematic of the proposed transmission line structure used to extract modeling parameters pertaining to the CNT networks under test is shown in Fig. 1. It consists of substrate and superstrate cut with identical length l, width S, and height h. The substrate shown in Fig. 1(b) has a groove cut along its length. Depth g and width W of the groove are designed such that $g << W$, $g << h$, and $W << S$. The superstrate shown in Fig. 1(a) has no groove and is used as the top cover. The dielectric

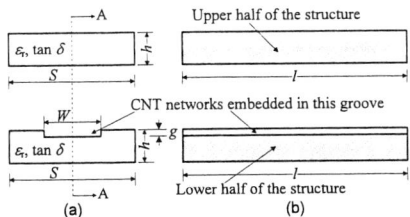

Fig. 1. Configuration of the proposed transmission line test structure. (a) Elevation view of the superstrate and substrate. (b) Side view at the cut A-A showing superstrate and substrates. Dimensions are not to scale.

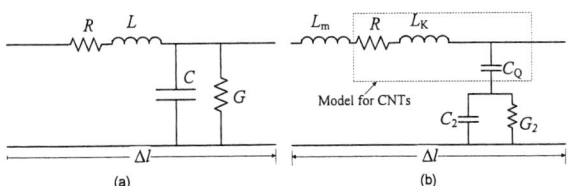

Fig. 2. Equivalent circuit model for CNT-based transmission line. (a) Well-known circuit model of transmission line. (b) Separated circuit to model CNT networks.

substrate used to hold the CNTs is Alumina 96% with thickness 1.016 mm(40 mil). Several samples of transmission line are fabricated. Their common features are as follows: substrate width $S = 31.25\,\text{mm}$, groove depth $g = 0.2\,\text{mm}$, and groove width $W = 1.45\,\text{mm}$. The samples have different length, $l \in \{2\,\text{cm}, 3\,\text{cm}, 4\,\text{cm}, 5\,\text{cm}\}$. CNT networks are obtained from commercial supplier in powder form. According to the supplier's data sheet, it is described as cleaned single wall nanotubes with purity $> 90\,\text{wt\%}$, ash $< 1.5\,\text{wt\%}$, diameter $1\,\text{nm}$ to $2\,\text{nm}$, and length $5\,\mu\text{m}$ to $30\,\mu\text{m}$. The CNTs are used as provided from the supplier. CNT networks are mechanically packed inside the groove. Mechanism of filling the groove consists of picking CNTs from their container then inserting gently inside the groove using tweezers. No further process is necessary. CNTs adhere efficiently to the groove surfaces due to Van Der Waal forces. To ensure the existence of a continuum of CNT networks in the groove, dc resistance between the input and output SMA connectors is always measured. The microwave measurements are carried using the S-parameters performance network analyzer (PNA): Agilent E8361A. Start and stop frequencies of the PNA are set to $10\,\text{MHz}$ and $400\,\text{MHz}$, respectively with a $0.975\,\text{MHz}$ frequency step. Input power to each port equals $-10\,\text{dB}$ and the intermediate frequency bandwidth is set to $20\,\text{Hz}$ to reduce effect of the random noise and reach the PNA maximum dynamic range.

III. CIRCUIT MODEL (1)

The well-known circuit model of transmission line is given in Fig. 2(a). Circuit transformations leads to the circuit model in Fig. 2 (b) which extracts the first proposed circuit model related to CNT networks from the total circuit model for transmission line. Practically, scattering parameters of the actual physical transmission line are measured. Values of lumped elements in the total circuit model are related to those measured scattering parameters. In Fig. 2(b), L_m is the magnetic inductance corresponding to magnetic energy storage due to current flow along the transmission-line trace, R represents the losses in CNT networks, L_K models the kinetic inductance of CNTs, C_Q is the quantum capacitance of CNTs, C_2 is the electrical capacitance representing storage of electric field in the substrate region between the transmission-line trace and ground plane, and G_2 models the substrate losses. The equivalent circuit of Fig. 2(b) indicates that the electrical C_2 and quantum C_Q capacitances contribute to the total capacitance C. The total conductance G and capacitance C in Fig. 2 (a) are related to the circuit elements shown in Fig. 2 (b) as

$$G = \frac{\omega^2 C_Q^2 G_2}{G_2^2 + \omega^2 (C_Q + C_2)^2}, \quad C = \frac{C_Q \left[G_2^2 + \omega^2 C_2 (C_Q + C_2)\right]}{G_2^2 + \omega^2 (C_Q + C_2)^2}. \tag{1}$$

R, L, C, and G are computed from microwave measurements of scattering parameters of the actual physical transmission line test structures. From (1), C_Q is obtained in terms of C_2, G_2, and C as follows:

$$C_Q = \frac{G_2^2 + \omega^2 C_2^2 - 2\omega^2 C C_2 + \sqrt{\left(G_2^2 + \omega^2 C_2^2\right)^2 - 4\omega^2 C^2 G_2^2}}{2\omega^2 (C - C_2)} \tag{2}$$

Mathematical studies of (2) indicate that the model given in Fig. 2 is valid if $C < C_2$ and $\omega C_2 > G_2$.

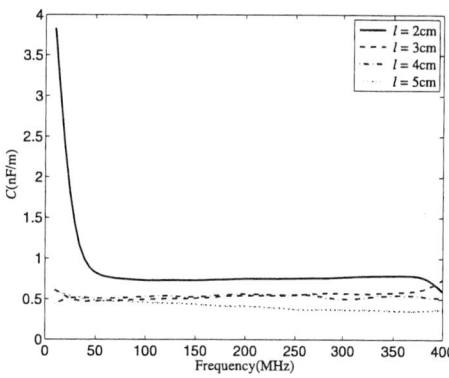

Fig. 3. Capacitance C per unit length in nF/m versus frequency in MHz for stripline structure. Dimensions of parameters are: $g = 0.2\,\text{mm}$, $W = 1.45\,\text{mm}$, $S = 31.25\,\text{mm}$, $h = 1.016\,\text{mm}$, $l = 50\,\text{mm}$, $\varepsilon_r = 9.4$, and $\tan\delta = 0.006$. The capacitance is extracted from measurements of scattering parameters.

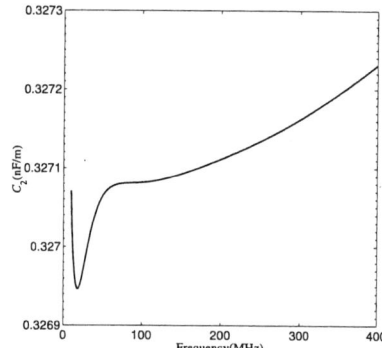

Fig. 4. Capacitance C_2 per unit length in nF/m versus frequency in MHz for stripline structure. Dimensions of parameters are: $g = 0.2\,\text{mm}$, $W = 1.45\,\text{mm}$, $S = 31.25\,\text{mm}$, $h = 1.016\,\text{mm}$, $l = 50\,\text{mm}$, $\varepsilon_r = 9.4$, and $\tan\delta = 0.006$. The results are obtained from HFSS simulations.

A. Results

Computation of C_Q based on (2) requires the knowledge of C, C_2, and G_2. C is extracted from measured scattering parameters [11]. C_2 and G_2 are obtained from full-wave simulations using HFSS [14] for the case of perfect conducting transmission-line trace. The results given herein correspond to the stripline test structure shown in Fig. 1 where dimensions of parameters are:

978-1-4244-4447-2/09 $25.00 © 2009 IEEE

$g = 0.2\,\text{mm}$, $W = 1.45\,\text{mm}$, $S = 31.25\,\text{mm}$, $h = 1.016\,\text{mm}$ and $l = 50\,\text{mm}$. Alumina 96% is used as substrate where its relative dielectric constant $\varepsilon_r = 9.4$ and loss tangent $\tan\delta = 0.006$.

The graph of C per unit length is shown in Fig. 3. The electrical capacitance C_2 per unit length is presented in Fig. 4. The graph shows that C_2 is quite a constant equal to 0.327 nF/m. The conductance G_2 per unit length is presented in Fig. 5. The given values of conductance show that substrate losses are quite negligible, yet increasing linearly with frequency. The computed values of C_2 and G_2 correspond to the condition $\omega C_2 > G_2$. The capacitance of CNTs computed using (2) is given in Fig. 6. The graph indicates that the computed quantum capacitance C_Q of CNTs is negative. At this point, the model need to be revisited to figure out the proper circuit model.

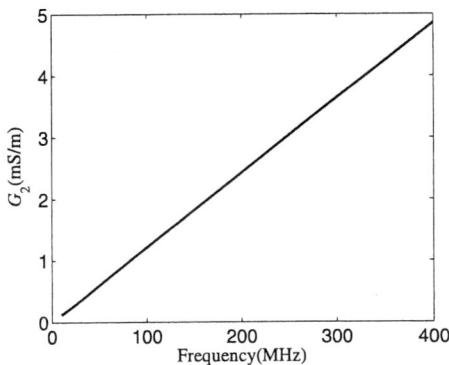

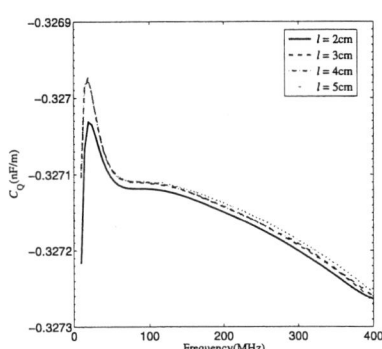

Fig. 5. Conductance G_2 per unit length in mS/m versus frequency in MHz for stripline structure. Dimensions of parameters are: $g = 0.2\,\text{mm}$, $W = 1.45\,\text{mm}$, $S = 31.25\,\text{mm}$, $h = 1.016\,\text{mm}$, and $l = 50\,\text{mm}$. The results are obtained from HFSS simulations.

Fig. 6. Extracted capacitance of CNTs C_Q per unit length in nF/m versus frequency in MHz for stripline structure. Dimensions of parameters are: $g = 0.2\,\text{mm}$, $W = 1.45\,\text{mm}$, $S = 31.25\,\text{mm}$, $h = 1.016\,\text{mm}$, and $l = 50\,\text{mm}$. Capacitance is computed using (2).

IV. CIRCUIT MODEL (2)

The equivalent circuit model given in Fig. 2 did not predict proper value for C_Q to represent the measured characteristics of CNT networks. In this section, different model is given as shown in Fig. 7. The circuit model consists of series and shunt branches. The series branch consists of serial connection of resistance R_1, inductances L_m and L_K, and capacitance C_1. The shunt branch consists of inductor L_2 in series with parallel combination of capacitance C_2 and conductance G_2. In this model L_K represents the kinetic inductance of CNT networks, C_2 represents electrical energy stored in substrate and G_2 is the conductance due to substrate losses. C_2 and G_2 are the same elements included in the circuit model of Fig. 2. The relations between the

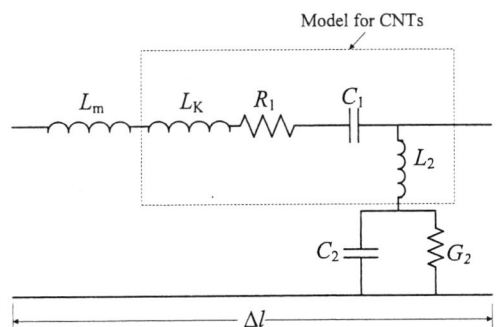

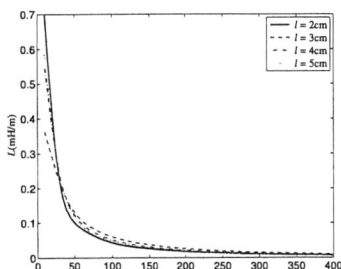

Fig. 7. Equivalent circuit model for CNT-based stripline. The series branch consists of a serial connection of resistance R_1, inductances L_m and L_K, and capacitance C_1. The shunt branch consists of an inductor L_2 in series with parallel combination of capacitance C_2 and conductance G_2.

Fig. 8. Extracted equivalent inductance L in the series branch of the equivalent circuit model shown in Fig. 7. Dimensions of parameters are: $g = 0.2\,\text{mm}$, $W = 1.45\,\text{mm}$, $S = 31.25\,\text{mm}$, $h = 1.016\,\text{mm}$, and $l = 50\,\text{mm}$.

elements of this circuit model and those equivalent elements shown in Fig. 2(a) are

$$R = R_1, \quad L = L_m + L_K - \frac{1}{\omega^2 C_1}, \quad G = \frac{G_2}{\left(1 - \omega^2 L_2 C_2\right)^2 + \omega^2 L_2^2 G_2^2}, \quad C = \frac{C_2\left(1 - \omega^2 L_2 C_2\right) - L_2 G_2^2}{\left(1 - \omega^2 L_2 C_2\right)^2 + \omega^2 L_2^2 G_2^2}. \tag{3}$$

L_2 is expressed in terms of C, C_2 and G_2 as given by the following equation

$$L_2 = \frac{-G_2^2 - \omega^2 C_2 \left(C_2 - 2C\right) + \sqrt{\left(G_2^2 + \omega^2 C_2^2\right)^2 - 4\omega^2 C^2 G_2^2}}{2\omega^2 C \left(G_2^2 + \omega^2 C_2^2\right)}. \tag{4}$$

Mathematical studies of (4) indicate that the model given in Fig. 7 is valid if $C > C_2$.

The graph of inductance L_2 per unit length is presented in Fig. 9. The best fitting curve is $L_2 = 7.74 \times 10^7 f^{-2}$ where L_2 is in H/m and f is in Hz. The capacitance C_1 in the series branch is computed using $C_1 = \frac{1}{\omega^2 (L_m + L_K - L)}$. Values for L_K are taken from the inductance curve extracted from microwave measurements shown in Fig. 8 at zero frequency. Those values are 0.508 mH/m, 0.4168 mH/m, 0.7639 mH/m, and 0.9434 mH/m corresponding to $l = 2$ cm, 3 cm, 4 cm, and 5 cm, respectively. The magnetic inductance $L_m = 291$ nH/m is obtained from HFSS and it is negligible compared to the kinetic inductance. The extracted capacitance C_1 is shown in Fig. 10. The best fitting curve for C_1 is $C_1 = 1.59 f^{-1.82}$ where C_1 is in F/m and f is the frequency in Hz. The resistance R_1 is given in [11] and it has the frequency dependence $R_1 = \frac{5 \times 10^8}{\sqrt{f}}$. The frequency-dependent lumped elements L_2 and C_1 added in the circuit model given in Fig. 7 predict properly electrical properties of CNT networks as shown in Figs. 9 and 10.

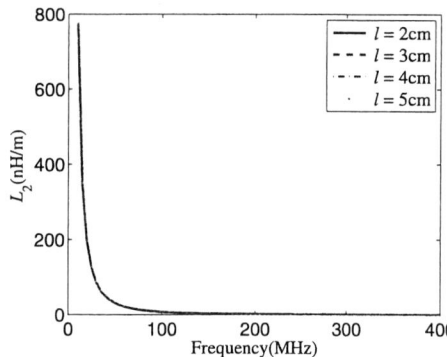

Fig. 9. Extracted series inductance L_2 in the shunt branch of the equivalent circuit model shown in Fig. 7. Dimensions of parameters are: $g = 0.2$ mm, $W = 1.45$ mm, $S = 31.25$ mm, $h = 1.016$ mm, and $l = 50$ mm.

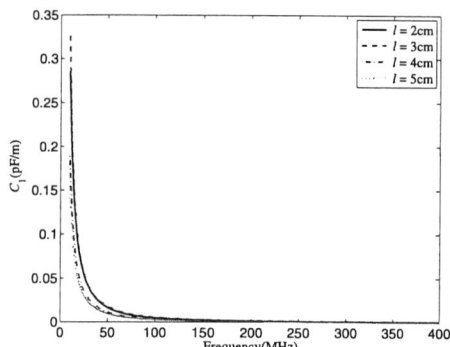

Fig. 10. Extracted series capacitance C_1 in the series branch of the equivalent circuit model shown in Fig. 7(b). Dimensions of parameters are: $g = 0.2$ mm, $W = 1.45$ mm, $S = 31.25$ mm, $h = 1.016$ mm, and $l = 50$ mm.

V. CONCLUSIONS

Frequency-dependent equivalent circuit models are developed for carbon nanotube networks. The models are extracted from microwave measurements of the scattering parameters of transmission line structures where CNT networks replace the planar metallization.

REFERENCES

[1] S. Iijima, "Helical microtubules of graphitic carbon," *Nature*, vol. 354, pp. 56-58, Nov. 1991.

[2] J. F. Davis, M. Bronikowski, D. Choi, L. Epp, M. Hoenk, D. Hoppe, B. Kowalczyk, F. Noca, E. Wong, B. Hunt, B. Chang, M. Jouzi, M. Tzolov, A. Yin, J. Xu, J. D. Adam, R. M. Young, J. Adams, B. Rogers, "High-Q mechanical resonator arrays based on carbon nanotubes," in *IEEE Conference on Nanotechnology*, San Francisco, CA, USA, Aug. 2003, pp. 635-638.

[3] J. Appenzeller, "Carbon nanotubes for high-performance electronics - progress and prospect," *Proceedings of the IEEE*, vol. 96, no. 2, pp. 201-211, Feb. 2008.

[4] Y. Wang, K. Kempa, B. Kimball, J. B. Carlson, G. Benham, W. Z. Li, T. Kempa, J. Rybczynski, A. Herczynski, and Z. F. Ren, "Receiving and transmitting light-like radio waves: antenna effect in arrays of aligned carbon nanotubes," *Applied Physics Letters*, vol. 85, no. 13, pp. 2607-2609, Sep. 2004.

[5] Q. Zhu, L. Wu, S. Sheng, Z. C. Mei, W. F. Liu, W. L. Cai, and L. Z. Yao, "Possibility of constructing microwave antenna with carbon nanotubes," *Journal of Vacuum Science Technology B: Microelectronics and Nanometers Structures*, vol. 25, no. 5, pp. 1630-1634, Sep./Oct. 2007.

[6] K. Kashiwagi, S. Yamashita, Y. Nasu, H. Yaguchi, C. S. Goh, and S. Y. Set, "Planar waveguide-type saturable absorber based on carbon nanotubes," *Applied Physics Letters*, vol. 89, 0812259(1-3), 2006.

[7] A. Naeemi, R. Sarvari, and J. D. Meindl, "Performance comparison between carbon nanotube and copper interconnects for gigascale integration (GSI)," *IEEE Electron Device Letters*, vol. 26, no. 2, pp. 84-86, Feb. 2005

[8] M. J. O'Connell, *Carbon Nanotubes Properties and Applications*, Taylor and Francis, Boca Raton, FL, 2006.

[9] P. J. Burke, "An RF circuit model for carbon nanotubes," *IEEE Trans. Nanotechnol.*, vol. 2, no. 1, pp. 55-58, Mar. 2003.

[10] G. W. Hanson, "Fundamental transmitting properties of carbon nanotube antennas," *IEEE Trans. Antennas Propag.*, vol. 53, no. 11, pp. 3426-3435, Nov. 2005.

[11] M. A. EL Sabbagh, S. M. El-Ghazaly, and H. A. Naseem, "Carbon nanotube-based planar transmission lines," in *IEEE MTT-S Int. Microwave Symp. Dig.*, Boston, MA, 7-12 June 2009, pp. 353-356.

[12] D. Stauffer and A. Aharony, *Introduction to Percolation Theory*, Taylor and Francis, Washington, DC, 1992.

[13] M. A. EL Sabbagh and S. M. El-Ghazaly, "Miniaturized carbon nanotube-based RF resonator," in *IEEE MTT-S Int. Microwave Symp. Dig.*, Boston, MA, 7-12 June 2009, pp. 829-832.

[14] Ansoft HFSS, Pittsburgh, PA, Version 11.1.3, 2008.

A two-level optimization scheme for bandwidth optimization of a microprocessor vertical interconnect

Arun V. Sathanur[1], Vikram Jandhyala[1], and Henning Braunisch[2]

[1]ACE Lab, Department of Electrical Engineering, University of Washington, Seattle, WA, USA

[2]Components Research, Intel Corporation, 5000 W. Chandler Blvd., Chandler, AZ, USA

E-mail: {arunsv,vj}@u.washington.edu, henning.braunisch@intel.com

Abstract: A parametric model of a generic differential ten-layer microprocessor package line with two motherboard layers has been developed. A dimensionality reduction scheme and a reusable, multi-dimensional look-up table precede the global optimization phase which is facilitated by a smooth interpolation scheme based on splines. An accelerated boundary element based full-wave electromagnetic solver has been used to construct a look-up table of S-parameters for a number of designs. The second phase features a custom local optimizer incorporating all the variables without any dimension reduction. This methodology has been applied to automated synthesis of the differential package line resulting in a significant improvement of the return loss performance.

1. Introduction

Presently, multi-core processor based computing systems are gaining popularity since they allow an increased system performance without the need for indefinitely increasing the frequency of operation of the central processing unit (CPU) cores. With the advent of the multi-core processors, an increased demand is being placed on the maximum input/output (I/O) data rates of the CPU package structures. The two different aspects of the electrical performance of a CPU package structure are related to power delivery and high-speed I/O bandwidth. The present work focuses only on the bandwidth optimization. The critical portion of the package is the vertical interconnect, comprising features such as plated through-holes (PTHs), micro-vias, pads, anti-pads, and lands. Appropriate design of the vertical interconnect is critical for channel impedance matching; degradation of matching results in increased reflections and contributes to reduced signaling speeds and data bandwidth. It is therefore imperative that schemes be developed to optimize the dimensions of the various features of the vertical portion of the package interconnect to achieve maximum possible bandwidth within the existing manufacturing realm before introducing any disruptive technologies.

The complex features of the vertical interconnect portion, along with the increased frequency of operation and reduced pitch, which results in increased mutual coupling, makes the full-wave analysis of package interconnects mandatory. The requirement to adequately capture all the feature sizes makes the simulations computationally very expensive. Traditional methods of connecting quasi-static models of the various portions of the interconnect in SPICE-like circuit simulators do not produce satisfactory results beyond a few GHz [1]. However, many commercial full-wave field solvers based on the Method of Moments (MoM) and Finite Element Method (FEM) suffer from high main memory requirements and increased time to solution. Accelerated Boundary Element Method (BEM) based electromagnetic solvers have been receiving considerable interest, and one such solver [2] is used here as the forward analysis tool in the optimization process.

Even with the use of fast solvers, the large number of parametric variables creates a high dimensionality design space and makes the overall optimization procedure based on using the electromagnetic solver in the loop extremely expensive. The work in [3] utilizes response surface models and a hierarchical scheme for optimization. Response surface models suffer from low accuracy and a hierarchical scheme for a large number of variables leads to an exponentially large number of

---

The authors wish to acknowledge the support of Intel Corporation and in particular thank Dr. Kemal Aygün and Dr. Zhichao Zhang for helpful discussions.

simulations. This paper presents a combination of techniques to make the problem tractable in a reasonable amount of time while circumventing the concerns of the previous methodology.

2. The parametric package model

To develop an automated framework for parametric sweeps, optimization and statistical analysis of CPU packages, the first step is the development of a parametric package model, in extension of [3]. A ten-layer differential package line is used as the test structure. It is comprised of three build-up layers on the bottom followed by four core layers and three build-up layers on the top giving rise to a 3-4-3 package structure with plated through-holes (PTHs) traversing the core layers. The micro-via from layer 4B lands directly on a large Ball Grid Array (BGA) pad in layer 5B, which then couples through a solder joint to the motherboard that is represented by two metal layers. A profile view of such a structure is shown in Fig. 1.

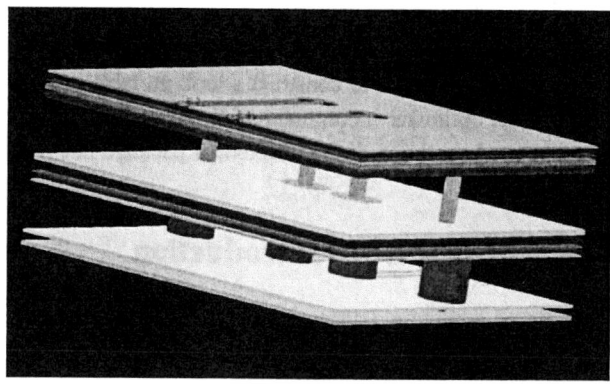

Figure 1: Profile view of the differential package line

3. The optimization process

The flowchart shown in Fig. 2 summarizes the steps followed in the optimization process. The process starts with reduction in the number of dimensions. Two approaches to dimensionality reduction are implemented. Parametric sweeps are conducted around a few random design points to assess the sensitivity of the differential return loss (henceforth called DS_{11}) with respect to each of the parameters. Only sensitive variables are selected for the first phase of the optimization. Several parameters, by their physical origin, have very similar effects on the objective function DS_{11}. Examples include the build-up voiding extensions (describing the extension of the void beyond the pad edge in the build-up layers 5B/5F, 4B/4F and 3B/3F), the core voiding extensions (extension of the void beyond the pad edge in the core layers 2B/2F and 1B/1F), the build-up pad extensions (extension of the void beyond the via edge in the build-up layers) and so on. Each of these groups of variables is represented by a single parameter per group. These two observations lead to a tremendous decrease in the number of simulations, without sacrificing accuracy. Sampling is done adequately along each dimension and a look-up table (LUT) is created for the S-parameters. Creation of such LUT is made possible by a combination of several factors such as decreased hardware cost, highly parallelized codes and acceleration of the BEM solvers themselves.

There are several advantages of the LUT approach. It is completely re-usable in the sense that it needs to be constructed only once per technology; the same LUT can be used as a starting point for incremental changes in the technology and it is very useful when objective functions change. The advantage becomes even more evident in the case of multi-objective optimization where a number of single objective optimization problems have to be solved.

Due to the limitations of traditional gradient based optimizers, Simulated Annealing, a physics based global optimizer, is utilized to find the global minimum of the approximate surface [4]. This stage is aided by utilization of splines for providing the interpolated values of the objective function in the points interior to each n-dimensional hypercube defined by the grid points. Since the function evaluations are done through a closed-form expression fitted to the LUT through spline interpolation, this step is computationally expedient and provides a better approximation to the global optimum as opposed to an inspection of the LUT. The next stage involves a custom local optimizer to refine the coarse global optimum obtained. In this stage, all the dimensionality reduction schemes are removed to allow complete exploration of the space around the coarse, global optimum. The choice of the algorithm for the local optimizer is governed by the fact that the Hessian matrix is very expensive to compute through finite differences (approximate $2n^2$ simulations for n parameters where n can easily reach 15). Hence the steepest descent algorithm which requires only the gradient computation is chosen. The flow for the local optimizer is shown in Fig. 3.

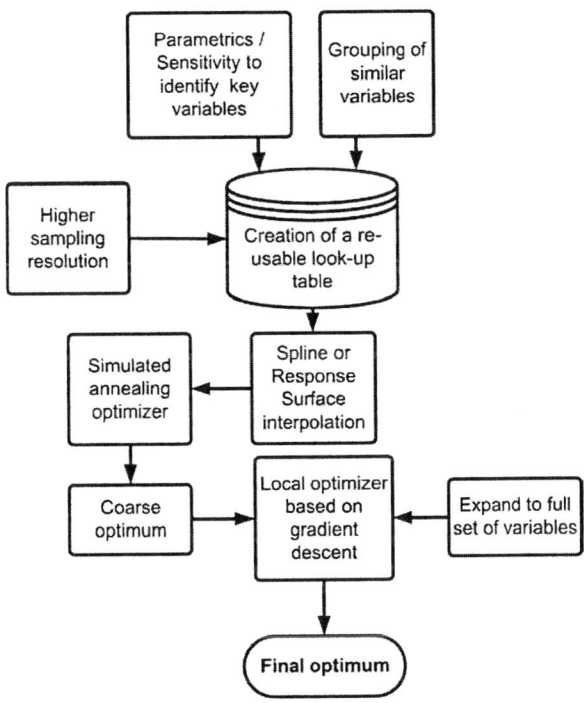

Figure 2: Flow diagram of the optimization process

4.   Results of numerical experiment

The optimization scheme developed has been applied to the differential package line shown in Fig. 1. The LUT involved three variables, the build-up layer void extensions tied together, the core layer void extensions tied together and the PTH drill diameter. It had 147 entries. All the simulations were carried out at 10 GHz on an eight-core machine while using seven of the eight cores. The local optimizer featured all the void sizes and pad sizes as unique parameters and the PTH drill diameter. The total time taken to construct the LUT was just under 9 hours and the local optimizer phase took 78 simulations and a time of 4 hours and 45 minutes for an overall optimization time of about 14 hours. The frequency sweeps for the three structures are shown in Fig. 4. The differential return loss for the final optimized structure is improved by 10–20 dB over the entire frequency range up to 12 GHz. The 10-dB differential return loss bandwidth is improved by more than 5x.

5.   Conclusions

A two stage optimizer was presented to facilitate optimization of microelectronic structures with large number of parameters requiring computationally expensive simulations. The process was applied to a representative differential microprocessor package interconnect line. Results indicate significant improvements in differential return loss. Work is concurrent on exploring trade-offs through multi-objective optimization and including statistical analysis.

978-1-4244-4447-2/09 $25.00 © 2009 IEEE

Expand to full set of variables

Set $\mathbf{x}_0 = \mathbf{x}_{\text{global}}$

(while !convergence)

{

Compute gradient ∇f

Compute $\beta_0 = \left(\dfrac{d_{\min}}{|\nabla f|^2} \right)$, $d_{\min}$ being the minimum decrease needed in the objective function

Start line search from $\beta = \beta_0$

$\mathbf{x}^1 \leftarrow \mathbf{x} - \beta \nabla f$

Compute $f(\mathbf{x}^1)$

Increase β until $\left| \beta |\nabla f|^2 - \left(f(\mathbf{x}^1) - f(\mathbf{x}) \right) \right| \le \alpha \left| f(\mathbf{x}^1) - f(\mathbf{x}) \right|$ and $\left(f(\mathbf{x}^1) < f(\mathbf{x}) \right)$, $\alpha = 0.05 - 0.1$

i.e. the function decreases and the predicted decrease differs from actual decrease by less than $5 - 10\%$

}

Figure 3: Algorithm for the local optimizer

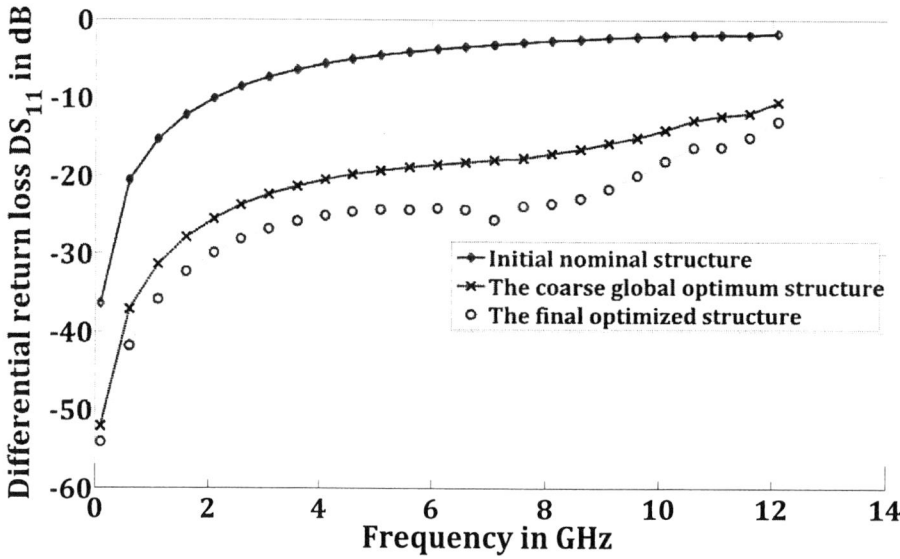

Figure 4: Comparison of frequency sweeps pre- and post-optimization of differential return loss

References

1. J. Mao, G. Fitzgerald, A. Kuo, and S. Wane, "Coupled analysis of quasi-static and full-wave solution towards IC, package and board co-design," in *Proc. IEEE Electrical Performance of Electronic Packaging (EPEP)*, Atlanta, GA, Oct. 29–31, 2007, pp. 111–114.

2. D. Gope and V. Jandhyala, "Efficient solution of EFIE via low-rank compression of multilevel predetermined interactions," *IEEE Trans. Antennas Propagat.*, vol. 53, no. 10, pp. 3324–3333, Oct. 2005.

3. A. V. Sathanur, V. Jandhyala, K. Aygün, H. Braunisch, and Z. Zhang, "Return loss optimization of the microprocessor package vertical interconnect," in *Proc. IEEE Electronic Components and Technol. Conf. (ECTC)*, San Diego, CA, May 26–29, 2009, pp. 1636–1642.

4. S. Kirkpatrick, C. D. Gelatt, and M. P. Vecchi, "Optimization by simulated annealing," *Science*, vol. 220, no. 598, pp. 671–680, May 1983.

Defining a Multi-Channel Infrastructure to enable MNA Formulation of Opto-Electronic Circuits

P. Gunupudi†, T. Smy†, J. Klein‡ and J. Jakubczyk‡

†Department of Electronics, Carleton University, Ottawa, Canada, ON K1S 5B6
email: {pavan, tjs}@doe.carleton.ca

‡Optiwave Systems Inc., 7 Capella Court, Ottawa, Ontario, K2E 7X1
email: {jackson.klein, jan.jakubczyk}@optiwave.com

Abstract

This paper presents a pre-calculation engine to enable formulation of opto-electronic systems incorporating multiple channels using a Modified Nodal Analysis (MNA) approach for self-consistent simulation in a SPICE-like single-engine simulator.

I. INTRODUCTION AND OVERVIEW

Rapid advances in microelectronics have pushed the borders of interconnect technology; chips and PCBs are now operating at higher frequencies and packages require higher pin-out densities. New system specifications pose significant challenges to the design community in producing reliable designs. In order to address these challenges, optical circuitry is being gradually introduced beside electrical components at all levels of the design hierarchy [1]–[3]. With both optical and electrical components present at the same level of design hierarchy, designers need a tool capable of simulating opto-electronic circuits reliably and efficiently. Recently such a simulation framework (*OptiSPICE*) was introduced where modifications to a Modified Nodal Analysis (MNA) approach [4], [5] and an infrastructure for optical nodes was suggested to facilitate simulation of optical and electrical components in a single-engine simulator [6].

One of the chief advantages of optical systems is the ability to carry information signals in separate channels within the same optical waveguide using wavelength division multiplexing (WDM). Inclusion of multiple channels in the formulation of MNA matrices for an opto-electronic system using the *OptiSPICE* infrastructure needs first the determination of channel topologies and then the evaluation of several quantities in a pre-simulation stage. *This paper discusses the necessity of evaluating these quantities before simulation and presents efficient ways of designing the pre-calculation engine that enables formulation of system equations using an MNA approach.*

The simulation framework presented in [6] introduces the concept of an optical node and illustrates the modeling of signal propagation between optical nodes. These signals are modeled using the complex envelope of electric field of each mode present at an optical node and are represented by the magnitude and phase of the envelope. The use of complex envelope improves simulation speeds as rapid variations of carrier frequency need not be captured. As optical components in general have several active modes, optical signals in this framework contain information related to each mode at every node of interest. A set of such modes, each consisting of magnitude and phase of the complex envelope (for both forward and reverse traveling waves) of electric field constitutes the optical signal for a single channel.

To extend the *OptiSPICE* infrastructure to multiple channels, additional information must be included in device models and optical nodes. For devices with independent channels, this simply requires duplication of the mode structure for each channel. Prior to simulation and formulation of system matrices, the number and type of modes present in each component and carrier frequencies of these modes have to be determined. Determining the mode structure (the number and particular mode shapes) depends on the optical device geometry. For simple device geometry, analytical mode shapes such as Bessel functions can be appropriate. For more complicated situations, numerical mode solvers based on 2D solutions of Maxwell's Equations can be used. Existing mode solvers available in the market can be coupled relatively easily with this framework and determining the modes of optical components is usually straight-forward; the primary exception being mirrors, optical interfaces and free-space elements involving diffraction over a distance. These elements have to be specially handled in the pre-calculation engine and will be discussed in detail.

At optical interfaces where components interact, the mismatch between mode structures will produce reflections, scattering and transmission. At each such interface, optical connector elements are inserted to take the modal information from the input device with a set of mode shapes and map this information to the modes of the output device. This mapping is done through a set of coefficients calculated using overlap integrals which will be described in this paper.

This paper discusses the above-mentioned issues and provides efficient solutions for determining modes and channels of optical components and their topological connections along with the evaluation of overlap integrals to establish mode mapping from one optical component to the other to enable an MNA type of formulation of opto-electronic circuits in a SPICE-like simulation engine [5].

978-1-4244-4447-2/09 $25.00 © 2009 IEEE

II. PRE-CALCULATION ENGINE FOR OPTO-ELECTRONIC SIMULATION

As mentioned above, for a multi-channel optical system several calculations have to be performed and channel topologies determined before performing simulation using the opto-electronic simulator described in [6]. These calculations are made within a pre-calculation engine; issues related to this engine are discussed in this section.

A. Channel Topology

The mode structure of devices, diffraction effects of elements based on free space propagation, and characterization of optical filters are all dependent on the carrier frequency. As such, the first task is to determine the channel topography of the optical system.

For the purpose of determining the channel topology, it is useful to categorize devices into two classes; 1) sources which generate an optical signal at a particular carrier frequency and 2) devices that operate on an input signal without generating a new carrier frequency. All optical devices are capable of being excited at several carrier frequencies. The particular carrier frequency or set of frequencies present in an optical device is determined by internal excitations (in the case of a source) and external signals incident from an optical element to which it is connected. The carrier frequencies present in this optical element, in turn depend on other optical elements connected to it. This optical path can be traced back all the way to optical sources. As such, in the pre-calculation stage, in order to create the optical channel topography, one has to efficiently traverse through all elements in the network to determine the particular route taken by channels corresponding to particular carrier frequencies. A wide variety of elements influence the topography of the channel network. Simple input/output devices have identical channel specifications at both ports. Some devices such as photodiodes and perfectly absorbing mirrors terminate channels. Multiple input/output devices such as splitters/joiners may have varying channel specifications on different ports. Optical filters selectively terminate and transmit channels. Reflections at optical connectors and mirrors complicate the tracing of optical channels because of the bi-directional nature of the optical signal. In addition, if there are multiple sources operating at the same carrier frequency care must be taken to merge channels and avoid duplication.

The problem of tracing the channel topography is essentially recursive. Each element has one or more ports. A channel present at one port is either transmitted to other ports or blocked. It is possible to trace channels from one element to another until a termination condition is found. Care must be taken to ensure that recursion is terminated and infinite loops are avoided. Shown in Algo. 1 is the pseudo-code for a recursive method used to determine the channel topology.

The channel topography is built by the function BUILDCHANNELTOPO which takes as input a list of all sources present in the circuit. Each source is then used to trace an optical channel by recursively calling the function TRACEOPTLINK which has as input the *ActiveChannel* to be traced and an optical element. The recursion is terminated when another element is encountered with the same channel as *ActiveChannel* or if the encountered element terminates the channel. This method establishes the channel topology for an optical circuit which can then be used to assign mode information for elements described in the next subsection.

B. Mode-calculation

After determining channel topography, the pre-calculation engine should determine mode shapes present in each optical device to allow for the calculation of device coupling. For most optical devices, mode shapes are determined by the physical geometry of the device and the carrier frequency. Modes of these components can be determined analytically, using a mode-solver or read from a library. A few elements need to be handled differently as the modes present are determined by adjacent devices. A mirror has only one port and the mode structure of the reflected signal is determined by its input device. The optical connector is not a physical device and represents the interface between two physical elements. It therefore has an input mode structure determined by the input device and an output mode structure determined by the output device. The free-space element is similar as it has input modes and output modes defined by neighbouring elements, however, it also has an internal field distribution that must be determined from diffraction of input fields. A complication can arise when a free-mode element is connected to another free-mode element. In view of this complication, it is necessary to propagate mode shapes through free-mode elements until all mode shapes are known.

C. Evaluation of Device Coupling

Once channel topology is determined and mode information for each optical component is evaluated, device coupling can be calculated. After this, MNA matrices can be formed and the system simulated.

For optical connectors, which represent the interface between two physical devices with specific mode structures, values of the connection matrix C must be calculated. This matrix relates the input (I_i) and output (O_i) complex envelopes for each mode such that $O_i = C_{ij}I_j$. Coefficients of this matrix represent the percentage of energy from the j'th input mode that excites the i'th output mode. This coefficient can be calculated using an overlap integral of the form [7]:

$$C_{ij} = \int \int \hat{O}_i(x,y)\hat{I}_j(x,y)dxdy \tag{1}$$

```
BUILDCHANNELTOPO (ListOfSources)
1  for source in ListOfSources

2      if (NEWCHANNEL (source.CarrierFreq))

3          ActiveChannel = ADDCHANNEL (Channels,source.CarrierFreq)
4          CREATECHANNEL(ActiveChannel)
5      else
6          ActiveChannel = GETCHANNEL (Channels,source.CarrierFreq)

7      TRACEOPTLINK(source.Output,ActiveChannel)

8  return


TRACEOPTLINK(OpticalElement,ActiveChannel)
1   if OpticalElement.type == PHOTODIODE or OpticalElement.type == SOURCE) return
2   if (ChannelExisted = UPDATEELEMENTCHANNELS(OpticalElement,ActiveChannel)) return

3   if (OpticalElement.type == MIRROR) TRACEOPTLINK(OpticalElement.InputPort,ActiveChannel)
4   if (OpticalElement.type == OPTICALCONNECTOR)

5       TRACEOPTLINK(OpticalElement.InputPort,ActiveChannel) // reflections
6       TRACEOPTLINK(OpticalElement.OutputPort,ActiveChannel)

7   if (OpticalElement.type == OPTICALFILTER)

8       If (OpticalElement.ChannelTransmitted)TRACEOPTLINK(OpticalElement.OutputPort,ActiveChannel)

9   if (OpticalElement.type == OPTICALSPLITTER)

10      for port in OpticalElement.OutputPorts TRACEOPTLINK(port,ActiveChannel)

11  if (OpticalElement.type == OPTICALJOINER)

12      for port in OpticalElement.InputPorts TRACEOPTLINK(port,ActiveChannel)

13  return
```

Algo. 1. Two pseudo-code functions for the recursive generation of channel topography.

where $\hat{I}_j(x,y)$ and $\hat{O}_i(x,y)$ represents the normalized mode shapes for input and output modes respectively. If input and output mode structures are same, then due to the orthonormal nature of modes the C matrix will be an identity matrix. As some of the optical energy may be reflected, absorbed or scattered the total output energy may be less then that of the input.

Free-space elements also require coupling information to be determined. However, the process is somewhat more complicated. The Fresnel diffraction integral can be used to determine field distribution at the output of the device [8],

$$D_t(r,z_0) = \sum_j \hat{D}_j(r,z_0) \quad \text{with} \quad \hat{D}_j(r,z_0) = \frac{k}{2\pi i z_0} \int \int \hat{I}_j(r',0) e^{ik\frac{(r-r')^2}{2z_0}} \, dr' d\phi \tag{2}$$

where D_t is the total electric field at the output and $\hat{D}_j$ represents the field distribution of the component due to the j'th input mode I_j of the input device. It should be noted that these field components are no longer orthonormal due to the diffraction process. The relationship between input modes and output modes is linear and can be represented by the scattering matrix S.

$$O_i = S_{ij} I_j \quad \text{where} \quad S_{ij} = \int \int \hat{O}_i(r) \hat{D}_j(r) dr d\phi \tag{3}$$

where the excitation of the output element's i'th mode by the j'th input mode can be determined by an overlap integral. Once S is determined, the free-space element is characterized and can be used in the formulation of MNA matrices. Calculation of the matrices C and S are quite computationally intensive, but typically independent of the actual system simulation. They can therefore be computed in the pre-calculation stage prior to simulation. Optionally they can be cached to disk using a unique identifier and read to be used later for subsequent runs.

III. RESULTS

The example presented in Fig. 1a shows a simple coarse WDM example. This example has four laser sources – each driven by a BJT circuit and at carrier frequencies spaced from $840\ nm$ to $900\ nm$, a four input optical mux, followed by a multimode fiber and then a demux connected to four detector circuits. The output of each mux has an optical filter that selects a particular optical channel. The lasers and drivers are mounted on a single substrate and thermally coupled through the substrate resulting in differential heating. The simulation involves three distinct energy domains; electrical, thermal and optical.

The first stage in the formulation of system equations for this example would be creation of the channel topography using the algorithm shown in Algo. 1. Starting with source *Laser-1*, the pre-calculation engine initiates the first channel (840 nm). As there is a mode mismatch between the laser and waveguide inputs of the mux element, an optical connector is needed to represent mode coupling and reflections present at the interface. The first channel is propagated through this optical connector and the Mux element. The next physical element it encounters is a multi-mode fiber. Once again as there is a mode mismatch at the interface an optical connector is needed. At this point, there are two paths to be followed by the algorithm: 1) to the output of the fiber and (if there are reflections) 2) back to all the input ports of the mux element. Following the 1st path, the channel will propagate through another optical connector and then to all the outputs of the demux. Following the algorithm through output port 1, the channel is passed to the detector. The detector includes an optical filter which terminates all channels except one which is passed to the photodiode which terminates the remaining channel. The photodiode is placed in an electrical circuit which provides biasing and amplifies the photocurrent. As this path is now complete, the algorithm completes tracing of channel 1 propagating a channel from input ports of the mux back to the other sources which terminate the channel. Following this procedure, the channel topography is traced for all four channels. The actual topography is dependent on the system description and the presence of reflections. The proposed algorithm minimizes channel lengths; only propagating channels into branches which are needed. This is important as the size of system equations can become extremely large for more complicated systems if all channels are propagated over all branches.

Once the channel topography has been determined devices can be fully characterized: mode structures can be determined for all elements; optical coupling matrices (C) can be calculated to characterize optical interfaces; stamps for electrical models and laser rate equation models can be formulated; and the other optical elements described by linear and non-linear models can be stamped. Finally, the global MNA matrices are formed and DC, transient or small-signal simulation can be performed.

Fig. 1b shows results of a transient simulation for this circuit, presenting optical power at detectors for the 1st and 2nd channels for anti-symmetrical bit streams. Apart from the apparent offset, a difference in the signals propagated in the two channels can be seen. This variation in bit-shapes is due to differences in the layout of the integrated laser and driver which creates different thermal resistances for the two sources and disparate device temperatures. The temperature sensitivity of laser rate equations results in differing output powers.

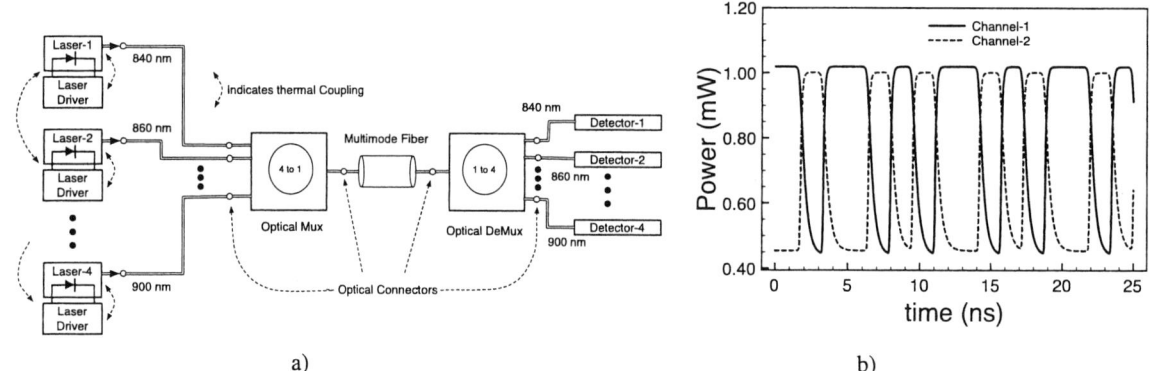

a)   b)

Fig. 1. Multi-channel optical circuit. a) System description. b) Output optical power.

REFERENCES

[1] L. Tsybeskov, D. Lockwood, and M. Ichikawa, "Silicon photonics: Cmos going optical," *Proceedings of the IEEE*, vol. 97, no. 7, pp. 1161–1165, July 2009.
[2] K. Ohashi, K. Nishi, T. Shimizu, M. Nakada, J. Fujikata, S. Ushida, S. Torii, K. Nose, M. Mizuno, H. Yukawa, M. Kinoshita, N. Suzuki, A. Gomyo, T. Ishi, D. Okamoto, K. Furue, T. Ueno, T. Tsuchizawa, T. Watanabe, K. Yamada, S.-i. Itabashi, and J. Akedo, "On-chip optical interconnect," *Proceedings of the IEEE*, vol. 97, no. 7, pp. 1186–1198, July 2009.
[3] A. Krishnamoorthy, R. Ho, X. Zheng, H. Schwetman, J. Lexau, P. Koka, G. Li, I. Shubin, and J. Cunningham, "Computer systems based on silicon photonic interconnects," *Proceedings of the IEEE*, vol. 97, no. 7, pp. 1337–1361, July 2009.
[4] C. Ho, A. Ruehli, and P. Brennan, "The modified nodal approach to network analysis," *Trans. on Circuits and Systems*, vol. 22, pp. 504–509, Jun. 1975.
[5] T. Quarles, A. Newton, D. Pederson, and A. Sangiovanni-Vincentelli, *SPICE 3 Version 3F5 User's Manual*, Dept. of EECE, Univ. of California, Berkeley.
[6] P. Gunupudi, T. Smy, J. Klein, and J. Jakubczyk, "Self-consistent simulation of optoelectronic circuits using a spice-like framework," in *IEEE 13th Workshop on Signal Propagation on Interconnects*, Strasbourg, France, May. 2009.
[7] C.-L. Chen, *Foundations for Guided-Wave Optics*. Wiley, 2006.
[8] M. Born and E. Wolf, *Principles of Optics: Electromagnetic Theory of Propagation, Interference and Diffraction of Light*. Cambridge, UK: Cambridge University Press, 1964.

Accelerated Frequency Domain Analysis by Susceptance-Element Based Model Order Reduction of 3D Full-wave Equations

Narayanan T.V., Sung-Hwan Min, Madhavan Swaminathan

Interconnect and Packaging Center
School of Electrical and Computer Engineering
Georgia Institute of Technology, Atlanta, GA
E-mail: {narayanantv, sunghwan.min}@gatech.edu, madhavan.swaminathan@ece.gatech.edu

Abstract — A circuit-equivalent frequency-domain three-dimensional electromagnetic simulation for package structures is proposed. A robust and passive susceptance-element based model order reduction is applied to the governing equation for accelerated simulation and proof-of-concept is shown with the examples of power-ground structure simulations.

I. INTRODUCTION

The analysis of power-ground structures presents a significant problem for signal/power integrity, as well as for electromagnetic interference (EMI) considerations [1] [2]. Such problems have been analyzed in the past by making use of time- and frequency-domain techniques. Most of these solvers can be classified, based on the degrees of freedom, as two-dimensional (2D), two-point-five dimensional (2.5D) and three-dimensional (3D). Though 3D solvers are the most accurate, they also impose a heavy penalty in terms of time and memory required for analysis. This problem can be alleviated by means of model order reduction approaches. The asymptotic waveform evaluation (AWE) [3] and Pade' via Lanczos (PVL) [4] algorithms have been previously applied for the analysis of interconnect structures. However, ensuring passivity was a bottleneck for these processes, which was addressed by the Passive Reduced-order Interconnect Macromodeling Algorithm (PRIMA) [5]. An analysis of the application of these techniques as applied to the 2D compact finite-difference frequency domain (FDFD) formulation is shown in [6]. However, PRIMA does not guarantee reciprocity among the ports. The Efficient Nodal Order Reduction (ENOR) [7] algorithm overcomes this problem by performing reduction of the governing equation in the nodal analysis form. It applies an orthogonal projection on the system based on moment-matching techniques in combination with Arnoldi-like orthogonalization [8].

This paper builds on the use of a full-wave simulation method by converting the Maxwell's equations into an electrical equivalent network [9], which has been validated. This offers the advantages of 1) making use of Spice-based circuit solvers to run full-wave simulations and 2) using circuit based numerical techniques to speed-up the simulation. An improved ENOR (imp-ENOR) algorithm [10] is then applied to the nodal formulation of this equivalent circuit resulting in a model-order reduced state equation. The system response can then be obtained in a fast memory-efficient manner as compared to the simulation of the full 3D problem. The paper is organized as follows: Section II describes the formulation of the equivalent circuit based simulator and the application of the imp-ENOR. Section III details the results and discussion.

II. FORMULATION

Consider the differential form of Maxwell's equation in the frequency domain

$$\nabla \times \mathbf{E} = -j\omega \mathbf{B} \tag{1}$$

$$\nabla \times \mathbf{H} = j\omega \mathbf{D} \tag{2}$$

where, $\mathbf{E}$ and $\mathbf{H}$ are the vector electric and magnetic fields
$\mathbf{D}$ and $\mathbf{B}$ are the vector electric and magnetic field densities
ω is the frequency in radians

Assuming an isotropic, lossless and homogeneous medium, the above equations can be written for a 2D transverse magnetic (TM) wave as:

$$j\omega\varepsilon E_z = \frac{\partial H_y}{\partial x} - \frac{\partial H_x}{\partial y} - J_z \tag{3}$$

$$j\omega\mu H_x = -\frac{\partial E_z}{\partial y} \tag{4}$$

$$j\omega\mu H_y = \frac{\partial E_z}{\partial x} \tag{5}$$

where, ε and μ are the material permittivity and permeability, respectively

E_p and H_p represent the electrical and magnetic field in the p-direction (p = x or y).

J_z is the external current source in the z-direction

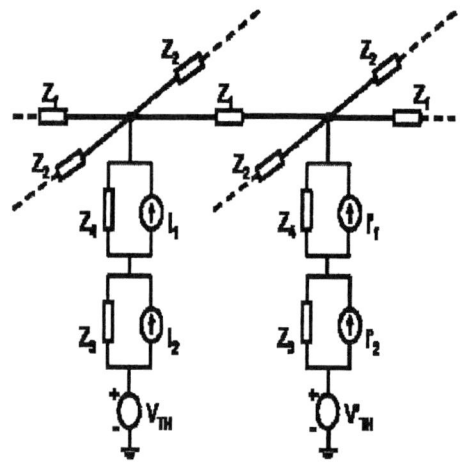

Fig.1 Equivalent-circuit unit-cell for 2D

Discretizing the above equations using the Yee-grid, so as to implicitly satisfy the divergence laws, we can form an electrical equivalent circuit for the resulting equations as shown in Fig. 1. The nodal voltages represent the electrical fields and the magnetic fields map to the branch currents. The circuit branch connected to ground can be further simplified to an equivalent Norton circuit, thus reducing the problem to one of solving only for the nodal voltages. The solution of the electrical network results in a linear equation of form $\mathbf{Ax} = \mathbf{b}$, where $\mathbf{A}$ is the sparse and banded nodal analysis (NA) amplification matrix, $\mathbf{x}$ is the vector of unknown nodal voltages and $\mathbf{b}$ is the vector containing external current sources.

The current-controlled current sources (CCCS) are converted into voltage-controlled current sources (VCCS). The controlled sources can also be viewed in terms of mutual susceptances between the voltage nodes. The circuit elements in the equivalent network in Fig.1 are given as:

Impedances: $\quad Z_1 = j\omega\mu\Delta x \quad Z_2 = j\omega\mu\Delta y \; Z_3 = \dfrac{1}{(j\omega\varepsilon)\Delta x} \quad Z_4 = \dfrac{1}{(j\omega\varepsilon)\Delta y}$ (6)

CCCS:

$$I_1 = j\omega\mu\,\Delta y\left(V_{i+\frac{1}{2},j+1} - V_{i+\frac{1}{2},j}\right) \quad I_2 = j\omega\mu\,\Delta x\left(V_{i,j+\frac{1}{2}} - V_{i+1,j+\frac{1}{2}}\right)$$

$$I'_1 = j\omega\mu\,\Delta y\left(V_{i+\frac{3}{2},j+1} - V_{i+\frac{3}{2},j}\right) \quad I'_2 = j\omega\mu\,\Delta x\left(V_{i+1,j+\frac{1}{2}} - V_{i+2,j+\frac{1}{2}}\right) \tag{7}$$

Dependent Voltage Sources:

$$V_{TH} = \frac{J_{z,ext}(i,j)}{j\omega\varepsilon} \qquad V'_{TH} = \frac{J_{z,ext}(i+1,j)}{j\omega\varepsilon} \tag{8}$$

where, Δx and Δy are the grid spacing along the X- and Y- directions, respectively. Perfect electric conductor (PEC) and perfect magnetic conductor (PMC) boundary conditions are enforced by shorting and opening the nodal points along the boundaries of the simulation domain, respectively. This can be easily extended for the 3D case. For a lossless structure with N unknowns and p ports, this results in an equation of the form:

$$\left(s\mathbf{C} + \frac{\Gamma}{s}\right)\mathbf{V}(s) = \mathbf{BI} \tag{9}$$

where, $\mathbf{C}, \Gamma \in \mathbf{R}^{N \times p}$ are the nodal capacitance and susceptance matrices, respectively; $\mathbf{V} \in \mathbf{C}^N, \mathbf{I} \in \mathbf{R}^p$ are the nodal voltages and port excitation currents, respectively and $\mathbf{B} \in \mathbf{R}^{N \times p}$ is the incidence matrix for current excitation.

The imp-ENOR algorithm is then used to construct an orthonormal basis $\mathbf{Q} \in \mathbf{C}^{N \times q}$ and project (9) onto this basis, resulting in the reduced-order system of the same form

$$\left(s\widetilde{\mathbf{C}} + \frac{\widetilde{\Gamma}}{s} \right) \widetilde{\mathbf{V}}(s) = \widetilde{\mathbf{B}}\mathbf{I} \tag{10}$$

where, $\widetilde{\mathbf{C}} = \mathbf{Q}^T \mathbf{C}\mathbf{Q}$, $\widetilde{\Gamma} = \mathbf{Q}^T \Gamma \mathbf{Q}$, $\widetilde{\mathbf{V}} = \mathbf{Q}^T \mathbf{V}$ and $\widetilde{\mathbf{B}} = \mathbf{Q}^T \mathbf{B}$. The above equations can then be solved very quickly to obtain the response over a wide range of frequencies.

III. RESULTS AND DISCUSSION

To verify the accuracy of the equivalent-circuit based full-wave simulation followed by the improved ENOR approach, a power-ground structure, as shown in Fig. 2, is simulated. Two 15mm x 15mm thin metal planes connected by a via are considered, placed in a homogeneous dielectric medium ($\varepsilon_r = 4.5$) in a PEC box of dimensions 25mm x 25mm x 80μm. A unit cell of 0.5mm x 0.5mm x 10μm was used to discretize the structure shown in Fig. 2.

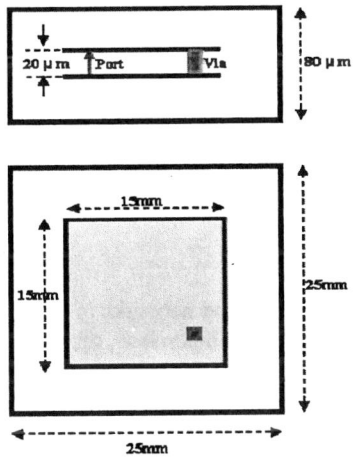

Fig.2 (a) Cross-section and (b) top-view of metal plane structure

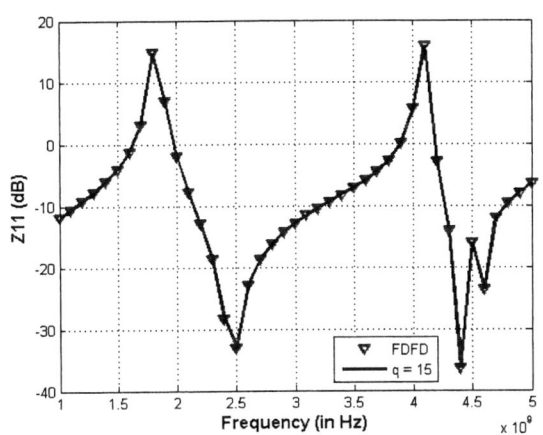

Fig.3 Z_{11} of the power-ground plane structure simulated using improved ENOR compared with full-wave 3D simulation [9]

The imp-ENOR algorithm is applied and the model-order is reduced to 15. Fig.3 shows a favorable comparison of the reduced model and the full-wave 3D simulation [9]. The average time per frequency point for the full 3D simulation was of the order of 120s, whereas the reduced-order system response took on the order of tenths of milliseconds per frequency point.

To demonstrate the convergence of the solution as the order of the reduced system is increased, a three-metal plane structure with aperture on the top plane is considered. The structure, with PEC boundaries and port placement is shown in Fig.4. The discretization of the 22mm x 22mm x 100μm volume is done using a unit cell of 0.5mm x 0.5mm x 10μm. The imp-ENOR algorithm is then applied, with the reduced order of the system varied as 2, 5 and 10. Fig.5 shows the impedance response of the two-port system for this set-up. As can be seen, as the reduced-order of the model is increased, the system response converges to the full-wave simulation. Whereas the 2-pole system is inadequate, the 5-pole system starts to diverge at frequencies above 4.5GHz. The 10-pole reduced order system shows a good match with the full-wave simulation results. The average time per frequency point for the full 3D simulation was of the order of 180s, whereas for the reduced-order system, it took on the order of tenths of milliseconds.

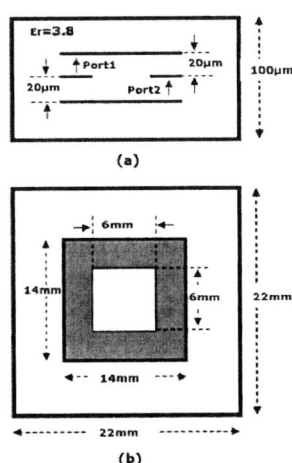

Fig.4 (a) Cross-section and (b) top-view of metal plane structure with aperture.

978-1-4244-4447-2/09 $25.00 © 2009 IEEE

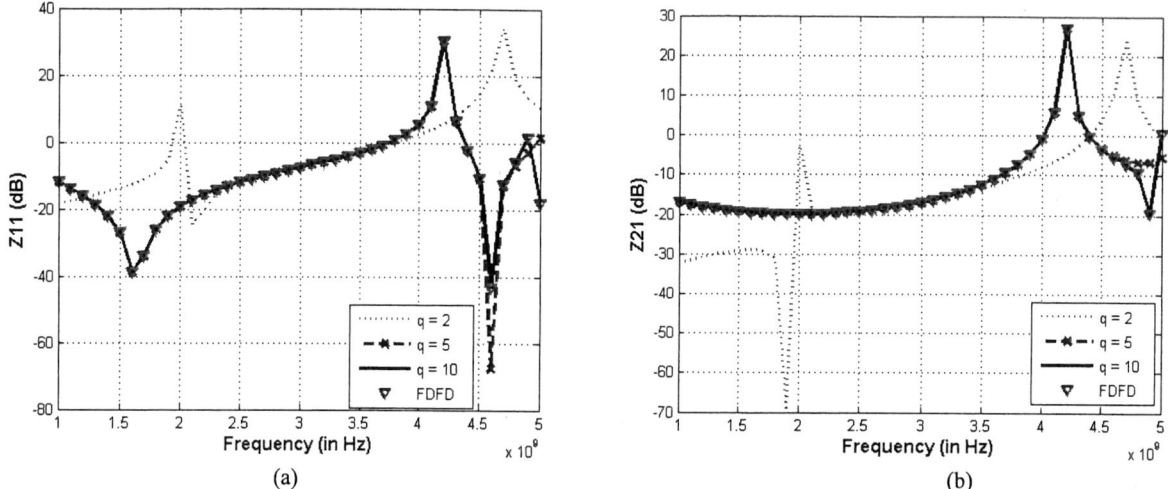

Fig.5 Impedance parameters (a) Z11 and (b) Z21, for the metal plane aperture structure
showing convergence as the order of the system is increased

In conclusion, the imp-ENOR approach has been applied to the 3D full-wave circuit-equivalent governing equation and the proof-of concept has been demonstrated by means of power-ground structure examples. The reduction in the order of the system from tens of thousands to double-digit values implies a significantly reduced simulation time. The inherent passive nature of the reduction ensures proper simulation without the need for additional passivity enforcement algorithms.

REFERENCES

[1] Frank Y. Yuan, "Electromagnetic modeling and signal integrity simulation of power/ground networks in high speed digital packages and printed circuit boards," *Proc. of 35th Annual ACM IEEE Design Automation Conference*, pp. 421-426, 1998.

[2] Madhavan Swaminathan, E. Engin, *Power integrity modeling and design for semiconductors and systems*, 1st ed., Prentice Hall PTR, 2007

[3] L. T. Pillage and R. A. Rohrer, "Asymptotic waveform evaluation for timing analysis," *IEEE Trans. Computer-Aided Design*, vol. 9, pp. 352–366, Apr. 1990.

[4] P. Feldmann and R. W. Freund, "Efficient linear circuit analysis by Pade' approximation via the Lanczos process," *IEEE Trans. Computer-Aided Design*, vol. 14, pp. 639–649, May 1995.

[5] A. Odabasioglu, M. Celik and L. Pileggi, "PRIMA: passive reduced order interconnect macromodeling algorithm," *IEEE Trans. Computer-Aided Design*, vol. 17, no. 8, pp. 645 - 653, August 1998

[6] Dagang Wu, Ji Chen, "Application of model order reduction techniques to compact FDFD method for guided wave structures," *Antennas and Propagation Society International Symposium*, pp. 636 – 639, Volume 2, Issue , 22-27 June 2003

[7] B. N. Sheehan, "ENOR: Model Order Reduction of RLC Circuits using nodal equations for efficient factorization," *Proc. of 36th Annual ACM IEEE Design Automation Conference*, pp. 17–21, 1999.

[8] M. Silveira, M. Kamon, I. Elfadel and J. White, "A coordinate-transformed Arnoldi algorithm for generating guaranteed stable reduced-order models of RLC Circuits," *Proc. of 33rd Annual ACM IEEE Design Automation Conference*, pp. 288-94, 1996

[9] T. V. Narayanan, K. Srinivasan, M. Swaminathan, "Fast memory-efficient full-wave 3D simulation of power planes ", *Proc. of IEEE Int. Symposium on Electromagnetic Compatibility, 2009 (to appear)*

[10] H. Zheng and L. Pileggi, "Robust and passive model order reduction for circuits containing susceptance elements," *Proc. of IEEE/ACM International Conf. on Computer-Aided Design*, pp. 761–766, 2002.

An Unconditionally Stable Time-Domain Finite Element Method of Significantly Reduced Computational Complexity for Large-Scale Simulation of IC and Package Problems

Houle Gan and Dan Jiao

School of Electrical and Computer Engineering, Purdue University
465 Northwestern Avenue, West Lafayette, IN 47907
Phone: 765-494-5240; Fax: 765-494-3371; e-mail: djiao@purdue.edu

ABSTRACT: An unconditionally stable and computationally efficient time-domain finite-element method is developed to solve large-scale IC and package problems. In this method, an analytical expression of the time dependence is developed for the field unknowns inside conductors. And hence any time step can be used to stably solve the system of equations inside conductors. A matrix solution is involved in the analytical expression. It is efficiently obtained by the time-domain finite-element reduction-recovery method, the factorization cost of which is $O(M)$, with M much less than the system matrix size N. The system of equations exterior to conductors is formed by a backward difference method, and hence is also unconditionally stable. The resultant system matrix is solved efficiently by an $\mathcal{H}$-matrix based direct sparse solver, which is shown to outperform the state-of-the-art direct sparse solver. The system exterior to the conductors and that interior to the conductors are then solved by a staggered marching scheme, the convergence of which is theoretically proved. Applications to on-chip problems have demonstrated a time step that is three orders of magnitude larger than what is permitted by an explicit time-domain scheme, with fast CPU run time, modest memory consumption, and without sacrificing accuracy.

KEYWORDS: Unconditionally stable schemes, fast solvers, electromagnetic simulation, time domain, on-chip, package

1. INTRODUCTION

The complexity of today's integrated circuit and package design requires electronic design automation tools to handle very large scale problems. To overcome the challenge of simulating very large scale problems, recently, a family of time-domain finite-element reduction-recovery (FE-RR) methods is developed [1-3]. In these methods, a linear system of equations $Ax = b$ is *rigorously* reduced to an orders of magnitude smaller system $A_r x_r = b_r$, with x_r being a subset of x. The reduction from A to A_r is achieved either via analytical means or with negligible computational cost. The reduced system $A_r x_r = b_r$ is solved in optimal complexity, i.e., linear complexity. The x_r solved from the reduced system is the same as that solved from the original system $Ax = b$. In addition, the rest of the unknown x can be recovered from x_r in linear complexity. The Hierarchical Finite-Element Reduction Recovery Method (HiFE-RR) [2] is applicable to any Manhattan-type multilayered structure; the Orthogonal Finite-Element Reduction Recovery Method (OrFE-RR) [3] further advances the method to solve any irregularly-shaped multilayer structure. The FE-RR methods have been successfully applied to solve large-scale IC and package design problems.

In essence, the FE-RR methods have successfully overcome the challenge of solving large-scale matrices encountered in analyzing IC and package problems. However, like many other time domain methods, the time step allowed by the FE-RR methods is subject to a certain constraint in order to maintain time-domain stability. For a central difference based FE-RR method, the time step required for stability needs to satisfy the following criterion [4]

$$dt \leq 2 / \sqrt{\rho\left(\mathbf{T}^{-1}\mathbf{S}\right)} \,, \tag{1}$$

where $\rho(\cdot)$ denotes the spectral radius of matrix $(\cdot)$, $\mathbf{T}$ is the mass matrix, and $\mathbf{S}$ is the stiffness matrix resulting from a finite-element-based analysis. For typical on-chip dimensions, (1) leads to a time step as small as 10^{-16} seconds. The objective of this work is to develop an unconditionally stable time domain scheme while still maintaining the computational efficiency of the FE-RR method. An unconditionally stable scheme allows for the use of any large time step without affecting the stability. In other words, any choice of the time step can make the time-domain simulation stable. The time step is thus only constrained by the accuracy requirement such as that imposed by the sampling theorem.

2. CHALLENGES OF DEVELOPING AN UNCONDITIONALLY STABLE TIME-DOMAIN FINITE-ELEMENT SCHEME OF REDUCED COMPUTATIONAL COMPLEXITY

The electric field $\mathbf{E}$ inside 3D integrated circuits and packages satisfies the second-order vector wave equation

$$\nabla \times [\mu_r^{-1}\nabla \times \mathbf{E}(\mathbf{r},t)] + \mu_0\varepsilon\partial_t^2\mathbf{E}(\mathbf{r},t) + \mu_0\sigma\partial_t\mathbf{E}(\mathbf{r},t) = -\mu_0\partial_t\mathbf{J}(\mathbf{r},t) \text{ in } V \tag{2}$$

subject to certain boundary conditions. A time-domain finite-element solution of (2) and its boundary condition results in the following system of ordinary differential equations [4]:

$$\mathbf{T}\frac{d^2u}{dt^2} + \mathbf{R}\frac{du}{dt} + \mathbf{S}u = f \,, \tag{3}$$

in which $\mathbf{T}$, $\mathbf{R}$, and $\mathbf{S}$ are square matrices, and u and f are column vectors. The matrix elements of $\mathbf{T}$, $\mathbf{R}$, and $\mathbf{S}$ are given by

$$\mathbf{T}_{ij} = \mu_0\varepsilon\left\langle \mathbf{N}_i, \mathbf{N}_j \right\rangle_V \,, \quad \mathbf{R}_{ij} = \mu_0\sigma\left\langle \mathbf{N}_i, \mathbf{N}_j \right\rangle_V \,, \quad \mathbf{S}_{ij} = \mu_r^{-1}\left\langle \nabla \times \mathbf{N}_i, \nabla \times \mathbf{N}_j \right\rangle_V \,, \tag{4}$$

---

*This work was supported by a grant from Intel Corporation, a grant from Office of Naval Research under award N00014-06-1-0716, and a grant from NSF under award 0747578 and award 0802178.

978-1-4244-4447-2/09 $25.00 © 2009 IEEE   145

where $\langle .,. \rangle_V$ and $\langle .,. \rangle_S$ denote inner products evaluated over a volume V, and a surface S respectively, $\mathbf{N}$ is the vector basis functions used to expand unknown electric field $\mathbf{E}$. Eqn. (3) is then solved in a time marching fashion.

There are a variety of time-domain differencing schemes that can render (3) unconditionally stable. Examples are a backward differencing scheme and a Newmark-beta method [4]. If the former is used, (3) becomes

$$\left(\mathbf{T} + dt\mathbf{R} + dt^2\mathbf{S}\right)u^{n+1} = \left(2\mathbf{T} + dt\mathbf{R}\right)u^n - \mathbf{T}u^{n-1} - f^{n+1}. \tag{5}$$

If the Newmark-beta method is used to discretize (3) in time, (3) becomes

$$\left(\mathbf{T} + 0.5dt\mathbf{R} + \beta dt^2\mathbf{S}\right)u^{n+1} = \left[2\mathbf{T} - (1 - 2\beta)\mathbf{S}\right]u^n - \left(\mathbf{T} - 0.5dt\mathbf{R} + \beta\mathbf{S}\right)u^{n-1} - \left[\beta f^{n+1} + (1 - 2\beta)f^n + \beta f^{n-1}\right], \tag{6}$$

where $\beta >= 0.25$. As can be seen from (5) and (6), both unconditionally stable schemes require the solution of a combined mass matrix $\mathbf{T}/\mathbf{R}$ and stiffness matrix $\mathbf{S}$. We have also investigated a variety of other differencing schemes that are unconditionally stable. All these schemes require a sum of $\mathbf{T}$, $\mathbf{R}$, and $\mathbf{S}$ to be solved.

The combined mass and stiffness matrix is a general sparse matrix, the direct solution of which is shown to have an optimal computational complexity of $O(N^{1.5})$ [5]; the iterative solution of which is not advantageous since it has to be repeated for each right hand side. In a time-domain simulation, the number of right hand sides to be solved is equal to the number of time steps, hence an iterative solution of (3) is expensive. As a result, it becomes highly challenging to develop a time domain finite element method which is not only unconditionally stable, but also computationally efficient.

A closer look at (3) reveals that the system matrix inside conductors and that outside of conductors each has a unique matrix property. Such a matrix property can be utilized to significantly speed up the matrix solution. For example, the system matrix inside conductors is dominated by $\mathbf{R}$ and $\mathbf{S}$, whereas that outside of conductors has only $\mathbf{T}$ and $\mathbf{S}$. However, when solving the system as a whole, the matrix property of each sub-system is overwhelmed. Based on this observation, we develop an unconditionally stable finite element method that has a significantly reduced computational complexity as follows.

3. PROPOSED METHOD

The proposed method has three components: a staggered time marching scheme, an efficient FE-RR-based analytical solution of the system matrix inside conductors, and an efficient $\mathcal{H}$-matrix-based solution of the matrix outside of conductors.

A. Staggered Time Marching

We divide the unknowns into two groups. One group consists of all the unknowns inside conductors; and the other group consists of the rest. The unknowns inside conductors are ordered layer by layer. The global system equation (5) is then rewritten as

$$\begin{bmatrix} \mathbf{A}_{ii} & \mathbf{A}_{io} \\ \mathbf{A}_{oi} & \mathbf{A}_{oo} \end{bmatrix} \begin{bmatrix} u_i \\ u_o \end{bmatrix} = \begin{bmatrix} b_i \\ b_o \end{bmatrix}, \tag{7}$$

where u_i denotes all the unknowns inside conductors, u_o denotes what is outside, $\mathbf{A}_{ii}$ is the sub-matrix formed by u_i, $\mathbf{A}_{oo}$ is the sub-matrix formed by u_o, $\mathbf{A}_{io}$ and $\mathbf{A}_{oi}$ represent the coupling between the u_i and u_o, and b_i and b_o are the corresponding right hand sides. To fully take advantage of the matrix property of $\mathbf{A}_{ii}$ and $\mathbf{A}_{oo}$, we solve (7) in a staggered manner

$$\begin{cases} \mathbf{A}_{ii}u_i^{k+1} = b_i - \mathbf{A}_{io}u_o^k \\ \mathbf{A}_{oo}u_o^{k+1} = b_o - \mathbf{A}_{oi}u_i^{k+1} \end{cases}, \tag{8}$$

where k denotes the iteration number. It is clear that when the iteration converges, $u^{k+1} = u^k$, and hence the solution of (8) is the same as that of (7).

Denoting the error of u_i and the error of u_o at the k-th iteration by ε_i^k, and ε_o^k respectively, we have

$$\varepsilon_i^{k+1} = \mathbf{A}_{ii}^{-1}\mathbf{A}_{io}\mathbf{A}_{oo}^{-1}\mathbf{A}_{oi}\varepsilon_i^k, \quad \varepsilon_o^{k+1} = \mathbf{A}_{oo}^{-1}\mathbf{A}_{oi}\mathbf{A}_{ii}^{-1}\mathbf{A}_{io}\varepsilon_o^k. \tag{9}$$

In order to guarantee the convergence of (8), it is required that

$$\left\| \mathbf{A}_{ii}^{-1}\mathbf{A}_{io}\mathbf{A}_{oo}^{-1}\mathbf{A}_{oi} \right\| < 1, \text{ and } \left\| \mathbf{A}_{oo}^{-1}\mathbf{A}_{oi}\mathbf{A}_{ii}^{-1}\mathbf{A}_{io} \right\| < 1. \tag{10}$$

In addition, the smaller the norm is, the faster the convergence of (8) is. For the system equation shown in (5) we have

$$\mathbf{A}_{ii} = \mathbf{T}_{ii} + dt\mathbf{R}_{ii} + dt^2\mathbf{S}_{ii}, \quad \mathbf{A}_{io} = \mathbf{T}_{io} + dt^2\mathbf{S}_{io}, \quad \mathbf{A}_{oi} = \mathbf{T}_{oi} + dt^2\mathbf{S}_{oi}, \quad \mathbf{A}_{oo} = \mathbf{T}_{oo} + dt^2\mathbf{S}_{oo}. \tag{11}$$

Since $\mathbf{A}_{ii}$ is dominated by $\mathbf{R}$, which is associated with conductors' conductivity, (10) is, in general, satisfied in on-chip and package applications. In addition, the norm has a small magnitude. For example, our numerical experiments show that for typical on-chip dimensions, (8) converges in a few iterations with a time step four orders of magnitude larger than what can be used in a conventional time-domain finite element method.

The convergence of (8) is determined by the following criterion

$$\left\| u^{k+1} - u^k \right\| / \left\| u^k \right\| < \delta, \tag{12}$$

where δ is a parameter used to control the accuracy.

It is worth mentioning that conductors in modern on-chip structures are heavily populated, and hence the sizes of $\mathbf{A}_{ii}$ and $\mathbf{A}_{oo}$ are approximately half the size of the original global matrix. If a direct sparse solver is used to decompose the system matrix, the optimal complexity is $O(N^{1.5})$. Therefore, even if no fast algorithm is developed to solve $\mathbf{A}_{ii}$ and $\mathbf{A}_{oo}$, the matrix decomposition time is naturally reduced by a factor of 1.4, compared to solving the global matrix as a whole. In the following two sections, we present two fast algorithms to solve $\mathbf{A}_{ii}$ and $\mathbf{A}_{oo}$ efficiently.

B. The FE-RR Based Analytical Solution of A_{ii}

Inside conductors, the displacement current is negligible compared to conduction current. Hence, (3) is reduced to

$$\mathbf{R}_{ii}\frac{du_i}{dt} + \mathbf{S}_{ii}u_i = f \ . \tag{13}$$

This is a first-order ordinary differential equation, the analytical solution of which is known as

$$u_i(t) = e^{-\mathbf{R}_{ii}^{-1}\mathbf{S}_{ii}t}\left[\int e^{\mathbf{R}_{ii}^{-1}\mathbf{S}_{ii}t}\mathbf{R}_{ii}^{-1}f(t)\,dt + C\right], \tag{14}$$

where C is the initial condition. Since the solution is analytical, there is no requirement on the time step. In other words, any time step can be used without rendering the scheme unstable. Next, we show how to efficiently evaluate (14).

The matrix exponential in (14) can be expanded as

$$e^{\mathbf{R}_{ii}^{-1}\mathbf{S}_{ii}t} = \mathbf{I} + \mathbf{R}_{ii}^{-1}\mathbf{S}_{ii}t + \frac{\left(\mathbf{R}_{ii}^{-1}\mathbf{S}_{ii}t\right)^2}{2!} + \frac{\left(\mathbf{R}_{ii}^{-1}\mathbf{S}_{ii}t\right)^3}{3!} + \cdots + \frac{\left(\mathbf{R}_{ii}^{-1}\mathbf{S}_{ii}t\right)^k}{k!} + \cdots \ . \tag{15}$$

It is shown in the literature that such an expansion always converges irrespective of the matrix properties of $\mathbf{R}$ and $\mathbf{S}$. When numerically computing (15), the sum can be truncated at the k-th term where

$$\left\|\mathbf{R}_{ii}^{-1}\mathbf{S}_{ii}t\right\| \ll k \ . \tag{16}$$

It is clear that by utilizing (14) and (15), the solution to (13) can be obtained by using $\mathbf{R}_{ii}^{-1}$ only. Since $\mathbf{R}_{ii}$ is a mass matrix, it can be efficiently solved by the FE-RR methods [1-3]. The factorization cost of $\mathbf{R}_{ii}$ is negligible; and the solution of $\mathbf{R}_{ii}$ scales linearly with the number of unknowns.

Though mathematically correct, a sum like (15) cannot be computed because when t is large, (15) exceeds what a computer can store. Thus, we developed the following scheme to evaluate (14) accurately and efficiently. At each time step, we use the u_i computed at the previous time step as the initial condition C. Thus, (14) becomes

$$u_i^{n+1} = e^{-\mathbf{R}_{ii}^{-1}\mathbf{S}_{ii}dt}C + e^{-\mathbf{R}_{ii}^{-1}\mathbf{S}_{ii}dt}\int_0^{dt}e^{\mathbf{R}_{ii}^{-1}\mathbf{S}_{ii}t}\mathbf{R}_{ii}^{-1}f(t)\,dt \ , \tag{17}$$

where C is u_i^n. In addition, to achieve a good accuracy, we evaluate the integral by using data at both the n-th time step and the $(n+1)$-th step. Hence, (17) becomes

$$u_i^{n+1} = \frac{\mathbf{R}_{ii}^{-1}f^{n+1}dt}{2} + e^{-\mathbf{R}_{ii}^{-1}\mathbf{S}_{ii}dt}\left[\frac{\mathbf{R}_{ii}^{-1}f^n dt}{2} + u_i^n\right]. \tag{18}$$

As can be seen from (18), only the second term in the right-hand side requires the calculation of the matrix exponential. In addition, this term only involves field values at the previous time step. Hence, it does not need to be evaluated at each iteration in the staggered marching process. It only needs to be done once at each time step.

C. The $\mathcal{H}$-Matrix Based Fast Solution of A_{oo}

The $\mathcal{H}$ matrix is a general mathematical framework that enables a highly compact representation and efficient numerical computation of dense matrices [6]. In [7-8], a fast $\mathcal{H}$-matrix-based solver is developed to efficiently store and compute the inverse of a finite-element-based matrix. It can be proven that $\mathbf{A}_{oo}$ can be represented by an $\mathcal{H}$ matrix without any approximation, and the inverse of this sparse matrix has a data-sparse $\mathcal{H}$-matrix approximation with error well controlled. Hence, the fast direct solver we have developed in [7-8] can be utilized to solve $\mathbf{A}_{oo}$ in almost linear complexity. In addition, when a large time step is used, the frequencies to be simulated are not high. Hence the rank required to represent $\mathbf{A}_{oo}$'s inverse is low, which further expedites the solution of $\mathbf{A}_{oo}$.

4. NUMERICAL RESULTS

In order to validate the proposed method, we simulated two on-chip circuits. The first one is a multi-conductor interconnect structure embedded in multilayer dielectrics. The cross section of the structure is plotted in Fig. 1, with permittivity of each layer shown. The conductivity of all the conductors is 5×10^7 S/m. The total length of the interconnect is 90 μm, with a 10 μm thick air layer attached at the near, and far end respectively. The air region is introduced to facilitate the construction of a first-order absorbing boundary condition. The interconnect is excited by a current source at the near end of a wire as shown in Fig. 1. The voltages at both ends of the excited wire are extracted.

To simulate this structure, a central difference based time-domain

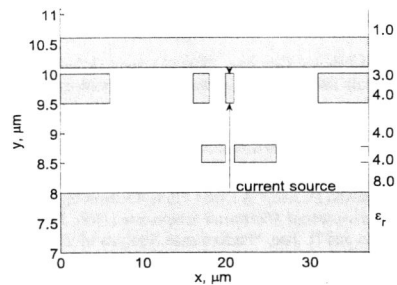

Fig. 1. Cross-sectional view of an on-chip interconnect.

978-1-4244-4447-2/09 $25.00 © 2009 IEEE   147

finite-element scheme requires a time step as small as 10^{-16} seconds in order to maintain stability. If an explicit FDTD-based method is used to simulate this example, a similar time step is required because the smallest space step is less than 10^{-7} m. The proposed method allows for a time step as large as 1×10^{-13} seconds, which is 1000 times larger than the time step allowed by a conventional time-domain finite-element scheme. The δ used in (12) is 10^{-4}, resulting in an average iteration number of only 5.9 for the convergence of the staggered marching. The number of terms used for the series expansion in (15) is only 8, and hence the computational overhead is minimal.

The brick elements are used to discretize the structure, resulting in 8783 unknowns inside conductors and 11662 unknowns outside of conductors. The time-domain simulation results are shown in Fig. 2. As can be seen clearly, the results generated by the proposed method show an excellent agreement with those generated by a conventional scheme employing a direct multi-frontal-based sparse solver. More importantly, the proposed method allows for a 1000 times larger time step, and hence is at least 1000 times faster to finish the simulation of a same time interval. In addition, the matrix factorization cost of the proposed method is much less than that of the conventional method in both memory consumption and CPU time. The detailed comparison can be seen in Table I.

Table I Performance Comparison

	Conventional method	Proposed method
Factorization CPU time	49.297 s	1.016 s
Factorization Memory cost	237 M	41 M

The second structure is a test-chip interconnect structure [9]. This interconnect involves 146 parallel returns (parallel to M2 lines) in the M3 layer, two M2 signal lines, and two M2 return lines, which are backed by a solid metal plane in M1. The structure is of length 100 mu. A current source is injected at the near end of one M2 wire and the voltages at this end, V1, and at the near end of an adjacent wire, V2, are sampled. Both conventional method and the proposed method were used to simulate this example. Once again, the proposed method allows for the use of a time step as large as 1×10^{-13} seconds. The average iteration number for the staggered marching is only 4.8. The number of expansion terms in (15)

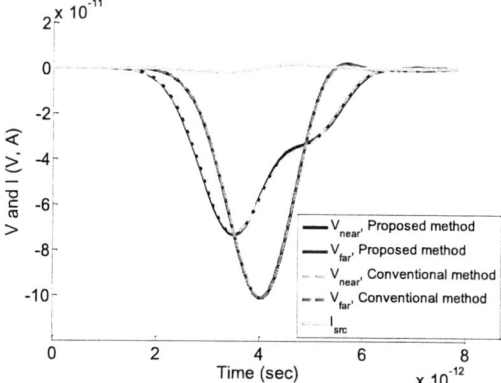

Fig. 2. Time-domain voltages of an on-chip interconnect.

is only 8. As can be seen from Fig. 3, the results generated by the proposed method and the conventional method are on top of each other. Due to the large time step, the proposed method is shown to be three orders of magnitude faster than a conventional scheme.

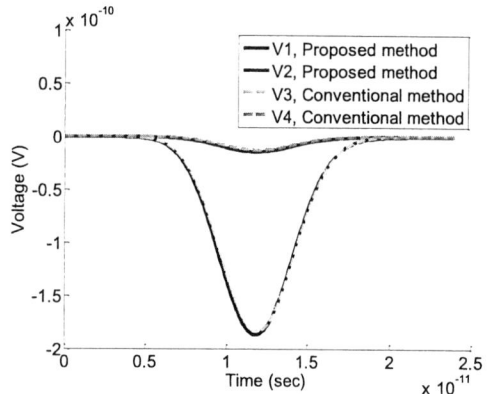

Fig. 3. Time-domain waveforms of a large-scale test-chip interconnect.

REFERENCES

[1] H. Gan and D. Jiao, "A Time-Domain Layered Finite Element Reduction Recovery (LAFE-RR) Method for High-Frequency VLSI Design," *IEEE Trans. Antennas Propagat.*, vol. 55, no. 12, pp. 3620-3629, Dec. 2007.

[2] H. Gan and D. Jiao, "Hierarchical Finite Element Reduction Recovery Method for Large-Scale Transient Analysis of High-Speed Integrated Circuits," to appear, *IEEE Trans. on Advanced Packaging*, 2009.

[3] Duo Chen and Dan Jiao, "Time-Domain Orthogonal Finite-Element Reduction-Recovery (OrFE-RR) Method for Electromagnetics-Based Analysis of Large-Scale Integrated Circuit and Package Problems," to appear, *IEEE Transactions on Computer Aided Design of Integrated Circuits and Systems*, 2009.

[4] D. Jiao and J. M. Jin, "Finite element analysis in time domain," in The Finite Element Method in Electromagnetics, New York: John Wiley & Sons, 2002, pp. 529-584.

[5] A. George, "Nested dissection of a regular finite element mesh," *SIAM J. on Numerical Analysis*, 10(2):345–363, April 1973.

[6] W. Hackbusch and B.Khoromaskij, "A Sparse Matrix arithmetic based on $\mathcal{H}$-matrices. Part I: Introduction to $\mathcal{H}$-Matrices," *Computing*, 62:89-108, 1999.

[7] H. Liu and D. Jiao, "A Direct Finite-Element-Based Solver of Significantly Reduced Complexity for Solving Large-Scale Electromagnetic Problems," 4 pages, *International Microwave Symposium (IMS)*, June 2009.

[8] H. Liu and D. Jiao, "Performance Analysis of the H-Matrix-Based Fast Direct Solver for Finite-Element-Based Analysis of Electromagnetic Problems," *IEEE International Symposium on Antennas and Propagation*, 4 pages, June 2009.

[9] M. J. Kobrinsky, S. Chakravarty, D. Jiao, M. Harmes, S. List, and M. Mazumder, "Experimental validation of crosstalk simulations for on-chip interconnects at high frequencies using S-parameters," *EPEP* 2003, pp. 329-332.

Design of a SIW-Based Data Communication System Using a SIW Six-Port Receiver

Abdulhadi E. Abdulhadi, Asanee Suntives and Ramesh Abhari

McGill University, Dept. of Electrical and Computer Engineering, 3480 University St., Montreal, QC, H3A 2A7, Canada
Phone: (514) 398-1451, Fax: (514) 398-4470, Emails: {abdulhadi.abdulhadi, asanee.suntives}@mail.mcgill.ca, ramesh.abhari@mcgill.ca.

Abstract: Substrate integrated waveguides (SIWs) are employed as an efficient low-cost wideband interconnect solution for high-speed signaling. The basic SIW data communication system includes mixers at the driver and receiver blocks in order to transmit baseband signal through the bandpass waveguide channel. Operational limitations are imposed due to limited bandwidth and non-ideal characteristics of the mixers. In this paper, an alternative technique utilizing a SIW six-port receiver is used to detect the transmitted data. The proposed system is designed for transmission of 5Gbps QPSK data and successfully evaluated through harmonic balance and envelope circuit simulations.

I. Introduction

Susceptibility of the conventional planar interconnects to crosstalk and electromagnetic interference along with the conductor and dielectric losses at high frequencies have driven research and development efforts to find new ways of interconnect design. Recently, a family of substrate integrated circuit structures has been introduced that offer numerous operational and system integration advantages [1]. Substrate Integrated Waveguide (SIW), i.e. a rectangular waveguide, is one of these embedded structures that has been used as a wide-band interconnect with excellent transmission and isolation characteristics in analog and digital applications [2]-[6]. The challenge in utilizing a SIW interconnect for data communication is adapting the baseband signaling system to support the bandpass nature of the waveguide. In the previous designs, digital signal is up-converted before and down-converted after the SIW interconnect by use of active mixers [3], [4]. These blocks replace the common driver and receiver stages of a serial data communication link and may add to system implementation costs. Use of mixers brings new system design concerns such as increased power consumption, LO-RF isolation and limited IF bandwidth.

A new approach to the design of communication system receivers has been introduced in [7]-[9] which utilizes a six-port detection system. This technique has been successfully implemented in direct down conversion and heterodyne receivers [8], [9].

In this paper the SIW based data communication system introduced earlier is redesigned by using the new concept of six-port detectors. The system transmits a QPSK modulated data at the rate of 5Gbps. The six-port junction is designed in SIW technology and optimized for the operation bandwidth of 7GHz around the centre frequency of 25GHz. Evaluation of the designed high-speed system is conducted through a number signal integrity indicator plots such as output constellation and eye diagrams demonstrating the successful operation of the data communication system using a bandpass channel.

II. Substrate Integrated Waveguide Data Communication System

Integrated waveguide interconnects provide a bandpass signaling channel. In the digital systems presented in [3], [4], simple forms of modulation and demodulation schemes, i.e. up and down conversion, are used to enable baseband signaling through SIWs. Moreover, a lowpass filter is needed at the receiver to block image and unwanted signals. The bandwidth of the system depends on the bandwidth of all system components including the SIW interconnect. Mainly, the dominant mode of the SIW, TE_{10}, is used for transmission. The bandwidth of the SIW is determined by its cutoff frequency and the excitation of higher order modes. The SIW structure considered in this paper is shown in Fig. 1. The physical design parameters of this SIW are the diameter (D) of the holes, the spacing (a) between the holes, the spacing (W) between the two rows and the thickness of the substrate. The values of these parameters which are based on the structure in [10] are: ε_r = 2.33, D = 0.76mm, a = 1.524mm, W1 = 6.258mm, h = 0.254mm, W2 = 2.286mm, W3 = 0.711mm, L = 3mm.

Fullwave simulations of this interconnect show a cutoff frequency of 16GHz, a return loss of more than 15dB and the insertion loss of less than 0.5dB over the entire bandwidth of 19–32GHz. However, when this SIW interconnect is incorporated in the system suggested in [4], the overall signaling bandwidth is reduced. For example, in the studied case presented in [4], the two mixers used in the system had the IF bandwidth of 8GHz which is much less than the bandwidth available by the SIW interconnect. In addition, the lowpass filter employed in the implement system [4] had a 3-dB cutoff of 7.6GHz. Hence, the transmission capacity of the realized data transmission link is further reduced. One way to increase the channel capacity as proposed in [11] is by using more elaborate modulation/demodulation and coding/decoding schemes. In fact, the modulation method incorporated in the basic system of [4] is BPSK. Implementation of the system with QPSK modulator and demodulators is the next step in improvement of data link capacity. A six-port detection system is inherently well-suited for the detection of QPSK data. The six-port receiver system is designed in the following sections.

978-1-4244-4447-2/09 $25.00 © 2009 IEEE

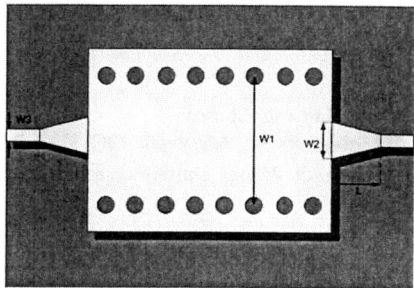

Fig. 1. Top view of the SIW interconnect.

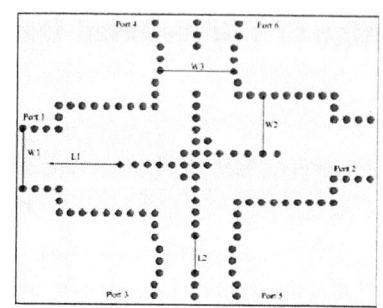

Fig. 2. Top view of the six-port SIW component used in the receiver of the high-speed system.

III. Design of the Six-Port Direct Conversion Block for the SIW Signaling System

To replace the receiver of the bandpass data communication system of [4] with the six-port detection system, a six-port junction should be designed. The centre design frequency of 25 GHz is considered that lies within the bandwidth provided by the SIW interconnect. It is intended to design a SIW six-port junction that offers a 7GHz bandwidth. The SIW six-port consists of two power dividers and two 90° hybrid couplers [12], [13]. To create a wideband junction, the power dividers and 90° hybrid couplers are individually designed and optimized using HFSS fullwave solver to minimize S_{11} and to maintain S_{21}, and S_{31} close to -3dB. Then the couplers and power dividers are put together and the overall six-port component is configured as shown in Fig. 2.

Next, the geometrical parameters of the overall SIW junction across the intended operating band are tweaked to minimize S_{11} and S_{22} and to keep S_{31} to S_{61} and S_{32} to S_{62} close to -6dB. The phase differences between the transmission coefficients should be multiples of 90° as indicated in [9]. The final dimensions and parameters obtained from fullwave optimization are the following: ε_r = 2.2, D = 0.8mm, a = 1.5mm, W1 = 5.898mm, W2 = 6.2491mm, W3 = 11.161mm, L1 = 7.193mm, L2 = 8.675mm, and h = 0.508mm.

The simulated scattering parameters are presented in Figures 3(a), 3(b), and 4(a), 4(b). It can be noted that the transmission coefficients S_{31} to S_{61} and S_{32} to S_{62} are close to -6 dB at the design frequency of 25 GHz. As well, the phase differences between the output ports are almost 90° or multiples of 90° over the wide frequency band of 7GHz. At the center frequency of 25 GHz the phase difference between the output ports are: Phase(S_{31}) − Phase(S_{51}) = 90.41°, Phase(S_{41}) − Phase(S_{61}) = 89.35°, Phase(S_{32}) − Phase(S_{52}) = 90.02°, and Phase(S_{62}) − Phase(S_{42}) = 87.09°.

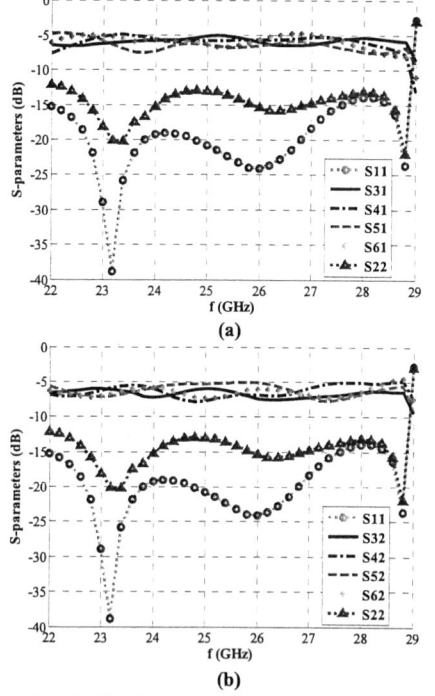

Fig. 3. (a) Simulated magnitudes of S_{31} to S_{61}. (b) Simulated magnitudes of S_{32} to S_{62}.

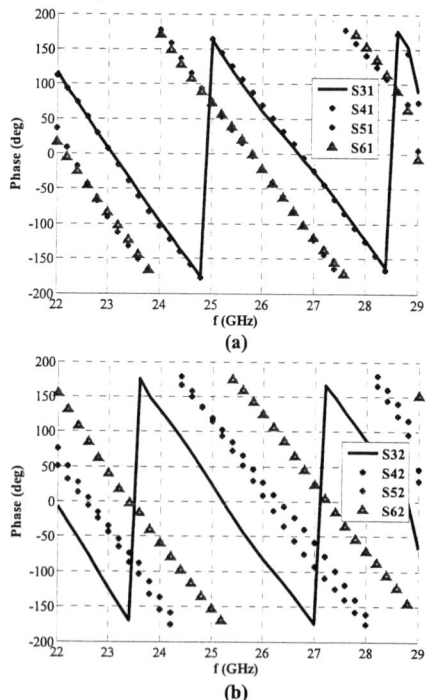

Fig. 4. (a) Simulated phases of S_{31} to S_{61}. (b) Simulated phases of S_{32} to S_{62}.

978-1-4244-4447-2/09 $25.00 © 2009 IEEE

In order to evaluate the performance of the SIW six-port junction for signal detection, as suggested in [9], harmonic balance simulations at the centre frequency are conducted using Agilent ADS. The LO is connected to port 2 and RF signal is fed to Port 1. Ports 3 to 6 are connected to power detectors. The phase of the signal at Port 2 which represents the LO signal in the system is fixed. The phase of the signal at Port 1 which is the incoming RF signal is swept from 0 to 360°. The voltage waveforms at the output ports are plotted in Fig. 5. By closely observing the minima of the output voltages in this figure, it can be observed that phase differences close to 90° are achieved. Moreover, the output voltages of Ports 3 and 5 and Ports 4 and 6 are anti-phase. The next step is adding the two differential amplifiers at the output ports of the SIW six-port junction to extract I/Q data.

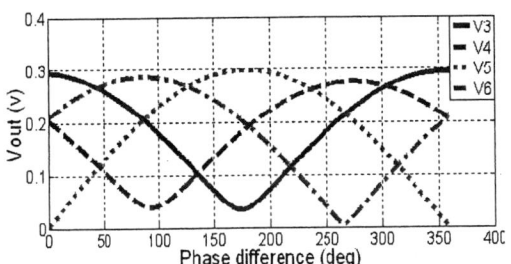

Fig. 5. Output voltages of the six-port junction vs. the phase of input RF signal.

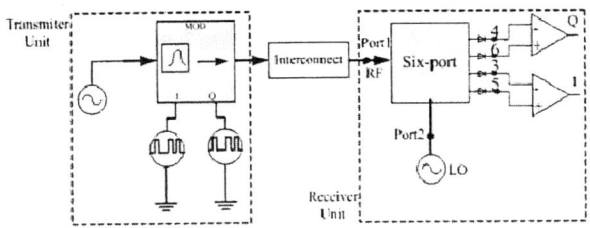

Fig. 6. SIW-based data communication system using a six-port receiver.

IV. High-Speed Digital System Simulations

The launched QPSK data into the SIW interconnect is generated using pseudorandom sequences with the rate of 5Gbps and rise/fall time of 36ps. In ADS simulation test bench the two data streams are input to an I/Q vector modulator with modulation frequency of 25 GHz. Envelop simulations are conducted using 20 ns long pseudorandom bit sequences. The input and output I/Q sequences are presented in Fig. 7. It can be observed that the demodulated output signals closely follow the same bit sequencing as the original input but with a lower magnitude.

Another way to verify the efficiency of the data communication system is by monitoring the output constellation diagram that is presented in Fig. 8 showing accurate positioning of the clusters of demodulate components in the I/Q plane.

One of the important measures of signal integrity performance of a data transmission system is the output eye diagram. Figures 9 and 10 depict the eye diagrams of the demodulated data stream on the Q and I branches. It can be observed that in both cases wide eye opening is achieved and no significant degradation of the rise/fall time is obtained. The eye diagram of the I branch as opposed to that of the Q branch shows ringing spikes that are indicative of the mismatch in the transmission path. In fact, the transmission coefficients of the output ports S_{31} to S_{61} and S_{32} to S_{62} are not exactly equal at the design frequency. They vary within a 2dB range from ideally needed value of -6dB as can be seen in Fig. 3. In addition, as mentioned earlier the phases of S_{31} to S_{61} and S_{32} to S_{62} are not exactly multiplies of 90°. These factors degrade the performance of the system.

V. Conclusions

Recently, Substrate Integrated Waveguide (SIWs) has been proposed as an alternative interconnect in high-speed communication systems. Data transmission in these bandpass channels is enabled by mixers at the driver and receiver end of the SIW interconnect. In this paper, the newly proposed six-port detection system is utilized instead of the mixer at the data reception end. A six-port SIW-based junction is designed and optimized for wideband operation around 25 GHz. The serial data communication system is evaluated in transmission of a high-speed QPSK modulated signal. The scattering parameters of the SIW interconnect and six-port junction are ported to ADS circuit solver and global system simulations are conducted. Envelop simulations demonstrate excellent quality in the demodulated I/Q output signals and attest to the efficiency of the designed bandpass data communication system in terms of maintaining signal integrity. The use of six-port detection eliminates the need for expensive mixers and enables wider signaling bandwidth due to its passive nature.

978-1-4244-4447-2/09 $25.00 © 2009 IEEE

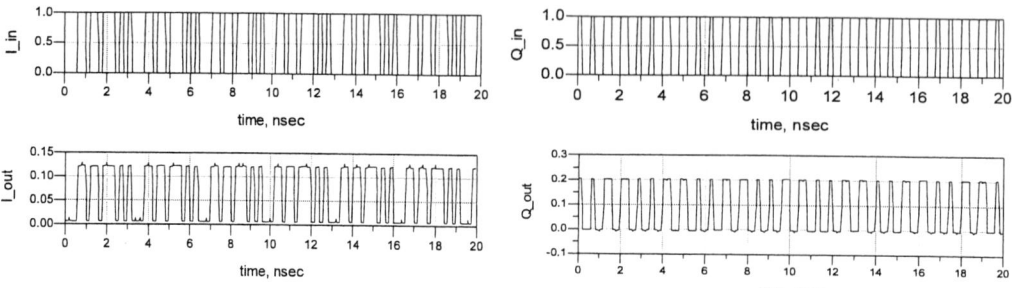

Fig. 7. Input and output 5Gbps data sequences in the I/Q branches.

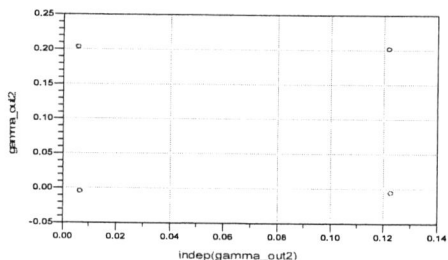

Fig. 8. Constellation diagram of demodulated 5 Gbps QPSK signal.

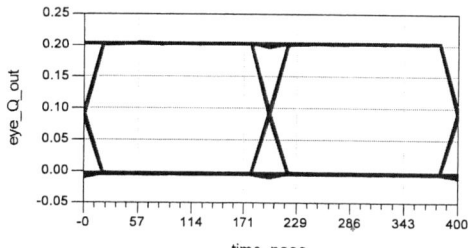

Fig. 9. Simulated output eye diagram of the system at the Q branch.

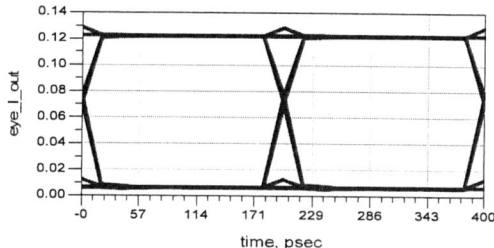

Fig. 10. Simulated output eye diagram of the system at the I branch.

References

[1] Ke Wu, D. Deslandes, Y. Cassivi, "The substrate integrated circuits - a new concept for high-frequency electronics and optoelectronics," *6th International Conference on Telecommunications in Modern Satellite, Cable and Broadcasting Service, TELSIKS* , vol.1., Oct. 2003.

[2] Deslandes, D., Ke Wu, "Single-substrate integration technique of planar circuits and waveguide filters," *IEEE Trans. Microwave Theory and Tech.,* vol.51, no.2, pp. 593-596, Feb. 2003.

[3] A. Suntives and R. Abhari, "Design and characterization of the EBG waveguide-based interconnects," *IEEE Trans. Adv. Packag.*, vol. 30, no. 2, pp. 163–170, May 2007.

[4] A. Suntives and R. Abhari, "Ultra-High-Speed Multichannel data transmission using hybrid substrate integrated waveguide," *IEEE Trans Microwave Theory and Tech..*, vol.56, no.8, pp.1973-1984, Aug. 2008.

[5] J. J. Simpson and A. Taflove, "Computational study of millimeter wave metal-pin photonic bandgap waveguides for use as ultra high-speed bandpass wireless interconnects," in *Proc. IEEE AP-S Int. Symp.*, Columbus, OH, vol. 4, pp. 871–874, Jun. 22–27, 2003.

[6] Davide Urbano, Emilio Arnieri, Gregorio Cappuccino, and Giandomenico Amendola, "Simulated and timing performances of integrated waveguides for ultra-high speed interconnect," *Microwave and Optical Tech. Letters*, vol. 50, issue 3, pp. 666 – 672, Jan. 2008.

[7] T. Hentschel, "The six-port as a communications receiver," *IEEE Trans. on Microwave Theory and Tech.,* vol.53, no.3, pp. 1039-1047, March 2005.

[8] S.O.Tatu, E. Moldovan, K. Wu, R. G. Bosisio, "A new direct millimeter-wave six-port receiver," *IEEE Trans. on Microwave Theory and Techniques*, vol. 49, no.12, pp.2517-2522, Dec 2001.

[9] S. O. Tatu and E. Moldovan, "V-Band Multiport Heterodyne Receiver for High-Speed Communication Systems," *EURASIP Jour. of Wireless Communications and Networking 2007*, Article ID 34358, 7 pages.

[10] Dominic Deslandes and Ke Wu, "Integrated microstrip and rectangular waveguide in linear form," *IEEE Microwave and Wireless Components letters*, vol. 11, no. 2, Feb. 2001.

[11] Asanee Suntives, High-Speed Data Transmission Using Substrate Integrated Waveguide-Type Interconnects. PhD Thesis, Dept. of Elect. and Comp. Eng., McGill University, 2009.

[12] S. Germain, D. Deslandes and K. Wu, "Development of substrate integrated waveguide power dividers," *IEEE CCECE Canadian Conference on Electrical and Computer Engineering,* vol.3, pp. 1921-1924, May 2003.

[13] X. Xinyu, R.G. Bosisio, and K. Wu, "A new six-port junction based on substrate integrated waveguide technology," *IEEE Trans Microwave Theory and Techniques*, vol.53, no.7, pp. 2267-2273, July 2005.

Model-to-Hardware Correlation of Disk Resonators for Via-Array Modeling in High-Speed PCBs

Arun Reddy Chada[#1], Young H. Kwark[^2], Xiaoxiong Gu[^3], and Jun Fan[#4]

[#] *Missouri S&T EMC Laboratory, Missouri University of Science and Technology (formerly University of Missouri-Rolla),*
Rolla, MO 65401, USA

[1]ac253, [4]jfan@mst.edu

[^] *IBM T.J. Watson Research Center, Yorktown Height, NY 10598, USA*

[2]kwark, [3]xgu@us.ibm.com

Abstract—**In high-speed printed circuit board (PCB) designs, vias play a critical role in determining signal and power integrity. This paper uses disk resonators to investigate modeling approaches for common via array structures. Boundary integral equation (BIE) and effective dielectric medium approaches are used for a perforated disk resonator with non-plated holes, using dielectric properties derived from a solid disk resonator study. Both approaches are compared and validated with measurements.**

I. INTRODUCTION

Vias necessary in multilayer high-speed printed circuit board (PCB) designs usually behave as discontinuities for high-speed signals (signal vias), or introduce unwanted parasitics for dc power distribution (power and ground vias). Many modeling approaches have been introduced for vias or via arrays in typical PCB structures. In this paper, a canonical problem of the return loss at a via port located at the center of a perforated circular parallel plane pair (a disk resonator) is studied.

The disk resonator constitutes a radial waveguide structure. Since the thickness is small compared to wavelengths in the frequency range of interest, only the TM_{z00} parallel plane mode is assumed to be excited in the structure. We also assume that the spacing between the non-plated holes is large enough so that no higher-order wave scattering occurs amongst the holes, and the via and hole diameters are small enough so that no significant φ-directional variation exists. The TM_{z00} parallel plane waves do get reflected from the disk perimeter, leading to resonances equally spaced in frequency.

The disk resonator is modeled in this paper using a physics-based equivalent circuit approach [1]. The via itself, which is connected to one of the planes as shown in Fig. 1, is modeled as a via-plane capacitance, obtained from the analytical expression developed in [2]. The rest of the plane pair including all the non-plated holes is modeled as a parallel-plane impedance Z_{pp}. A boundary integral equation method (BIE) [3] is used herein to obtain the Z_{pp} where the non-plated holes are accounted for either by using the radial-wave scattering parameters built in the BIE formulation [3], or by using an equivalent dielectric medium that is derived based on the Maxwell Garnett formula [4]. The properties of the dielectric medium in the disk resonator are derived from the study of a solid disk resonator without non-plated holes, built from the same dielectric material.

The modeling approaches including the BIE and the equivalent dielectric medium methods are discussed in section II. The procedure used to obtain the dielectric properties from measurements of a solid disk resonator are detailed in section III. Section IV discusses the model-to-hardware correlation of the perforated disk resonator, with conclusions and directions in Section V.

II. MODELING APPROACHES FOR Z_{pp}

Various methods, including the finite element method [5], method of moments [6], finite difference method [7]-[8], and cavity model with segmentation [9]-[12], have been widely used to obtain the impedance Z_{pp} of a plane pair. However, they are either not suitable, or inefficient, when dealing with a plane pair with many holes. A boundary integral equation (BIE) method was introduced in [3], which is particularlly suitable for modeling a plane pair with multiple holes having perfect magnetic conductor (PMC) boundaries. Since the plane pair thickness is very small, all the non-plated holes in the perforated disk resonator in this paper can be assumed to be PMC holes. In the BIE formulation, radial-wave scattering parameters S^R_{pp} are introduced and the Z_{pp} is obtained by transforming the radial scattering parameters into impedance parameters. The introduction of the radial-wave scattering parameters as well as the transforms greatly simplifies the calculation of the Z_{pp} for a perforated plane pair, since only the plane pair boundary needs to be discretized as unknowns in the integral equation formulation. The hole circumferences do not need to be discretized, making modeling much more efficient.

The non-plated holes in the perforated disk resonator can also be accounted for by using an effective dielectric medium. With the effective dielectric medium, the perforated disk resonator can be modeled as an equivalent solid disk resonator, where the impedance Z_{pp} can also be obtained using the BIE method except that there is only one port at the centrally-located via location in this case. To obtain the effective dielectric medium, the non-plated holes are considered as air dielectric media. Then the extended Maxwell Garnett formula is applied to find the effective dielectric material properties from the mixture of air and FR-4 dielectric media in [4] is applied. For convenience, the related expressions are listed below as

$$\varepsilon_{eff} = \varepsilon_b + v_i.\varepsilon_b.\frac{(\varepsilon_i - \varepsilon_b)}{\varepsilon_b + (1-v_i).(\varepsilon_i - \varepsilon_b).F_z}, \qquad (1)$$

$$F_z = \left(\frac{1}{a}\right)^2 \ln a, \qquad (2)$$

$$a = \frac{l}{d}, \qquad (3)$$

$$F_x + F_y + F_z = 1, \qquad (4)$$

$$F_x = F_y = \frac{1 - F_z}{2}, \qquad (5)$$

where l is the height of the cylindrical inclusion of the air dielectric in the FR-4 base medium, d is the diameter of the cylindrical inclusion, ε_b is the permittivity of the base material, ε_i is the permittivity of the inclusion material, v_i is the volume fraction of the inclusions, and F_x, F_y, F_z are the depolarization factors along the x, y, and z directions.

III. SOLID DISK RESONATOR STUDY

To obtain the dielectric properties of the FR-4 material used in the perforated disk resonator, a solid disk resonator made from the same dielectric material was studied first. As shown in Fig. 1, the solid disk resonator is a circularly-shaped plane pair with one via located at the center, which is shorted to the bottom plane. The return loss at this centrally-located via port was measured using a vector network analyzer as well as modeled using the BIE method overviewed earlier. Then the dielectric properties were derived by fitting the modeled impedance results to the measurement results.

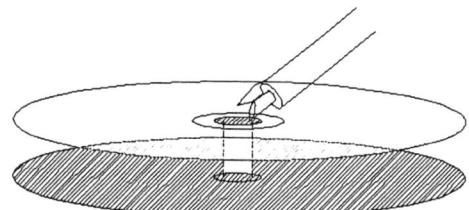

Fig. 1. Side view of the solid disk resonator with a measurement probe.

The diameter of the solid disk is 3.017 inches. The radius of the centrally located via is 6 mils. In addition, the via has a pad and an antipad in the top plane with the corresponding radii of 11.5 mils and 17 mils, respectively. The dielectric thickness is 58.5 mils, and the thickness of the top and bottom metal planes is 2 mils each. Simulations were performed in the frequency range from 100 MHz to 15 GHz.

The first order Debye model including the dielectric material conductivity is used to describe the dielectric material as

$$\varepsilon_r = \varepsilon_\infty + \frac{(\varepsilon_s - \varepsilon_\infty)}{(1 + j\omega\tau)} - \frac{j\sigma_e}{\omega\varepsilon_0}, \qquad (6)$$

where $\varepsilon_0 = 8.85 \times 10^{-12}$ F/m, ε_s is the static relative dielectric constant, ε_∞ is the optic relative permittivity, τ is the relaxation time, and σ_e is the conductivity of the dielectric material.

From (6), the dielectric constant and loss tangent terms can be computed as

$$\varepsilon_r = real(\varepsilon_r), \qquad (7)$$

$$\tan \delta = -\frac{imag(\varepsilon_r)}{\varepsilon_r}. \qquad (8)$$

The initial values of the dielectric properties were chosen based on the manufacturer's datasheet. Then the parallel plane-pair impedance Z_{pp} was obtained using the BIE method. In addition, the via-plane capacitance was calculated as 144.3 fF using the expression in [2] for the given via/pad/anti-pad geometries. An equivalent circuit model [1] was assembled in ADS as shown in Fig. 2, to calculate the return loss at the centrally-located via port.

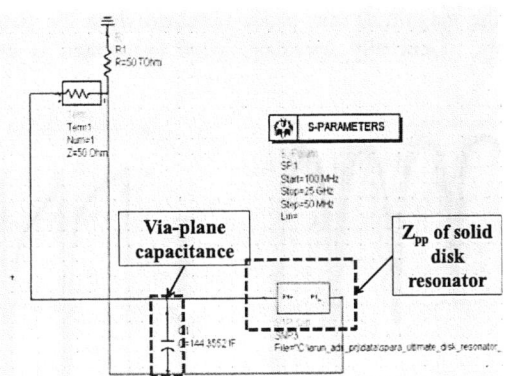

Fig. 2. ADS schematics to calculate the return loss at the via port for the solid disk resonator.

The simulation results obtained from ADS were then compared with the measured ones. The values of the dielectric properties were adjusted and the modeling procedures repeated until the simulated and measured results had a good fit. The final values of ε_s, ε_∞, τ, and σ_e were determined to be 4.4943, 4.315, 3.33 x 10^{-11} s, and 0.002068 S/m, respectively, for this particular dielectric medium used in the disk resonators.

Fig. 3 shows the comparisons of the magnitude and phase of the measured and simulated S_{11} results. Good agreement has been achieved for the frequencies up to 15 GHz.

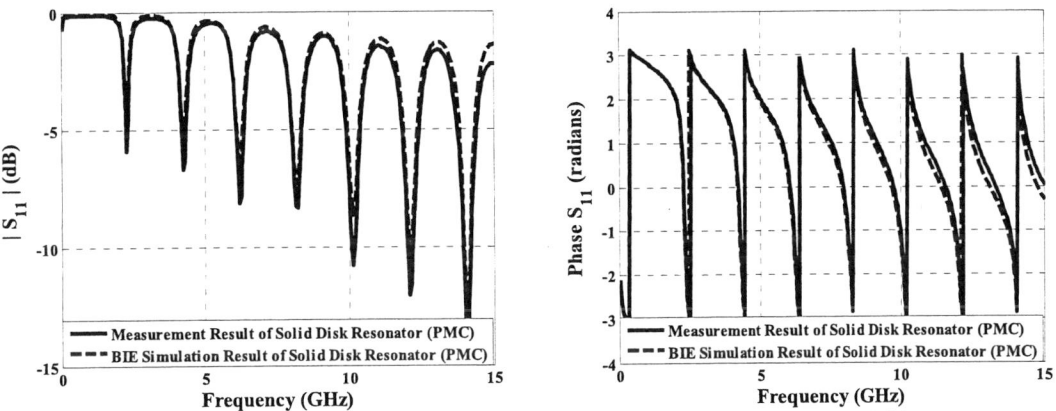

Fig. 3. Comparison of magnitude and phase of the measured and simulated S_{11} results for the solid disk resonator.

IV. MODEL-TO-HARDWARE CORRELATION FOR PERFORATED DISK RESONATOR

After obtaining the dielectric properties of the dielectric medium used in the disk resonators, the perforated disk resonator was studied using the same approach. The geometry of the perforated disk resonator is illustrated in Fig. 4. It has a via located at the center that is shorted to the bottom plane, as well as 4500 non-plated holes in a 1 mm X 1 mm grid. The diameter of the non-plated holes is 18.3 mils. All other geometrical dimensions remain the same as the solid disk resonator studied in Section III.

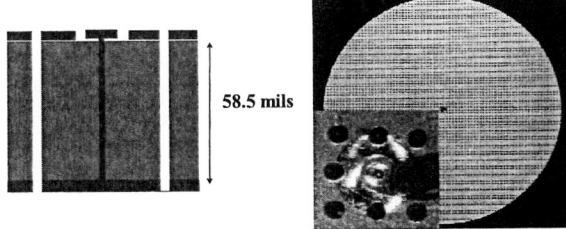

Fig. 4. Central via cross-section with only two non-plated holes illustrated (left), top view of perforated disk resonator (right)

The return loss at the centrally-located via port was simulated similarly in ADS from an equivalent circuit for the perforated disk resonator. The parallel plane-pair impedance Zpp in this case, however, was calculated using either a pure BIE formulation or an effective dielectric medium approach. The first order Debye model derived from the solid disk resonator study was used in the simulations.

978-1-4244-4447-2/09 $25.00 © 2009 IEEE   155

Fig. 5 plots the comparison of the magnitude and phase obtained from the pure BIE approach, the effective dielectric medium approach, and measurements. Generally speaking, good agreement is demonstrated up to 15 GHz with better agreement at lower frequencies.

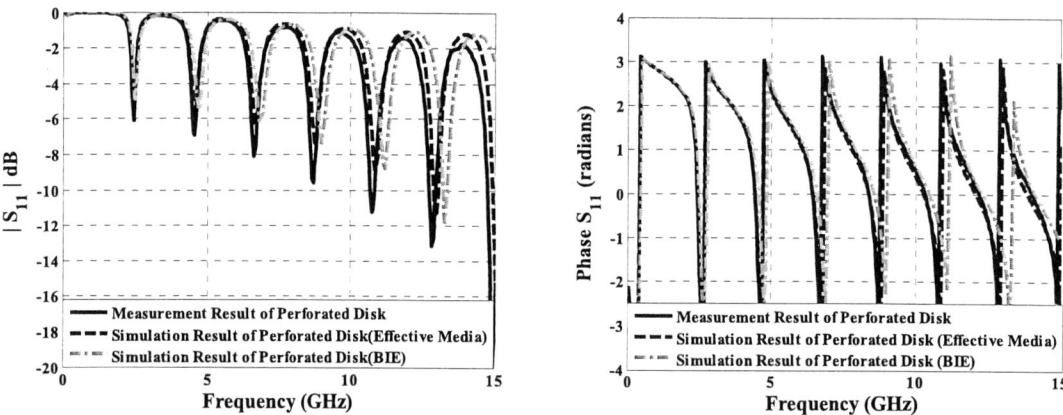

Fig. 5. Comparison of Magnitude and Phase of the S_{11} results for the perforated disk resonator.

At higher frequencies, the divergence can be partly attributed to neglect of higher order wave scattering among the non-plated holes. A more refined formulation that includes multiple scattering [13] may further improve the high-frequency modeling accuracy, but with considerably more complexity. Interestingly, the effective dielectric medium approach seems to work better than the pure BIE approach at high frequencies.

V. CONCLUSION

Model-to-hardware correlation for a perforated disk resonator with 4500 non-plated holes is presented in this paper. The procedure used to obtain accurate dielectric properties from a solid disk resonator constructed from the same dielectric material, as well as the modeling techniques to handle an array of non-plated holes embedded in the parallel plane geometry, have been demonstrated to be effective in the frequency range up to 15 GHz despite the assumptions of a single TM_{z00} mode and the neglect of higher order wave scattering amongst the holes, Future studies may include the multiple scattering formulation for improved accuracy at high frequencies.

ACKNOWLEDGEMENT

The measurements in this study were supported by the Defense Advanced Research Projects Agency under its Agreement No. HR0011-07-9-0002.

REFERENCES

[1] C. Schuster, Y. Kwark, G. Selli, and P. Muthana, "Developing a 'Physical' Model for Vias," DesignCon 2006, Santa Clara, CA USA, Feb. 6 – 9, 2006.

[2] Y. Zhang, J. Fan, G. Selli, M. Cocchini, and F. De Paulis, "Analytical evaluation of via-plate capacitance for multilayer packages or PCBs," *IEEE Transactions on Microwave Theory and Techniques*, Vol. 56, No. 9, pp. 2118-2128, September 2008.

[3] Y. Zhang, G. Feng, J. Fan, "Radial Waveguide Interpretation of the Impedance of Power/Ground Planes and its Evaluation by Boundary Integral Method," *IEEE Trans. Microwave Theory Tech.*, submitted, 2008.

[4] A. Sihvola, Electromagnetic mixing formulas and applications, IEE, London, UK, 1999.

[5] C. Guo, T. H. Hubing, "Circuit models of power bus structures on printed circuit boards using hybrid FEM-SPICE method," *IEEE Trans. Adv. Packaging.*, vol. 25, no 3, pp. 441-447, Aug 2006.

[6] W. Shi, and J. Fang, "New efficient method of modeling electronics packages with layered power/ground planes," *IEEE Trans. Adv. Packaging.*, vol. 25, no. 3, pp. 417-423, Aug. 2002.

[7] T. -K. Wang, S.-T. Chen, C.-W. Tsai, S.-M. Wu, J. L. Drewniak, and T.-L. Wu, "Modeling noise coupling between package and PCB power/ground planes with an efficient 2-D FDTD/lumped element method," *IEEE Trans. Adv. Packag.*, vol. 30, no. 4, pp. 864-871, Nov.2007.

[8] A.E. Engin, K. Bharath, and M. Swaminathan, "Multilayered finite -difference method (MFDM) for modeling of package and printed circuit board planes," *IEEE Trans. Electromagn. Compat.*, vol. 49, no. 2, pp. 441-447, May 2007.

[9] Z. L. Wang, O. Wada, Y. Toyota, and R. Koga, "Convergence acceleration and accuracy improvement in power bus impedance calculation with a fast algorithm using cavity modes," *IEEE Trans. Electromag. Compat.*, vol. 47, no. 1, pp. 2-9, Feb. 2005.

[10] P. Liu, and Z. –F Li, "An efficient method for calculating bounces in the irregular power/ground plane structure with holes in high speed PCBs," *IEEE Trans. Electromag. Compat.*, vol. 47, no. 1, pp. 889-898, 2005.

[11] Y. Joeong, A. C. Lu, L. L Wai, W. Fan, B. K. Lok, H. Park, and J. Kim "Hybrid analytical modeling method for split power bus in multilayered package," *IEEE Trans. Electromag. Compat.*, vol. 48, no. 1, pp. 82-94, Feb. 2006.

[12] C. Wang, J. Mao, G. Selli, S. Luan, L. Zhang, J. Fan, D. J. Pommerenke, R. E. DuBroff, and J. L. Drewniak, " An efficient approach for power delivery network design with closed-form expressions for parasitic interconnect inductances," *IEEE Transactions on Advanced Packaging*, Vol. 29, No. 2, pp. 320-334, May 2006.

[13] Y. Zhang, J. Fan, A. R. Chada, J. L. Drewniak, "A concise multiple scattering method for via array analysis in a circular plate pair," EDAPS 2008 Electrical Design of Advanced Packaging and Systems Symposium, pp. 143-146, Dec 2008.

978-1-4244-4447-2/09 $25.00 © 2009 IEEE

A Precise Analytical Eye-diagram Estimation Method for Non-ideal High-Speed Channels

Jeonghyeon Cho, Eakhwan Song, Jongjoo Shim, Jiseong Kim, and Joungho Kim

Terahertz Interconnection and Package Laboratory, Department of Electrical Engineering, KAIST, Daejeon, Republic of Korea
Phone) +82-42-879-9866, Fax) +82-42-869-8058, E-mail) caleb1@eeinfo.kaist.ac.kr; teralab@ee.kaist.ac.kr

In this paper, we propose an analytical eye-diagram estimation method for a channel of a pair of differential microstrip traces on PCBs with arbitrary source and load terminations. The closed-form equation of the voltage transfer function for the given channel structure is derived and the method to deduce the worst case data patterns by considering the asymmetric and finite slew rates of the input signals is introduced. The validity of the proposed method was verified through comparison with the DDJ and eye-opening voltage values obtained by using HSPICE simulations.

I. INTRODUCTION

Most high speed signal lines on PCBs are in the form of a pair of differential microstrip traces with arbitrary source and load terminations as shown in Fig. 1. For these channels, reflection as well as frequency-dependent loss are unavoidable and seriously deteriorate the signal integrity of the received signal, which limits the maximum achievable data transmission rate through the channel [1]. There have been numerous previous studies to estimate the eye-diagram in analytical ways [2]-[4]. In these methods, the output response can be obtained through the inverse fast Fourier transform (IFFT) of the product of the output responses of a channel in the frequency domain and the voltage transfer function of the channel. Meanwhile, the eye-diagram of a channel is extracted from the worst case data patterns that are derived through the linear combination of the output response of a channel for the unit pulse signal, which is also known as the Peak distortion inter-symbol interference analysis (PDA) method.

In this paper, an analytical eye-diagram estimation method for a channel of a pair of differential microstrip traces on PCBs with arbitrary source and load terminations is presented. For this purpose, a closed-form equation to obtain the voltage transfer function of the given channel structure is derived. Moreover, a modified PDA method to account for the asymmetric and finite rise/fall times of the input signal is developed. The acquisition of the voltage transfer function and the understanding of the worst case data patterns constitute the basis of the eye-diagram estimation. It was successfully demonstrated that the proposed method can provide fast and precise eye-diagram estimation for the given channel structure through a comparison with the eye-diagram obtained by using an HSPICE simulation.

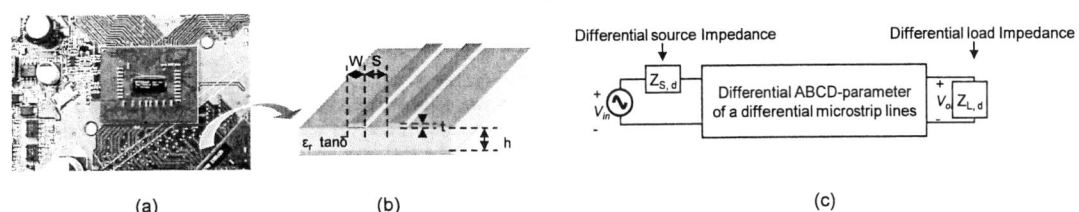

(a)   (b)   (c)

Fig. 1 (a) High speed differential microstrip lines on PCBs, (b) the cross section of a pair of differential microstrip traces for their network function extraction, and (c) the schematic block diagram of a pair of differential microstrip traces with arbitrary source and load terminations.

II. VOLTAGE TRANSFER FUNCTION FOR A CHANNEL OF A PAIR OF DIFFERENTIAL MICROSTRIP TRACES ON PCBs

The differential mode ABCD parameter of a pair of differential microstrip traces on PCBs can be written by:

$$\begin{bmatrix} A_d & B_d \\ C_d & D_d \end{bmatrix} = \begin{bmatrix} A_o & 2B_o \\ \dfrac{C_o}{2} & D_o \end{bmatrix} = \begin{bmatrix} \cosh\gamma_o l & 2Z_{0,o}\sinh\gamma_o l \\ \dfrac{\sinh\gamma_o l}{2Z_{0,o}} & \cosh\gamma_o l \end{bmatrix} \tag{1}$$

where $Z_{0,o}$ and γ_o are the odd mode characteristic impedance and propagation constant of a pair of microstrip traces on PCBs, respectively. For a given differential microstrip structure, they can easily be extracted from the empirical equations in [5].

For a channel of a pair of differential microstrip traces on PCBs with arbitrary source and load terminations, the total ABCD parameter is given by:

$$
\begin{bmatrix} A_{T,d} & B_{T,d} \\ C_{T,d} & D_{T,d} \end{bmatrix} = \begin{bmatrix} 1 & Z_{S,d} \\ 0 & 1 \end{bmatrix} \begin{bmatrix} \cosh\gamma_o l & 2Z_{0,o}\sinh\gamma_o l \\ \dfrac{\sinh\gamma_o l}{2Z_{0,o}} & \cosh\gamma_o l \end{bmatrix} \begin{bmatrix} 1 & 0 \\ \dfrac{1}{Z_{L,d}} & 1 \end{bmatrix}
$$
$$
= \begin{bmatrix} \cosh\gamma_o l + \dfrac{Z_{S,d}\cosh\gamma_o l}{2Z_{0,o}} + \dfrac{2Z_{0,o}\sinh\gamma_o l + Z_{S,d}\cosh\gamma_o l}{Z_{L,d}} & 2Z_{0,o}\sinh\gamma_o l + Z_{S,d}\cosh\gamma_o l \\ \dfrac{\sinh\gamma_o l}{2Z_{0,o}} + \dfrac{\cosh\gamma_o l}{Z_{L,d}} & \cosh\gamma_o l \end{bmatrix}
$$

(2)

Hence, the voltage transfer function for this channel can be obtained by:

$$
H(f) = \frac{1}{A_{T,d}} = \frac{2Z_{0,o}Z_{L,d}}{4Z_{0,o}{}^2\sinh\gamma_o l + \left\{2Z_{0,o}\left(Z_{S,d}+Z_{L,d}\right)+Z_{S,d}Z_{L,d}\right\}\cosh\gamma_o l}
$$

(3)

The estimated and the simulated voltage transfer function for the given channel structure are compared in Fig. 2.

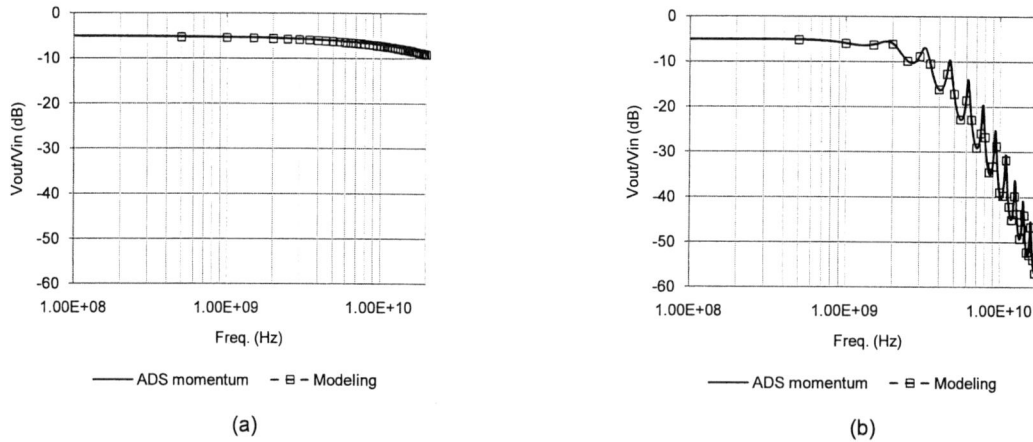

(a)   (b)

Fig. 2 Comparison between the obtained voltage transfer functions of two kinds of channels that have a pair of 5 cm long differential microstrip traces – (a) with 100 ohm resistors for both source and load terminations and (b) with a series of a 100 ohm resistor and a 2 nH inductor and a parallel 0.75 pF capacitor for source termination and a series of a 100 ohm resistor and a 4 nH inductor and a parallel 1 pF capacitor for source termination, respectively, through the ADS momentum method (solid line) and Eq. (3) (dashed line).

III. Modified Peak Distortion Inter-symbol Interference Analysis (PDA) Method

The conventional PDA method assumes that the rise/fall times of an input signal are infinite, or at least symmetric. Therefore, it is sufficient to derive the worst case '1' and '0' data patterns only with the output response of a unit pulse signal, where the unit pulse signal is a signal which has a single data '1' period during its whole time period, and worst case '1' and '0' data patterns are the input data patterns which produce the lowest high and highest low output levels. However, practical digital signals are subject to asymmetric and finite slew rates due to the asymmetry and finite turn-on resistances of pull-up and pull-down output drivers. The asymmetry of the rise/fall time may bring about considerable duty cycle distortion (DCD), and DCD is one of the major components of DDJ. Hence, for precise DDJ estimation, the asymmetry of the rise/fall time must be taken into consideration [6] and, thus, a modified PDA method is proposed.

For the modified PDA method, four kinds of unit pulse signals are defined as the first step. To consider the asymmetry of signal pull-up and pull-down characteristics, the worst case '1' and '0' data patterns must be derived from the output response of a unit pulse '1' and '0', respectively, where the unit pulse '1' signal is the same signal defined as the unit pulse signal in the conventional method, and the unit pulse '0' signal is a signal that has a single, unit time interval, data '0' period during its whole time period. Moreover, a couple of unit pulse signals must additionally be defined to consider data patterns consisting of consecutive '1's or '0's. As shown in Fig. 3, when the rise/fall times of an input signal are asymmetric and finite, the consecutive '1's or '0's data patterns cannot be represented as the combination of two consecutive unit pulse '1' or '0' signals, and thus, the preceding unit pulse signals must be replaced by the auxiliary unit pulse signals, named unit pulse '1^c' and '0^c' signals in this paper. The unit pulse '1^c' and '0^c' signals are similar to the original unit pulse '1' and '0' signals, respectively, but they are different in that they have common rise/fall times. That is, the unit pulse '1^c' is a signal which has a single data '1' period during its whole time period and common rise/fall times that are the same as the rise time of the unit pulse '1' or '0' signal. On the other hand, the unit pulse '0^c' is a signal which has a single data '0' period during its whole time period and common rise/fall times that are the same as the fall time of the unit pulse '1' or '0' signal. Then, the worst case data patterns are deduced from the output responses of four kinds of unit pulse input signals and through their linear combination. The complete equations for the output responses of worst case '1' and '0' data patterns, which are which determine the inner contour of eye-diagrams, are given by:

$$s_1(t) = y_1^*(t) + \sum_{\substack{k=-\infty \\ k \neq 0}}^{\infty} y_1^*(t - kT)\Big|_{y_1^*(t-kT)<0} \tag{4}$$

$$s_0(t) = y_0^*(t) + \sum_{\substack{k=-\infty \\ k \neq 0}}^{\infty} \left(y_0^*(t - kT) - Y_1 \right)\Big|_{y_0^*(t-kT)>Y_1} \tag{5}$$

where $y_1^*(t)$ and $y_0^*(t)$ are the output responses of unit pulse '1' or '1^c' and '0' or '0^c' signals, and Y_1 is the ideal output logic '1' level.

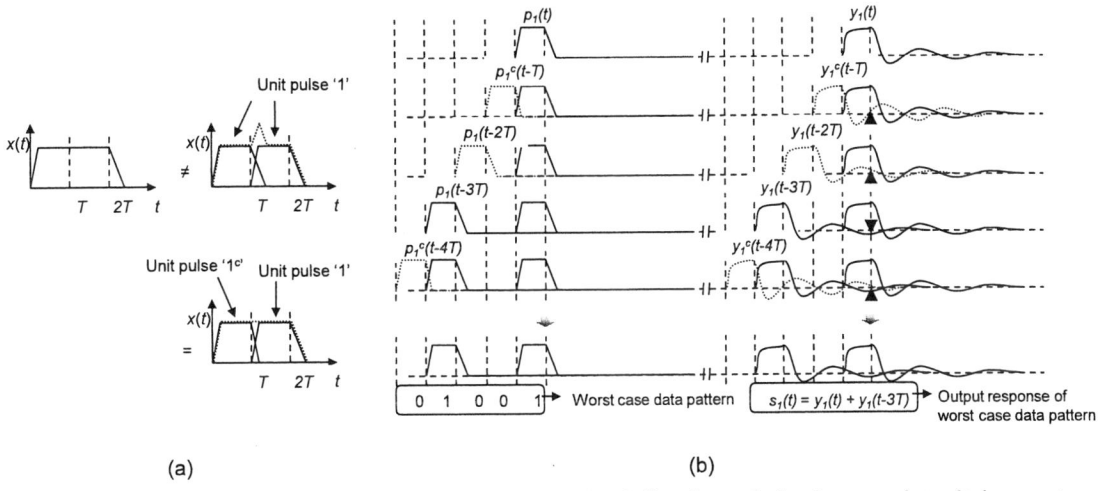

(a)   (b)

Fig. 3 (a) Consecutive '1's and their decomposition, (b) proposed modified peak distortion method and an example to obtain a worst case '1' data pattern through the proposed method.

IV. SIMULATION-BASED VERIFICATION

In this section, we report simulation-based verification results on the proposed eye-diagram estimation method. To start with, the transfer function of a channel was modeled with the proposed modeling method described in Section 1. Then, the output responses for four unit pulse signals were obtained, either through a series of mathematical computations using the FFT of four unit pulse signals, as shown in Section 2, multiplication of them by the transfer function of a channel, and the IFFT of them or through time domain convolution of four unit pulse signals and the impulse response of the transfer function of a channel in the time domain. Next, by using the modified PDA method described in Section 3, the output responses of worst case '1' and '0' data patterns were extracted. For the channel in Fig. 2 (b), they are determined as '11010010001' and '0010101110' for each case. The overall procedures are summarized in Fig. 4. Finally, the estimated DDJ and eye-opening voltage values determined by using these short data patterns were compared with those values determined by using a 2^{10}-1 bit PRBS data pattern, which is shown in Fig. 5. As shown in this figure, the proposed method can successfully present fast and precise eye-diagram estimation.

978-1-4244-4447-2/09 $25.00 © 2009 IEEE

V. CONCLUSION

In this paper, an analytical eye-diagram estimation method for a channel of a pair of differential microstrip traces on PCBs with arbitrary source and load terminations is presented. The voltage transfer function of the given channel structure is derived in the form of a closed-form equation. Moreover, to consider the asymmetric and finite slew rates of an input signal, a modified PDA method is introduced. In this procedure, a couple of worst case data patterns, which determine the inner contour of an eye-diagram, are deduced from the output responses of four kinds of unit pulse input signals and through their linear combination. Finally, the DDJ and eye-opening voltage values for the given channel are estimated from the eye-diagram formed by these worst case data patterns. It was successfully confirmed that the proposed method can provide fast and precise eye-diagram estimation for the given channel structure through comparison with the DDJ and eye-opening voltage values obtained by using HSPICE simulations.

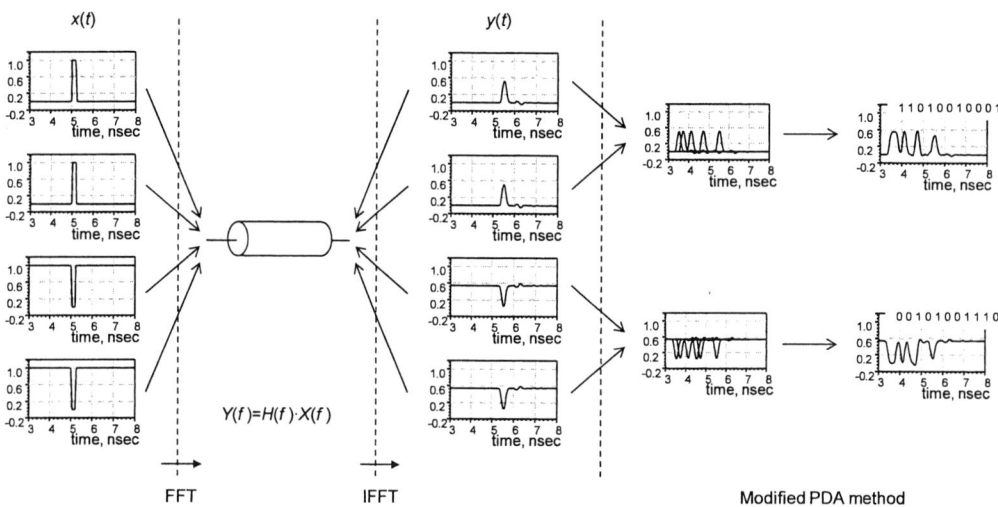

Fig. 4 Overall procedures to obtain worst case '1' and '0' data patterns by using the proposed methods in Section 2 and 3 for the channel in Fig. 2 (b).

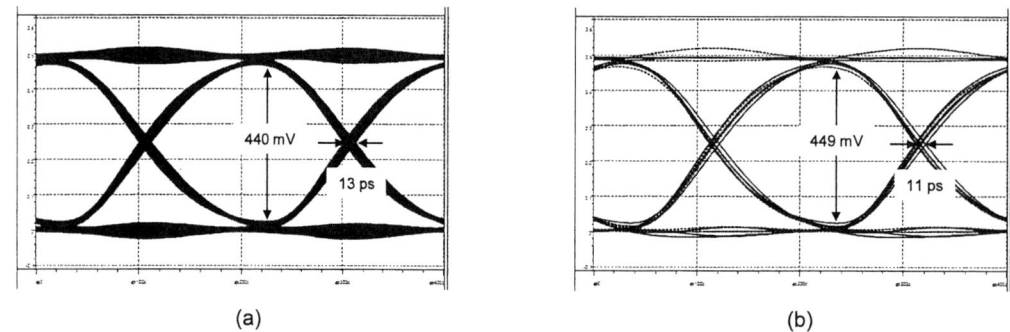

Fig. 5 Estimated eye-diagram (a) by using a 2^{10}-1 bit PRBS data pattern and (b) by using both worst case '1' and '0' data patterns

REFERENCES

[1] J. Liu and X. Lin, "Equalization in high-speed communication systems," *IEEE Circuits Syst. Mag.*, vol. 4, no. 2, pp. 4-17, 2nd quarter, 2004.
[2] B. K. Casper, M. Haycock, and R. Mooney, "An accurate and efficient analysis method for multi-Gbps chip-to-chip signaling," in *Proc. IEEE Symp. VLSI Circuits*, pp. 54-57, Jun. 2002.
[3] H. Zhu, C. Cheng, A. Deutsch, and G. Katopis, "Prediction and optimizing jitter and eye-opening based on bitonic step response," in *Proc. IEEE EPEP 2007*, pp. 155-158, Oct. 2007
[4] R. Shi et al., "Efficient and accurate eye diagram prediction for high speed signaling," *IEEE/ACM ICCAD 2008*, pp. 655-661, Nov. 2008.
[5] K.C.Gupta, Ramesh Garg et al, "Microstrip lines and slotlines", Artech House, 1996
[6] J. Cho et al., "A precise analytical estimation method of data-dependent jitter for high speed serial links with the consideration of finite slew rate of input signal," *KJJC-AP/EMC/EMT 2009*, pp. 215, May 2009

A Fast Methodology for the Synthesis of Dispersive Multi-Port Equivalent Circuit Model of Multiple Coupled Bond Wires

J. H. Chung[1], V. Okhmatovski[2], and A.C. Cangellaris[1]

[1] ECE Department, University of Illinois, Urbana-Champaign, 1406 W. Green St., Urbana, IL 61801
[2] ECE Department, University of Manitoba, 75A Chancellor Circ., Winnipeg, MB R3T 5V6, Canada

Abstract

A methodology is presented for the fast synthesis of a SPICE-compatible, dispersive, multi-port equivalent circuit model of a group of coupled bond wires. The expediency of the proposed method stems from its reliance upon resistance and inductance matrices for the coupled wires extracted at low and high frequencies only, which are then combined with a rational function interpolation process to obtain $\mathbf{R}(\omega), \mathbf{L}(\omega)$ over the entire frequency bandwidth of interest.

Introduction

Due to their electrically small lengths, the electrical modeling of bond wires for signal and power integrity analysis is customarily done making use of three-dimensional capacitance, resistance and inductance extractors. Such an approach has been shown to lead to accurate models up to frequencies at which the wire electrical length (length in wavelengths) remains below 0.25. For such R,L,C, modeling of the bond wires, the primary cost of the development of the equivalent circuit is associated with the calculation of the frequency-dependent resistance and inductance matrices. In particular, in order to capture accurately the frequency dependence of these matrices, the field solver used must comprehend accurately the field penetration inside the wires. Irrespective of the type of solver used (integral equation based or finite element based) the two frequency regimes at which resistance and inductance extraction is the fastest are: a) frequencies low enough for the skin depth in the metallization to be comparable to or larger than the wire cross-sectional dimensions, and b) frequencies high enough at which the skin depth is much smaller than the cross-sectional dimensions of the wire. For the former, a coarse discretization of the wires suffices for accurate resistance and inductance extraction while for the latter use of surface impedance boundary condition eliminates the need for discretizing the interior of the wires. This observation motivates the possibility of relying on extracted values for $\mathbf{R}(\omega), \mathbf{L}(\omega)$ obtained only at these two frequency regimes to describe the dispersive properties of the synthesized circuit over the entire frequency bandwidth of interest. The feasibility of such an approach is examined in this paper.

Methodology

Figure 1 depicts a representative geometry of a group of bond wires. Quasi-static, integral equation-based solvers are used for the extraction of the capacitance, inductance and resistance matrices of the coupled bond wires. For the case of the magneto-quasi-static problem, solved for the calculation of the inductance and resistance matrices, we distinguish between two frequency regimes: a) a low-frequency regime that extends over the range of frequencies for which skin depth is larger or at most of the order of the wire diameter, and b) a high-frequency regime that encompasses the range of frequencies for which the skin depth is a small fraction of the wire diameter. For frequencies in the low-frequency regime use of a filament discretization of the wires for the discretization of the volumetric current density $\mathbf{j}$ in the wires is economical. The pertinent magneto-quasi-static integral equation is [1]

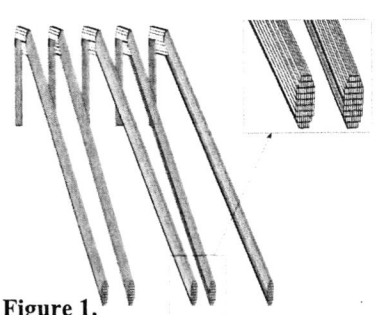

Figure 1.

$$\frac{\mathbf{j}(\mathbf{r})}{\sigma} + \frac{j\omega\mu}{4\pi} \int_V \frac{\mathbf{j}(\mathbf{r}')}{|\mathbf{r}-\mathbf{r}'|} dv' = -\nabla\Phi, \quad \mathbf{r} \in V \tag{1}$$

Through a standard Galerkin process the discrete forms of (1) leads to a matrix equation of the form

$$\left([R]+j\omega[L]\right)[I]=[V], \text{ where it is } [R]_{ij}=\frac{l_i\delta_{ij}}{\sigma a_i}, [L]_{ij}=\frac{j\omega\mu}{4\pi V_j}\int_{V_j}\frac{\mathbf{l}_i\cdot\mathbf{l}_j}{|\mathbf{r}_i^c-\mathbf{r}'|}dv', [V]_i=\Phi(\mathbf{r}_i^a)-\Phi(\mathbf{r}_i^b),$$

$i,j=1,...,N$, and δ_{ij} is the Kronecker's delta and N the number of filaments. The quantity $[V]_i$ denotes the difference of the nodal potentials at the end points of the i th filament. Augmenting the above linear systems with the discrete form of the charge conservation equation, $\nabla\cdot\mathbf{j}=0$, enforced at each one of the n nodes formed by the junction of different conductor segments closes the system for the calculation of the N filament currents and the n nodal potentials [1].

In the high-frequency regime if the wires are discretized volumetrically according to skin-depth the model may result in a matrix equation of a very large size. To circumvent this problem at the frequencies when the skin-depth is substantially smaller than crossectional dimensions of the wires one can reduce the volumetric integral equation (1) to its surface counterpart with the use of appropriate impedance boundary condition

$$Z_s\mathbf{J}(\mathbf{r})+\frac{j\omega\mu}{4\pi}\int_S\frac{\mathbf{J}(\mathbf{r}')}{|\mathbf{r}-\mathbf{r}'|}ds'=-\nabla\Phi, \quad \mathbf{r}\in S, \tag{2}$$

where $Z_s=(1+i)\sqrt{\omega\mu_0/(2\sigma)}$ is the Leontovich's surface impedance [2, p. 533]. This integral equation

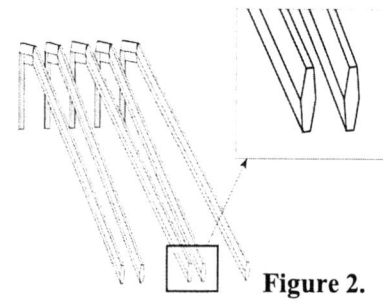

Figure 2.

can be discretized using the boundary element method of moments. As depicted in Fig. 2, confinement of the unknown functions to the surfaces of the wires results in a matrix equation of substantially smaller size, thus allowing for expedient extraction of the R and L parameters at high frequencies.

The extracted values of $\mathbf{R}(\omega),\mathbf{L}(\omega)$ at the low- and high-frequency regimes are used as input to a rational function-fitting algorithm [3] to yield a rational function interpolation over the bandwidth of interest, including the intermediate frequency band over which $\mathbf{R}(\omega),\mathbf{L}(\omega)$ were not computed through the solution of the magneto-quasi-static equation. More specifically, since it is the diagonal elements of the impedance matrix for the coupled bond wires that exhibit the strongest dependence with frequency, rational function interpolations are generated for these terms in the form [4]

$$Z_{kk}(s)=R_{kk}(s)+sL_{kk}(s)=R_0^{(kk)}+sL_0^{(kk)}+\sum_{m=1}^{Q_1}R_m^{(kk)}\left(s-P_m\right)^{-1} \tag{3}$$

where $s=j\omega$. An equivalent circuit representation of (3), depicted in Fig. 2 by each branch formed by the series connection of a resistor, an inductor and Q_1 $R\|L$ circuits, follows immediately from (3) by noting that it can be recast in the form

$$Z_{kk}(s)=\overline{R}_0^{(kk)}+sL_0+\sum_{m=1}^{Q_1}\overline{R}_m^{(k)}s\left(s-P_m\right)^{-1}; \overline{R}_0^{(kk)}=R_0^{(kk)}-\sum_{m=1}^{Q_1}\left(R_m^{(kk)}/P_m\right), \overline{R}_m^{(kk)}=R_m^{(kk)}/P_m, \overline{L}_m^{(kk)}=-R_m^{(kk)}/P_m^2$$

In the above equation, expressions for the elements of the equivalent circuit in terms of the terms in the rational function interpolation in (3) are provided. As far as the mutual impedances are concerned, the values computed at the highest frequency of interest are used for the definition of the equivalent circuit. These then, together with the extracted capacitances are used to complete the multi-port, equivalent circuit for the coupled bond wires depicted in Fig. 3.

978-1-4244-4447-2/09 $25.00 © 2009 IEEE

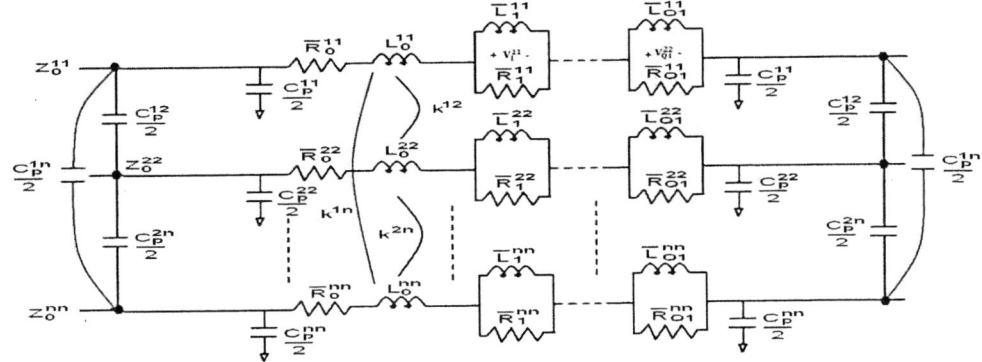

Figure 3: Equivalent circuit representation of n coupled bond wires.

Numerical Examples

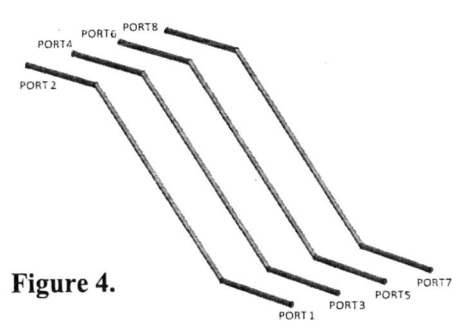

Figure 4.

To demonstrate the proposed methodology we consider the geometry of four copper bond wires depicted in Fig. 4. The wire radius is 10^{-2} mm, each wire bond is 1.1983 mm long, and the axis-to-axis distance between adjacent wires is 0.2 mm. A ground plane is present below the wires, placed at a distance of 1 mm from the bottom side of the wires. Depicted in Fig. 5 is the comparison of the resistance and reactance versus frequency for one of the bond wires over the 5 GHz frequency bandwidth of interest calculated using the proposed method, with those obtained using the 3D extractors. Very good agreement is observed.

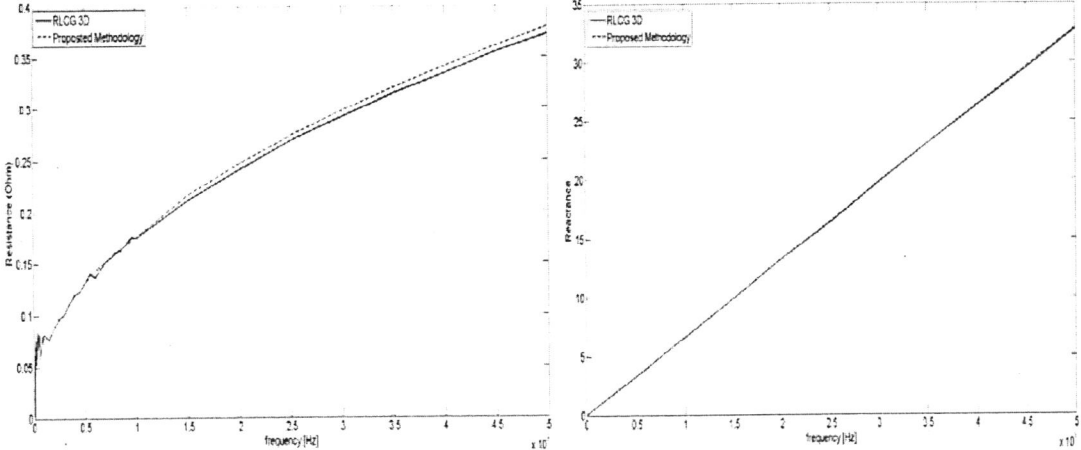

Figure 5: Calculated frequency-dependent resistance (left), and reactance (right) for one of the four wire bonds of Fig. 4.

Next, the synthesized equivalent multiport is used with a set of source and load condtions to perform a transient simulation using SPICE. With reference to Fig. 4, a volatge source with input impedance of 50 ohms is connected at Port 1. All remaining ports are terminated at 50 ohms. The excitation waveform is a pulse train of period 2 ns, pulse width 1 ns, rise and fall times of 0.2 ns and amplitude of 2 V. The resulting waveforms at several of the ports are depicted in Figs. 6 and 7. The generated responses have been computed using two different synthesized circuits. One of the circuits was synthesized using the proposed methodology. The other was synthesized using resistance and inductance matrix data computed by the 3D quasi-static extractors over the entire frequency band. The two sets of responses are in very good agreement for both the driven and the victim wires.

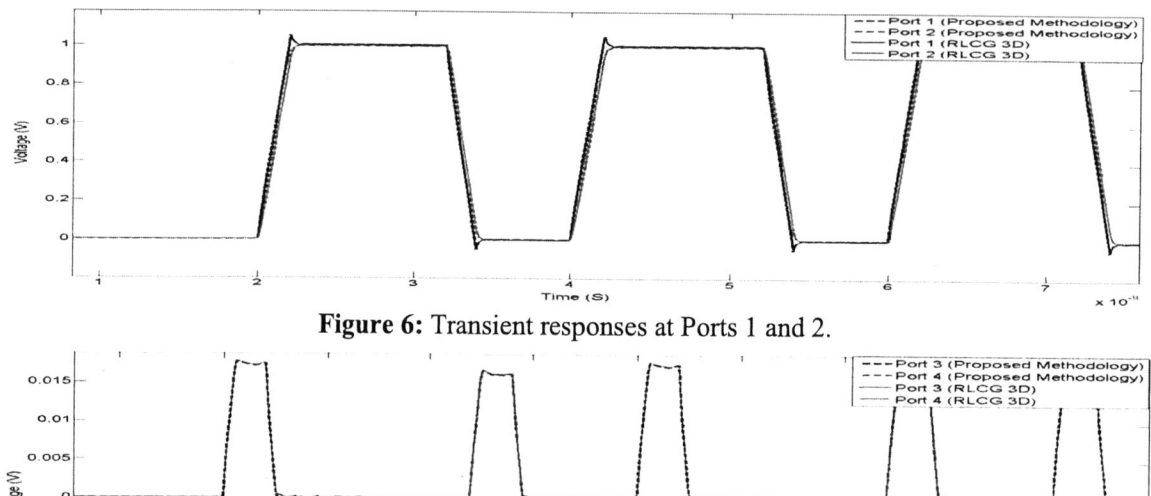

Figure 6: Transient responses at Ports 1 and 2.

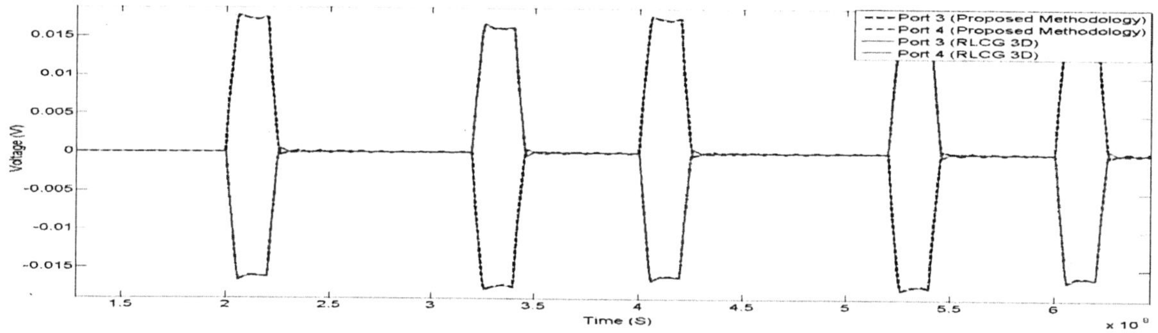

Figure 7: Transient responses at Ports 3 and 4.

Concluding Remarks

In summary, we have proposed a computationally efficient scheme for the development of equivalent circuit representation of coupled bond wires of arbitrary configuration and number, the computation cost associated with the extraction of the frequency-dependent resistance and inductance matrices through the solution of a magneto-quasi-static, three-dimensional boundary value problem. This is achieved by combining extracted values for $\mathbf{R}(\omega), \mathbf{L}(\omega)$ using a low-density volumetric grid at a subset of low frequencies at which the skin depth is larger than or at most comparable with extracted values at a subset at the upper end of the frequency bandwidth of interest where the skin depth is much smaller than the wire diameter and thus a surface impedance can be imposed to eliminate the need for volumetric discretization. These sets of values are then used in conjunction with a rational function fitting algorithm to interpolate these values over the entire frequency bandwidth of interest and, subsequently, generate the desired multi-port equivalent circuit.

References

[1] M. Kamon, M.J. Tsuk, and J.K. White, "FASTHENRY: A multipole-accelerated 3D inductance extraction program," *IEEE Trans. Microwave Theory Tech.*, vol. 42, no. 9, pp. 1750-1758, Sept. 1994.

[2] J.A. Stratton, *Electromagnetic Theory*. New York: McGraw-Hill, 1941.

[3] B. Gustavsen and A. Semlyen, "Rational approximation of frequency domain responses by Vector Fitting," *IEEE Trans. Power Delivery*, vol. 14, no. 3, pp. 1052-1061, July 1999.

[4] K.M. Coperich, J. Morsey, V.I. Okhmatovski, A.C. Cangellaris and A.E. Ruehli, "Systematic development of transmission-line models for interconnects with frequency-dependent losses," *IEEE Trans. Microwave Theory Tech.*, vol. 49, no. 10, pp. 1677-1685, Oct. 2001.

A New EBG Structure for Low Frequency Power Plane Noise Mitigation

A.Ciccomancini Scogna[#], G.Romo[#]

[#]CST of America, 492 Old Connecticut Path, #505, Framingham (MA), 01701, USA - antonio.ciccomancini@cst.com, gerardo.romo@cst.com

Abstract— A new electromagnetic bandgap structure for power plane noise mitigation in the low frequency range is proposed. It consists of an improved asymmetric square patch with meander lines, optimized inductance and high-K dielectric material for the substrate. An example of PWR/GND plane pairs is analyzed where fundamental stop band is demonstrated over a frequency range of less than 300 MHz to about 1 GHz.

Key words: Power Integrity, Signal Integrity, SSN and EBG.

I. INTRODUCTION

Electromagnetic Band Gap (EBG) structures are one of the most recent and alternative solutions for noise mitigation in high-speed printed circuit boards (PCBs) and packages [1-3]. Great interest has been given to the application of EBG for the reduction of simultaneous switching noise (SSN).

Most of the works are focused on how to extend the stop band for more efficient noise suppression (in high frequency), however it appears difficult to lower the effective frequency ranges of the EBGs, infact not many papers are available in literature focusing on this aspect. In [4] an alternative hybrid EBG layer is proposed with consistent noise suppression below 200MHz, nevertheless in the paper the Signal Integrity aspect (SI) is not treated. Due to the patterned EBG layer the return current path can produce multiple impedance variation (clearly observed in the TDR waveform), consistent signal distortion (time signal analysis) as well as degradation (insertion loss) [5].

The present is on the line of power integrity (PI) analysis and SSN mitigation in low frequency range by means of a new 2 dimensional (2D) EBG layer as well as in the line of SI for high speed interconnects which is very important for a successful application of EBGs. The proposed geometry is similar to the one described in [6] and it is an enhancement of the standard EBG structure with square patches and meander lines.

The main difference of the proposed EBG is the increased inductance due to asymmetric meanders on the edges of the square patches in order to optimize the routing of the PWR layer and therefore improve the conduction at DC as well as the noise isolation in the low frequency range. It is demonstrated how the present EBG allows noise suppression of -30dB from 300MHz when standard fr4 dielectric material is used, however by using high K dielectrics, the noise isolation can be lowered to less than 100Mhz, therefore pushing the usage of EBGs in frequency where decoupling capacitors are typically used.

The structure of the paper is the following: in the next section the proposed EBG design is illustrated along with some relevant dimensions. In section III the PI is studied by means of noise coefficient and in section IV the SI of a single-ended (SE) and differential (DIFF) 90 degree bend microstripline is analyzed. Section IV offers some final remarks and design recommendations.

II. PROPOSED EBG STRUCTURE

The shape of the considered EBG is illustrated in Fig.1 along with the relevant dimensions: it is a 2D square patch with an improved asymmetric meander lines.

The shape is similar to the model already proposed in [6]; however in this case the space around the metal patches is optimized by using different lengths for the meanders in the right/left side and top/bottom side.

The main idea is to increase the inductance of the meander lines and therefore to provide noise mitigation in the low frequency range when multiple unit cells are cascaded. The simulated test board is a 2 layers (PWR/GND) board whose dimensions are W = 9.6 cm, H = 7.2 cm, h = 0.02 cm (see Fig.2). The dielectric material is FR4 (ε_r = 4.4 and tan δ = 0.02 at 0.5 GHz).

978-1-4244-4447-2/09 $25.00 © 2009 IEEE

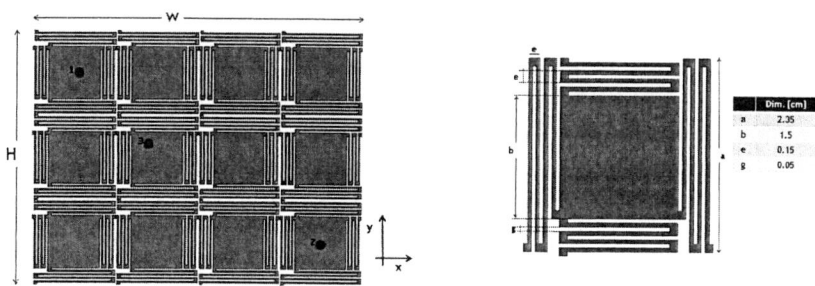

Fig. 1 - Top view of the PCB and port location (1 is the input port and 2, 3 are the output ports), proposed patch for the EBG layer and relevant dimensions (in [cm])

III. POWER INTEGRITY ANALYSIS

A Finite Integration Technique (FIT) based numerical code [7] is used for the numerical simulations. The excitation of the system is given by a lumped source placed in position x=1, y=6cm (Fig.1). Fig.2a-b illustrates the noise coefficient on ports 2 and 3 for the proposed EBG structure; the comparison with the board with a similar EBG layer (as proposed in [6]) is also illustrated for comparison.

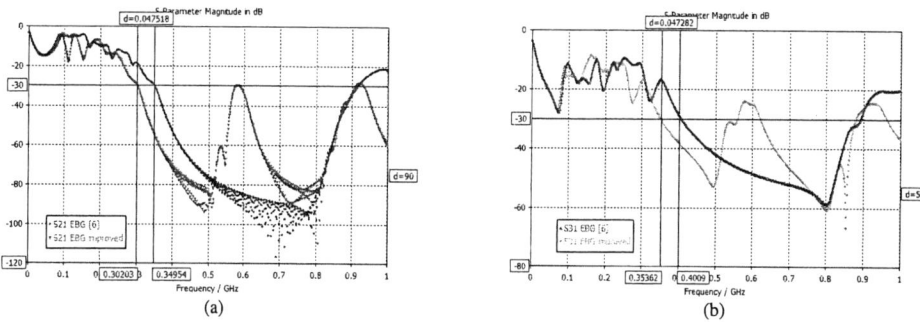

Fig. 2 - Noise coefficients on a): port 2 and b): port 3

A figure of merit of -30 dB is used to compare the different results and a reduction of 40MHz in terms of noise mitigation for both S_{21} and S_{31} is achieved by adopting the proposed EBG structure vs. the one described in [6].

As next step the same geometry is simulated by employing a high-K [8] dielectric material with $\varepsilon_r=30$ and the new results are illustrated (along with the same results already commented) in Fig.3a-b. It this case the noise mitigation is further lowered and the same figure of merit can be extended to 280 MHz for S_{21} and 110 MHz for S_{31}. Furthermore S_{31} also presents a narrow bang gap in the very low frequency range 10-25MHz.

The different behavior of S_{21} and S_{31} in bandgap and level of noise attenuation is probably related to the asymmetric nature of the proposed EBG patch whose meander has a different length for the top/bottom and right/left edge.

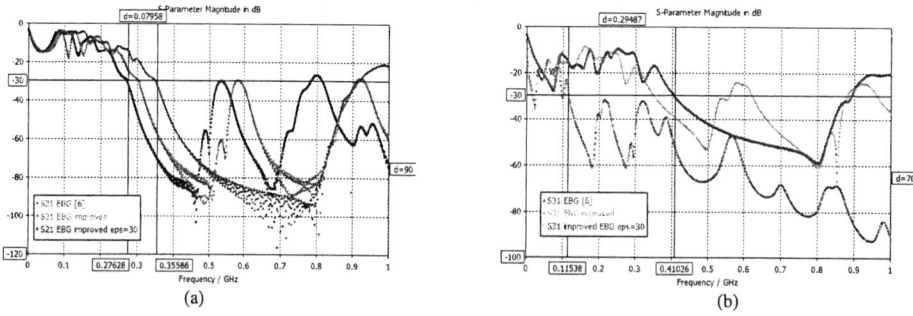

Fig. 3 - Noise coefficients on a): port 2 and b): port 3 for configuration presented in [6], improved EBG and improved EBG with high k dielectric

978-1-4244-4447-2/09 $25.00 © 2009 IEEE

To further investigate the non uniform behavior of the EBG layer in correspondence of the ports 2-3 (according to the location in Fig.2) a second configuration is studied. In order to analyze the worst case scenario, 4 patches only (2x2) are considered and the ports are placed in the middle of each patch, as depicted in Fig.4.

In this way the noise coefficients on ports 2-3 will represent the coupling mechanism between 2 patches only. It should be noted that in this study all the dimensions are kept the same as in Fig.1 and a permittivity of $\varepsilon_r=30$ is used for this analysis.

Results are reported in Fig.5 and if the same figure of merit of -30dB is considered, stop bands are detected in the frequency range 100MHz-180MHz, 230MHz-300MHz and 380MHz-1GHz for both S_{21} and S_{31}, therefore proving the effectiveness and the consistency of the proposed EBG pattern in the low frequency range.

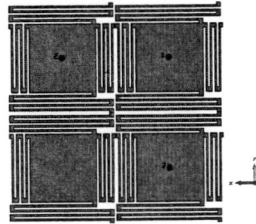

Fig. 4 – Test vehicle for the PI analysis

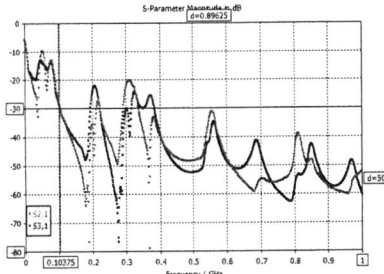

Fig. 5 - Noise coefficient for configuration in Fig.4

The understanding of the behaviour of the EBG structure can be also widened by looking at the spatial distribution of the surface current on the EBG layer. Fig.6 illustrates the surface current for two different frequencies: 100MHz and 320MHz. The current variation is represented by a color contrast in these figures and the unit in the color bars is in [A/m]. Isolation is of course desirable between the input port 1 and the output ports 2/3.

Fig. 6 shows that the EBG structure does not provide good isolation at 320 MHz since this frequency is still in the pass band region, however at 100 MHz (a frequency value in the stop band region) the noise generated by the current source on the input port cannot propagate to the other metal patches characterizing the EBG structure which means that eventual noise generated by digital circuits cannot propagate to the RF circuits located at the output ports.

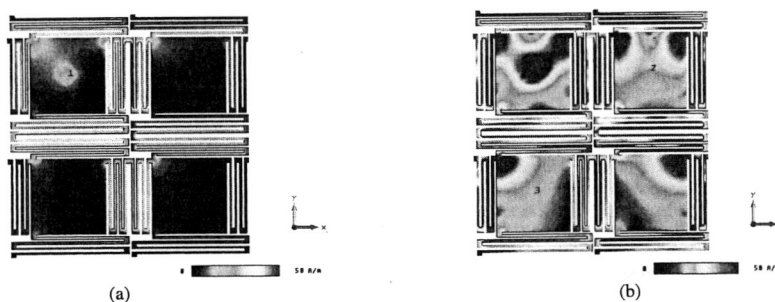

(a)   (b)

Fig. 6 Current distribution on the EBG layer at a): 100MHz and b): 320MHz.

IV.   SIGNAL INTEGRITY ANALYSIS OF SE AND DIFF SIGNALS

Although the proposed PWR/GND plane design shows excellent performance on eliminating SSN noise at broad band frequency ranges, the presence of the EBG structure might degrade the quality of signals propagating on traces. For this reason a 90 degree bend microstripline is modeled in order to study the SI performance in presence of the EBG layer. In particular two different configurations for the microstriplines are considered: SE and DIFF, with a second line placed adjacent to the first one with an offset of 0.01cm. The simulated structure is depicted in Fig.7 where the top view (with port location) and perspective view of the board are illustrated.

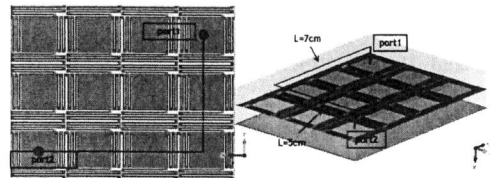

Fig. 7 - Top and perspective views with port location for the SE 90 degree micro strip line.

A modification of existing S-parameter theory is needed to measure DIFF-mode and common-mode circuit performance. The modified theory is based on the mixed propagation mode and describes the common-mode and differential-mode conversion. If one is interested in the performance of this structure at differential mode propagation the introduction of a mixed mode scattering matrix S_{mm} [9] is needed.

Fig.8 compares the insertion loss for the SE line and the DIFF lines. In the low frequency range (up to 400 MHz) some resonances are detected in the insertion loss due to the SE configuration, however in the S_{dd} (corresponding in this case to S_{dd21}) those dips disappear and a less degraded behavior is observed. A standard -3dB figure of merit used for the detection of insertion loss quality would in this case allow increasing the bandwidth up to 0.8GHz.

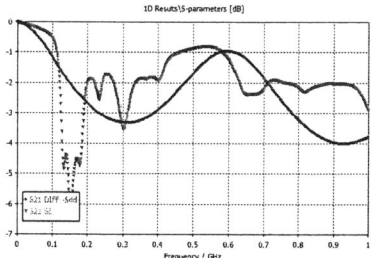

Fig. 8 – Insertion loss: SE and DIFF.

V. CONCLUSIONS

In this paper noise mitigation in the low frequency range by means of an improved 2D EBG with square patches, asymmetric meander lines and high-K dielectric material for the substrate is presented. The calculated noise coefficient reveals good noise suppression in the frequency range 100MHz-1GHz. Decoupling capacitors can be completely replaced by this EBG structure. The worst case scenario is investigated by considering 2x2 square patches only and consistency in the results is found.

The SI is also studied by considering a 90 degree bend microstripline and 2 different configurations are analysed: SE and DIFF signals. Frequency domain results are provided which prove the effectiveness of the proposed design in terms of quality of transmitted signal (insertion loss) when the DIFF excitation is used.

REFERENCES

[1] J. Qin and O. M. Ramahi, "Ultra-Wideband Mitigation of Simultaneous Switching Noise Using Novel Planar Electromagnetic Bandgap Structures", *IEEE Microwave and Wireless Components Letters,* Vol. 16, No. 9, pp. 487-489, September 2006-9.

[2] A.Ciccomancini Scogna, M.Schauer, "A Novel Electromagnetic Bandgap Structure for SSN Suppression in PWR/GND plane pairs", *Proceedings of IEEE International Symo. ECTC,* Arizona, USA, May 2007

[3] T. Lin Wu, Y.H. Lin, T.K.Wang, et Al, "Electromagnetic band gap power/ground planes for wideband suppression of ground bounce noise and radiated emission in high speed circuits", *IEEE Trans. On Microwave Theory and Tech*, vol. 53, no. 5, pp. 2935–2942, September 2005.

[4] W. McKinzie, "A Low Frequency hybrid EBG Structure for Power Plane Noise Suppression", on *Proc. of EPEP*, 2006.

[5] A. Ciccomancini Scogna, A.Orlandi, V.Ricchiuti, "Signal Integrity analysis of single-ended and differential signaling in PCBs with EBG structure", submitted for *Proc. of 2009 IEEE International Symposiun on EMC*, Detroit, August 2008.

[6] A. Ciccomancini Scogna, " SSN mitigation by means of a 2D EBG structure with square patches and meander lines", *in IEEE Proceedings of EPEP*, San Jose', CA, USA, October 07.

[7] CST STUDIO SUITE 2009™, Computer Simulation Technology, www.cst.com

[8] A. Tambawala, E.Engin, M. Swaminathan and Al, "Compact Electromagnetic bandgap Structures for Power Plane Isolation using High-K dielectrics", on *Proceedings of DesignCon 2007*, CA, USA

[9] E. Bockelman, W. R. Eisenstandt, " Combined Differential and Common Mode Scattering Parameters: Theory and Simulation ", *IEEE* Trans. Microwave Theory & Technique, vol. 43, no. 7, pp. 1530-1539, July 1995.

978-1-4244-4447-2/09 $25.00 © 2009 IEEE

Sensitivity Analysis of Lossy Transmission Lines based on the Passive Method of Characteristics

Amir Beygi and Anestis Dounavis

Department of ECE, University of Western Ontario, London, Ontario, N6A 5B9

Tel: (519) 661-2111 ext: 81255; Fax: (519) 850-2436; Email: adounavis@eng.uwo.ca; abeygi@uwo.ca

ABSTRACT: This paper describes an efficient approach for time domain sensitivity analysis of lossy transmission lines in the presence of nonlinear termination. Sensitivity information is derived using the recently developed Passive Method of Characteristics (PMoC) macromodel. An important feature of the proposed method is that the derivatives of the sensitivity network are obtained analytically from the PMoC macromodel, leading to significant computational advantages.

I. INTRODUCTION

The rapid decrease in feature size and the associated growth in circuit complexity and higher operating speeds has made the analysis of interconnects a critical aspect of system reliability, speed of operation and cost. As operating frequencies increase, interconnects behave as transmission lines and effects such as signal delay, distortion, attenuation, ringing and crosstalk are observed [1]. As a result, circuit designers must consider these effects at the early stages of the design cycle to ensure circuit performance and reliability.

To model distributed interconnects in the framework of nonlinear circuit simulators macromodeling algorithms are required to convert the partial differential equations describing interconnects into ordinary differential equations [1]-[12]. Among the most commonly used algorithms are those based on the Method of Characteristics (MoC) [3]-[12]. The efficiency of MoC is derived by extracting the propagation delay which allows the attenuation function to be approximated with a low-order rational function.

In [11]-[12], a passive MoC macromodel is described where the curve fitting to realize the macromodel depends only on per-unit-length (p.u.l.) parameters and not on the discretization of the macromodel. Thus, with the knowledge of the rational functions derived by the p.u.l. parameters, the MoC is formulated in a closed form manner for any line length while ensuring passivity (referred to as PMoC). In this paper, a new algorithm to perform sensitivity analysis of nonlinear circuits with distributed interconnects is presented. The method is based on the PMoC used to model distributed interconnects as well as to derive the sensitivity network with respect to any interconnect parameter. A major advantage of the proposed algorithm is that the sensitivity network is derived analytically from the rational approximation of the original system.

II. REVIEW OF THE PASSIVE METHOD OF CHARACTERISTICS

The frequency domain solution of Telegrapher's equations of a two conductor transmission line can be expressed as [10]

$$I_1(s) = Y_0(s)V_1(s) - J_1(s); \qquad I_2(s) = Y_0(s)V_2(s) - J_2(s)$$

$$J_1(s) = H(s)[Y_0(s)V_2(s) + I_2(s)]; \quad J_2(s) = H(s)[Y_0(s)V_1(s) + I_1(s)]$$

$$H(s) = e^{-\Gamma(s)d}; \qquad \Gamma(s) = \sqrt{(G+sC)(R+sL)}; \qquad Y_0(s) = \sqrt{(G+sC)/(R+sL)} \qquad (1)$$

where s is Laplace transform variable; ($I_1(s)$, $V_1(s)$) and ($I_2(s)$, $V_2(s)$) are the current-voltage pairs at the near-end and far-end ports respectively; $J_1(s)$ and $J_2(s)$ are current controlled current sources; $\Gamma(s)$ is the propagation function; $Y_0(s)$ is the characteristic admittance; G, C, R and L are the p.u.l. parameters and d is the length of the line. The solution of Telegrapher's equations cannot be expressed in the time domain as ordinary differential equations, which makes it difficult to interface with nonlinear SPICE circuit simulators.

The PMoC is derived by extracting the propagation delay from $H(s)$ as

$$H(s) = e^{-sT}e^{-(d\Gamma(s)-sT)} \approx e^{-sT}Q(s) \qquad (2)$$

where T is the extracted delay and $Q(s) \approx e^{-(d\Gamma(s)-sT)}$ corresponds to the delayless propagation operator that takes into the account the effects due to line dispersion and attenuation. One of the main challenges in constructing the MoC in a closed form manner is the curve fitting for $Q(s)$, which depends on the line length d. If the transmission line is analyzed at different line lengths, then the rational approximation for $Q(s)$ needs to be re-evaluated. To address this issue, a new formulation is described to make the curve fitting of $Q(s)$ independent of the line length. In the proposed scheme, a rational approximation is performed on $\Gamma(s)$ instead of $Q(s)$ as

$$\Gamma(s) = \sqrt{(G(s) + sC(s))(R(s) + sL(s))} \approx sT + P(s) \tag{3}$$

where $P(s)$ is approximated as a rational function. In addition, the characteristic admittance $Y_0(s)$ is also approximated as a rational function. It should be noted that the curve fitting for both $\Gamma(s)$ and $Y_0(s)$ depend only on the p.u.l. parameters and is independent of the line length d and the discretization of the MoC macromodel. To approximate $Q(s)$ using $P(s)$, a closed form Padé approximation is used as [11]-[12],

$$Q(s) = \left(\sum_{i=0}^{M} \frac{(M+N-i)!M!}{(M+N)!i!(M-i)!} (-P(s)d)^i \right) \Big/ \left(\sum_{i=0}^{N} \frac{(M+N-i)!N!}{(M+N)!i!(N-i)!} (P(s)d)^i \right) \tag{4}$$

Thus to analyze transmission lines at different line lengths, the rational function $P(s)$ is multiplied by the line length d and the appropriate order M and N is selected to obtain the desired frequency domain accuracy for $Q(s)$. Unlike other MoC macromodels, the advantages of this formulation is that with the knowledge of the rational functions $\Gamma(s)$ and $Y_0(s)$, the PMoC can be formulated in a closed form manner for any line length. In addition, the passivity of the macromodel is assured for any line length, provided that the rational approximations of $\Gamma(s)$ and $Y_0(s)$ satisfy the conditions of the passivity theorem [11]-[12]. The next section describes the sensitivity network of transmission lines using the PMoC.

III. Sensitivity Analysis

To derive the sensitivity network, let the system of (1) be expressed as

$$AX = Bu \tag{5}$$

where

$$A = \begin{bmatrix} Y_0(s) & 0 & -H(s) & 0 \\ 0 & Y_0(s) & 0 & -H(s) \\ -Y_0(s) & 0 & 0 & 1 \\ 0 & -Y_0(s) & 1 & 0 \end{bmatrix}; \quad X = \begin{bmatrix} V_1 \\ V_2 \\ K_1 \\ K_2 \end{bmatrix}; \quad B = \begin{bmatrix} 1 & 0 \\ 0 & 1 \\ 1 & 0 \\ 0 & 1 \end{bmatrix}; \quad u = \begin{bmatrix} I_1 \\ I_2 \end{bmatrix} \tag{6}$$

Differentiating (5) with respect to a transmission line parameter λ yields

$$Ax^s = Bu^s - \frac{\partial A}{\partial \lambda} X \tag{7}$$

where

$$X^s = \begin{bmatrix} V_1^s \\ V_2^s \\ K_1^s \\ K_2^s \end{bmatrix} = \begin{bmatrix} \frac{\partial V_1}{\partial \lambda} \\ \frac{\partial V_2}{\partial \lambda} \\ \frac{\partial K_1}{\partial \lambda} \\ \frac{\partial K_2}{\partial \lambda} \end{bmatrix}; \quad u^s = \begin{bmatrix} I_1^s \\ I_2^s \end{bmatrix} = \begin{bmatrix} \frac{\partial I_1}{\partial \lambda} \\ \frac{\partial I_2}{\partial \lambda} \end{bmatrix}; \quad \frac{\partial A}{\partial \lambda} X = \begin{bmatrix} \Psi_1 \\ \Psi_2 \\ \Psi_3 \\ \Psi_4 \end{bmatrix} = \begin{bmatrix} \frac{\partial Y_o}{\partial \lambda} V_1 - \frac{\partial H}{\partial \lambda} K_1 \\ \frac{\partial Y_o}{\partial \lambda} V_2 - \frac{\partial H}{\partial \lambda} K_2 \\ -\frac{\partial Y_o}{\partial \lambda} V_2 \\ -\frac{\partial Y_o}{\partial \lambda} V_1 \end{bmatrix} \tag{8}$$

Equation (7) represents the sensitivity network with respect to λ, where V_1^s and V_2^s are the sensitivities of the voltages; I_1^s and I_2^s are the sensitivities of the currents; and K_1^s and K_2^s are the sensitivities of the K_1 and K_2 variables, respectively. The sensitivity network of (7) is similar in form to the original system, with the exception that the sensitivity network contains an additional derivative term of $(\partial A/\partial\lambda)X$. From the above analysis, it can be shown that the original and sensitivity networks have the same admittance matrices and that $(\partial A/\partial\lambda)X$ is modeled with additional voltage and current sources. As a result, Modified Nodal Analysis (MNA) matrices of the original and sensitivity networks are the same and the solution of the sensitivity network does not require additional Lower-Upper (LU) decompositions to invert the sensitivity network matrices. The time domain representations for both the original and sensitivity networks are obtained by approximating $\Gamma(s)$ and $Y_0(s)$ as rational functions and using (2)-(4) to approximate $H(s)$. For the sensitivity network X is known form the solution of the original network and $\partial A/\partial\lambda$ is obtained by differentiating Y_o and H with respect to an electrical parameter λ, where the product of $(\partial A/\partial\lambda)X$ is modeled as voltage and current sources.

IV. COMPUTATIONAL RESULTS

In this example, a seven-transmission line network with nonlinear CMOS inverters is shown in Fig. 1. The p.u.l. parameters of each line are R=8.26Ω/m, L=361nH/m, C=140pF/m, G=0.0 and the length of each line is 10cm. The input voltage is a trapezoidal pulse of amplitude 5V with rise/fall time of 0.2ns, pulse width of 5ns and a period of 10ns. The transient response at node V1 using the PMoC and the conventional lumped model [2] is shown in Fig 2. Fig. 3 shows the sensitivities with respect to the electrical p.u.l. parameters of the resistance, inductance, capacitance, and line length. The results of the proposed method are compared with the perturbation of the conventional lumped segmentation model. Both the proposed method and the SPICE perturbation results are in good agreement.

It is to be noted that using the proposed PMoC provides the following advantages. i) Using the PMoC provides significant CPU advantage compared to lumped segmentation model. For this example the PMoC required 3.7 seconds to obtain the transient response while the conventional lumped model required 36 seconds. ii) Perturbation based techniques can lead to inaccurate results depending on the magnitude of the perturbation. iii) In addition the perturbed network must be solved separately for every parameter of interest. However, in the proposed approach, the sensitivity information with respect to all the parameters can be essentially obtained from the solution of the original network since both the original and sensitivity network have the same admittance matrices and additional LU decompositions to invert the sensitivity network matrices are not required.

REFERENCES

[1] R. Achar and M. Nakhla, "Simulation of high-speed interconnects," *Proc. IEEE*, vol.89, no. 5, pp. 693-728, May 2001.

[2] C. R. Paul, *Analysis of Multiconductor Transmission Lines*, New York, NY: Wiley, 2008.

[3] F. H. Branin, Jr., "Transient analysis of lossless transmission lines," *Proc. IEEE*, vol. 55, no. 11, pp. 2012–2013, Nov. 1967.

[4] A. J. Gruodis and C. S. Chang, "Coupled lossy transmission line characterization and simulation," *IBM J. Res. Development*, vol. 25, pp. 25–41, Jan. 1981.

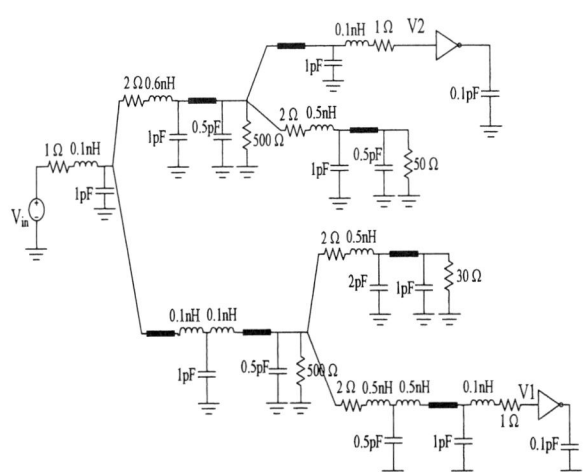

Fig. 1. Transmission line network with nonlinear termination

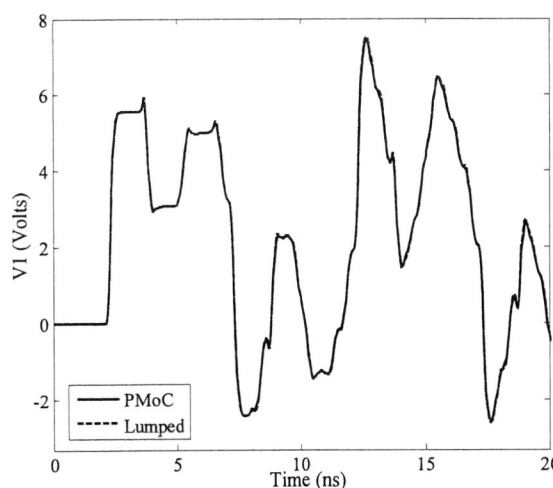

Fig. 2. Transient response of V1.

978-1-4244-4447-2/09 $25.00 © 2009 IEEE

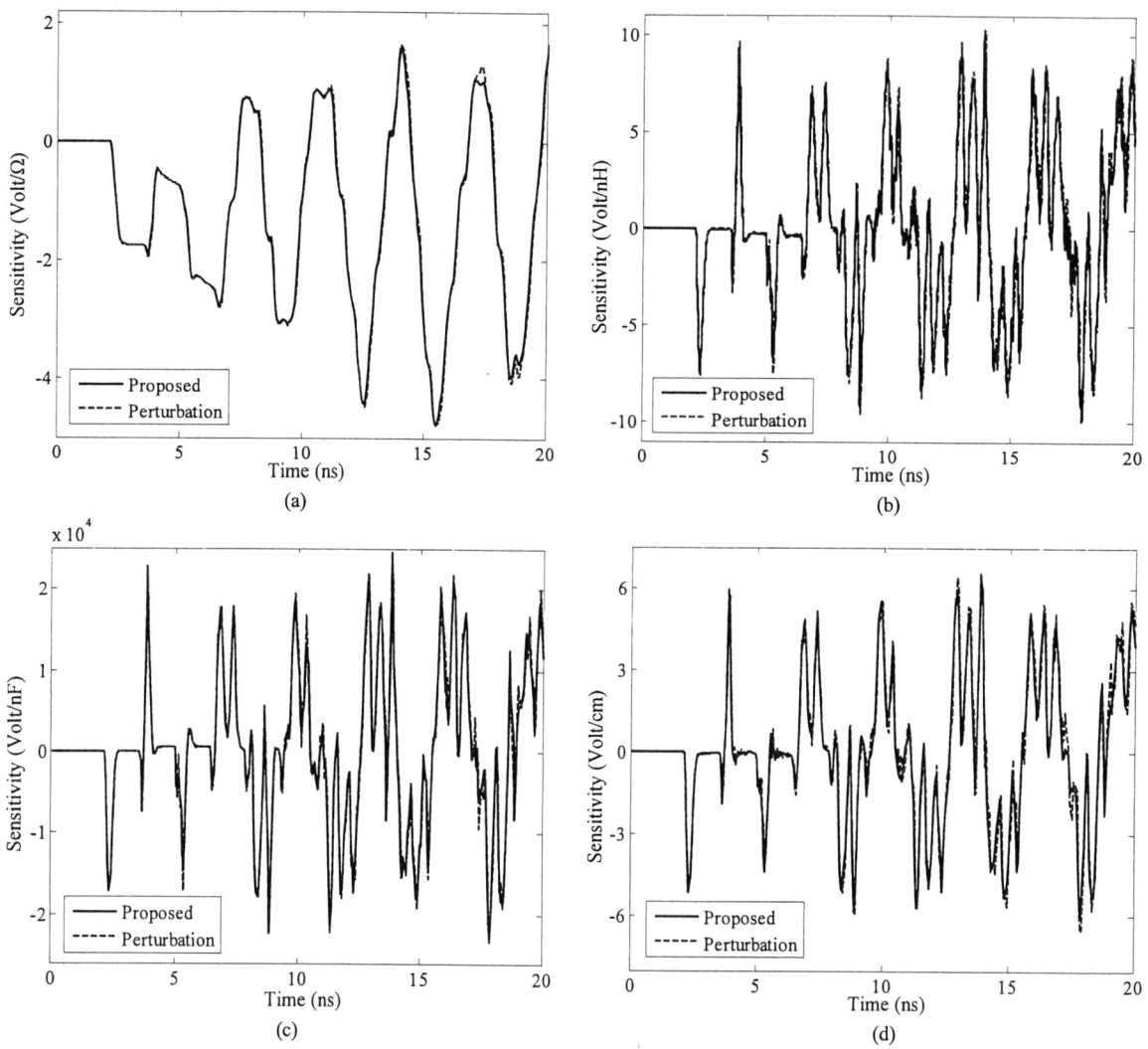

Fig. 3. Sensitivity of V1 with respect to (a) p.u.l resistance (b) p.u.l inductance (c) p.u.l capacitance and (d) line length.

[5] F. Y. Chang, "The generalized method of characteristics for waveform relaxation analysis of lossy coupled transmission lines," *IEEE Trans. Microw. Theory Tech.*, vol. 37, pp. 2028-2038, Dec. 1989.

[6] S. Lin and E. S. Kuh, "Transient simulation of lossy interconnects based on the recursive convolution formulation," *IEEE Trans. Circuits Syst. I*, vol. 39, no. 11, pp. 879–892, Nov. 1992.

[7] D. B. Kuznetsov and J. E. Schutt-Ainé, "Optimal transient simulation of transmission lines," *IEEE Trans. Circuits Syst. I*, vol. 43, no. 2, pp. 110–121, Feb. 1996.

[8] J. F. Mao; E. S. Kuh, "Fast simulation and sensitivity analysis of lossy transmission lines by the method of characteristics," *IEEE Trans. Circuits Syst. I*, vol. 44, no. 5, pp. 391–401, May 1997.

[9] I. M. Elfadel, H.-M. Huang, A. E. Ruehli, A. Dounavis, and M. S. Nakhla, "A comparative study of two transient analysis algorithms for lossy transmission lines with frequency-dependent data," *IEEE Trans. Adv. Packag.*, vol. 25, no. 5, pp. 143-153, May 2002.

[10] S. Grivet, Talocia, H-M. Huang, A. E. Ruehli, F. Canavero, and I. Elfadel, "Transient analysis of lossy transmission lines: An efficient approach based on the method of characteristics," *IEEE Trans. Adv. Packag.*, pp. 1-12, Feb. 2004.

[11] A. Dounavis, V. A. Pothiwala, "Passive closed-form transmission line macromodel using method of characteristics," *IEEE Trans. Adv. Packag.*, Vol 35, pp. 190-202, Feb. 2008.

[12] A. Dounavis, V. A. Pothiwala, A. Beygi, "Passive macromodeling of lossy multiconductor transmission lines based on the method of characteristics," *IEEE Trans. Adv. Packag.*, Vol 32, pp. 184-198, Feb. 2009.

Fast Full Wave Analysis of PCB Via Arrays with Model-to-Hardware Correlation

Xiaoxiong Gu[*], Boping Wu[+], Christian Baks[*] and Leung Tsang[+]

[*] IBM T. J. Watson Research Center, Yorktown Heights, NY 10598, USA

{xgu, cbaks @us.ibm.com}

[+] University of Washington, Seattle, WA 98195, USA

{bennywu, tsang1 @u.washington.edu}

Abstract

This paper applies a methodology based on a mostly-analytical Foldy-Lax full wave scattering model to analyze via arrays in a multilayered printed circuit board. We have demonstrated good model-to-hardware correlation up to 20 GHz by simulating and measuring 8x8 via arrays including 29 signal vias and 35 ground vias in a 18-layer test board. The required CPU in the 64-via case is 4 seconds which is more than four orders of magnitude faster than the market leading general-purpose 3D full wave solver.

I. INTRODUCTION

Via arrays in the connector fields such as LGA (land grid arrays) and BGA (ball grid arrays) are common vertical interconnect structures between a module (package) and a printed circuit board. Aggregate bandwidths are increasing rapidly for on-board communication links [1], requiring ever greater numbers of high-speed vias. It is currently a design challenge to optimize the placement of all the high-speed signal and ground vias in a timely fashion in order to maximize signal transmissions and to minimize the cross-talk between the signals, especially considering that most commercial three-dimensional (3D) general-purpose full wave electromagnetic solvers are not fast enough to compare a variety of configurations. In this paper, we applied a fast full wave 3D interconnect modeling method [2–6] to simulate massively-coupled multiple vias in multilayered PCBs. The admittance parameters are efficiently calculated by using equivalent magnetic current sources at the via-void. The magnetic fields inside the parallel-plate waveguide are expressed by the dyadic Green's functions in terms of waveguide modes in the vertical direction and vector cylindrical wave expansions in the horizontal direction. We have demonstrated our modeling approach using a university-developed MATLAB-based software tool [7] by simulating 8-by-8 via arrays each including 29 signal vias and 35 ground vias in a 18-layer board with layered dielectrics. The numerical results of signal reflection and crosstalk coupling are well correlated with frequency domain measurement data given by a 2-port network analyzer. The simulation time is about more than four orders of magnitude faster than the widely used 3D general-purpose EM solver HFSS[TM].

II. MODELING APPROACH BASED ON FOLDY-LAX MULTIPLE SCATTERING EQUATIONS

Vias in multilayered PCBs are considered essentially as vertical PEC cylinders through a multilayered substrate with layered dielectrics. The Foldy-Lax method is a full 3D characterization of the fields due to the multiple scattering among the vias in between the parallel planes excited by the magnetic current sources on the anti-pad aperture of each via (Fig. 1(left)). Fig. 1(right) shows the magnetic current on one aperture assuming a static symmetric transverse electromagnetic (TEM) field distribution. V is the port voltage across the via antipad. Such magnetic sources excite cylindrical waves between two parallel planes. Multiple scattering among vias occurs when multiple vias are present. The Foldy-Lax multiple scattering equations shown in (1) are derived by using cylindrical expansion of the excited wave (2) and requiring the total field to meet PEC

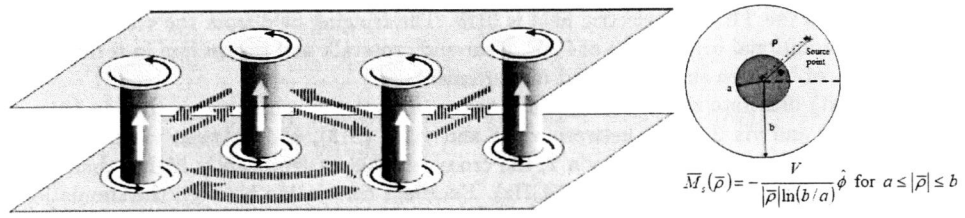

$$\overline{M}_i(\bar{\rho}) = -\frac{V}{|\bar{\rho}|\ln(b/a)}\hat{\phi} \text{ for } a \le |\bar{\rho}| \le b$$

Fig. 1. Multiple via scattering between planes (left) and equivalent magnetic current source at the via-hole (right).

978-1-4244-4447-2/09 $25.00 © 2009 IEEE

boundary conditions on each via surface [2]. The Foldy-Lax multiple scattering equations state that the final exciting field of cylinder q is equal to the incident field plus the scattering fields from all other cylinders, such that we can include couplings among all the vias between two parallel-plates. Since the waveguide modes are decoupled into the equation, the solution of the Foldy-Lax equation can be calculated for each waveguide mode separately.

$$w_{l,n}^{TM(q)} = a_{l,n}^{TM(q)} + \sum_{p=1, p \neq q}^{N} \sum_{m=-\infty}^{\infty} H_{n-m}^{(2)} \left(k_{\rho l} \left| \bar{\rho}_p - \bar{\rho}_q \right| \right) e^{j(n-m)\phi_{\bar{\rho}_p \bar{\rho}_q}} T_m^{TM} w_{l,m}^{TM(p)} \tag{1}$$

$$\bar{H}_{ex}^{(p)} = \sum_{m,l} w_{l,m}^{TM(p)} Rg \bar{H}_m^{TM} \left(k_{\rho l}, k_{zl}, \bar{\rho} - \bar{\rho}_p, z \pm d/2 \right) \tag{2}$$

Here, $a_{l,n}^{TM(q)}$ is the incident field of the magnetic current source onto cylinder q and can be calculated by using the dyadic Green's function derived for a single via case. $w_{l,m}$ are unknown coefficients solved by the Foldy-Lax multiple scattering equations. T_m^{TM} are then known T-matrix coefficents for PEC cylinders. η is the characteristic impedance of the medium surrounding the via, a is the radius of the via, $k_{zl} = l\pi/d$ and d is the separation between the reference planes, $k_{\rho l} = \sqrt{k^2 - k_{zl}^2}$ and k is the wave number in the dielectric medium. $\bar{H}_m^{(2)}$ is the mth order Hankel function of the second kind, $\phi_{\bar{p}}$ is the azimuthal angle from origin to the center of via p.

The final solution of the interior problem are given in terms of the currents going into or out of vias on the top and bottom sides of the parallel planes and the port voltages across each via anti-pad. For example, with a unit voltage across the anti-pad, the z-dependence and the waveguide modes can be seen from the following expression for the surface current on cylinder p shown in (3).

$$\bar{I}^{(p)} = \hat{z} \sum_{m,l} w_{l,m}^{TM(p)} \frac{4\cos\left[k_{zl}\left(z \pm d/2\right)\right]}{\eta H_m^{(2)}\left(k_{\rho l} a\right)} e^{-jm\phi_{\bar{p}}} \tag{3}$$

We see that the current is a summation over all the waveguide modes represented by l and azimuthal modes represented by m. The output of the analysis can be also summarized in an admittance Y matrix of size $2N$ x $2N$ where N is the number of vias. For multiple interior layers, the Y matrices can be cascaded by either first converting into ABCD matrices or directly by cascading in a Spice-like simulator. Scattering parameters are given as the final results. All equations above have been implemented in a standalone MATLAB-based software tool which the authors used to perform simulations for the via array experiment summarized in the next section.

III. Model-to-Hardare Correlation Results

Previous validation of the analytic via model with measurement results, in terms of scattering parameters, was reported in [4]. However, the previous test sites did not emphasize on the multiple scatterings of the via array designs. Also, these test sites contained both vias and transmission lines which deviate our focus and strength on the vertical interconnect problem. In this paper, in order to make a direct comparison with the via modeling approach outlined in section II, we have built via array structures without any trace connection in a 18-layer test board. Fig. 2(left) shows the measured layer stack of the board. All vias in the array go through from the top plane to the bottom plane. Fig. 2(middle) illustrates the pattern of an 8-by-8 via array with 80-mil pitch. There are 29 signal vias and 35 ground in one array. The radius of via drill is 5 mils. The antipad radius is 15 mils. There are only circular pads on the top and bottom planes with a radius of 10 mils. All ground planes have the same size as the original board panel (21.8-inch by 16.8-inch) except for the top and bottom planes which have a much smaller square shape with a size of 600-mil by 600-mil. The dielectric material of the board is Nelco4000-13 with a dielectric constant of 3.7 and loss tangent of 0.009. Three vias labeled in Fig. 2(right) were measured using network analyzer with GS probes (225um pitch) from the surface of the top plane as shown in Fig. 2(right). Except for those contacted with the probes, both ends of all the signal vias were left open in the measurement set-up. In the simulation, a capacitance of 93fF was used as the termination to account for the non-ideal capacitive-dominant open port. Here, the equivalent capacitance across the antipad due to the TEM-like electric field is 31fF. The fringing field from the via pad in the air was estimated to cause an additional capacitance of 62fF. Near-end crosstalk and return loss in terms of scattering parameters are compared between simulation and measurement.

Fig. 3 and Fig. 4(left) illustrate the simulated and measured near-end cross-talk in the via array with 80-mil pitch between via 1 and via 2 (S12), between via 1 and via 3 (S13), and between via 2 and via 3 (S23). Because of the ground via between via 1 and via 2, the cross-talk S23 is significantly higher than S12 and S13 (e.g., S23=-6.8dB, S12=-25dB, S13=-31.7dB at 10GHz). Excellent correlation between the simulation and the measurement in these cases were obtained up to 20GHz. We also measured the near-end crosstalk S23 for a 40-mil pitch via array with exactly the same signal/ground pattern and via geometries. What is really interesting

978-1-4244-4447-2/09 $25.00 © 2009 IEEE

is that if we compare two S23 curves in Fig. 4(left) and Fig. 4(right), the crosstalk S23 is actually higher for the larger pitch array than for the smaller pitch array (e.g., at 10GHz, S23=-6.8dB for the 80mil pitch array whileas S23=-12dB for the 40mil pitch array). This observation would be reversed if there were no other vias but via 2 and via 3 on the test site. The explanation of this observation resides in the fact that the multiple wave scattering amongst all neighboring vias significantly affects the crosstalk. Depending on the geometries, there can be more energy coupled between two vias with larger distance. Such effect is also well captured by the scattering-based simulation which shows very good correlation with measurement in Fig. 4(right).

The measured and simulated return loss of the three vias on the top plane (S11,S22,and S33) are shown in Fig. 5. The resonance frequencies around 17GHz and above 20GHz are due to the via length (vias acting like hanging stubs in this case) and they are also affected by the surrounding vias as previously observed in [4]. Here, via 2 and via 3 have very similar local environment so that their return loss curves look similar as well (Fig. 5(left)). On the other hand, via 1 has more neighboring ground vias which cause the stub resonant frequency to shift higher (Fig. 5(right)). Simulation results also demonstrate such a trend to a similar degree.

For the numerical simulation of the 80-mil pitch via array structure, our simulation using the MATLAB-based tool took only about 4 seconds of CPU time and 20M of RAM to finish for running 20 frequency points from 1 GHz to 20 GHz on an Intel T2600 2.16 GHz processor, whereas it was reported that HFSS^{TM}took 28 hours and 23 minutes of CPU time and 5.19G of RAM to analzye 64 vias in a 8-layer board for the same frequencies using an Intel Xeon Quad-core 3.0 GHz processor [7]. Here, our multiple scattering solution demonstrates a speed-up factor of more than 10 thousand times faster than the popular 3D general-purpose electromagnetic solver.

IV. Conclusions

Multilayered PCB via array structures have been designed, fabricated and measured. Good correlation between hardware measurement and numerical simulation results demonstrates that the scattering based Foldy-Lax via modeling approach can significantly reduce the requirement of computational resources, making it suitable for the signal integrity analysis of a large number of vias and various via structures.

Acknowledgments

The authors would like to thank Miroslav Kotzev, Renato Rimolo Donadio, Christian Schuster from Technische Universitt Hamburg-Harburg, Germany and Young Kwark, Ki Jin Han from IBM for their help and discussion on designing and testing the printed circuit board, as well as Mark Ritter for his management support. All the hardware and measurement performed in this study were supported by the Defense Advanced Research Projects Agency under its Agreement No. HR0011-07-9-0002.

References

[1] D. G. Kam, M. B. Ritter, T. J. Beukema, J. F. Bulzacchelli, P. K. Pepeljugoski, Y. H. Kwark, L. Shan, X. Gu, C. W. Baks, R. A. John, G. Hougham, C. Schuster, R. Rimolo-Donadio, and B. Wu. Is 25 Gb/s on-board signaling viable? *IEEE Transactions on Advanced Packaging*, pages 328–344, May 2009.

[2] H. Chen, Q. Li, L. Tsang, C.C. Huang, and V. Jandhyala. Analysis of a large number of vias and differential signaling in multilayered structures. *IEEE Transactions on Microwave Theory and Techniques*, 51:818–829, March 2003.

[3] L. Tsang, H. Chen, C. C. Huang, and V. Jandhyala. *Methods for modeling interactions between massively coupled multiple vias in multilayered electronic packaging structures, U.S. Patent 7 149 666*. 2006.

[4] X. Gu and M. Ritter. Application of Foldy-Lax multiple scattering method to via analysis in multilayer printed circuit board. In *Proc. of DesignCon Conference*, Santa Clara, CA, February 2008.

[5] B. Wu and L. Tsang. Signal integrity analysis of package and printed circuit board with multiple vias in substrate of layered dielectrics. *IEEE Transactions on Advanced Packaging*, in press, 2009.

[6] B. Wu and L. Tsang. Modeling multiple vias with arbitrary shape of antipads and pads in high speed interconnect circuits. *IEEE Microwave and Wireless Comp. Lett.*, Vol. 19 ,1:12–14, January 2009.

[7] B. Wu and L. Tsang. IC package and PCB analysis tool for 3D full-wave modeling of multi-stack layered-media multi-core designs. In *SRC/GRC annual review*, Atlanta, GA, June 2009.

978-1-4244-4447-2/09 $25.00 © 2009 IEEE

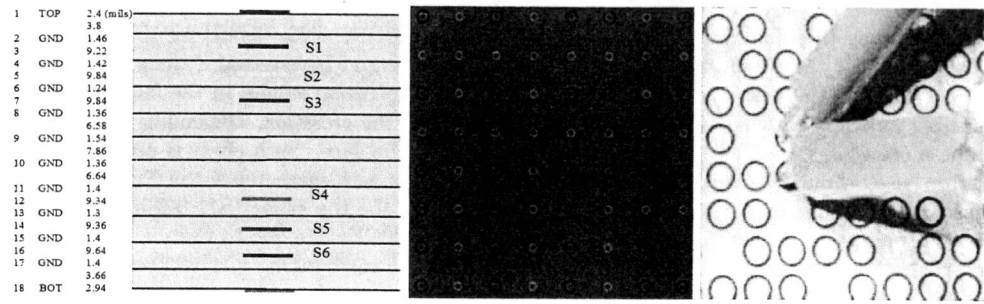

Fig. 2. Measured layer stack of the test board (left), top-view of the via array with 80-mil pitch(middle) and probing measurement on the board surface (right).

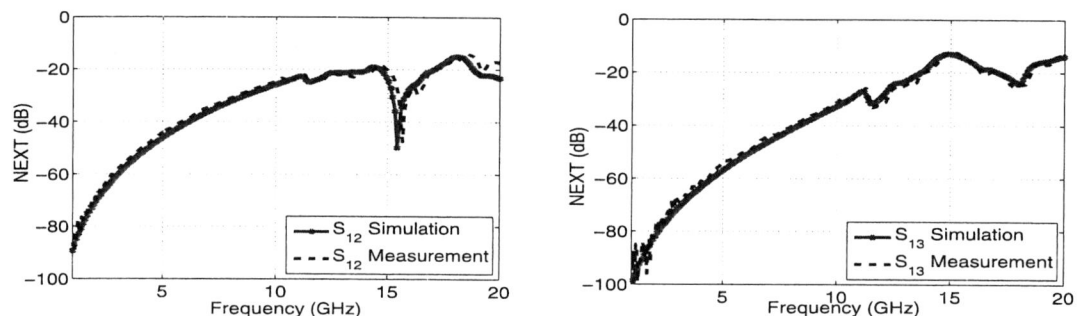

Fig. 3. Near-end crosstalk (80-mil pitch): between via 1 and via 2(left) and between via 1 and via 3 (right).

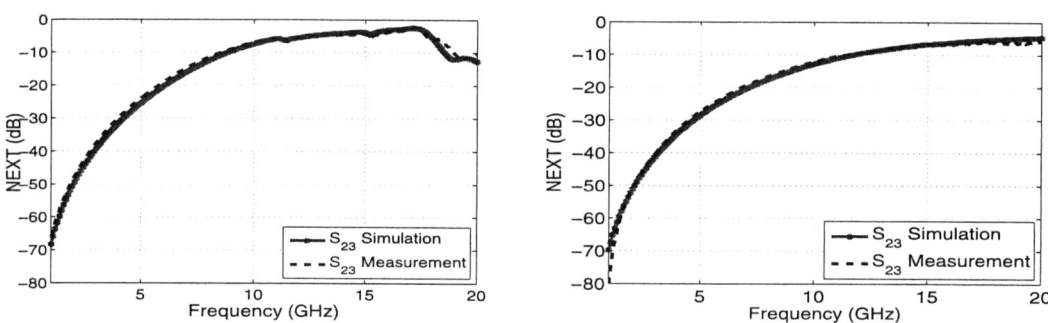

Fig. 4. Near-end crosstalk between via 2 and via 3: array with 80-mil pitch(left) and array with 40-mil pitch (right).

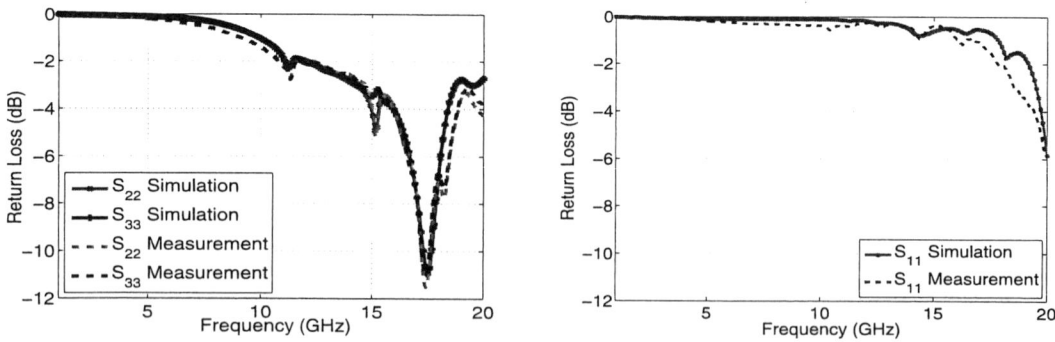

Fig. 5. Left: Return Loss on via 2 and via 3 (80-mil pitch). Right: Return Loss on via 1 (80-mil pitch).

978-1-4244-4447-2/09 $25.00 © 2009 IEEE

Efficient Capacitance Solver for 3D Interconnect Based on Template-Instantiated Basis Functions

Yu-Chung Hsiao [yuchsiao@mit.edu], Tarek El-Moselhy [tmoselhy@mit.edu], Luca Daniel [luca@mit.edu]

Massachusetts Institute of Technology

Abstract

In this paper we show how the highly restrictive design rules of the recent sub-micro to nano-scale Integrated Circuit technologies allow to use a limited number of pre-computed surface charge distributions as a set of fundamental template basis functions in an efficient integral equation based 3D capacitance solver. Several examples verify that our solver can achieve final accuracies of less than 2% using 5× to 30× fewer unknowns than standard piecewise constant basis functions for the same accuracy, resulting in up to 25× speedups.

I. Introduction

The state-of-the-art in efficient capacitance extraction methods for integrated circuits involves 2D cross section scanning, determining wire adjacency, calculating 2D capacitance in a table lookup approach, and then reconstructing quasi-3D capacitance. Such approach is indeed fast, yet it is accurate only for 2D structures. Full 3D structures (e.g. crossing wires in adjacent metal layers) need the accuracy of electrostatic field solvers such as [1]–[4]. The most efficient of such tools are based on solving integral equations using piece-wise constant basis functions combined with standard collocation testing and iterative techniques. Such solvers are typically accelerated by fast matrix-vector products, which have a significant computational overhead, but scale almost linearly with the number of conductors. Hence they are ideal for very large scale examples.

On the other hand, improving time and memory requirements by the use of higher order basis functions such as piece-wise linear and quadratic bases is a common practice in almost all numerical communities when solving differential and integral equations. Sometimes even more efficient solvers are obtained by employing or developing specialized basis functions with "built-in" known physical properties such as sinusoidal bases for high frequency resonating antenna problems [5], loop-star bases for diverge-free unknowns [6], conduction mode bases for Helmholtz current distributions inside conductors [7]–[11], and edge and corner bases for surface charge density in capacitance extraction problems for microelectromechanical devices [12].

As in [12], this paper investigates the use of specialized basis functions to represent effectively the surface charge density distributions in integral equation based capacitance extraction solvers. However, the key idea in this paper is to exploit the charge distributions properties due to the highly restrictive design rules of the recent sub-micro to nano-scale integrated circuit and packaging technologies, as highlighted in Section III-A. As we will demonstrate in the example session, in this scenario the edge and corner bases introduced in [12] are not required to achieved accuracies of about 5%, typically required by integrated circuit and packaging applications. On the other hand, charge distributions and fringing fields induced by adjacent crossing wires, when neglected, can easily generate unacceptable errors in the 20% range. Pre-computed surface charge distributions shapes (defined in Section III-B) will be used in this work as specialized basis functions (Section III-C) to represent such induced charge distributions. A similar idea was introduced in [9] for proximity effect induced *currents*, as opposed to *charges*. An additional difference in this work is the idea of assembling the basis functions a priori and "on the fly" from just two basic building blocks. In this way analytical formulas and numerical tabulation of the Galerkin coefficients for our limited number of template building blocks can effectively limit the setup overhead as shown in Section III-D, obtaining fast simulation times and affordable memory requirements as demonstrated in the examples in Section IV.

II. Background

A standard way to extract the capacitance matrix for a *n*-conductor system embedded in a uniform medium with dielectric constant ε is to solve the integral equation

$$\int_S \frac{\rho(\mathbf{r}')}{4\pi\varepsilon\|\mathbf{r}-\mathbf{r}'\|}d\mathbf{r}' = \Phi(\mathbf{r}) \tag{1}$$

for the surface charge density ρ, given the electric potential $\Phi(\mathbf{r})$. By expressing the charge density $\rho(\mathbf{r}') = \Sigma_j \rho_j \psi_j(\mathbf{r}')$ in a linear combination of N basis functions ψ_j and by using the standard Galerkin testing method, (1) becomes

$$\left[\int_{Si}\int_{Sj} \frac{\psi_i(\mathbf{r})\psi_j(\mathbf{r}')}{4\pi\varepsilon\|\mathbf{r}-\mathbf{r}'\|}d\mathbf{r}'d\mathbf{r}\right]\rho_j = \int_{Si}\psi_i(\mathbf{r})\Phi(\mathbf{r})d\mathbf{r} \tag{2}$$

where the integration in the brackets forms a system matrix, and ρ_j is a vector of N unknowns corresponding to each basis function.

978-1-4244-4447-2/09 $25.00 © 2009 IEEE

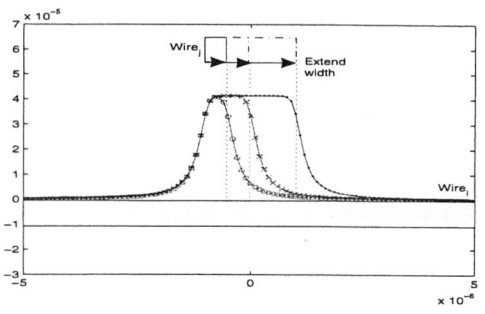

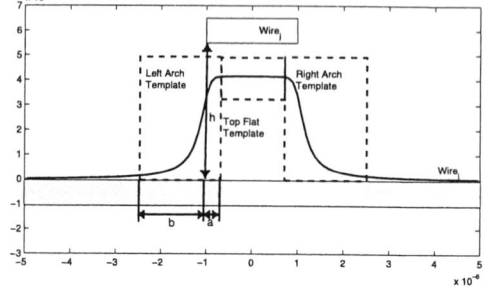

(a) The shape and slope of the induced charge density (cross-section) of the bottom wire are not affected by the width of the crossing wire.

(b) Basis function instantiation and assembly process: a single basis is constructed connecting one flat and two arch templates.

Fig. 1. Capturing charge density induced by crossing wires.

III. TEMPLATE INSTANTIATED BASIS FUNCTIONS

A. Observed Charge Density Properties for Typical IC Interconnect

One key observation is that the surface charges accumulating on conductor corners and edges, as well as the charges induced on a conductor surface due to a nearby conductor, in typical IC interconnect geometries are generally quite confined. A second key observation is that corner and edge charge accumulations affect the wire capacitances by no more than a few percent, and therefore can be safely ignored for typical IC target accuracies of about 5%, as the first example in Section IV will verify. A third key observation is that the shape and slope of the charge density induced by a nearby wire is the same regardless of the width of the crossing wire as seen on Fig. 1(a). From the above observations, one can conclude that in order to represent most, if not all, charge distribution scenarios in IC/package applications, specialized basis functions can be easily instantiated a-priori from a very small number of pre-defined templates as shown for instance in Fig 1(b) for the two crossing wires, and as described in more details for other cases in Section III-C below. One can further conclude that given the small number of template building blocks, the coefficients of the Galerkin system matrix in eq. (2), representing the interaction between different bases, can be either tabulated and retrieved very efficiently, or computed partially analytically as shown in Section III-D, therefore avoiding expensive setup costs.

B. Definition of Charge Density Building Block Templates

We first define two simple 1D shape templates as follows:
- "Flat curve" template: $T_F(\cdot) = 1$, a 1D constant function.
- "Arch shape" template: $T_A(\cdot, a, b, h)$, a family of 1D decaying functions. Arch templates are characterized by the three parameters (a, b, h) defined in Fig. 1(b).

Using the two 1D template shapes above we define 2D building blocks:
- "Flat Building Block" : $B_F(u, v) = T_F(u) \cdot T_F(v)$,
- "Arch Building Block" : $B_{A,u\pm}(u, v) = T_A(\pm u) \cdot T_F(v)$ or $B_{A,u\pm}(u, v) = T_A(\pm v) \cdot T_F(u)$, for decaying in $\pm u$ and $\pm v$ directions, respectively,

where (u, v) are local coordinates on each conductor face. In our original formulation we had defined other building blocks such as corner and edge templates [12] shown in Fig. 3(b) and other blocks such as $T_A(\pm u) \cdot T_A(\pm v)$. In our experimentations we have however made the critical observation that only the Flat and Arch building blocks defined above are essential and sufficient to achieve the target 5% accuracies of typical IC capacitance extraction.

C. Instantiation and Assembly of Charge Density Basis Functions from Building Block Templates

Fig. 1(b) shows an example of the instantiation and assembly process for a basis function solely responsible to capture local charge accumulation induced on the bottom conductor by a nearby crossing conductor. A left arch building block, a right arch building block, and a flat building block are first instantiated to fit the appropriate dimensions of the neighboring wires. The three blocks are then connected together to constitute a *single* basis function, hence they will contribute to a single unknown in the final system (2). Each additional crossing wire will contribute a single extra basis function to the bottom conductor, hence contributing a single extra unknown to the system.

Another typical example of such instantiation and assembly process is illustrated in Fig 2(b) which shows one single basis function constructed on the fly by the solver by instantiating and connecting three arch building blocks and one flat building block in order to fit the wire dimensions shown in Fig. 2(a). The total number of basis functions used to represent the charge density of all surfaces of the bottom conductor is 7, i.e. one flat basis covering completely each face of each conductor, plus

978-1-4244-4447-2/09 $25.00 © 2009 IEEE   178

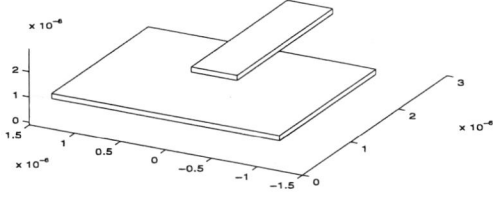

(a) Wires' geometry.

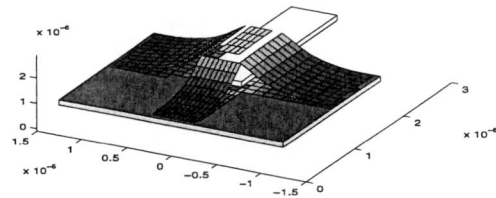

(b) The single basis function representing the induced charge density is assembled using three arch building blocks and one flat building block.

Fig. 2. Instantiation and assembly process for partially overlapping wires

the basis function shown in Fig. 2(b). One can notice how additional building blocks of the form $T_A(\pm u) \cdot T_A(\pm v)$ could have been used to capture corner fringe. However as mentioned multiple times the simple basis function as shown in Fig. 2(b) is necessary and sufficient for the target 5% accuracy of this problem.

A simple algorithm implementing the ideas illustrated in the two examples above has been developed (and cannot be included because of space limitations) to instantiate and assemble basis functions from our two building blocks for any given collection of wires in a Manhattan layout with rectangular wires.

D. Efficient System Matrix Assembly

In order to reduce the time required to calculate the Galerkin integrals for each of the system entry in eq. (2), we adopt a partially numerical - partially analytical scheme, summarized in Table below. In order to further increase efficiency we truncate and use a piecewise linear approximation for the arch shape.

Building Block Interaction Type	Integration Schemes
Flat with *Flat*	3D analytical and 1D numerical
Flat with *Arch*	3D analytical and 1D numerical
Arch with *Arch* (different directions)	2D analytical and 2D numerical
Arch with *Arch* (same direction)	Summation of *Flat* with *Arch*

IV. EXAMPLES

Figure 3(a) shows the parametric sweep of aspect ratio and area for a single conductor solved using only one single flat basis function over each face, combined with the standard Galerkin testing approach. This simple setup achieves less than 3% relative error. Including additional basis functions representing edge and corner singularities [12] as shown in Fig. 3(b) achieves a significantly smaller relative error of 10^{-3}%. This example however demonstrates that for the 5% accuracy required by integrated circuit designs, edge and corner basis functions do not need to be included.

In the Table below we summarize the performance of several examples where we used the basis functions described in Section III-C with a standard Galerkin testing, and we compare them to piecewise constant (PWC) basis with collocation testing in uniform discretization. In both methods, systems are solved by standard Gaussian elimination. All our examples have been run in Matlab on a desktop computer with a Xeon 2.93GHz CPU.

Example	Partially overlapping wires Fig. 2(a)			7 by 7 buses Fig. 4(a)			Routing wires between modules Fig. 4(b)		
Relative Error	1.6%			2.1%			1.7%		
	This work	PWC	Improvement	This work	PWC	Improvement	This work	PWC	Improvement
Unknown Number	17	572	33×	966	4688	4.8×	120	1754	14.6×
Filling Time (sec)	0.03s	0.75s	25×	14.1s	13.3s	0.94×	0.35s	1.9s	5.4×
Solving Time (sec)	< 0.1ms	0.015s	> 150×	0.05s	3.3s	60.7×	< 1ms	0.24s	> 240×
Total Time (sec)	0.03s	0.76s	25.3×	14.2s	16.6s	1.2×	0.35s	2.2s	6.1×

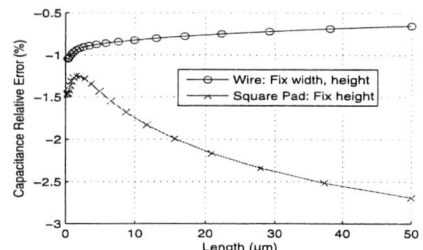

(a) Parametric sweep for capacitance error neglecting edge, corner singularity basis functions

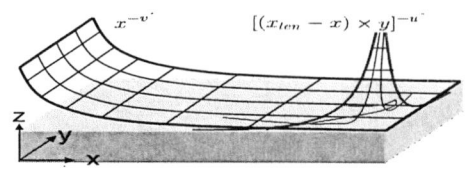

(b) Edge and corner singularity basis functions

Fig. 3.   Verification for neglecting singularity basis functions

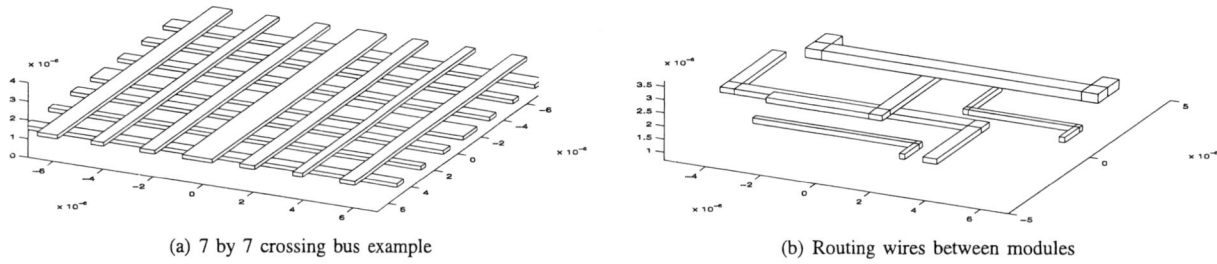

(a) 7 by 7 crossing bus example

(b) Routing wires between modules

Fig. 4.   Two larger examples

V. Conclusions

In this paper we have presented an integral equation based capacitance solver which instatiates on the fly a small number of specialized basis functions to capture charge distributions induced by nearby conductors. In a medium size example, our solver used a total of just 120 unknowns, obtaining a worst-case relative error less than 2% compared to the result extracted by piecewise constant basis in a very fine discretization with tens of thousands of unknowns. Furthermore, the piecewise constant basis method requires 1754 unknowns to produce the same 2% error in a coarser discretization. Hence, for the same 2% accuracy, our algorithm requires approximately 14.6× fewer unknowns, resulting in an overall 6× speedup.

Acknowledgment

Funding for this project has been provided by Mentor Graphics, AMD, and the Semiconductor Research Corporation.

References

[1] K. Nabors, J. K. White, "FASTCAP A multipole-accelerated 3D capacitance extraction program" *IEEE Transactions on Computer-Aided Design of Integrated Circuits and Systems*, pp. 1447–1459, 1991.

[2] R. B. Iverson, Y. L. Le Coz, "A Stochastic Algorithm for High Speed capacitance Extraction in Integrated Circuits," *Solid-State Electronics*, Vol. 35, No. 7, pp. 1005–1012, 1992.

[3] J. R. Phillips, J. K. White, "A precorrected-FFT method for electrostatic analysis of complicated 3D structures" *IEEE Transactions on Computer-Aided Design of Integrated Circuits and Systems*, pp. 1059–1072, 1997.

[4] S. Kapur, D. Long, "IES3: Efficient Electrostatic and Electromagnetic Simulation" *IEEE Computational Science & Engineering*, pp. 60 – 67, 1998.

[5] W. A. Imbriale, "On numerical convergence of moment solutions of moderately thick wire antennas using sinusoidal basis functions", *IEEE Transactions on Antennas and Propagation*, v Ap-21, n 3, p 363-6, May 1973.

[6] "Loop-Star Decomposition of Basis Functions in the Discretization of the EFIE", *IEEE Trans. on Antennas and Propagation* Vol. 47, No. 2, Feb 1999.

[7] P. Silvester, "Model network theory of skin effect in flat conductors", *Proceedings of IEEE*, 54(9):1147-1151, September 1966.

[8] L. Daniel, A. Sangiovanni-Vincentelli, J. White, "Interconnect Electromagnetic Modeling using Conduction Modes as Global Basis Functions", *IEEE 9th Topical Meeting on Electrical Performance of Electronic Packaging*, 2000.

[9] L. Daniel, A. Sangiovanni-Vincentelli, J. White,"Proximity Templates for Modeling of Skin and Proximity Effects on Packages and High Frequency Interconnect", *IEEE/ACM International Conference on Computer Aided Design*, San Jose, Nov 2002.

[10] S. Ortiz, R. Suaya, "Fullwave volumetric Maxwell solver using conduction modes", *IEEE/ACM International Conference on Computer-Aided Design*, 2006.

[11] K.J. Han, E.Engin, M. Swaminathan, "Cylindrical Conduction Mode Basis Functions for Modeling of Inductive Couplings in System-in-Package (SiP)", *IEEE Topical Meeting on Electrical Performance of Eletronic Packaging*, 2007.

[12] Y. Su, E. T. Ong, K H Lee, "Automatic classification of singular elements for the electrostatic analysis of microelectromechical systems", *Journal of Mecromechanics and Microengineering*, Vol 12, pp.307-315, 2002.

978-1-4244-4447-2/09 $25.00 © 2009 IEEE

An Ultra Compact Electromagnetic Band Gap filter for GHz Power Noise Suppression Using LTCC technology

Yu-Wen Huang[#1], Ting-Kuang Wang, Tzong-Lin Wu[#2]

Department of Electrical Engineering and Graduate Institute of Communication Engineering,
National Taiwan University,Taipei, 10617, Taiwan
E-mail: wtl@cc.ee.ntu.edu.tw

Abstract

A new structure to suppress simultaneous switching noise using an electromagnetic band gap filter is proposed. Based on periodic structure concept, this structure is one dimensional periodic structure with three unit cells. The size is comparable with noise suppression component such as capacitors, ferrite beads and π filters. But unlike these components, the proposed structure would not suffer from noise problems at higher frequencies. The special characteristic is its flexible operation frequency. The band gap can be designed by using proper geometrical structure parameter. To make use of the parameter in this work, the noise beyond limitation of the π filter could be suppressed by this structure. In this work, the size of proposed structure is 1.2mm×3.8mm×0.728mm. The stop band is from 2GHz to 5.5GHz and is validated both by simulation and experiment. Over 25dB noise reduction in the stop band could be achieved by this structure.

Keywords : Power integrity (PI), Simultaneous switching noise (SSN), Electromagnetic band gap (EBG), Periodic structure

I. INTRODUCTION

Simultaneously switching noise (SSN) has been a serious concern in recent year because of the tendency of faster data rate and lower power supply voltage in the IC design. With a rapid switching of current, the inductive of the power supply wire and the bonding wire causes the voltage level at a position in a plane cavity to fluctuate. This kind of noise could be an electromagnetic source which propagates to other positions by the power/ground resonance and it would couple to the nearby circuit and interfere with the operation of other devices by the interconnections. Therefore, the switching noise must be under control to ensure reliable circuit, device, and system operation. There have been numerous studies of design and analysis methodologies to reduce the SSN. Decoupling capacitor is the approach we usually used, but the high frequencies performance is limited by its equivalent series inductive. To keep good power integrity (PI) in high frequencies, several methods are proposed. One of the methods is electromagnetic band gap (EBG) structures such as long period coplanar EBG (LPC-EBG) which suppress noise by design the periodic patterns directly on the power or ground metal planes [1]. Another method based on a photonic crystal substrate concept was proposed to efficiently isolate the power noise by periodically inserting the high-DK rods into the substrate between the power and ground planes [2]. Furthermore, [3] demonstrated that how to use the artificial-substrate EBG (AS-EBG) broaden the bandwidth. Power-plane segmentation is also often used for dc power-bus noise isolation in multilayer printed circuit board (PCB) designs. To achieve desirable RF noise isolation, different power-plane segmentations can be used. In the [4], the noise isolation with several power-plane segmentation designs including power islands, and totally segmented power planes is studied. For the DC connection, the segmentation provides a high series impedance in the power plane reducing the conducted noise currents, thus providing isolation of a noise source. In addition to the connecting bridge, ferrite bead also has the same effect. The π filter is the combination of bead and lumped capacitors which could be used to filter noise. But this method suffers from limited bandwidth in noise reduction because of the equivalent series inductance for capacitor and the saturation of frequency response for ferromagnetic material.

In this work, a novel method is proposed to impede GHz SSN. The idea is using periodic structure concept to design electromagnetic band gap filter. The proposed structure would not suffer from the problem of working at higher frequencies because the band gap can be designed by using proper

978-1-4244-4447-2/09 $25.00 © 2009 IEEE

geometrical structure parameter. From the co-simulation result, the proposed structure can be observed that the noise suppression is better than π filter. Over 40dB noise reduction could be achieved.

II. PROPOSED STRUCTURE AND ITS CHARACTERISTIC

Periodic structures support slow-wave propagation and have pass band and stop band characteristics similar to filters. Based on the concept above, this electromagnetic band gap filter is designed. The proposed structure consists of three unit cells, and the unit cell can be seen in Fig. 1(a). It is composed of shunt multilayer capacitor and series transmission line. The width, length and height of unit cell structure are denoted by W, L and H, respectively, and L is equal to W in the unit cell of our design. The side view of unit cell structure is shown in Fig. 1(b) where P_n means n_{th} power planes and G_n is n_{th} the ground plane. As could be seen, the unit cell structure contains N power and ground planes. Each power plane or ground plane is connected by vias. Series transmission line is located at the bottom layer, and the reference ground is G_N. With the help of via and transmission line, the power plane of different unit cells could be cascaded periodically as shown in Fig 1(c).

In this work, it is assumed that the structure parameters (W, H, N) = (1.2mm, 0.728mm, 7) and transmission line is 3.2 mm long, the total length of this design is 3.8mm. Just as mentioned before, this design is flexible when it is used to suppress noise. It is unlike π filter which only works at lower frequency. We could adjust the band gap to where the noise is. Fig.2 shows the variation of band gap with the different layers N in the structure.

III. THEORETICAL MODEL AND BAND GAP PREDICTION

In order to efficiently predict the band gap, the analysis of infinite periodic structure is necessary. The modelling is set up under the assumption where the structure in infinitely long or perfectly matched at the end. Fig. 3 shows the equivalent model of unit cell structure. It is composed of multilayer capacitor, via and the transmission line. However, attention should be paid when constructing the model of multilayer capacitor. It can not be viewed as a big capacitor anymore. The mutual inductive coupling between plane and plane in the multilayer capacitor should be considered [5]. The insertion loss S_{21} obtained from modeling and full-wave simulation tool HFSS is shown in Fig. 4. Great consistency is seen.

IV. PI PERFORMANCE AND CO-SIMULATION

The performance of our design is implemented by the LTCC process with relative permittivity 7.8, relative permeability 1, and dielectric loss tangent 0.005. The reasonably good agreement between the measurement and simulation is seen in Fig. 5. The bandwidth is defined as the frequency range in which $|S_{21}|<-25$dB. It is observed that the stop band is between 2 to 5.5 GHz. At least 25 dB of SSN suppression can be achieved within the stop band in the designs.

To demonstrate the performance of proposed structure, co-simulation environment is established by a system of C-band LNA and 16 inverters in package as shown in Fig. 6. All circuit use TSMC 0.18 μm model. The DC power of 1.8V is transported by the power trace. The trace is represented by a large inductor (L_{DC} = 5 nH) and DC resistance (R_{DC} = 0.2 Ω). For LNA, the input signal is a 5 GHz sin wave with amplitude 0.01 V , and its resistance load is 50 Ohm. 16 CMOS inverters switch at the same time with the periodical pulse train with period of 0.208 ns, rising and falling time of 0.035 ns, and duty cycle of 50%. The corresponding fundamental frequency is 4.8 GHz. After setting up the co-simulation, it would be found that 5 GHz input signal frequency of LNA is too close to the 4.8 GHz SSN generated by 16 inverters to work normally. Three cases for device under test (DUT) to test its noise suppression capability. One is the reference case that no circuit is used to impede power noise. Another is the proposed structure. The third is a π filter, which the modeling method of ferrite bead is from [6], and the equivalent model of both two capacitors GRM0335C1E120JD01 are released by MURATA. Fig. 7 shows transient simulation in ADS. For the reference case, the power of noise is -10dBm. It is obvious that about 50dBm suppression is achieved by the proposed structure as the π filter does nothing. The suppressive ability is better than π filter at higher frequencies.

978-1-4244-4447-2/09 $25.00 © 2009 IEEE

V. CONCLUSION

In this work, we propose a novel method for GHz SSN coupling suppression using a compact electromagnetic band gap filter. Compared with traditional components we use to impede noise, it could work at higher frequencies and the band gap could be decided by different structure parameters. The equivalent lumped model is also established, and good agreement is seen between the modeling and simulation. The measurement also has good consistency. By using chip-package co-simulation, the suppressive ability is verified. From the output power spectrum of LNA, it is observed that over 50dBm noise suppression could be achieved by proposed structure. It can be observed that it is better than π filter. This structure can be used to solve noise problem at higher frequencies.

REFERENCES

[1] Tzong-Lin Wu, Yen-Hui Lin, Ting-Kuang Wang, Chien-Chung Wang, and Sin-Ting Chen, "Electromagnetic Bandgap Power/Ground Planes for Wideband Suppression of Ground Bounce Noise and Radiated Emission in High-Speed Circuits", *IEEE Trans. Microwave Theory and Tech.*, vol. 53, No.9, pp. 2935-2942, Sept. 2005.

[2] Tzong-Lin Wu and Sin-Ting Chen, "A Photonic Crystal Power/Ground Layer for Eliminating Simultaneously Switching Noise in High-Speed Circuit", *IEEE Trans. Microwave Theory and Tech.*, vol. 54, No.8, pp. 3398-3406, Aug 2006.

[3] T. K. Wang, T. W. Han, and T. L. Wu, "A novel EBG power plane with stopband enhancement using artificial substrate," *IEEE Trans. Microwave Theory and Tech*, vol. 56, No.5, pp. 1164-1171, Sept. 2008.

[4] W. Cui, J. Fan, Y. Ren, H. Shi, J. L. D rewniak, and R. E. DuBroff, "DC power-bus noise isolation with power-plane segmentation," *IEEE Trans. Electromagn. Compat.*, vol. 45, pp. 436–443, May 2003.

[5] J. Kim and M. Swaminathan,"Modeling of multilayered power distribution planes using transmission matrix method," *IEEE Trans. Adv. Packag.*, vol. 25, no. 2, pp.189-199, May 2002.

[6] Tae Hong Kim, Junho Lee, Hyungsoo Kim, and Joungho Kim, "3 GHz wide frequency model of ferrite bead for power/ground noise simulation of high-speed PCB," Electrical Performance of Electronic Packaging, 2002. pp.217-220.

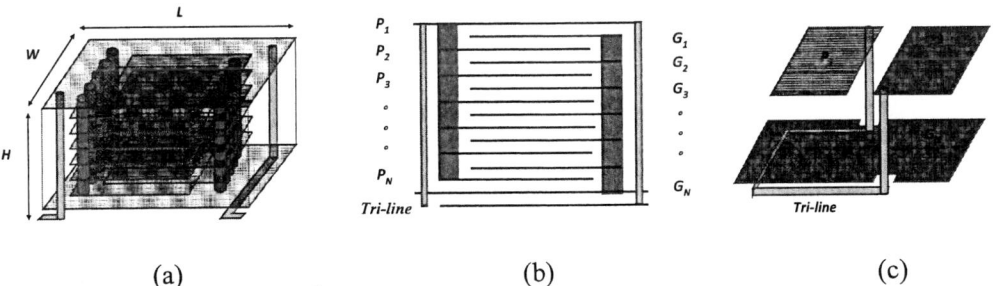

(a) (b) (c)

Fig. 1 (a) Unit cell of proposed structure (b) The connection between unit cells (c) Side view of unit cell structure

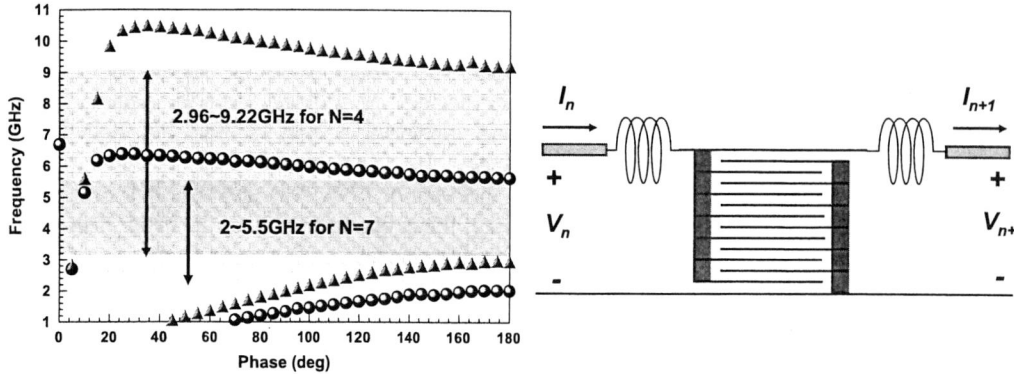

Fig. 2 The variation of band gap depended on the total layers of structure N

Fig. 3 Equivalent model of unit cell of proposed structure

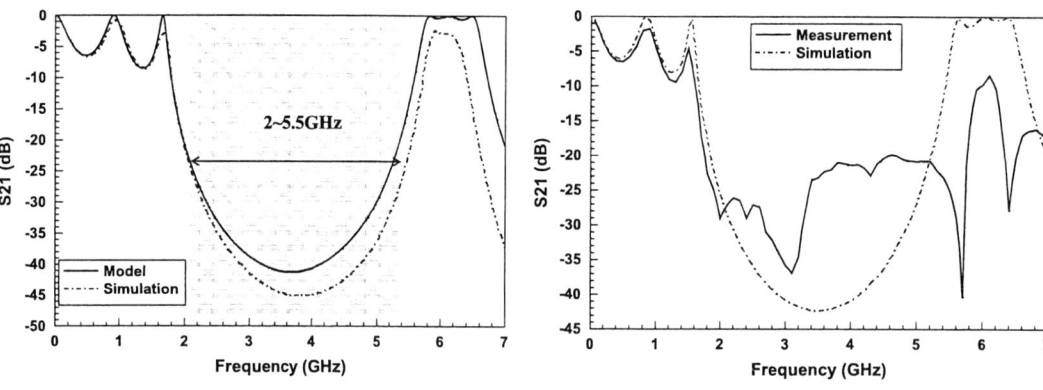

Fig. 4 S_{21} from HFSS and model

Fig. 5 S_{21} from measurement and simulation

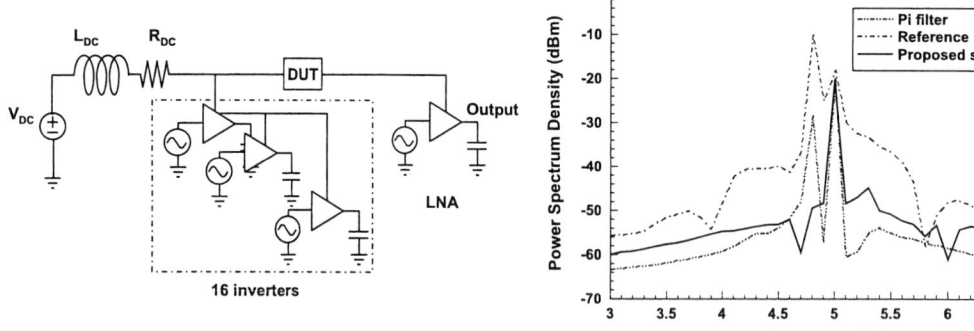

Fig. 6 Co-simulation environment

Fig. 7 Power spectrum density of LNA output

A New Extraction Method of Characteristic Parameters of a Coupled Transmission Line

Minwoo Kang[1], elano@ramrec.sch.ac.kr, Daehoon Jang[1], hb17894@hanmail.net, Kwangsik Park[1], p0403@hanmail.net, Chilhyeun Gwon[1], 050713kch@hanmail.net, Kwisoo Kim[1], kwisoo@ramrec.sch.ac.kr, Jongsik Lim[1], jslim@sch.ac.kr, Kwansun Choi[1], cks1329@sch.ac.kr, Dal Ahn[1], dahnkr@sch.ac.kr

[1]Dept. of Electrical Communication System Engineering, Soonchunhyang, University, Rep. of KOREA
Tel : 82 – 41 – 530 – 1607, FAX : 82 – 41 – 530 – 1609

Abstract

This paper suggests a new extraction method of the characteristic parameters of a coupled transmission line, which is structured like a microstrip line. A coupled line coupler that has -20dB coupling at the center frequency 1.5GHz is used for the extraction of the characteristic parameters. First, it extracts the characteristic parameters using the simulation results of the coupler. The paper shows the accuracy of this extraction method by comparing the results of EM simulation with the results of circuit simulation using the extracted characteristic parameters.

Keywords: coupled transmission line, characteristic parameters, EM simulation, circuit simulation

1 Introduction

Wireless and information technology industries have been growing rapidly due to the development of compact high performance versatile communication devices. Consequently, the demand has grown for microwave planar devices, such as a filter, directional coupler, power divider, which dominantly comprise the RF and microwave systems, providing high performance and small size at reduced cost. These devices consist of the coupled transmission line. [1]

The characteristic of the coupled transmission line is decided by the impedance of the even/odd-mode and the effective permittivity.
When circuit is designed, it is important to predict the characteristic parameters so that the circuit can yield the desired characteristic of the device. However, it is too difficult to know the even-odd mode characteristic impedance and effective permittivity in a coupled line. [2]-[5]

In this paper, a new method that uses the s-parameters of the coupled line to provide a practical prediction of the even/odd-mode characteristic impedance and effective permittivity is proposed.
It is advantageous to know the exact characteristic parameters of the coupled line based on the S-parameter.

2 Analysis

Figure 1 is an equivalent circuit of a coupled line under even/odd-mode excitation. The electronic-magnetic wave is represented as (1) at each port. [6]-[8]

The transmission coefficients and reflection coefficients are obtained from (2) using (1).

A 2-port network can be represented by using the results of a 4-port network as in (3). $A_{e(o)}$ and $B_{e(o)}$ can be calculated by using a conversion formula between ABCD-parameter and S-parameter. [6]

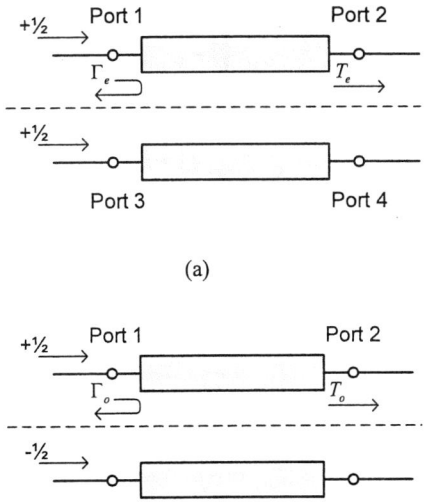

(a)

(b)

Figure 1   Equivalent circuit of even/odd-mode

$$S_{11} = \frac{1}{2}(\Gamma_e + \Gamma_o)$$

$$S_{21} = \frac{1}{2}(T_e + T_o)$$

$$S_{31} = \frac{1}{2}(\Gamma_e - \Gamma_o) \qquad (1)$$

$$S_{41} = \frac{1}{2}(T_e - T_o)$$

$$\Gamma_e = S_{11} + S_{31}$$

$$T_e = S_{21} + S_{41}$$

$$\Gamma_o = S_{11} - S_{31} \qquad (2)$$

$$T_o = S_{21} - S_{41}$$

$$\begin{pmatrix} A_{e(o)} & B_{e(o)} \\ C_{e(o)} & D_{e(o)} \end{pmatrix} = \begin{pmatrix} \cos\theta_{e(o)} & jZ_{0e(0o)}\sin\theta_{e(o)} \\ jY_{0e(0o)}\sin\theta_{e(o)} & \cos\theta_{e(o)} \end{pmatrix} \quad (3)$$

$$A_e = \frac{\left(1 - \Gamma_e^2\right) + T_e^2}{2T_e}$$

$$B_e = Z_0 \frac{\left(1 + \Gamma_e\right)^2 - T_e^2}{2T_e}$$

$$A_o = \frac{\left(1 - \Gamma_o^2\right) + T_o^2}{2T_o} \qquad (4)$$

$$B_o = Z_0 \frac{\left(1 + \Gamma_o\right)^2 - T_o^2}{2T_o}$$

Equation (5) gives the even/odd-mode characteristic impedance.

$$Z_{0e} = \frac{B_e}{\sqrt{1 - A_{(e)}^2}}$$

$$Z_{0o} = \frac{B_o}{\sqrt{1 - A_{(o)}^2}} \qquad (5)$$

Equation (6) and (7) represent the wave length λ_0 and effective permittivity ε_{eff} of a microstrip line.[9] λ_0 is the wavelength at the frequency at which the value of $A_{e(o)}$ is zero. And λ_0 has the maximum coupling

coefficient at quarter wavelength, and it is represented by (7).

$$\lambda_0 = \frac{c}{f_0} \qquad (6)$$

$$\varepsilon_{eff} = \left(\frac{\lambda_0}{4 \times l_p}\right)^2 \qquad (7)$$

In this case, c is the velocity of light in free space and l_p is the physical length of a microstrip line. ($c = 3 \times 10^8$ m/sec)

3 The Calculation And Verification

We designed a directional coupler for a structure like the microstrip line to prove the accuracy of this extraction method of the characteristic parameters. The specification of the directional coupler is shown below and we carried out an EM simulation of this case. The substrate that was used in the EM simulation was RT duroid 5880 of height 0.7874mm (-Rogers Corporation)

- Specification
- Center frequency : 1.5GHz
- Coupling coefficient : -20dB
- In/Out put impedance : 50Ω

The even/odd-mode characteristic impedances were 55.277 Ω and 45.227 Ω, based on the width and length of this directional coupler.
Figure 2 shows the magnitude of the frequency response of the ABCD-parameter, which is the calculated S-parameter of the EM-simulation using (4).

From (3), if the electrical length is 90°, the magnitude of A_e and A_o must be zero.
Therefore, the frequency at which the magnitudes of A_e and A_o become zero is shown in Fig. 2 (a). Exactly, the magnitude of A_e and A_o are zero at 1.43GHz and 1.51GHz, respectively.
Also, the values of B_e and B_o are 55.96 and 45.83, respectively.

Substitute $A_{e(o)}$ and $B_{e(o)}$ in (5), and then calculate the practical even/odd-mode characteristic impedance. Next, the extraction method of effective permittivity of a coupler is presented.
The wave length λ_0 and effective permittivity ε_{eff} can calculate by using (6) and (7), respectively.
At this time, l_p has a quarter wave length at 1.5GHz, and the physical length is 36.595mm. And f_0 is the frequency where the electrical length of an equivalent circuit becomes 90°. Exactly, the value of

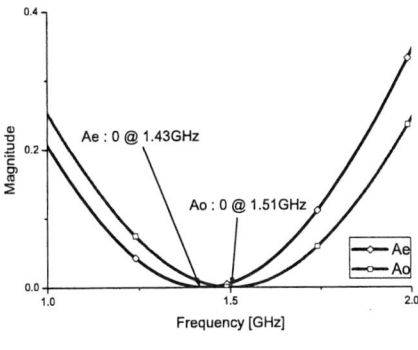

(a)

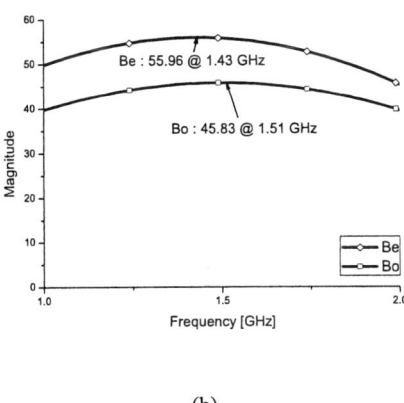

(b)

Figure 2   Frequency response of the magnitude of
ABCD-Parameter of even/odd-mode
(a) the magnitude of A_e and A_o (b) the magnitude
of B_e and B_o

Table 1 The values of characteristic parameters of
even/odd mode

Even-mode	Odd-mode
l_p = 36.595mm	l_p = 36.595mm
Z_{0e} = 55.962 Ω	Z_{0o} = 45.827 Ω
$\varepsilon_{eff(e)}$ = 2.056	$\varepsilon_{eff(o)}$ = 1.856

A_e is zero. Accordingly, we can calculate the values of even/odd-mode wave length and effective permittivity.

Table 1 shows the values of the characteristic parameters obtained by this procedure.
Figure 3 compares the result of EM simulation of a coupler with the results of circuit simulation using extracted characteristic parameters. We used HFSS ver. 11.2 and Designer 4.0 of Ansoft Corporation for the

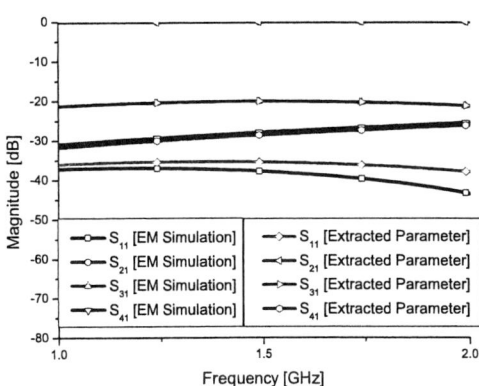

(a)

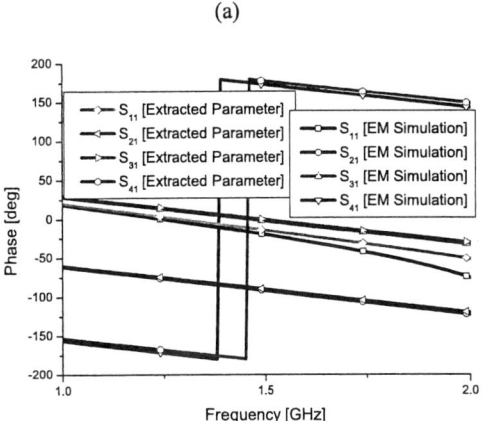

(b)

Figure 3   The comparison of EM simulation results and
circuit simulation results using extracted characteristic
parameters
(a) S-parameters (b) phases

EM and circuit simulations. Figure 3 (a) compares the S-parameter property and the results are almost accurate. And Figure 3 (b) compares the phase property and the results are almost accurate, too. Table 2 and Table 3 show the values of the S-parameter and phase at the center frequency and the results are almost accurate.
Accordingly, we proved the accuracy of this extraction method of characteristic parameters.

4 Conclusion

This paper proposes a new extraction method of characteristic parameters of a coupled transmission line structured like a microstrip line. We extracted the characteristic parameters from a coupler that has

Table 2 The values of the S-parameters @ 1.5GHz

	EM Simulation [dB]	Extracted Parameter [dB]
S_{11}	-0.074	-0.059
S_{21}	-37.806	-35.579
S_{31}	-20.088	-19.949
S_{41}	-27.833	-28.546

Table 3 The values of the phases @ 1.5GHz

	EM Simulation [deg]	Extracted Parameter [deg]
S_{11}	-19.895	-14.666
S_{21}	-92.065	-89.892
S_{31}	-0.216	-0.203
S_{41}	173.035	177.73

-20dB coupling coefficient at center frequency of 1.5GHz.

Table 2 and Table 3 show the values of the S-parameter and phase at the center frequency. The results of the EM simulation of a coupler and the results of circuit simulation using the extracted characteristic parameters were very close agreement.

The extraction method of the characteristic parameters will be very useful for the design of RF circuits and analysis of RF systems.

5 References

[1] R. Mongia, I. Bahl and P. Bhartia, joh, "RF and Microwave Coupled-Line Circuits Advanced," Artech House Publishers, boston, pp. 1-13

[2] M. K. Krage and G. I. Haddad," Chracteristics of coupled microstrip transmission lines-I : Coupled-mode formulation of inhomogeneous lines," *IEEE Trans. Microwave Theory Tech.,* vol.MTT-18, pp. 217-222, Apr.1970.

[3] S. D. Shamasundara and N. Singh, "Design of coupled microstrip lines," *IEEE Trans. Microwave Theory Tech.*, vol.MTT-25, pp.232-233, Mar.1977.

[4] S. Akhtarzad et al.,, "The design of coupled microstrip lines," *IEEE MTT Trans. Microwave Theory Tech.*, vol.MTT-23, pp.486-492, June.1975.

[5] E. G. Cristal and L. Young, "Theory and Tables of Optimum Symmetrical TEM-Mode Coupled-Transmission-Line Directional Couplers," *IEEE Trans. Microwave Theory and Tech.*, pp.543-558, 1965.

[6] David M. Pozar, Microwave Engineering 3rd ed., MA John Wiley & Sons, Inc., pp. 337-345, pp. 174-188, 2005.

[7] Jonguk Kim, Jong-sik Lim, Kwangsoo Kim, and Dal Ahn, "An Equivalent Circuit Model for Multi-port Networks," *European Microwave Conference 2007,* pp. 901-904, October 2007.

[8] Jinjoo Choi, Microwave Engineering 2nd ed., INFINITYBOOKS, Inc., pp. 393-403, 2007.

[9] G. Gonzalez, Microwave Transistor Amplifiers Analysis and Design 2nd ed, Prentice Hall, Inc., pp. 141-152, 1997

Design and Testing of a High Speed Module Based Memory System

Ravi Kollipara, Ming Li, Don Mullen, Wendemagegnehu Beyene, Chris Madden, Chuck Yuan,
Hideki Kusamitsu* and Toshiyasu Ito*
Rambus, Inc.
4440 El Camino Real, Los Altos, CA 94022 USA
ravik@rambus.com
*Yamaichi Electronics Co., Ltd.
3-28-7 Nakamagome Ohta-ku, Tokyo, Japan 143-8515

Abstract

An LCP flex based interconnect and a mating area array connector are used to increase the bandwidth of the module based memory systems. Simulations show that data rates in the range of 6.4 Gbps to 16.0 Gbps are possible depending on the memory system configuration. Tested prototype memory systems confirmed simulation predicted data rates of 16 Gbps and 12.8 Gbps for flex interconnect lengths of 6" and 12", respectively.

I. Introduction

The main memory for desktops, laptops, workstations and servers typically resides on modules. With the adoption of multi-core architectures by the computing industry and with the new computing trend of virtualization, the demand for memory performance is growing. However, the maximum data rate of module based DDR3 systems is limited to 1600 Mbps [1]. Even for a point-to-point based DQ link, a number of factors limit the data rate of the traditional module based memory system. The through-hole DIMM connectors have high crosstalk and reflections due to the high impedance of the connector. The balls and vias of the controller BGA package and the through-hole vias of the motherboard and the module PCBs also introduce crosstalk and reflections due to the impedance discontinuities present at these locations. The motherboard and module PCBs are fabricated typically from low-cost FR-4 dielectric material which has a high dielectric loss tangent leading to increased attenuation in the multi-GHz region. The FR-4 based PCBs may also contribute fiber-weave induced skew which limits the data rates. In addition, the FR-4 based PCB design rules, the DIMM connector pin pitch and the controller package ball pitch also limit the interconnect density leading to longer interconnects and increased channel attenuation or more routing layers. In this paper, a way to achieve data rates of up to ten times the highest data rate of DDR3 memory system is described and tested.

II. Design of a Flex Based Memory Interconnect System

High speed flex interconnects are based on low loss dielectric materials and two layer flex is becoming standard for higher interconnect density and signal integrity reasons. The most commonly used dielectric material for high speed applications is polyimide (PI) [2]. Though PI has low dielectric loss tangent, it is typically used with an adhesive as a bonding medium to form the metal clad laminate or cover layer. These adhesive materials have high dielectric loss tangent which increases the overall loss. However, special processing has shown that an all-polyimide-and-copper construction is possible which makes the overall loss the lowest possible for a PI based flex interconnect [3]. Another material that is getting serious consideration for high speed applications is liquid crystal polymer (LCP). When compared to PI, it has lower loss tangent and lower dielectric constant that are also much less sensitive to moisture, temperature and frequency variations. In addition, adhesiveless lamination is easily possible with LCP.

The challenge with using flex as a high speed interconnect is coming up with an interfacing method with the other components that is easily assembled and reliable. One major consideration is whether one or both ends of the interface need to be a detachable connection or can be a permanent one. Typically, permanent connections are more robust and can be excellent from a signal integrity point of view. However, systems may need to be flexible from assembly point of view which may force one or both ends of the interface to be a detachable one. This is where proper interface design becomes crucial. Both detachable and permanent connections were successfully tested in chip-to-chip [3], backplane [4] and memory applications [5-6]. In these applications, the flex interface is used to route high speed signal traces to bypass the package balls and vias, motherboard/line card vias, and the backplane connector's press-fit pin plated through hole vias or memory connector's pins and vias. As a result, package cost can

978-1-4244-4447-2/09 $25.00 © 2009 IEEE

be lowered by reduced ball count, skew is reduced as flex uses a homogeneous dielectric material, interconnect length is shorter, and interconnect density can be increased for a given loss. Most importantly, signal integrity is improved as the number of interfaces where reflections and crosstalk may occur is reduced.

A detachable, two-piece area array connector was evaluated for interfacing with flex on the memory controller side [5-6]. When the connector housing is fully populated with mating pins, the pin count is 216, resulting in 72 (12X6) differential pairs using one ground pin per diff pair. When populated with half the pin count, the diff pair count is reduced to 36 (12X3). The dimensions of the connector are ~ 10 mm x 15 mm x 3 mm. The 3-D EM simulation results of the two connector footprints are shown in Figure 1 in both frequency and time domain. An LCP based flex interconnect was also designed and evaluated for memory applications [5-6]. The two cross sections of the flex that were tested are shown in Figure 2. The higher interconnect density flex (140 diff pairs/inch) was designed to interface with the 216 pin count flex connector and the lower interconnect density flex (70 diff pairs/inch) was designed to interface with the 108 pin count flex connector. The measured insertion loss of the flex for the two cross sections as well as the simulated far-end crosstalk (FEXT) for 6" and 12" flex lengths are shown Figure 3. The measured insertion loss indicated an rms surface roughness of ~ 0.5 μm for the flex copper foil which is in the expected range of 0.45 to 0.55 μm indicated by the flex vendor.

III. Assembly of the Test System

A flex based memory module system can be put together in many different ways depending on the data rate, memory capacity, ease of assembly and cost considerations. One possible configuration that is geared towards driving higher data rates with a detachable flex interface on the memory controller side is shown in Figure 4. The DRAM package is directly soldered on to the top (ground) plane side of the flex. Copper bumps in the LCP flex are used to transition the DRAM power and ground as well as signals to the bottom layer of the flex. Power, ground and low speed DRAM signals have balls attached on the bottom layer of the flex and the flex with DRAM is soldered to the memory module. The high speed differential signals are routed on the bottom layer of the flex to the memory controller side and are transitioned to the top side of the controller package through the detachable mating flex connector. Solder balls are not attached to the DRAM differential signal pads on the bottom layer of the flex to avoid capacitive loading. The assembled components with 6" and 12" flex interconnect lengths are also shown in Figure 4.

A technology initiative focused on memory signaling for future generations of game consoles and graphics memory systems was introduced with a goal of delivering 1 TB/s of memory bandwidth [7]. Full implementation details of this initiative were disclosed at various technical conferences [8-10]. The memory interface used a point-to-point, bi-directional differential topology to ensure high data rates and high scalability. The memory controller and the memory were implemented in the TSMC 65 nm process technology. The design of the memory emulated the 40 nm DRAM. The DRAMs were soldered on an FR-4 motherboard resulting in FR4 trace lengths in the range of 2 to 3 inches. The maximum data rate of this memory system is 16 Gbps. The command/address (C/A) is also a point-to-point link running at the same data transfer rate of the DQ links. Both the WRITE and the READ equalizations are performed on the memory controller side using transmit pre-emphasis and receive continuous-time linear equalizer, as shown in the block diagram of Figure 5. All the key link parameters are verified and correlated with the system models [11-13] and the verified key link parameters are also listed in Figure 5.

The same test chips that were developed to test the motherboard soldered memory with 3" of FR4 PCB traces are used to test the flex based memory module system. The original controller flip-chip BGA package substrate is redesigned to accommodate the flex connector on the top layer of the package. The same original two-layer wirebonded chip scale DRAM package is used without modification. The same mother board that was used to test the soldered memory system is used to test the flex based memory system using sockets and is shown in Figure 4. The sockets do not degrade the performance as the high speed signals bypass the sockets and pass through the flex. The previously routed differential signal pairs in the motherboard do not interfere with the testing as they are not accessed either on the DRAM side or on the controller side.

IV. Simulation and Measurement Results

A MATLAB based BER simulation tool based on statistical approach [14] is used to simulate the possible data rates for the memory system. The key active component parameters listed in Figure 5 and the channel passive S-parameter models are used as inputs to the simulation tool. The MATLAB based system simulations showed that the 6" and 12" flex interconnects with the lower loss stack-up can handle data rates of 16 Gbps and 12.8 Gbps, respectively, at a BER of 1E_20 [6]. The 12" flex interconnect passes at 16 Gbps in the WRITE direction but not in the READ direction. In general, The memory performance is limited by the READ case as there is less Tx swing and limited Rx gain. Both the 6' and 12" flex interconnects are tested with the prototype system shown in Figure 4. The measured READ BER timing bathtub plots for the 6" flex and 12" flex interconnects at 16 Gbps and 12.8 Gbps, respectively, are shown in Figure 5. The measurements confirm that there is passing timing margin left at a BER of 1E-20.

V. Conclusion

One possible way of improving the performance of module based memory systems using a flex interconnect is shown. A detachable memory connector with a flexible footprint was developed and tested. Two LCP based flex stack-ups were considered and evaluated. Data rates in the range of 6.4 Gbps to 16.0 Gbps are possible depending on the memory system configuration.

References

[1] "DDR3 SDRAM Specification," JEDEC Standard Document JESD79-3B, April 2008.

[2] Joseph Fjelstad, Flexible Circuit Technology, Third Edition, BR Publishing, Inc. 2006.

[3] H. Braunisch et al, "High-Speed Flex Circuit Chip-to-Chip Interconnects," IEEE Trans. on Advanced Packaging, vol. 31, No. 1 (2008), pp. 82-90.

[4] K. Grundy et al. "Designing Scalable 10G Backplane Interconnect Systems Utilizing Advanced Verification Methodologies," 8-WP2, DesignCon2006, Santa Clara, CA.

[5] R.T. Kollipara et al., "Evaluation of High Density LCP Based Flex Interconnect for Supporting >1 TB/s of Memory Bandwidth", 58th Electronic Components & Technology Conference, May 2008, Orlando, FL.

[6] R.T. Kollipara et al., "Evaluation of a Module Based Memory System with an LCP Flex Interconnect", 59th Electronic Components & Technology Conference, May 2008, San Diego, CA.

[7] Steven Woo, "Computing Trends And Applications Driving Memory Performance," Rambus Developer Forum, Tokyo, Japan, November 28, 2007.

[8] K. Chang, et al., "A 16Gb/s/link, 64GB/s Bidirectional Asymmetric Memory Interface," Symposium on VLSI Circuits, Honolulu, HI, June 2008.

[9] N. Nguyen, et al., "A 16-Gb/s Differential I/O Cell with 380fs RJ in an Emulated 40nm DRAM Process", Symposium on VLSI Circuits, Honolulu, HI, June 2008.

[10] T. Wu, et al., "Clocking Circuits for a 16Gb/s Memory Interface," IEEE Custom Integrated Circuits Conference, San Jose, CA, September 2008.

[11] W. T. Beyene, et al. "Design and Analysis of a TB/sec Memory System," Proceedings of IEEE 17th Topical Meeting on Electrical Performance of Electrical Packaging (EPEP), San Jose, CA, October, 2008.

[12] W. T. Beyene, et al. "The Design and Signal Integrity Analysis of a TB/sec Memory System," DesignCon, Santa Clara, CA, February 2009.

[13] W. T. Beyene et al., "Advanced modeling and accurate characterization of a 16 Gb/s memory interface," IEEE Trans. on Advanced Packaging, Vol. 32, No.2, pp. 437-659, May 2009.

[14] V. Stojanovic, Channel Limited High-Speed Links: Modeling, Analysis and Design, PhD Thesis, Stanford University, 2004

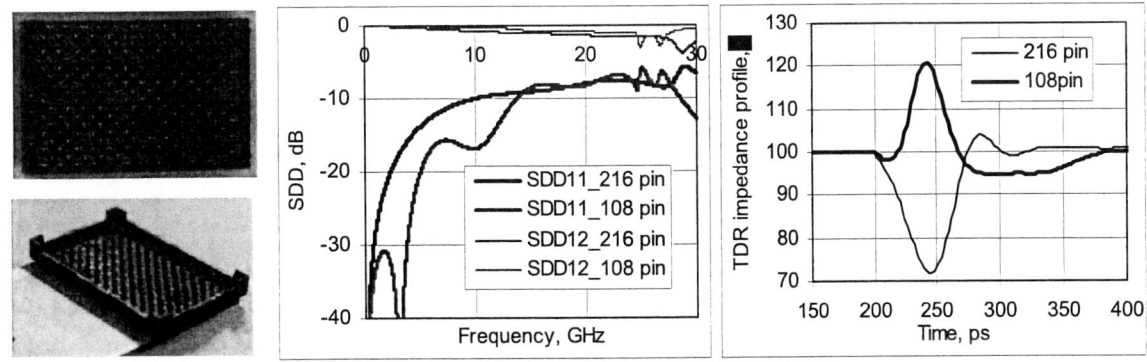

Figure 1 Modeled S-parameters and TDR impedance profile (Tr (20-80%) = 20 ps) of the flex connector.

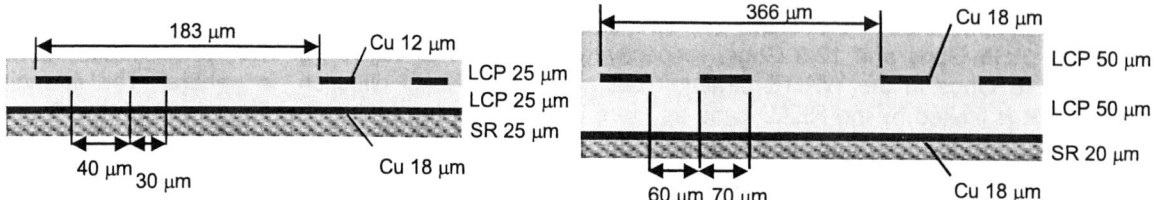

Figure 2 The stack-ups of the two fabricated flex insterconnects.

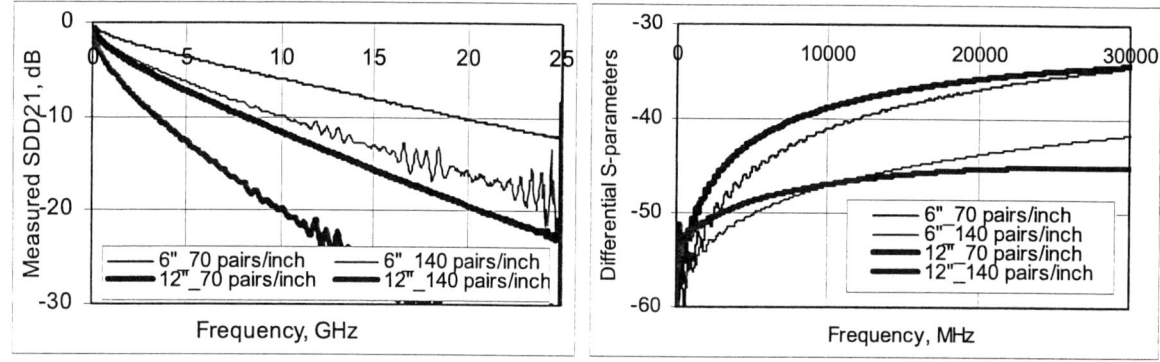

Figure 3 Measured differential insertion loss and simulated differential FEXT of the two flex stack-ups.

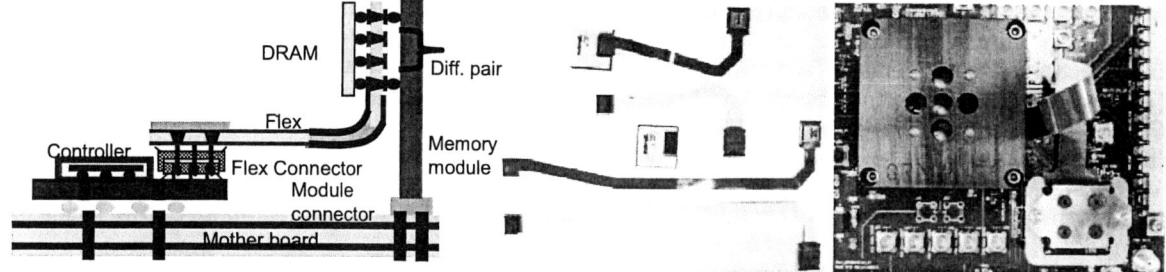

Figure 4 Proposed high speed memory module configuration, test components and the test system.

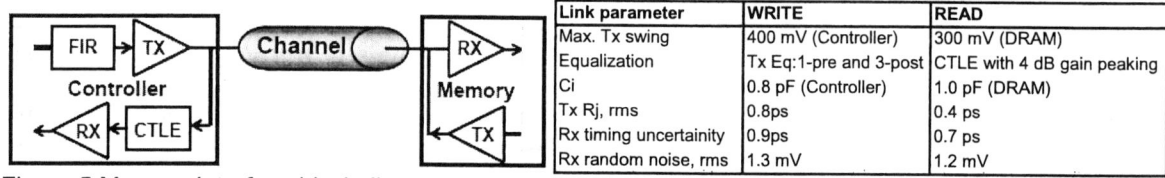

Link parameter	WRITE	READ
Max. Tx swing	400 mV (Controller)	300 mV (DRAM)
Equalization	Tx Eq:1-pre and 3-post	CTLE with 4 dB gain peaking
Ci	0.8 pF (Controller)	1.0 pF (DRAM)
Tx Rj, rms	0.8ps	0.4 ps
Rx timing uncertainity	0.9ps	0.7 ps
Rx random noise, rms	1.3 mV	1.2 mV

Figure 5 Memory interface block diagram and the memory system's key link parameters.

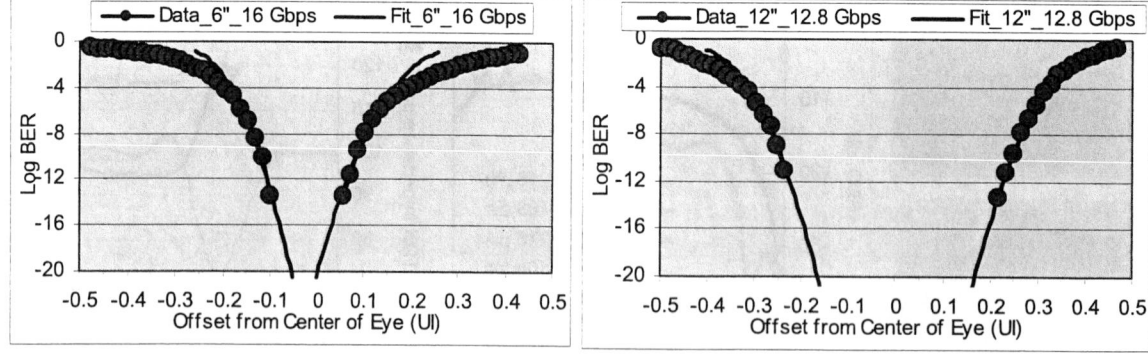

Figure 6 Measured READ timing bathtubs at 16Gbps for 6" flex and at 12.8 Gbps fo12" flex interconnect.

978-1-4244-4447-2/09 $25.00 © 2009 IEEE

The Extraction and Measurement of On-Die Impedance for Power Delivery Analysis

Xiaoping Liu, xiaoping.a.liu@intel.com, Yi-Feng Liu, yi-feng.liu@intel.com

Mailstop: CH7-231
5000 W. Chandler Blvd, Chandler, AZ 85226
Intel Corporation
Tel: (480) 554-4742, Fax: (480) 552-8466

Abstract

Power delivery system noise, current and impedance are the key performance factors for successful chipsets/CPUs design. Power delivery noise is significantly affected by on-die impedance and current, i.e., on-die power grid equivalent resistance (Rdie), capacitance (Cdie), and Icc(t). Rdie and Cdie have been challenging in pre-silicon extraction and post-silicon measurement. This paper focuses on Rdie and Cdie extraction and silicon correlation. We carried out Rdie and Cdie extraction comparison study for Intel products that contain several PHY blocks by using home-brew flow, a commercial tool, and lab measurement, to achieve good correlation. The work boosts our confidence in the accuracies of internal flow and commercial tool and therefore facilitates the adoption of the commercial tool to greatly improve die grid modeling effort.

I. Introduction

Power delivery becomes a big issue in the design of today's microprocessor, chipsets, communication chips, and other semiconductor products. The market demand for higher speed, more functions, and lower power consumption results in the significant increase in operating frequency and the number of transistors, and the decrease of voltage. The consequent large current and low voltage necessitate a low-impedance path from the power supply to the die. Low-impedance in power delivery loop reduces not only DC IR voltage drop, but also power delivery noise. This is critical in today's low voltage and large current silicon design, especially for the circuits with power management features and lower active Vcc margins, such as some IO (Input/Output) analog circuitry. To save power consumption, circuits like PCIe will be powered down during low idle state; when several circuits power on again, potential large di/dt current can generate excessive supply noise which can cause signal timing issue or even functional failure. Hence, power integrity should be ensured so that power delivery droop/noise can be reasonably small to meet more stringent targets. As a part of the low-impedance power delivery network, on-die low impedance means low resistance Rdie and high capacitance Cdie. New technology requirements even push the on-die impedance into the order of milliohm or less. Accurate modeling of Rdie and Cdie is essential to this effort.

The accuracy of die grid modeling is important in better prediction of voltage droop/noise, silicon decap quantity, package decap type/quantity and package power/ground planes design, and even IO lane-to-lane staggering control. These can alleviate overdesign or underdesign in silicon and package which results in good quality and cost reduction, and avoids design reiteration. Furthermore, in the fast-paced IT market nowadays, a modeling flow/tool possessing both accuracy and high efficiency/capacity is even more robust and in demand. It can shorten turnaround time so that the silicon/package designers can try different design options and do analysis to optimize designs quickly. This helps to shorten design to market time with high-performance and low-cost products.

The team developed the home-brew On-Die PDA toolset about 4 years ago [1] before the commercial solution was widely adopted in the design team. Since then, it has been used to deliver die grid model for several Intel products. To validate the two pre-silicon extraction toolsets, we correlated their results with lab measurement. Good correlation was observed in core Cdie [2][3]. For continuation, this paper extended to Rdie correlation for core/IOs and Cdie correlation for IOs, both of which have not been found in publications. We applied the two extraction flows to extract lumped RdieCdie for several PHY blocks as well as core. VNA (Vector Network Analyzer) was used in lab to measure lumped RdieCdie. Reasonably good correlation among the three approaches was observed.

II. Internal Tool: On-Die PDA (Power Delivery Analysis)

978-1-4244-4447-2/09 $25.00 © 2009 IEEE

On-Die PDA is an internal extraction/simulation tool [1][2] to generate on-die power delivery models, i.e. Rdie, Cdie, and Icc(t). Cdie extraction was described in detail in [1]. After Cdie calculation and power/ground grid extraction, the entire extracted Vcc/Vss network with all on-die decaps connected at their corresponding locations is plotted in Figure 1. By shorting all Vcc bumps together and shorting all Vss bumps together in a functional block, and running AC simulation, we can obtain the equivalent Rdie for this block.

III. Commercial Tool

The commercial toolset we used is developed **Error! Reference source not found.**for power integrity analysis solutions [4]. For IO, there are two types of tools: one is used to characterize IO circuitry, while the other to generate on-die models. When utilized together, they are capable of generating frequency-dependent on-die models for IO blocks. For core, another tool can be used to generate on-die models for core partitions. The on-die model extraction of the tools is based on the principle: transistors are modeled as current sources with ESR and ESC, decap cells are modeled as capacitors/ESR, and the power/ground grid is extracted as RLC network.

IV. VNA Lab Measurement

To measure the low impedance of power delivery network in frequency domain, we adopted a robust 2-port VNA metrology introduced [5]in 1999 [5]. The Device Under Test (DUT) was with die and package together. Package decoupling capacitors were removed to reduce the number of parameters in on-die impedance measurement. Probing was conducted at package probing pad due to the difficulty to probe directly at tiny silicon bumps. VNA testing yielded S-parameters for DUT, hence the impedance of DUT (Z_{DUT}). To extract RdieCdie, the assumption was made that the die and package could be modeled as the equivalent circuit in Figure 2, and curve-fitting technique was utilized [2][3].

V. Correlation Results

The above three approaches were applied to two Intel products. Table1 is for an IO interface from one product (Chip1) with post-silicon measurement unavailable yet. Table 2 and 3 are from another product (Chip2) which has lab measurement: the former is for IO, the latter is for core. The results from two simulation toolsets are close for both IO and core. The largest difference on Rdie is 20% and Cdie 17%. It also can be seen that, for Cdie, the extracted data agreed well with VNA lab data for both IO and core, with the max delta of 17%. However, for Rdie, bigger discrepancy was observed between extraction and lab measurement with VNA number about 7 times extraction number in the worst case. But note that the Rdie absolute value is quite small in this case.

Several factors might contribute to the discrepancies in RdieCdie correlation among the three techniques:
1. In real system, the equivalent Rdie and Cdie are functions of frequency, and they are not fixed numbers, so the lumped models of Rdie and Cdie with fixed values are definitely approximations. Besides, the model assumption for lab measurement in Figure 2 is highly simplified.
2. In lab measurement, when DC voltage is supplied to the silicon, the internal circuit states were unknown. This cannot be simulated, thus affecting Cdie and Rdie correlation. If the circuits were to be properly initialized and put into known state, correlation could have been better.
3. The lab measurement may not be very accurate, especially when measuring small Rdie for a power domain with big die area. The probing technique and the chosen port location affect measurement results. Package testing point should be properly designed and the relative locations of probes should be studied.
4. Curve-fitting scheme can cause discrepancy between simulated data and original data, especially when Rdie is small (like several miliohms), the delta can be big, due to the big range of the impedance curves. Highly zooming in the valley of the impedance curve is needed for proper Rdie fitting.
5. In On-Die PDA flow, curve-fitting scheme was used to extract RdieCdie, so the accuracy is also affected by package model accuracy. More accurate package model will help.
6. The Cdie from commercial tool is equivalent Cdie from AC simulation, but On-Die PDA obtains Cdie by summing up all capacitance components. The internal flow can be improved to match commercial tool.
7. In the two extraction tools, the assumption that all Vcc bumps are shorted and all Vss bumps are shorted is also a simplified one, which usually results in smaller Rdie.
8. Only two typical skew chips were measured in the lab. More parts are needed to establish the statistic profile (the mean and standard deviation) of Rdie and Cdie.

Regarding the Rdie discrepancy of several times when Rdie is small, a study was conducted to investigate its impact to power delivery noise. In a spice simulation test case, at platform power delivery system, Rdie was swept from 1mΩ (close to Chip2 entire core Rdie)/15mΩ/30mΩ/60mΩ while the other parameters remained unchanged. The system-level impedance Z(f) profiles in Figure 3 show that, as Rdie increases, the impedance Z at the 1st peak resonance frequency decreases, but impedance at higher frequencies increases. Figure 4 demonstrates the power supply noise waveforms Vcc(t) with Vcc peak-to-peak (Vccp-p) recorded in Table 4. In this test case, Vccp-p seems not sensitive to Rdie for small Rdie. For example, when Rdie increases from 1 mΩ to 15 mΩ, Vccp-p only increases by 7%. For Chip2, platform-level simulations with core Rdie sweep showed that either 0.7 mΩ or 5 mΩ did not make significant difference in Vccp-p. Therefore, the Rdie difference between extraction and lab measurement is acceptable. However, this might not be a general rule. The Rdie discrepancy origin will be further investigated in future work.

Finally, the efficiencies of On-Die PDA and commercial tool are also compared. For an interface such as Chip2 IO or core, On-Die PDA usually took an engineer about two weeks to extract Rdie and Cdie, with most of time in manual work, such as data preparation and step-by-step operation. Commercial tool normally took one or two days in data preparation and then the runtime was about one day.

VI. Conclusions

In this work, on-die grid modeling comparisons were performed for core and IOs by using internal On-Die PDA flow, commercial tool, and VNA lab measurement. The validation revealed that similar accuracies were achieved among PDA and commercial solution. In commercial tool, the feature of frequency-dependent die model can be useful for what-if type of power delivery solution explorations. But need to sanity check the results for validity. On-Die PDA is free but requires manual intervention and simulations; strong dependency on circuit design environment which is constantly changing with product generations or project team specific setup; and complexity.

VII. References

[1] Jung S. Kang, Peter P. Jeng, Michael M. DeSmith, Ke W. Wang, "On-die Model Methodology and its Application to RHSL Chipset Product on P1263", Internal conference paper, 2005.

[2] Yi-Feng Liu, Brian Wang, Mingming Xu, Xiaoping Liu, Jie Zhu Chen, Michael Desmith, "Correlation of On-Die Capacitance for Power Delivery Network", EPEP, 2008.

[3] PCG Low Power Design (LPD) Team, "Correlation of Chip2 Cdie Between Measurement, Commercial Tool, and On-Die PDA (Internal Script)", internal documentation, March 20, 2008.

[4] Emre Kulali, Evgeny Wasserman and Ji Zheng, "Chip Power Model - A New Methodology for System Power Integrity Analysis and Design", IEEE EPEP, proceedings pp. 259-262, Oct. 29-31, 2007.

[5] Istvan Novak, "Probes and setup for measuring power-plane impedances with vector network analyzer", DesignCon, 1999.

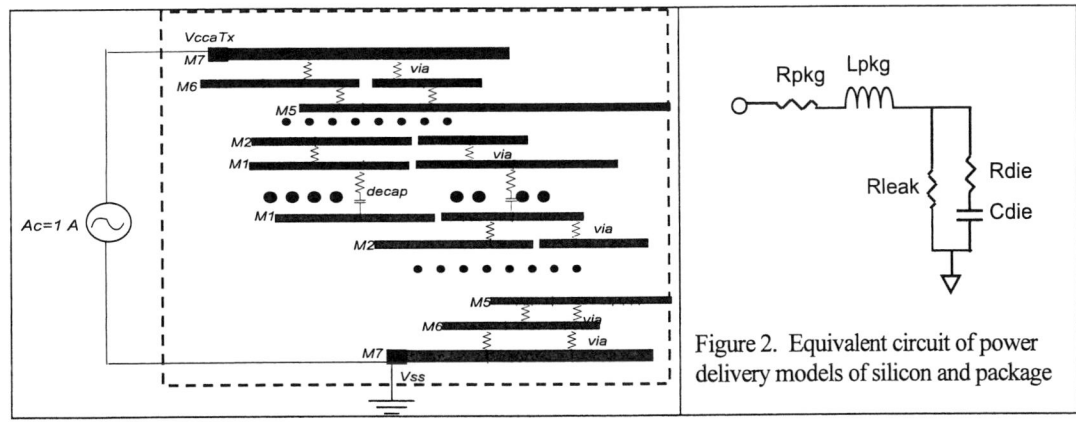

Figure 2. Equivalent circuit of power delivery models of silicon and package

978-1-4244-4447-2/09 $25.00 © 2009 IEEE

Figure 1. The entire extracted on-die power delivery network

PHY block	Temperature	Commercial Tool		On-Die PDA		Difference	
		Rdie (Ω)	Cdie (pF)	Rdie (Ω)	Cdie (pF)	Rdie	Cdie
Blk1	50C	0.622	60.3	0.735	61.6	18%	2%
Blk2	110C	0.100	340.2	0.105	363.9	5%	7%
Blk3	50C	0.330	251.5	0.303	229.7	-8%	-9%
Blk4	50C	0.063	368.8	0.058	360.6	-8%	-2%

Note: Typical,1.1V. Difference: use commercial tool as reference.

Table 1. Comparison between two extraction flows for Chip1 IO

PHY block	Commercial Tool		On-Die PDA		Lab VNA msmt	
	Rdie (mΩ)	Cdie (nF)	Rdie (mΩ)	Cdie (nF)	Rdie (mΩ)	Cdie (nF)
Blk1	5.1	20.3	4.4	23.8	13.0	23.2
Blk2	6.0	20.4	5.5	22.8	15.0	21.0

Note: Typical, 1.0V, 25C

Table 2. Comparison among three techniques for Chip2 IO

Partition	Commercial Tool		On-Die PDA		Difference	
	Rdie (m Ω)	Cdie (nF)	Rdie (m Ω)	Cdie (nF)	Rdie	Cdie
Par1	13.61	2.7	10.87	2.56	-20%	-5%
Par2	22.69	2.23	18.8	2.08	-17%	-7%

Note: Difference: use commercial tool as reference.

Table 3a. Comparison between two extraction flows for Chip2 core partitions

Commercial Tool		Lab VNA msmt	
Rdie (m Ω)	Cdie (nF)	Rdie (m Ω)	Cdie (nF)
0.7	89.4	5	105

Note: Typical, 0.86V, 25C.

Table 3b. Comparison between two techniques for Chip2 entire core

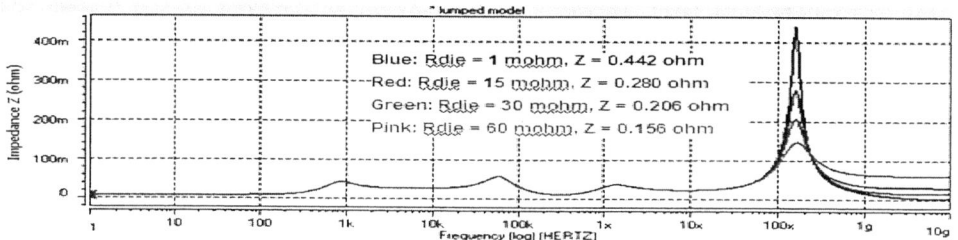

Figure 3. Platform-level power delivery impedance Z(f) profile for different Rdie

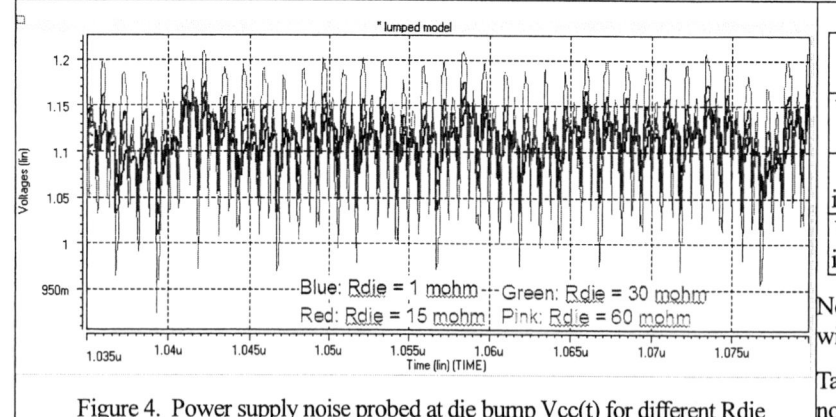

Figure 4. Power supply noise probed at die bump Vcc(t) for different Rdie

Rdie (mΩ)	1	15	30	60
Vccp-p (mV)	140	150	180	296
Rdie increase	0	140%	100%	100%
Vccp-p increase	0	7%	20%	64%

Note: "increase" means: compare with previous column

Table 4. Comparison of power noise Vccp-p for different Rdie

978-1-4244-4447-2/09 $25.00 © 2009 IEEE   196

Bit-Pattern Sensitivity Analysis and Optimal On-Die-Termination for High-Speed Memory Bus Design

Evelyn Mintarno[1], Steven Yun Ji[2]

[1] PO BOX 17116, Stanford University, Stanford, CA 94309, USA, [2]JF2-54, 2111 NE 25th Ave., Intel Corporation, Hillsboro, OR 97006, USA
[1]evemint@stanford.edu, [2]steven.yun.ji@intel.com, [1]Phone: (650)-387-8695, [2]Phone: (503)-712-1455

ABSTRACT

System IO power and performance are critical computer platform design parameters. IO power forms a significant portion of the overall power while its performance is often the bottleneck in achieving overall performance specification. This paper experimentally demonstrates, for the first time, optimal on-die-termination (ODT) schemes for DDR3-800MT/s and DDR3-1067 MT/s, revealed by thorough bit-pattern sensitivity analysis. Optimal ODT at IO receiver pads is proposed as a new critical design knob to achieve optimized power-performance trade-offs, dramatically improving signal integrity and power consumption. The thorough bit-pattern sensitivity analysis was found to be 100% more accurate than traditional approach. Up to 50% reduction in power consumption, 100% increase in timing margin, and 100% increase in voltage margin were demonstrated as the impact of the choice of ODT. It is also promised to become more important in the future at higher data rate.

INTRODUCTION

Demands for lower power consumption, higher performance, and smaller platform form-factor drive more complex platform memory bus designs. The design challenges are paramount, since fast edge rates and compact routing leads to more ISI (inter-symbol interference) and crosstalk. High-density interconnect can reduce the routing distance but it is a costly alternative. Design needs to tolerate process and operating condition variations which cause transmitter, channel and receiver imperfections. Therefore, transmitter and receiver have error tolerance and setup/hold-times requirements. Data transfer through channel is also distorted by ISI, crosstalk, and SSN. ISI arises from channel impedance mismatch resonance, reflections, return path discontinuity resonance, or split VSSP/VSS resonance. Crosstalk from nearby signals is affected by bit-pattern mode and aggressor/victim combination. Overall, sufficient signal integrity is needed to guarantee reliable data transfers in the presence of those imperfections.

In this paper, we propose that designing with optimal ODT (on-die termination), which can further be tuned autonomously post-manufacturing adaptive to manufacturing variations, is a cost effective approach to meet power-performance target as defined by user or system. ODT strongly determines power dissipation through DC paths from the pads to supply and ground. In the nominal case, larger ODT has significant power advantage and possible increase in eye height from self reflection at the receiver. However, as will be shown later, larger ODT results in worse timing margin, i.e. narrower "Vref" eye width due to increased sensitivity to bit patterns as a result of poor receiver side terminations and reflections. Such power-performance trade-offs associated with the choice of ODT and its sensitivity to input-pattern will be presented in this paper. This has a strong ROI (return on investment) on low power consumer products, e.g. notebook and small form-factor notebook, where battery life is one of the most critical design targets. Both simulation and post-silicon data of system margins have been reported on DDR2-533/667/800MT/s and DDR3-800MT/s [1]-[2]. This paper is intended to further address these important questions, for the first time: the scalability and pitfall of larger ODT, and potential solution space at higher data rate (DDR3-1067MT/s and DDR3-1333MT/s). Additionally, as will be shown later in results section, DDR3 timing/voltage margin were found to be a strong function of bit pattern stressing the bus. This indicates thorough data pattern sensitivity studies are needed in both simulation and validation. This work also suggests that with careful design of system channel and choice of memory type to support, higher data rate is achievable with larger ODT, which results in significant power saving.

EXPERIMENTAL SETUP

A common approach to quantify DDR3 margin is to measure DQ/DQS eye diagram of the received signal. Distortion in received data is illustrated by the fuzziness of the eye diagram. Voltage margin is defined as the difference between amplitude of received signal at sampling time and the receiver's threshold voltage. We use high-side and low-side center eye height rather than peak eye height to avoid misleading margin in the case when the eye is horizontally asymmetric, which we found to be the case with many bit patterns, especially with unequal low phase and high phase. Eye width is computed AC-DC following JEDEC definition, i.e. by adding minimum value of setup and hold times of high side and low side, of DQS rising and falling.

Fig. 1 shows DDR3 DQ signalling scheme. MCH (memory controller hub) is a CMOS driver. ODT is a series-resistor voltage divider to bias the receiver at half VDDQ. DDR3 DQ routing topology from MCH to SDRAM consists of MCH package routing, followed by breakout from MCH package balls, motherboard breakout, main routing, and break-in to DIMM pin. Fig. 2 shows the routing topology of Raw Card A and Raw Card C. As shown in Fig. 2, capacitive load of Raw Card A is twice larger than that of Raw Card B. Analysis indicates that DQ write cycles are more susceptible to error than read cycles for this particular design studied. Considering that ODT of 20ohm or lower is not allowed during DQ write, we investigate 4 representative ODT schemes: 40 ohms, 60 ohms, 120 ohms, and no ODT.

BIT PATTERN SENSITIVITY OF VARIOUS ON-DIE-TERMINATIONS

DDR3 defined by JEDEC spec is a bus without encoding scheme. In real applications, DDR3 bus may exercise all possible bit patterns. The margin either simulated or measured was found to be a strong function of bit pattern. This is of research

978-1-4244-4447-2/09 $25.00 © 2009 IEEE

interest for compact mobile device designs, where crosstalk is severe due to tight routing constraints. Ideally every possible combination of bit-pattern needs to be used in validation to nail down the worst-case eye. Channel memory, which becomes effectively longer at higher data rate, determines the number of bits per line needed to excite the worst-case-pattern. The number of possible input pattern combinations is roughly $2^{(number\ of\ victim/aggressor\ lines\ *\ number\ of\ bits)}$, which leads to long validation time.

There is no guarantee that a limited number of bit-patterns would capture the true worst case eye. Nevertheless, Table 3 shows a set of 17 different bit-patterns which we found to stress DDR3 eye diagram margin to an asymptotic level. We define the eye margin as the minimum among this set of test patterns and results provide a lot of insights to the trend. Even- and odd-mode refers to the phase relationship between victim and aggressive bits. LP and HP denote low-phase and high-phase, respectively. LP<n>_HP<m> refers to <n> bit "0" follow by <m> bit "1". Different LP and HP combinations can inject different frequency spectrums into the system. The highest frequency combination is LP1_HP1, which is a clock pattern.

Fig. 3-5 show that the quality of the eye diagram is highly dependent on input bit pattern used to drive the channel. More than 300ps (~100%) timing margin and 0.3V (~100%) high-side center eye-height differences were observed between the set of bit patterns and traditional clock pattern approach. They also show that larger ODT is more sensitive to bit pattern, due to less well terminated bus. Additionally, we found that higher data rate and capacitive load are more sensitive to stimulus.

Although bit pattern sensitivity varies with ODT, Raw Cards, and speeds, "odd" and "maxfreq" bit patterns (6-13) stress timing and voltage margins considerably. Unequal low phase and high phase patterns impact the slow rising slew rate, asymmetric high/low SSN, and cause horizontally asymmetric eye. Simply using clock pattern to produce eye diagram is not sufficient to get accurate timing and voltage margins. Additionally, often times clock pattern produce horizontally and vertically asymmetric eye. Bit pattern that stress timing margin stongly may not be the same as those that are stressful to voltage margin. These results shred important insights, especially for IO interfaces with building-in training capability, e.g. DDR3,

- The victim and aggressor bit patterns should be independent.
- Random pattern is not necessarily the best or most effective choice.
- Clock pattern is a poor choice to train the bus, e.g. to align strobes

Low power design has strong motivation to use larger ODT. However, validation should be done very carefully as larger ODT is more sensitive to bit patterns as shown in Fig. 5. Flatter channel bandwidth response typically has less variation in eye diagrams to bit streams. Therefore, optimized channel design may help widen the solution space of using larger ODT.

POWER-PERFORMANCE TRADE-OFF WITH ON-DIE-TERMINATION

Table 1 shows that power dissipation is a strong function of DRAM ODT scheme and is roughly constant with data rate. In Table 1, we assume: 2 channels with 1 SO-DIMM per channel (1DPC), MCH ODT of 120 Ohm, 60% bus utilization and 60/40 read/write split. As a design parameter, power can be optimized based on ODT, Raw Card type and speed. Although bit UI (unit interval) shrinks quickly, channel budget does not scale linearly as shown in Table 2. Bit pattern sensitivity analysis of various ODT reveals power-performance trade-off of various ODT schemes. In this section, the worst-case eye of all patterns is used to measure performance. In Fig. 6, we overlapped eye diagrams of all bit patterns for Raw Card A at 1067 MT/s, for various ODT. It appears that the optimal ODT (in this case 120-Ohms) generally has the most horizontally and vertically symmetric eye diagram, which helps increase voltage and timing margins.

Optimal ODT may have different trend for voltage and timing margin. Fig. 7 shows that optimal ODT for maximum timing margin is 60 ohms for Raw Card C at 1067 MT/s, 120 ohms for Raw Card A at 1067 MT/s and Raw Card C at 800 MT/s, no-ODT for Raw Card A at 800 MT/s. Fig. 8 shows that optimal ODT for maximum center eye height is 120 ohms for Raw Card A at 1067 MT/s, and no-ODT for other scenarios. Raw Card C timing and voltage margins are generally ~ 150ps+ and 50~100 mV better than that of Raw Card A, respectively. For the same configuration, 800 MT/s margin is on average 300ps+ and 50~100 mV better than that of 1067MT/s. It implies that Raw Card C is approximately one speed bin higher than Raw Card A if voltage margin is the limiter. At higher data rate the trend is to use stronger ODT (smaller value) as channel timing noise doesn't scale proportially with UI shrink. At 1066 MT/s, RC-A with no-ODT has significant votlage/timing performance degradation while RC-C still has adequate margin.

Fig. 9 shows the voltage margin of three kinds of raw cards from three different SDRAM vendors. It clearly indicates raw card type has a dominant effort on the margin. RC-D is dual-die package thus has the highest capacitive load thus has the worst voltage margin. RC-C has the lightest loading thus margin is in general the best. We did observe significant performance variation for the same raw card type but from different SDRAM vendors. The reason is that JEDEC spec has a wide range to accommodate design and manufacturing variations from many vendors. So this is as expected as SDRAM vendors tune their designs and processes independently. This also give a clear indication that performance, power and ODT may be tuned by carefully selecting raw-cards and vendors, if interoperability (freely swap DIMMs) is not a concern, e.g. in memory down designs where SDRAMs are soldered down on motherboards.

CONCLUSIONS AND FUTURE WORK

High speed memory bus designers are expected to create the most energy-efficient design while meeting performance constraints. Further requirement for robust design in the presence of manufacturing and stimulus variations increases design challenges. We have presented optimal ODT as a very powerful and cost-effective technique to achieve such robust energy-efficient design. Optimal ODT scheme involves a tradeoff among power, timing and voltage margins. The optimal ODT value is influenced by speed, raw card type, SDRAM vendor, platform design and optimization. Our investigation suggests that low-power termination using larger ODT can be scaled to 1066MT/s or higher in mobile 1DPC platform if the right tradeoff is made. Design and validation must account for bus margin high sensitivity to bit-pattern. DDR3 margin is also a strong function

978-1-4244-4447-2/09 $25.00 © 2009 IEEE

of raw card types and vendors. For light loading raw cards (RC-B/C), we may trade-off margins for larger ODT to save power. For heavy loading raw cards (RC-A/D/F), we need to use either strong ODT or scale down frequency to meet performance target. Our future work includes adaptive high speed links validations and robust high speed links optimization.

ACKNOWLEDGMENT

The authors wish to acknowledge the help and support of Alan Hatfield, Ricky Nguyen, Alex Thomas, Warren Barto, Eric Gantner, Nick Bouris, Paul Yang, Ranjan Sahoo, and Shivraj Thakare at Intel Corporation.

REFERENCES

[1] R. Das, S. Ji, S. Peterson, J. L. Chen, C. Pan, "Signal Integrity Impact of Ultra Low Power IO Initiatives," *EPEP*, 2006.
[2] R. Das, J. Iyer, Suchitha V., Y Praveen K., M. T. Tran, S. A. Thomas, S. K. Gupta, and W. Kraipak, "Novel Bus Termination Schemes to Reduce IO Power Consumption on Low Power Intel Small Form Factor Platforms," *ECTC*, 2008.

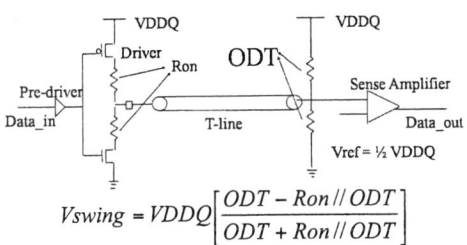

$$Vswing = VDDQ\left[\frac{ODT - Ron\,//\,ODT}{ODT + Ron\,//\,ODT}\right]$$

Fig. 1. MCH-SDRAM signalling scheme.

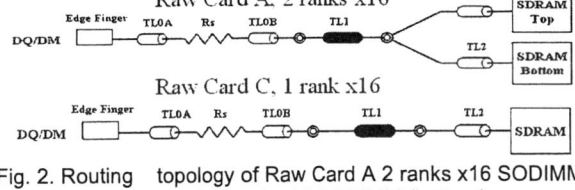

Fig. 2. Routing topology of Raw Card A 2 ranks x16 SODIMM (top) and Raw Card C 1 rank x16 SODIMM (bottom).

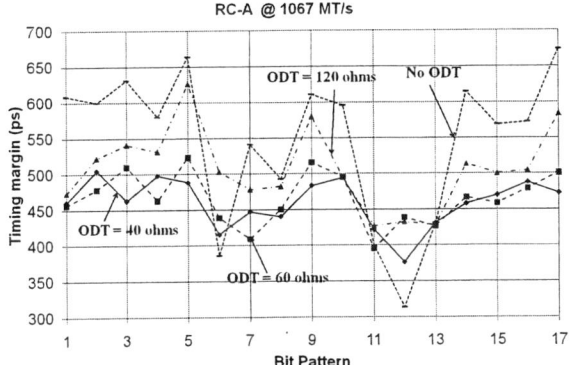

Fig 3. Bit pattern sensitivity of Raw Card A timing margin at 1067MT/s with various ODT.

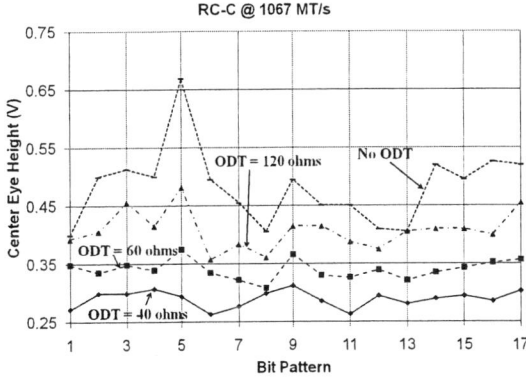

Fig. 4. Bit pattern sensitivity of Raw Card C high side center eye height at 1067MT/s with various ODT.

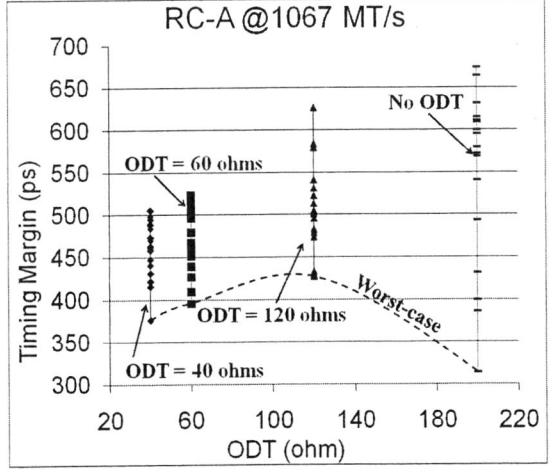

Fig 5. Bit pattern sensitivity of Raw Card A timing margin at 1067MT/s with various ODT.

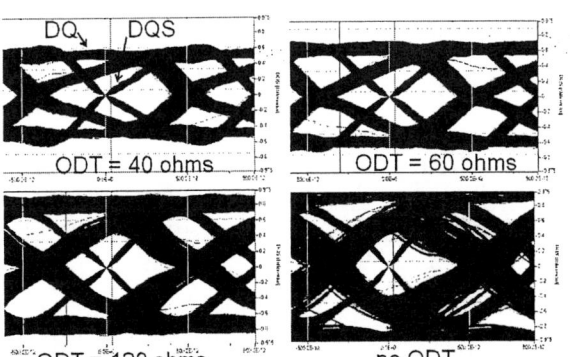

Fig. 6. All patterns eye diagrams of Raw Card A at 1067MT/s with various ODT.

978-1-4244-4447-2/09 $25.00 © 2009 IEEE

ODT (ohm)	800 MT/s	1067 MT/s
40	1.46 W	1.48 W
60	1.21 W	1.24 W
120	1.00 W	1.05 W
infinity	0.90 W	1.00 W

Table 1. Power consumption vs. ODT at various speeds.

Speed (MT/s)	AC threshold (mV)	DC threshold (mV)	UI (ps)	SDRAM tDS+TDH	MCH error (0.34 UI)	Channel budget
800	750+/- 175	750+/- 100	1250	0.18UI (225ps)	0.34 UI (425 ps)	0.48 UI (600 ps)
1067	750+/- 175	750+/- 100	938	0.13UI (125 ps)	0.34 UI (319 ps)	0.53 UI (497ps)

Table 2. JEDEC DDR3 eye-height and eye-width targets show that channel budget scales quickly.

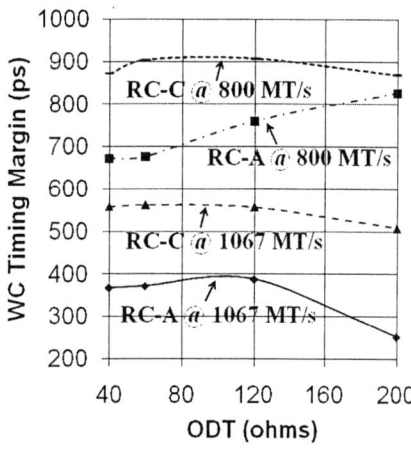

Fig. 7. Worst-case timing margin vs. ODT for various Raw Cards and speeds.

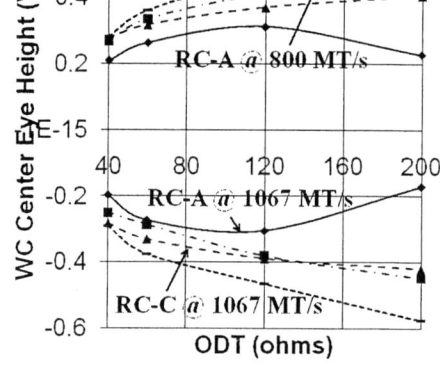

Fig. 8. Worst-case center eye height vs. ODT for various Raw Cards and speeds.

Table 3. Bittern-patterns used to stress system eye margin

No.	Pattern name	Victim	Aggressor
1	random	random	random
2	Even-mode LP1_HP1	010101...	010101...
3	Even-mode LP1_HP2	011011...	011011...
4	Even-mode LP2_HP1	001001...	001001...
5	Even-mode LP2_HP2	00110011...	00110011...
6	Odd-mode LP1_HP1	010101...	101010...
7	Odd-mode LP1_HP2	011011...	100100...
8	Odd-mode LP2_HP1	001001...	110110...
9	Odd-mode LP2_HP2	00110011...	11001100...
10	MaxFreq LP1_HP1	101010...	101010...
11	MaxFreq LP1_HP2	101010...	011011...
12	MaxFreq LP2_HP1	101010...	001001...
13	MaxFreq LP2_HP2	101010...	00110011...
14	Single-bit LP1_HP1	101010...	111111...
15	Single-bit LP1_HP2	011011	111111...
16	Single-bit LP2_HP1	001001...	111111...
17	Single-bit LP2_HP2	00110011...	111111...

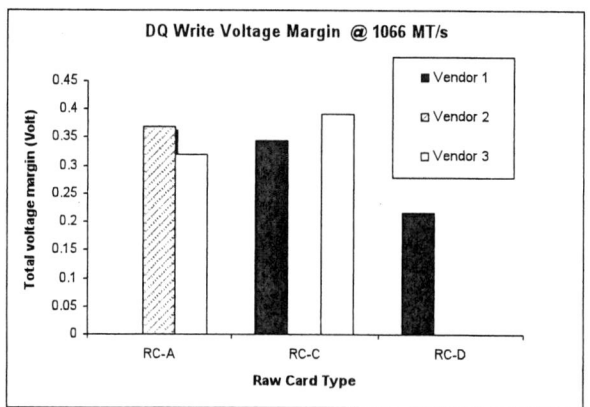

Fig. 9. Voltage margin vs. different vendor and Raw Card types.

An LCP Package Model for Use in Chip/Package Co-Design of an X-band SiGe Low Noise Amplifier

Chung Hang John Poh, Tushar K. Thrivikraman, Swapan K. Bhattacharya, Chad E. Patterson,
John D. Cressler, and John Papapolymerou

School of Electrical and Computer Engineering, 777 Atlantic Drive, NW,
Georgia Institute of Technology, Atlanta, GA 30332-0250, USA
E-mail: johnpoh@gatech.edu

Abstract

Abstract— We present the modeling of a liquid crystal polymer (LCP) package for use with an X-band SiGe HBT Low Noise Amplifier (LNA). The package consists of a 2 mil LCP laminated over an embedded SiGe LNA, with vias in the LCP serving as interconnects to the LNA bondpads. An accurate model for the packaging interconnects has been developed and verified by comparing to measurement results, and can be used in chip/package co-design.

I. INTRODUCTION

Understanding the electrical characteristics of a high-frequency package and its interconnects is important in order to optimize circuit performance. At X-band (8 – 12 GHz), the parasitics due to the package and interconnects can lead to significant degradation of electrical performance at these high frequency and hence need to be explicitly taken into consideration during the design stage. One way to model the parasitics of the interconnections is to perform a full electromagnetic (EM) simulation for the package; however, this is time consuming and increases the complexity for circuit designers. Therefore, it is highly desirable to implement an equivalent circuit model for the package that can be used in circuit design kits for circuit optimization in order to achieve better overall system performance.

In a packaged monolithic microwave integrated circuits (MMIC), the interconnections are usually made using wire bonding [1] which tends to add large impedances [2], degrading circuit performance. Liquid crystal polymer (LCP) has been used for the MMIC packaging because of its excellent high frequency electrical properties, such as low dielectric constant and low loss tangent [3]-[5]. The package design here consists of a 2 mil LCP which is been laminated onto the top surface of a SiGe HBT LNA die. The CPW lines of the LCP are connected directly to the bondpads of the SiGe die through 50 μm via interconnections. The details of the LCP fabrication are discussed in [6]. Using this approach, the large parasitics usually introduced by the bondwires interconnects in a conventional package are greatly reduced.

The SiGe LNA is shown in Figure 1 was fabricated in a commercially-available 200GHz, 0.13 μm BiCMOS SiGe technology and the measured on-wafer results of the LNA is shown in Table I. The details of the LNA design and measured results are discussed in [7].

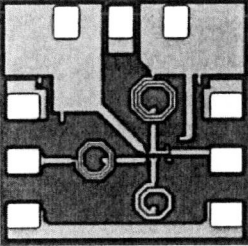

Figure 1. Photograph of the X-band SiGe LNA die.

TABLE I
SiGe LNA SPECIFICATIONS

Frequency (GHz)	9.5 – 10.5
Vcc (V)	1.5
Gain (dB)	10
Return loss (dB)	>10
Noise Figure (dB)	2
Power dissipation (mW)	< 2

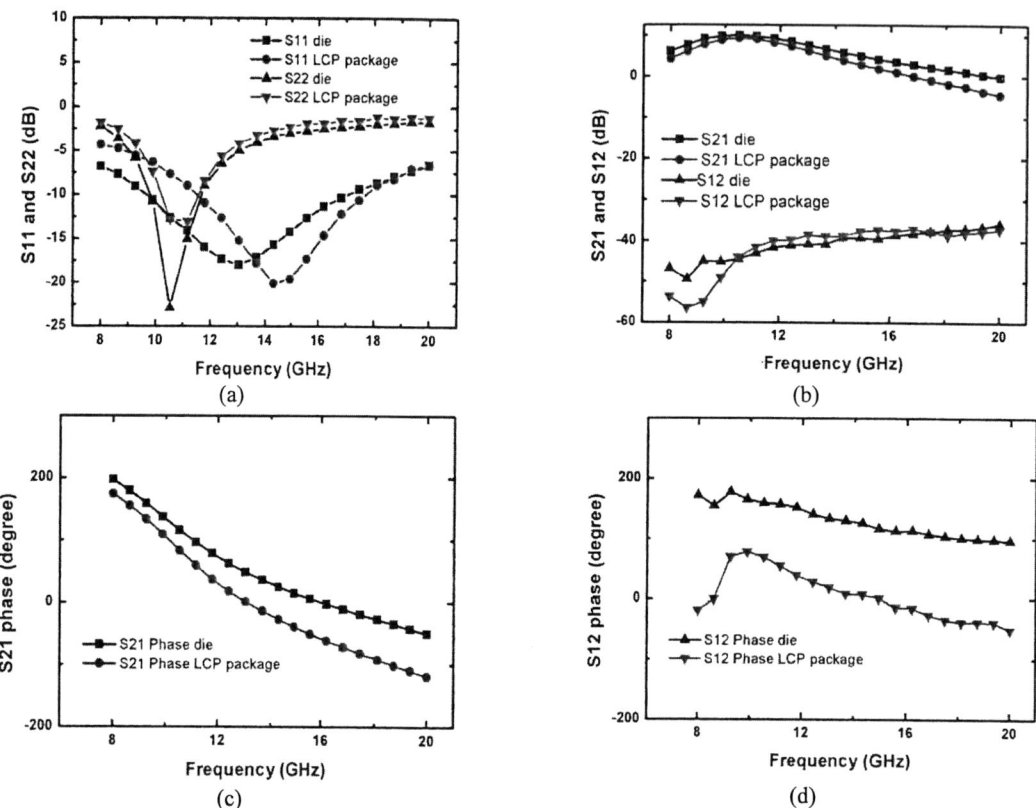

Figure 2. Measured S-parameter results, both before and after LCP packaging: (a) S11 and S22, (b) S21 and S12, (c) S21 phase and (d) S12 phase.

In Figure 2, the LCP-packaged SiGe LNA was measured and was compared with the on-wafer measurement result.

It can be seen that after the SiGe LNA had been packaged, there is a shift in the input matching and a slight degradation of less than 1 dB in the magnitude of S21. All these changes are due to the mismatch introduced by the package parasitics. In the next section, the effects of the LCP package on the LNA performance is studied and modeled.

II. SIMULATIONS AND PARASTIC EXTRACTION

The package was divided into five components, as shown in Figure 3 as in [8].

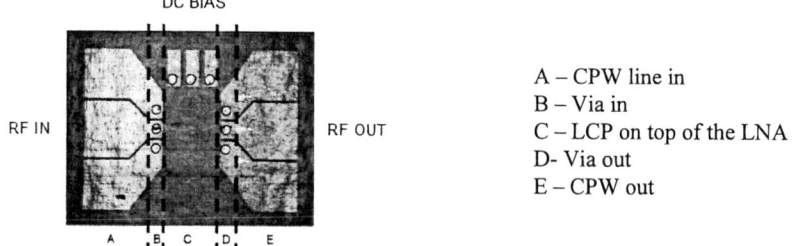

A – CPW line in
B – Via in
C – LCP on top of the LNA
D- Via out
E – CPW out

Figure 3. Top view of the LCP packaged SiGe LNA.

Each of the components was simulated in Ansoft HFSS and the equivalent circuit models were derived based on the simulated S-parameter results from HFSS. Agilent's Advanced Design System (ADS) was then used to optimize the lump elements in the equivalent circuit model to correlate the HFSS simulation results and the models. The CPW line of the LCP package shown in Figure 4 was represented by the circuit topology shown in Figure 5. In Figure 5, the L_{cpw} represents the self-inductance of the CPW line, while C_{cpw} represents the coupling between the signal and the ground pad in the CPW transmission line

978-1-4244-4447-2/09 $25.00 © 2009 IEEE

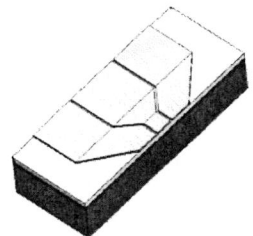

Figure 4. CPW line of the LCP package.

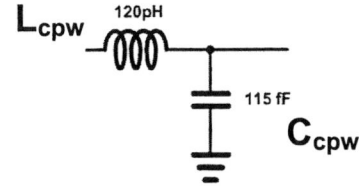

Figure 5. Equivalent circuit model for the CPW line.

As shown in Figure 6, the package consists of vias and LCP on the top surface of the SiGe LNA die. The equivalent circuit model for the vias [9] and the LCP are shown in Figure 7 and Figure 8, respectively. In Figure 8, the resistance R1 models the dielectric loss in the LCP material, both C1 and C2 model the coupling to the ground vias and, C3 models the coupling between the input and output vias.

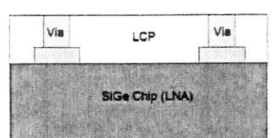

Figure 6. Cross-sectional view of the vias in the LCP package.

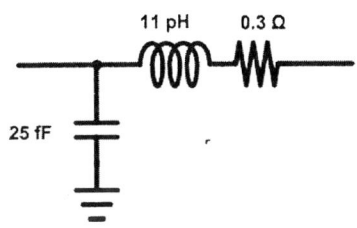

Figure 7. Equivalent circuit model for the interconnection via.

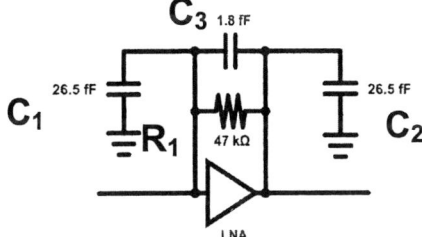

Figure 8. Equivalent circuit model for the LCP on top of the SiGe LNA die.

III. MODEL AND MEASUREMENT VERIFICATION

The equivalent circuit models for the individual components in the LCP package in Figure 9 were extracted and then added to the SiGe LNA for re-simulation (Figure 10) to verify the model and to compare with measured results.

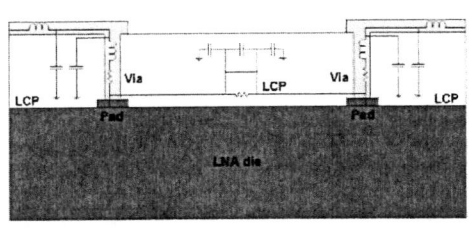

Figure 9. Cross-sectional view of the packaged SiGe LNA with the parasitic models.

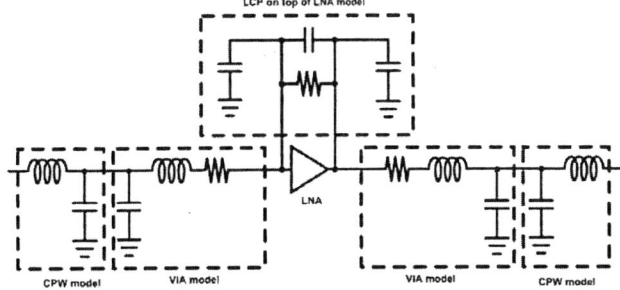

Figure 10. Schematic view of the packaged SiGe LNA with the parasitic models used for circuit simulation.

Figure 11 shows the measured results of the packaged LNA, together with the simulation results of the LNA with the LCP model included. It can be seen from the comparison that the simulations of the LNA which include the package model give good agreement with measured data.

978-1-4244-4447-2/09 $25.00 © 2009 IEEE 203

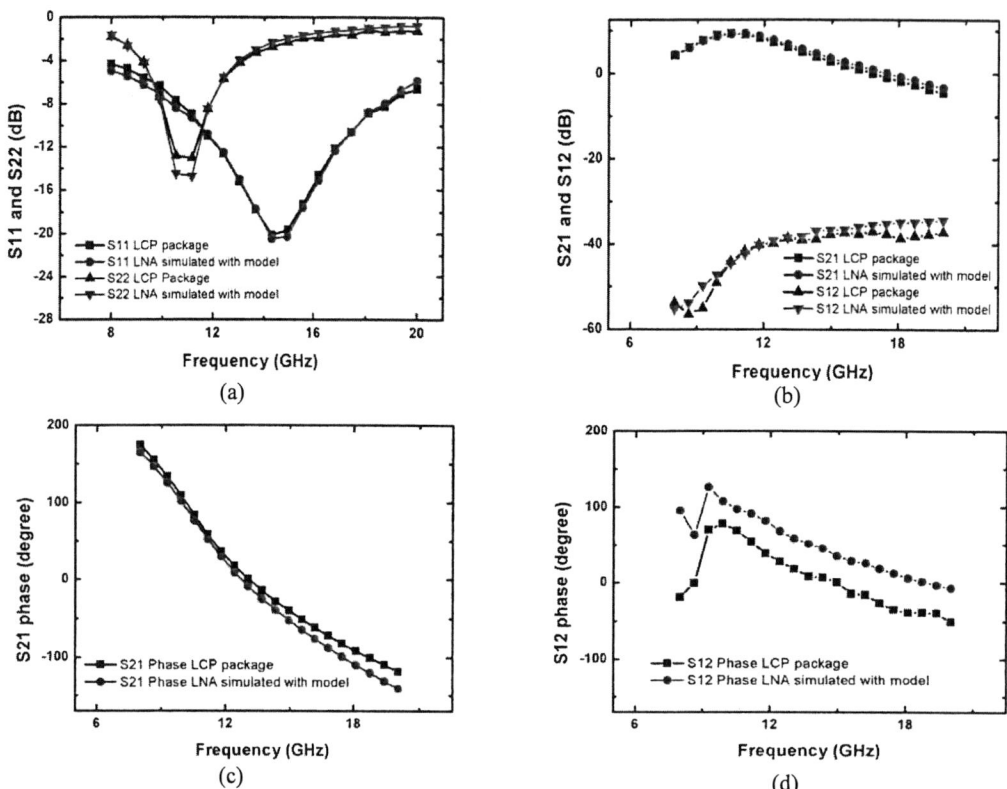

Figure 11. Measured and modeled S-parameters of the packaged LNA. (a) S11 and S22, (b) S21 and S12, (c) S21 phase and (d) S12 phase.

IV. SUMMARY

This paper has presented a model for an LCP packaged SiGe HBT LNA, and simulation results match closely with measurements. These results imply that the LCP package has a significant impact on the LNA performance due to its parasitics, and therefore it is important that all these parasitics be considered during the circuit design stage. With this LCP package model, circuit designers will be able to performance more accurate optimization for envisioned MMIC applications which require the use of embedded SiGe die in a LCP package matrix.

ACKNOWLEDGMENT

This work was supported by NASA under grant #NNX08AN22G. We are grateful to the GTRI NASA radar team for their contributions.

REFERENCES

[1] A. Sutono, N. G. Cafaro, J. Laskar and M. M. Tentzeris, "Experimental Modeling, Repeatability Investigation and Optimization of Microwave Bond Wire Interconnects", *IEEE Trans. on Advanced Packaging,* vol. 24, issue 4, pp. 595-603, Nov. 2001.
[2] H. Y. Lee, "Wideband characterization of typical bonding wire for microwave and millimeter-wave integrated circuits," *IEEE Trans. Microwave Theory Techniques,* vol. 43, pp. 63-68, Aug. 1994.
[3] D. C. Thompson, O. Tantot, H. Jallageans, G.E. Ponchak, M. M. Tentzeris and J. Papapolymerous, "Characterization of liquid crystal polymer (LCP) material and transmission lines on LCP Substrates from 30 to 110 GHz," *IEEE Trans. on Microwave Theory and Techniques,* vol. 52, no.4, pp. 1343-1352, Apr. 2004.
[4] D.C. Thompson, N. Kingsley, G. Wang, J. Papapolymerous, M. M. Tentzeris, "RF Characterization of thin film liquid polymer (LCP) packages for RF MEMS and MMIC integration," *IEEE MTT-S International Microwave Symposium Digest,* pp. 857-860, June 2005.
[5] D. C. Thompson, M. M. Tentzeris and J. Papapolymerous, "Packaging of MMICs in multilayer LCP substrates," *IEEE Microwave and Wireless Components Letters,* vol.16, no.7, pp. 410-412, July 2006.
[6] C. E. Patterson, T. K. Trivikraman, S. Bhattacharya, C. H. Poh, J. D. Cressler and J. Papapolymerou, "Organic wafer scale packaging for X-banc SiGe low noise amplifier", *39th European Microwave Conference,* Oct. 2009.
[7] T. K. Thrivikraman, W. M. L. Kuo, J. P. Comeau, A. K. Sutton, J. D. Cressler, P. W. Marshall and M. A. Mitchell, "A 2 mW, sub-2dB noise figure, SiGe low-noise amplifier for X-band high-altitude or space-based radar applications," IEEE RFIC Symposium Digest, pp. 629-632, July 2007.
[8] H. Wada, C, Makihara and C. Park, "High frequency package design technology using s-parameter synthesize method," *Wireless Communication Conference Digest,* pp. 151-155, Aug. 1997.
[9] A. Pham, J. Laskar, V.B. Krishnamurthy, H. S. Cole and T. Sitnik-Nieters, "Ultra low loss millimeter wave multichip module interconnects," *IEEE Trans. on Components, Packaging and Manufacturing Technology- Part B,* vol. 21, no. 3, pp. 302-308, Aug. 1997.

On Adding Metalization to Improve Via Performance on PCBs

Albert E. Ruehli, Xiaoxiong Gu and Mark B. Ritter
IBM T. J. Watson Research Center, Yorktown Heights, NY 10598, USA
{ruehli xgu mritter} @us.ibm.com

Abstract

The design of printed circuit boards evolves continuously in density, price and performance. In this short paper, we propose an approach where the performance of a via is altered by adding additional metalizations to enhance the ground return path for signals.

I. INTRODUCTION

A continuous change can be observed in the design processes for printed circuit boards (PCBs) and other package structures with the ever increasing density, decreasing price and higher performance requirements for electronic systems. The modeling of signal paths along via interconnects has in recent years been the source of much research [1–7]. The design improvement of the vias can contribute considerably to the reduction of loss, distortion and crosstalk of high speed signals on PCBs. This work has lead to multiple improvements like the placement of several ground vias near each signal via.

The optimization of the vias such that the discontinuities are minimized is an important issue. For example, in our earlier work [6] we considered the computation of the characteristic impedance of a via using both the transmission (S_{21}) and the reflection (S_{11}). A via optimization routine can be used to adjust the characteristic impedance of the via by changing its physical structure. However, it is difficult to match the impedance in a wide frequency range due to the frequency dependent nature of via impedance as seen in [6]. In this paper, we propose an approach where the signaling performance of the vias is altered by including additional specific metalizations in strategic locations. Here, we not only adjust the via parameters, but also add additional metalizations to enhance the ground return path and to improve the signaling performance by having a more constant broadband via impedance. We designed two modeling experiments to study the application of this method and to demonstrate its region of validity. The first example we show in the paper emulates the geometries typically available in a high-end PCB, with many closely located ground planes and ground vias next to a signal through via. The results show that a designed via with enough ground planes performs well so that no additional metalization is needed.

On the other hand, low-cost PCB designs may have less ideal ground return paths. For this reason, we considered the geometry shown in the second example where fewer ground planes and ground vias are available. We also assumed thicker dielectrics between the planes. In this case, additional metalization can be used to improve the signaling performance of the via by providing a better impedance match, as well as by increasing the inter-plane capacitive coupling near the signal via such that the return path is enhanced.

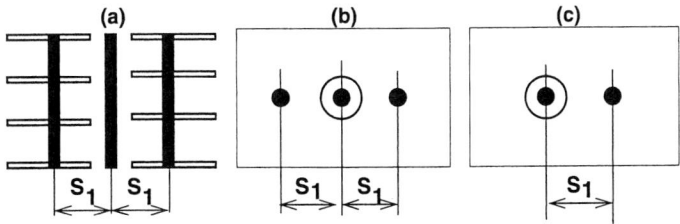

Fig. 1. (a): Side view of a via with two ground via (b) Top view. (c): Top view of a via with one ground via.

II. MODELING EXPERIMENTS

Today we can easily validate proposed designs with numerical electromagnetic solvers. In this paper, we used a 3D general purpose electromagnetic solver (HFSS^{TM}) and a 3D quasistatic extraction tool (Q3D^{TM}) to analyze our designs by computing via loss and via capacitance, respectively. As shown in Fig. 1, we consider two modeling experiments, i.e., one signal via with two adjacent ground vias which are shown in (a) and (b), and one signal via with one ground via shown in (c). In both cases, we used a dielectric with an $\epsilon_r = 4.4$ and a $\tan \delta = 0.02$. For each experiment, we first simulated the structure to obtain its frequency response (transmission and reflection) in terms of scattering parameters with and without additional metalization on the ground return path. We then extracted the characteristic via impedance Z_0 from the simulation data using method presented in [6]. In general, we would like to limit the frequency dependence of Z_0 as much as possible.

A. A high performance via with two ground vias

The first structure we modeled has two ground vias as shown in Fig. 2. There are totally 9 reference planes including 5 ground planes and 4 voltage planes in an alternating fashion. The plane size is 250-mil by 250-mil. The vertical plane-to-plane spacing is 10 mils. The distance between the centered signal via and two adjacent ground vias is 32 mils. The via drill is 4.4 mils and the antipad radius is 15 mils. We included additional ground circular pads to the ground vias between every two adjacent planes as shown in Fig. 2(right). Besides this, we also considered a coax like structure where the grounding pads are extended to connect both ground vias as shown in Fig. 5(left). For both vias the ground pad size is 18 mils in radius with a 15-mil coax hole radius.

Fig. 3 illustrates the return loss (S_{11}) of the signal via on the left and the insertion loss (S_{21}) on the right. The comparison is between a conventional via structure (Fig. 2(left)), one with circular pads attached to the ground vias (Fig. 2(right)), and one with the ground coax-like joint pads (Fig. 5(left)). Notice that the reflection is very small, i.e., lower than -30dB below 15 GHz for the configuration without additional metalization, indicating that the close-by ground vias and ground planes already provide an excellent return path. The plot of the characteristic impedance Z_0 of the signal via also confirms such a good via performance. Fig. 4(left) illustrates that without any additional metalization, the impedance Z_0 is well controlled in the range between 50.5ohms and 52.5ohms up to 20GHz. Adding more metalization on the ground vias slightly increases the capacitance between the signal via and ground, thus pulling down the impedance curve. In this particular case, the additional metalizations also show marginal effects on reducing the insertion loss of the signal via, e.g., about -0.2 dB reduction at 20 GHz as shown in Fig. 3(right).

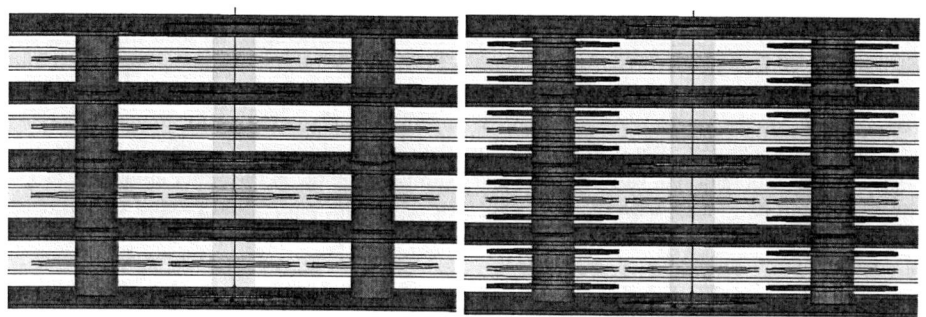

Fig. 2. Left: Signal via with two ground vias.   Right: Two grounds with circular paths added.

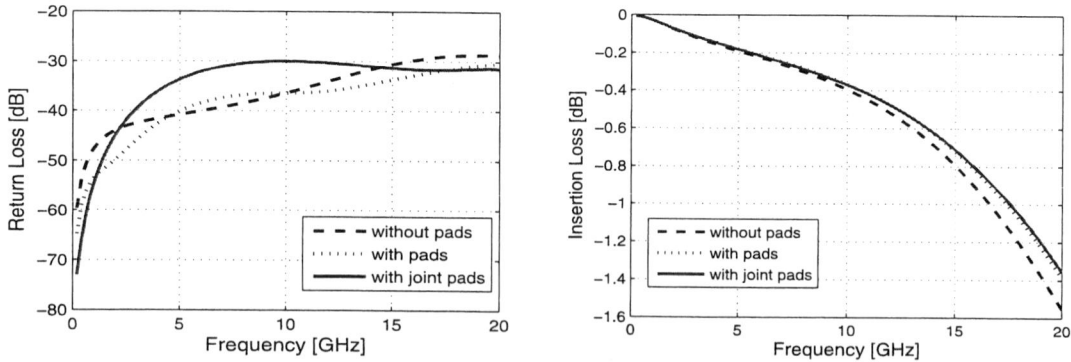

Fig. 3. Left: Return loss S_{11} for the signal via with two ground vias.   Right: Insertion loss S_{21}.

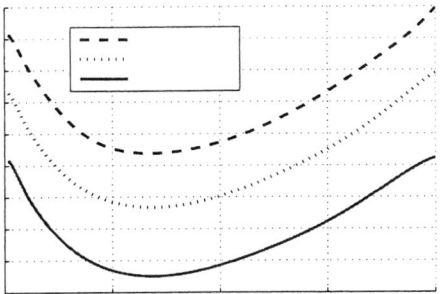 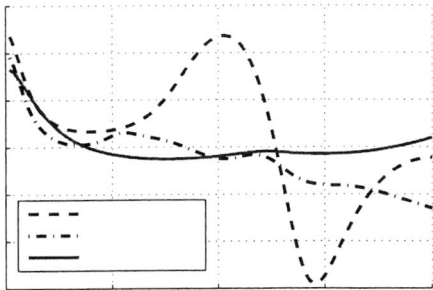

Fig. 4. Characteristic impedance: with two ground vias (left) and with one ground via (right).

B. A low performance via with one ground via

The second modeling experiment shown in Fig. 6 has only one neighboring ground via at a distance of 32 mils from the signal via. The plane to plane spacing is increased to 20 mils. The plane size remains the same as 250-mil by 250-mil. As for the layer stack, only the top and bottom planes are considered as ground planes. The three internal planes are voltage planes which are disconnected to the ground vias by antipads with a radius of 15 mils. Again, we augmented the basic model with additional metalization attached to the ground via, i.e., the circular pads shown in Fig. 7(right) and the coax-like pads shown in Fig. 5(right), as an attempt to improve the via performance.

Fig. 4(right) illustrates the signal via impedance Z_0 as a function of frequency. It is clear that the additional metalizations in this case enhance the ground return path over a wide frequency range so that the frequency dependence of Z_0 gets reduced. Notice that without metalization on the ground vias, the variation of the via impedance is much greater, i.e., from 40ohms to 67ohms. In contrast, with the coax-like ground pads, the via impedance curve tends to saturate especially at high frequencies above 5GHz. Table I lists the capacitance values of the voltage planes and the signal via with respect to the ground extracted by the quasi-static EM tool for the case with and without the additional metalization on the ground via. The results show that the capacitance of the signal via increases by about 5.6 percent with circular pads and 8.2 percent with coax-like pads. The capacitance of the voltage planes increases by about 2.9 percent with circular pads and 5.2 percent with coax-like pads. Simulation results of the reflection and transmission are plotted in Fig. 7. Compared with the previous two-ground via case, the return loss and insertion loss are much higher (e.g., S11=-18dB and S21=-1.25dB versus S11=-38dB and S21=-0.4dB at 10GHz) because of the worse ground path as well as the thicker dielectric layers. With additional metalization, the reflection loss and insertion loss also decrease for most of the frequencies up to 20GHz, e.g., about 2.5dB and 0.2dB reduction at 10GHz on S11 and S21 respectively for the case with coax-like additional metal pads on the ground via. For the signal via with and without circular pads in Fig. 6, we added microstrip lines (125-mil long, 6-mil wide, 1-mil thick) to connect to the signal via at the top and the bottom over a 5-mil thick dielectric. Simulation results based on an internal link simulator show that with circular pads, the eye opening of pseudorandom data through the trace-via-trace structure increases from 81.5 percent to 88.3 percent at 25 Gb/s.

III. Conclusions

This paper points towards a new direction for the improvement of the performance of PCB vias. The results show the performance of a via can be impacted with additional metalization on the ground return path. Specifically, we show how the performance of a via can be improved in terms of its characteristic impedance and losses with additional metal pads attached to a ground via. Research in this direction can improve other parts of a PCB or package. For example, the vias with the additional ground pads also provide better coupling between voltage planes and the ground planes which could reduce the impedance of the power delivery circuit. Further work which includes hardware measurement is currently in progress.

References

[1] Q. Zheng, E. Yang, and M. A. Tassoudji. Modeling and analysis of vias in multilayered integrated circuits. *IEEE Transactions on Microwave Theory and Techniques*, pages 206–215, February 1993.

[2] H. Chen, Q. Li, L. Tsang, C.C. Huang, and V. Jandhyala. Analysis of a large number of vias and differential signaling in multilayered structures. *IEEE Transactions on Microwave Theory and Techniques*, 51:818–829, March 2003.

[3] C. Schuster, Y. Kwark, G. Selli, and P. Muthana. Developing a physical model for vias. In *Proc. of DesignCon Conference*, Santa Clara,CA, February 2006.

978-1-4244-4447-2/09 $25.00 © 2009 IEEE   207

[4] Y. Jeng, W. Guo, G. Shiue, C. Lin, and R. Wu. Reflecitionless design for differential-via transitions using neural network-based approach. In *Digest of Electr. Perf. Electronic Packaging*, pages 143–146, Atlanta,GA, October 2007.

[5] X. Gu and M. Ritter. Application of Foldy-Lax multiple scattering method to via analysis in multilayer printed circuit board. In *Proc. of DesignCon Conference*, Santa Clara, CA, February 2008.

[6] X. Gu, A. E. Ruehli, and M. Ritter. Impedance design for multilayer pc-boards. In *Digest of Electr. Perf. Electronic Packaging*, San Jose,CA, October 2008.

[7] B. Wu and L. Tsang. Signal integrity analysis of package and printed circuit board with multiple vias in substrate of layered dielectrics, in press. *IEEE Transactions on Advanced Packaging*, 2009.

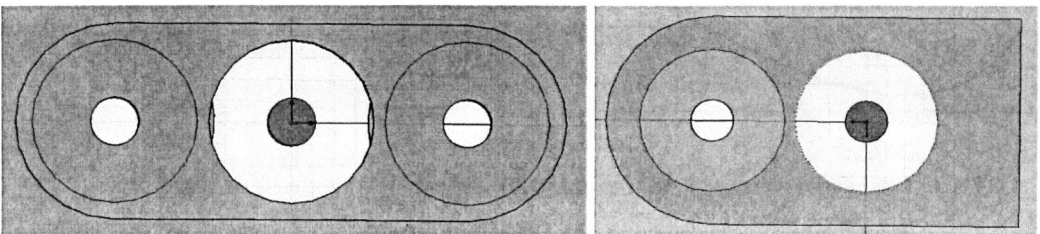

Fig. 5.   Left: Coax type ground to ground connection.   Right: Single ground coax type connection.

TABLE I

VIA AND PLANE CAPACITANCE VALUES EXTRACTED BY Q3D^{TM}

	signal via	top voltage plane	middle voltage plane	bottom voltage plane
no additional metalization	0.366pF	6.79pF	6.79pF	6.74pF
with circular pads on the ground via	0.388pF	6.94pF	6.93pF	6.92pF
with coax-like pads on the ground via	0.396pF	7.13pF	7.14pF	7.12pF

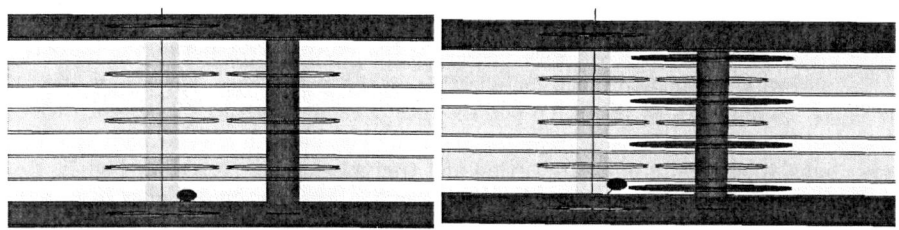

Fig. 6.   Left: Simple one ground via model.   Right: Improved via model with circular ground pads.

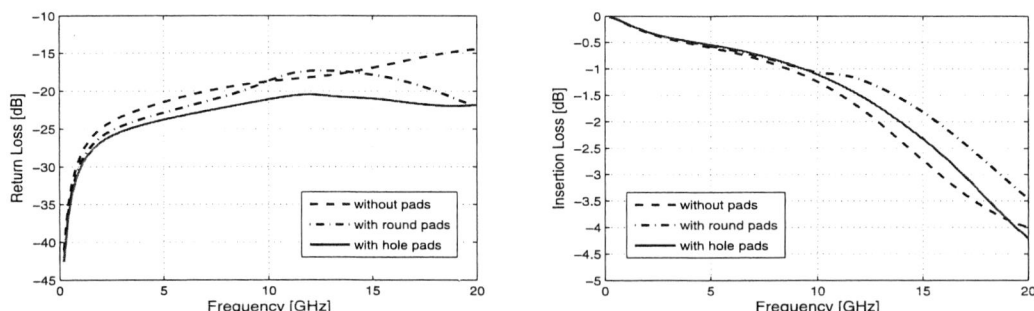

Fig. 7.   Left: Return loss S_{11} for the signal via with one ground via.   Right: Insertion loss S_{12}.

Next Generation I/O Power Delivery Design through SIPD co-analysis & Comprehensive Platform Validation

Yee Hung See Tau[1], Marcus Chan[2]
Intel Microelectronics (M) Sdn. Bhd.
Penang Design Center (PDC), Bayan Lepas FTZ Phase 3, 11900 Penang, Malaysia
yee.hung.see.tau@intel.com; marcus.chan@intel.com

Abstract

This paper illustrates different approaches in solving I/O power delivery noise issues and walk through pre-silicon design solution. It covers circuit and architectural design influence, on silicon and on board decoupling solutions selection and package and platform design optimization. SIPD co-simulations and appropriate package return path are the main topics to discuss and certainly impedance (Z) profile and transient analysis will be performed to observe the noise frequency and accurately address the root cause. All the above will be verified through comprehensive validation data.

1. Introduction

Design of a solid, economically sensible power delivery system for an IO interface to ensure that the SSO is within minimal the allowable levels required by the IO circuitry is not a simple task. Single ended IO interface 7:1:1 bump array design is not an optimize power delivery design if package capacitor is required and it is difficult to provide good return path for the package capacitor. Ground bumps which provided return path are located at inner of the bump array which enables signals to escape through top and 2nd layer of the package. This hybrid signal design had created long VSS fingers which contributed high loop inductance to package capacitor return path. In other hand, top layer ground plane has connections constraint to inner row of VSS bump which is critical to overall return path. Bad return path design will increase total loop inductance of power delivery network (PDN) and causing high VCC noise. Silicon cost reduction becomes a focus by avoiding bump array design change from 7:1:1 to 4:1:1 for single ended IO interface and it will be causing die size grow. Full blown analysis is carried out by including circuit transistor model, signal integrity model and power delivery network in one single network. It is new era of mind set for power delivery methodology. SSO timing analysis is the most important factor to decide good power delivery design.

2. Power Delivery analysis

Low speed memory uses a highly multiplexed 8-bit bus to synchronously transfer command, address and data. It operates at 83MHz, there are all together 16 channels of data lines and 2 clock channel as the buffer strength. Worst case analysis assumption is assuming all channels turn on at the same time. In power delivery analysis, it starts up with package & motherboard model extraction which will be used as power delivery network (PDN) based on the 3D simulation tool. The current profile extraction is done without SI model included which is pessimistic. Typically the Iccmax for one buffer is known or can be approximating based on the buffer impedance, and the Vcc and the target trace impedance as calculated as below

Iccmax per buffer = Vccmax/ (Z buffer + Zo of trace) (a)

Time & frequency domain are being carried out with Spice tool by hooking up the current profile to silicon of at the extracted PDN. Seeing the AC noise ~ 1.8Vp-p with this pessimistic assumption as shown in Figure 1.

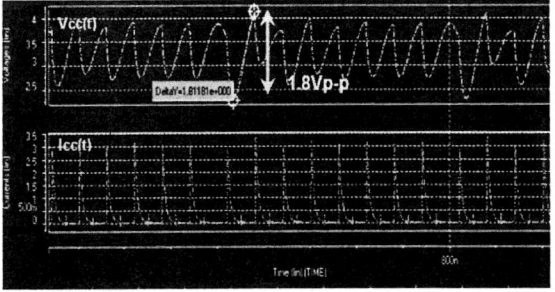

Figure 1: VCC noise using pessimistic Icc(t)

3. Determining power usage model in SIPD co-analysis

An appropriate SIPD co-simulation is needed and included SI model in full transistor simulation in real system environment. Other than that, correct usage model needs to redefine rather than assuming all data lines turn on at the same time. Realistic usage model is assuming data bits toggle randomly, there will be multiple buffers turn on and follow by 2 CLKs with 90 degree phase shift. To generate the 16 channels of current profile based on this single channel circuit test bench, 15X of multiplier is used to duplicate total of 16 channels current & connect the receiver side to lossless transmission line & Cload connected to each channel. The PDN will then be connected to the circuit deck accordingly to the die input node & the SI deck is connected to output node. Appropriate SIPD model connection is currently fulfill the formula (a) which represents the Icc(t) behavior profile as previous power delivery analysis is not included SI model which is pessimistic while generating current profile. Figure 2 is showing the full blown test bench including SIPD model.

Besides that, it is common that lossless transmission line is used in the past for SIPD co-analysis. It is a common practice where lossless transmission line (T-Line) model and perfect capacitors value are used in transistor model analysis. These assumptions, however, are found necessary to be changed in order to achieve better correlation with validation data. Lossless T-line model will draw more current compare to lossy T-line model, causing the power delivery droop appears higher. Additional decoupling capacitors will be required to dampen the noise due to this assumption. In the mean time, although the termination specification calls for 40pF termination per channel, this shall not be translated to putting 40pF value in the model. The proposal here is to use an IBIS receiver model as the termination

With this methodology, seeing the AC noise has reduced to 1.3V for the interface which operates at 1.8V & data timing push out (perfect source data signal vs. with PDN data signal) of 717ps in Figure 3.

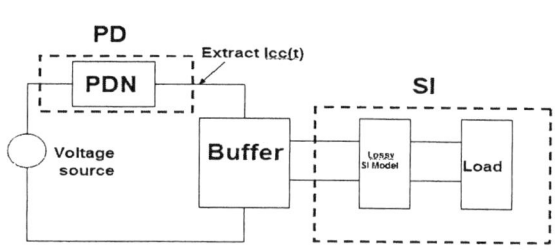

Figure 2: Appropriate SIPD model represent Icc(t) behavior profile

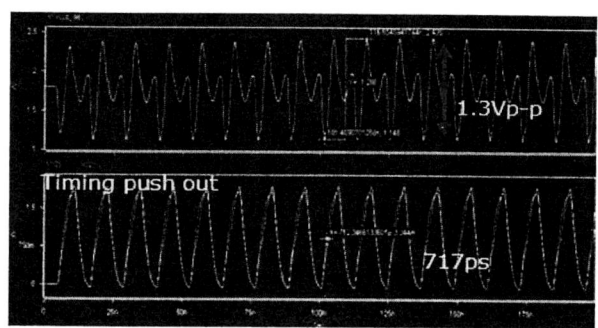

Figure 3: Vcc noise in full blown analysis including SIPD model

4. Optimization of Power Delivery Design

By doing the FFT based on the noise & current profile, seeing that sensitive frequencies are at about 80MHz & above. Package layout improvement is needed in order to improve package loop inductance and it also help to move the resonance frequency which is currently park at 83MHZ, which is critical for power delivery. Most of the VCC noise is contributed by the resonance noise at 83MHZ and it is proven in the FFT analysis is Figure 4

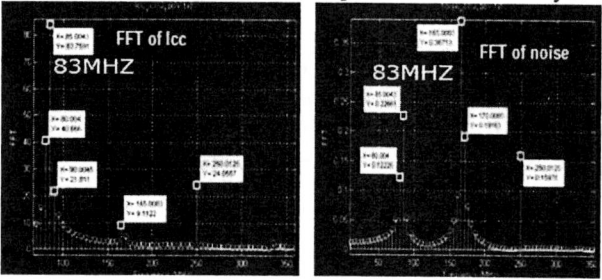

Figure 4: FFT analysis for both Icc(t) and Noise profile

Significant change at package layout return path is needed in order for package capacitor to provide effective/shortest current flow from die to package capacitor and vice versa. Package return path is pretty poor at top layer and 2nd layer from the DSC to die level if it is design incorrectly on the bump out assignment at the first place. Poor return path may also cause by the longer VSS fingers for giving way to signal routing at top layer. Figure 6 and 7 showed the before and after package return path improvement for package capacitor.

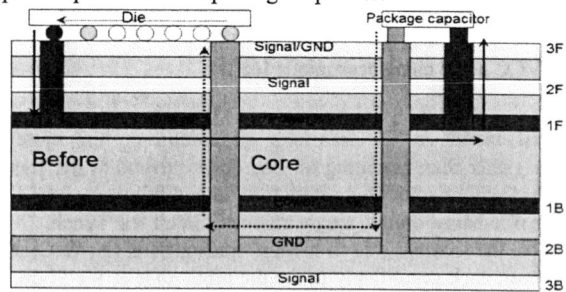

Figure 6: Before package return path improve

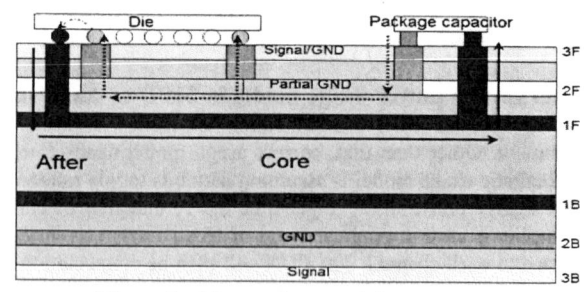

Figure 7: After package path improve

978-1-4244-4447-2/09 $25.00 © 2009 IEEE

With those improvements being done on package layout, seeing the VCC AC noise is ~ 422mV, reduced > 200% from the 1.3Vp-p before improving package return path. Resonance frequency had move away from operating frequency due to lower loop inductance after improving package return path. Latest package layout is able to push resonance frequency from 166MHZ to ~300MHZ as shown in the Figure 8.

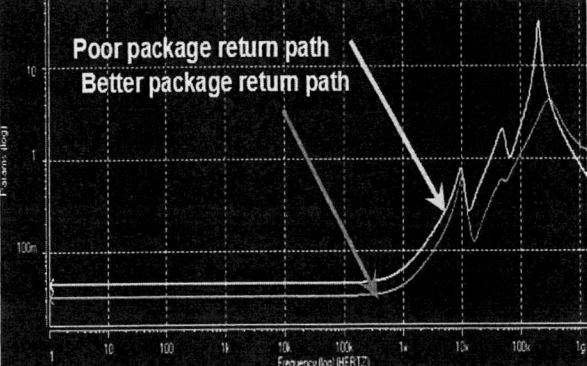

Figure 8: Impedance profile comparison between before and after package improvement.

5. Silicon on die modeling

To further reduce VCC noise, suitable silicon resistance connection need to redefine in order to reduce resonance peak and better loop inductance as show in Figure 9. Different values of R-connect had been analyzed and found that 0.1ohm is most suitable. Clearly seeing the inductance at 80MHz & above is significantly reduced with R-connect 0.1ohm.

By implemented appropriate SIPD co-analysis, return path improvement for package capacitor and proper R-connect value, VCC noise has been reduced to 283mV. The SSO is measure to find the relative delay difference between rising/falling edges of data signal and clock (90 degree phase shifted) due to the power delivery noise. Timing push out is improved and meeting timing spec with lower VCC noise as shown in Figure 10.

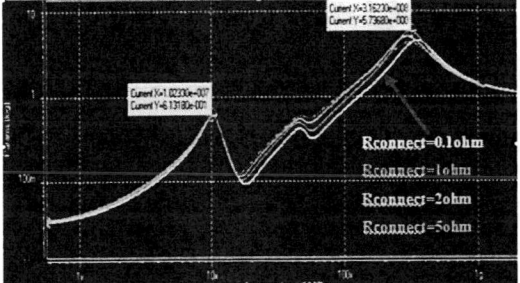

Figure 9: Different R-connect comparison in Z profile   **Figure 10**: SSO timing push out result vs AC noise

6. Lab Validation

Silicon level measurement is carried out through FIB (Focused Ion Beam) process for probing noise measurement purpose. The measurement is taken with the motherboard booted, interface is stressed with suitable software under different worst case bit pattern.

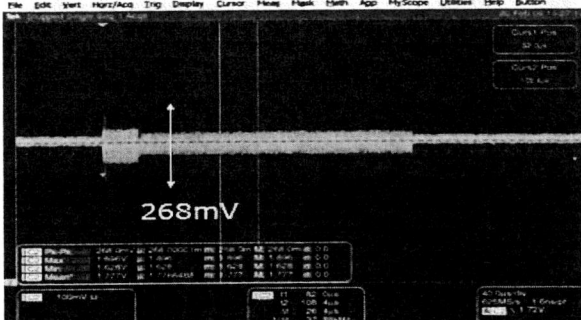

Figure 11: On-die noise measurement

Figure 11 showed 268mVp-p on-die noise and this correlates well with the predicted simulation data of 283mV. Zooming into the worst peak to peak noise to get the frequency will result in 83MHz. This result can be predicted through the Impedance

978-1-4244-4447-2/09 $25.00 © 2009 IEEE   211

profile whereby 83MHz is the sensitive frequency that is operating in and lower VCC noise can be achieved with reducing of the resonance peak.

The validation result proved that in order to reduce the VCC noise, shifting power delivery network resonance to a non-critical frequency by improving package loop inductance or adding decoupling capacitors on the package are needed. This will shift the resonance peak away from the sensitive frequency of the IO design and thus reducing VCC noise that will eventually affect the jitter and SSO timing specifications.

In this particular IO Power delivery design a package decoupling capacitor is recommended to reduce VCC noise. A simple experiment conducted in the lab which is to remove package capacitor to see the effectiveness of it.

Figure 12 showed that with the removal of the package capacitor, VCC noise measured had increased from 268mVp-p to 380mVp-p. There is an increase of 122mV which makes this interface failed SSO timing specifications. Again, it proved that higher VCC noise will directly influence timing performance of the interface. This significant increase of VCC noise caused malfunction of the device and reconfirmed that the package capacitor is important for this IO power delivery analysis. Other than probing at silicon level for noise measurement, signal probing is done at the device for package capacitor removal. Signal is distorted and doesn't reach full voltage swing with the removal experiment as shown in Figure 13.

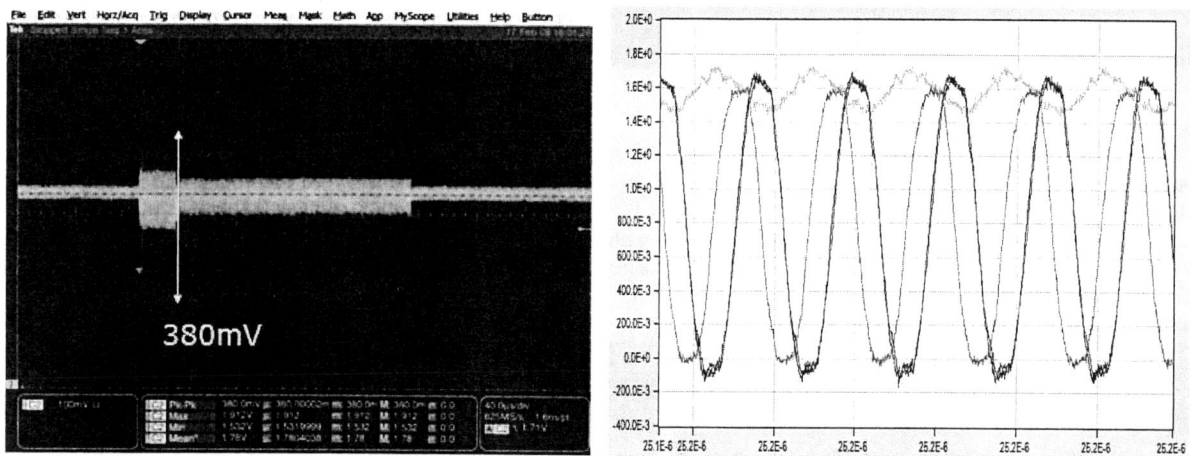

Figure 12: On-die noise measurement with removal of package capacitor

Figure 13: Distorted signal line with package capacitor removal

Further study is needed to improve the package routing for this interface in order to reduce package cost by removing this package capacitor. This includes having wider power planes, adding more Punch through Hole (PTH). Design of top-down structure is also an advantage when it comes to power delivery design with condition of proper bump pattern and appropriate ball location.

7. Summary

This paper describes innovation solutions in solving the power delivery issues without adding additional silicon and package cost. Appropriate SIPD co-analysis, package capacitor implementation, package returns path improvement and R-connect are the main key contributors for power delivery design. Well correlation between simulation and measurement proved that all the above solutions are valid.

8. Acknowledgements

The author would like all contributors who have spent time and effort on this analysis and validation work.

9. References

[1] Sanjiv Soman, "CPE Power Delivery Analysis Methodology Document", Intel internal document 2003
[2] Huey Lin Cheah, " Ibexpeak Fast Flash Power Delivery analysis", Intel internal document, 2007.
[3] Marcus Chan, "Ibexpeak NAND power delivery validation", Intel internal document, 2009
[4] Fei Deng, "Ibexpeak NAND interface review", Intel internal document, 2007
[5] Balsha R.Stanisic, Rob A.Rutenbar, L.Richard Carley, "Synthesis of Power Distribution to manage signal integrity in mixed–signal ICs", Kluwer Academic Publishers, pp.88-103

978-1-4244-4447-2/09 $25.00 © 2009 IEEE

Perturbation Based Modeling Strategy for Weakly Coupled Interconnects

Hao Shi

Apple Inc.

5 Infinite Loop, Cupertino, CA 95014
Phone: (408)862-0460, FAX: (408)974-4432, hshi@apple.com

Abstract — Coupling between two general interconnect structures, each viewed as the aggressor and victim sub-systems, respectively, are modeled with a 4-port network. A new method is introduced to estimate the coupling effects on the victim sub-system with known activities in the aggressor sub-system. It is verified with an idealized example, and is illustrated with a real-world PCB design scenario for noise prediction.

Index Terms — **coupled interconnect, crosstalk, perturbation.**

I. INTRODUCTION

Printed circuit board (PCB) design of modern electronic systems becomes more challenging as the signaling rate increases while the feature size decreases. Signal integrity (SI) problems are often investigated using models extracted by electromagnetic (EM) solvers. Full-wave EM solvers, yielding accurate numerical solutions to the Maxwell's equations to the full extent, are generally unsuited for solving the entire structure of a real-world design due to overwhelming demand for computational resources. An alternative approach, sometimes known as the hybrid approach, employs specialized EM solution techniques to extract models for separate categories of PCB interconnect structures. For instance, multi-conductor transmission-line solvers are used for traces, and resonant-cavity type of analytical modeling for vias and planes [1]. The hybrid approach, exemplified by some commercial simulation packages [2], captures the dominant SI effects inside a practical PCB design. However, coupling effects crossing the category boundaries, e.g. trace and via, are not systematically captured. Typical coupling between a trace and a nearby via is very small, below -40 dB in terms of the S-parameters. Surprisingly, such a weak coupling can still cause significant problems. One example is that the RF ringing on a DC-DC converter circuit may induce sizable noise onto a sensitive signal [3], and impair the system. Thus, it is desirable to complement the hybrid solution approach with additional terms accounting for the weak coupling scenarios. A novel strategy is proposed here consisting of the following steps: (a) solve the system voltage/current variables using a system-level circuit simulator with models from the hybrid approach, (b) capture the weak interconnect couplings, that are missed by the hybrid approach, using a full-wave EM solver, (c) calculate the additional terms as a first-order perturbation to the existing solution. First, derivations are given in Section II. Next, an idealized example is given in Section III to validate the formulation. Further demonstration with a practical example is shown in Section IV. Lastly, Section V gives the summary.

II. FORMULATION OF THE METHOD

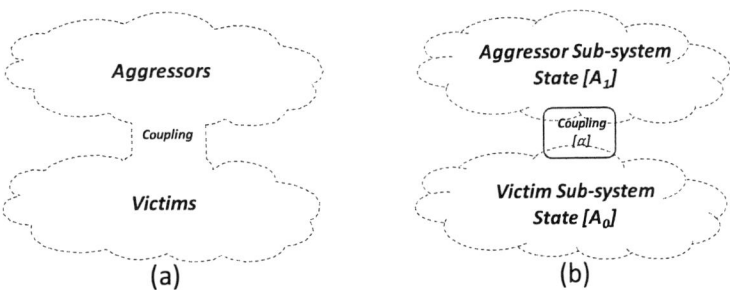

Figure 1: Coupled aggressor and victim sub-systems are (a) included in a larger system, or (b) divided into separate systems.

For ease of discussion of the coupling effects, the overall interconnect models can be divided into two sub-systems: aggressor and victim. In principle, the entire system can be solved with the full-wave EM solution approach, thus the coupling between aggressors and victims are readily captured, as illustrated by Figure 1(a). In reality, the particular coupling mechanism might be ignored, leaving the two sub-systems being independent from each other. If the aggregate of the voltage/current variables for the victim and aggressor systems are denoted by $[A_0]$ and $[A_1]$. respectively, then, the coupling structure can be isolated and modeled with a full-wave EM solver, with results symbolically represented by $[\varepsilon]$. The overall victim state shall be shown as:

$$[A] = [A_0] + [\alpha][A_1] \quad ...(1)$$

978-1-4244-4447-2/09 $25.00 © 2009 IEEE   213

Attaining the solution of $[\mathbf{A_0}]$ is straight forward, a matter of using an SPICE-like circuit simulator, when there are voltage/current excitations within the victim sub-system. The focus of this paper is to determine the first-order perturbation term $[\boldsymbol{\alpha}][\mathbf{A_1}]$, which is the effects on the victim sub-system due to activities in the aggressor sub-system.

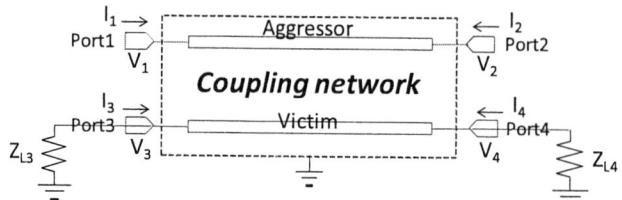

Figure 2: A generic model of the coupling network involving two interconnects.

A generic interconnect is represented by a 2-port network, therefore a 4-port network is adequate to capture the coupling between two interconnects, as shown in Figure 2. As indicated earlier that the coupling network can be solved by a full-wave EM solver, thereby the general outcome are scattering parameters, or S-parameters. By definition, the S-parameters relate the reflected waves (b's) to the incident waves (a's) by

$$b_n = \sum_{k=1}^{4} S_{nk} a_k, n = 1,2,3,4 \qquad a_n = (V_n + Z_0 I_n)/(2\sqrt{Z_0}), b_n = (V_n - Z_0 I_n)/(2\sqrt{Z_0})$$

When the b's and a's are replaced by the voltage/current variables, along with the termination conditions, the equations for the voltages in the victim sub-system are:

$$\begin{cases} V_3 - Z_0 I_3 = S_{31}(V_1 + Z_0 I_1) + S_{32}(V_2 + Z_0 I_2) + S_{33}(V_3 + Z_0 I_3) + S_{34}(V_4 + Z_0 I_4), V_3 = -I_3 Z_{L3} \\ V_4 - Z_0 I_4 = S_{41}(V_1 + Z_0 I_1) + S_{42}(V_2 + Z_0 I_2) + S_{43}(V_3 + Z_0 I_3) + S_{44}(V_4 + Z_0 I_4), V_4 = -I_4 Z_{L4} \end{cases} \quad ...(2)$$

which can be solved to yield

$$\begin{cases} V_3 = \alpha_{31}(V_1 + Z_0 I_1) + \alpha_{32}(V_2 + Z_0 I_2) \\ V_4 = \alpha_{41}(V_1 + Z_0 I_1) + \alpha_{42}(V_2 + Z_0 I_2) \end{cases} \quad ...(3)$$

where

$$\alpha_{31} = Z_{L3}[Z_0(S_{31} + S_{31}S_{44} - S_{34}S_{41}) + Z_{L4}(S_{31} - S_{31}S_{44} + S_{34}S_{41})]/\Delta$$

$$\alpha_{32} = Z_{L3}[Z_0(S_{32} + S_{32}S_{44} - S_{34}S_{42}) + Z_{L4}(S_{32} - S_{32}S_{44} + S_{34}S_{42})]/\Delta$$

$$\alpha_{41} = Z_{L4}[Z_0(S_{41} - S_{31}S_{43} + S_{33}S_{41}) + Z_{L3}(S_{41} + S_{31}S_{43} - S_{33}S_{41})]/\Delta$$

$$\alpha_{42} = Z_{L4}[Z_0(S_{42} - S_{32}S_{43} + S_{33}S_{42}) + Z_{L3}(S_{42} + S_{32}S_{43} - S_{33}S_{42})]/\Delta$$

and

$$\Delta = Z_0^2(1 + S_{33} + S_{44} + S_{33}S_{44} - S_{34}S_{43}) + Z_0 Z_{L3}(1 - S_{33} + S_{44} - S_{33}S_{44} + S_{34}S_{43})$$

$$+ Z_0 Z_{L4}(1 + S_{33} - S_{44} - S_{33}S_{44} + S_{34}S_{43}) + Z_{L3}Z_{L4}(1 - S_{33} - S_{44} + S_{33}S_{44} - S_{34}S_{43})$$

The above results map to the matrix form $[\boldsymbol{\alpha}][\mathbf{A_1}]$. Note that Z_0 is the constant reference impedance for all ports, which is 50.

III. VERIFICATION WITH AN IDEALIZED EXAMPLE

Now, we have a recipe to compute the noise voltages on the victim sub-system related to the activities in the aggressor sub-system. The validity of the technique shall be verified through an idealized circuit example.

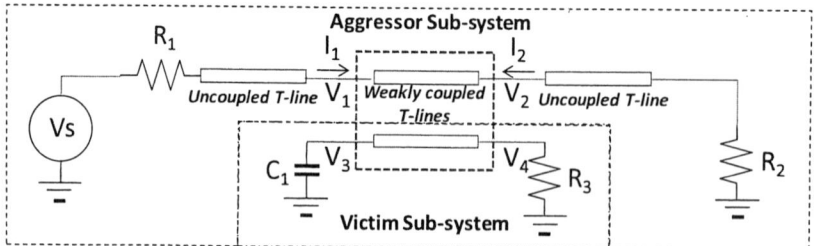

Figure 3: An ideal circuit network used to validate the perturbation based method.

As shown in Figure 3, an idealized circuit network is rendered in ADS [4], which includes two sub-systems that are linked with a weakly coupled transmission line model. The aggressor sub-system contains an AC voltage source, frequency swept from 10 MHz to 5 GHz, with magnitude of 1V. The source impedance R_1=50. A segment of coupled traces (widths are 0.25mm, edge separation is 1mm, length is 5mm) is sandwiched between two segments of uncoupled traces (width is 0.15mm, lengths are 20mm and 30mm). The traces are routed on a microstrip layer with 64um above a ground plane, with thickness of 31um, in FR4 dielectric material. The victim sub-system shares the coupled trace segment, and is simply terminated with a capacitor (5 pF) and resistor (120ohm) on two ends. The verification procedure consists of the following steps:

- Simulate the entire circuit, including the coupling effects between the two traces of the middle segment, save the $\{V_3,V_4\}$ solution as reference, labeled $\{V_{3r},V_{4r}\}$
- Solve the aggressor sub-system for $\{V_1,V_2,I_1,I_2\}$ by ignoring the coupling effects in the middle trace segment.
- Compute the S-parameters of the coupling network, which is modeled by the segment of two coupled traces
- Compute the $\{V_3,V_4\}$ using the recipe given in Section II, using $\{V_1,V_2,I_1,I_2\}$ as stimuli
- Compare the two sets of solutions: $\{V_3,V_4\}$ versus $\{V_{3r},V_{4r}\}$

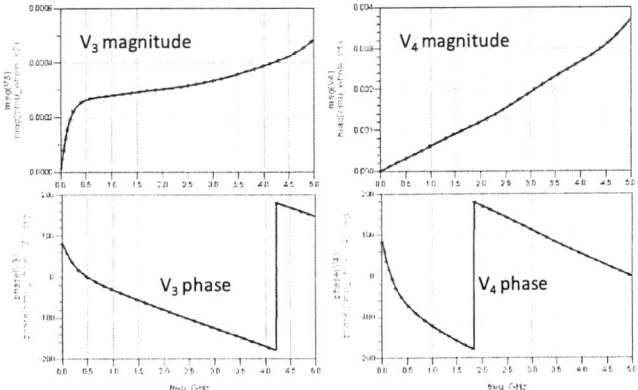

Figure 4: Victim voltages from the full circuit simulation (in solid red) and the perturbation approach (in blue circle).

As shown in Figure 4, the voltages in the victim sub-system computed with the perturbation approach are almost the same as the results obtained from the full circuit simulation. The perturbation based technique is thus verified.

IV. APPLICATION TO A REAL-WORLD COUPLING PROBLEM

In modern electronic systems, many functional circuits are consolidated into general-purpose ICs, each requires a set of DC voltages. Consequently, a significant portion of the system board real estate is dedicated to DC voltage regulator modules (VRMs). It is found in practical PCB designs that certain sensitive signal traces inadvertently get close-by to the metals associated with VRMs. Some of the VRMs are known to have RF ringing. The coupling path from a VRM metals to a signal traces can be subtle in many cases, and often not readily captured by an EM analysis using the hybrid modeling approach.

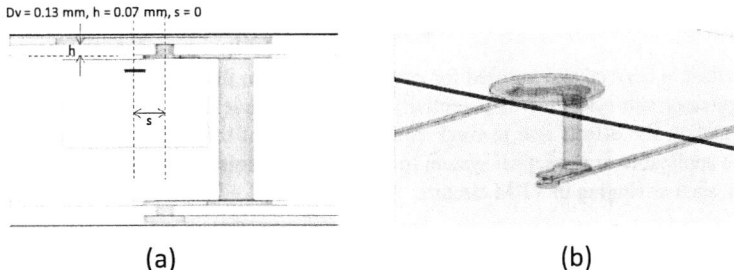

Figure 5: (a) Layer stack-up for the PCB structure of concern; (b) Two blind vias coupling to a stripline trace in Layer 3.

As shown in Figure 5, the coupling scenario involves an aggressor with two blind vias (each with drill diameter of 0.13mm) traversing between Layers 1 and 2 of a 4-layer board, and connected to a circular patch of 1.5mm in diameter on Layer 1. Traces in Layer 2 (primarily ground plane) connect the blind vias to a core via (drill diameter 0.25mm), which is tied with a trace segment on Layer 3, linking to another blind via, and emerges at the bottom side. The PCB dielectric material is FR4. The thickness of the core dielectric layer is 0.96mm, the distance between Layers 1-2, 3-4 are both 0.07mm. Trace width is 0.12mm

978-1-4244-4447-2/09 $25.00 © 2009 IEEE   215

on Layer 1 and 4, and 0.3mm on Layer 2 and 3, with keep-out distance to plane being 0.075mm. The victim sub-system includes simply a stripline trace in Layer 3, which is directly underneath the center-center line of the two blind vias.

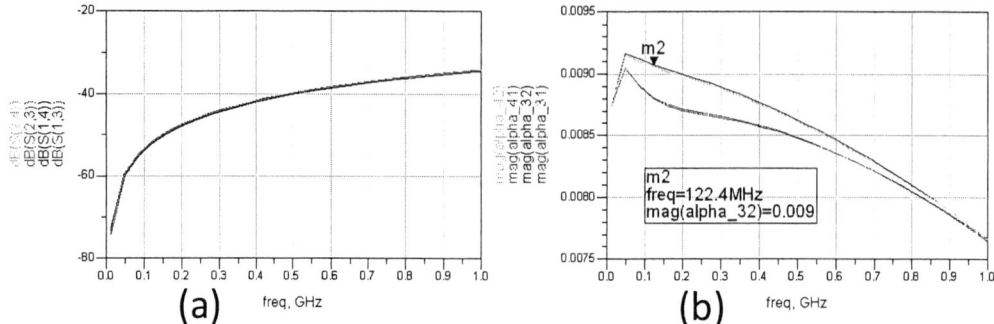

Figure 6: (a) Coupling S-parameters (in dBs), and (b) coupling coefficients (in magnitudes) for case depicted by Figure 5.

The above structure is modeled with HFSS [5], with Ports 1, 2 assigned at Layer 1, 4 of the aggressor, and Ports 3, 4 at two ends of the stripline in Layer 3. The coupling S-parameters, S_{31}, S_{32}, S_{41}, S_{42}, are plotted in Figure 6(a), which are all well below -40 dB around 125 MHz. The magnitude of the coupling coefficients, with two 3pF victim loads, are plotted in Figure 6(b).

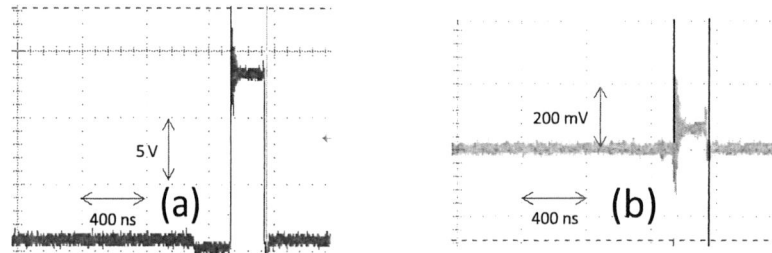

Figure 7: Scope shots of waveforms (a) on the phase node of a VRM, and (b) on a DDR3 DQ signal trace.

A similar coupling scenario, as discussed above, was seen in a real-world system prototype board. The RF ringing on the phase node of a VRM (buck converter from 12V to 1.1V), manifested as a decaying sinusoidal voltage waveform of 125 MHz, was superimposed on the rising edge of the 400 kHz rectangular switching waveform. The magnitude of the ringing is about 8 V, shown in Figure 7(a). We would use Eqn.(3) to estimate the noise voltages on the victim. Assuming the aggressor structure is electrically short such that $V_1 = V_2$ and $I_2 = -I_1$. When the four coupling coefficients are almost the same, shown in Figure 6, the contributions from the aggressor currents are cancelled, leaving $V_3 = V_4 = 2\alpha V_1$. The magnitude of α at 125 MHz, read from Figure 6(b), is about 0.009. Thus, $|V_3| = |V_4| = 2|\alpha V_1| = 2 \times 0.009 \times 8 = 0.144V$, which is consistent with the measured voltage noise peak amplitude (about 0.14V), shown in Figure 7(b), on a DDR3 DQ trace passing underneath the two blind vias.

V. SUMMARY

A perturbation based method is derived and verified for estimating noise on the victim sub-system due to weak coupling from known activities in the aggressor sub-system. It is potentially useful to complement the solutions from some of the PCB-level hybrid EM simulators by adding the effects due to weak coupling, on top of the simulation results without accounting for the weak coupling. It is also be applicable to a practical system in predicting the noise contamination on sensitive signal traces due to known aggressor activities, such as ringing in VRM circuits.

REFERENCES

[1] Xiaoxiong Gu, and *et al*, "Fully Analytical Methodology for Fast End-to-End Link Analysis on Complex Printed Circuit Boards including Signal and Power Integrity Effects," *DesignCon 2009 13-TA2,* Feb. 2009, Santa Clara, CA.
[2] Sigrity Corp, "PowerSI User's Guide," 2009.
[3] Amy Luoh, Gene Garrison, and Jon Powell, "Switching Voltage Regulator Noise Coupling Analysis for Printed Circuit Board Systems," *DesignCon 2009 10-TH,* Feb. 2009, Santa Clara, CA.
[4] Agilent Technologies, "Advanced Design System Documentation," 2009.
[5] Ansoft (Ansys Inc), "High Frequency Structure Simulator V11 User's Guide," 2009.

978-1-4244-4447-2/09 $25.00 © 2009 IEEE

The Impact of Guard Trace with Open Stub on Time-Domain Waveform in High-Speed Digital Circuits

Po-Wei Chiu, and Guang-Hwa Shiue

Department of Electronic Engineering, Chung Yuan Christian University, Taoyuan, Taiwan, R. O. C.

Advanced Packaging and EMC Lab

Tel: +886-3-2654627   Fax: +886-3-2654699   E-mail: ghs@cycu.edu.tw

Abstract

Grounded guard traces are increasingly utilized to mitigate the crosstalk noise interference of the printed circuit board or package. This article discusses how the couple microstrip line inserted guard trace with stub affects the crosstalk noise. The guard trace with an open stub seriously degrades the performance of the coupled line with respect to signal integrity (SI) while far-end crosstalk noise will still occurs for stripline structure. Additionally, an analytical formula is proposed to approximate the extra crosstalk noise which is caused by guard trace stub.

I. Introduction

Crosstalk noise interference seriously degrades the performance of high-speed digital circuits, especially with an increasing signal rise time. Such interference is generally represented in term of near-end crosstalk (NEXT) and far-end crosstalk (FEXT). In a typical PCB layout, the designer places a guard trace which is grounded by shorting via between adjacent parallel lines; doing so can effectively diminish crosstalk noise [1]. Additional, the spacing of shorting vias determines the resonance in frequency response, with more vias, in which the first resonance frequency shifts away from the bandwidth of interest [2]. In the time domain waveform, resonance related to spacing of the shorting via might lead to even more significant ringing noise on the victim line, which largely explains the FEXT on the guard trace [3].

Either the guard trace or ground plane with stub always occurs in practical package or PCB layout. Figure 1(a) illustrated a practical package layout. When the guard trace is close to the chip owing to that the dimension is restricted, the shorting vias can not keep the grounded from one end of guard trace to the other end, subsequently becoming an open stub between the couple line. Additionally, Fig. 1(b) shows the other practical printed circuit board (PCB), which has a guard trace with stub due to electromagnetic interference (EMI) related issues, i.e. increasing the ground plane or ground trace. For high-speed digital circuits, it can not be regarded as an ideal ground line or a lumped circuit component. This study demonstrates the above phenomenon and the feasibility of an approximate formula of extra large noise interference by guard trace stub. Eye diagrams for coupled microstrip and strip line inserting guard trace with/without the stub are also discussed.

II. Analyses of the guard trace stub

Figure 2 shows a typical microstrip couple line with n vias guard trace scheme, in which $\ell = 60mm$, $\ell_{stub} = 40mm$, $W = 3mm$, $W_g = 3mm$, $T = 0.035mm$, $H = 1.5mm$, loss tangent=0.02, $\varepsilon_r = 4.4$, and $S = 5mm$. In the frequency domain, Fig. 3 illustrated the relationship between resonance and number of vias, as simulated by HFSS [5]. The resonance is attributed to the vias spacing equivalent half-wavelength, and the stub incurs the resonance when the stub length is equivalent to a quarter-wavelength. When the guard trace has five vias, the first resonance shifts to a higher frequency. Although reducing crosstalk by shorting vias is preferred, with one via, doing so implies that once becoming a stub, the fist resonance frequency complies with the total stub length. The stub length is shorter, while the fist resonance frequency is higher. Hence, the shortest stub length or zero length can be maintained for guard trace stub.

During simulation of the time domain, the driver and load resistances are the chosen matched resistance value, while the rise time and amplitude of the ramp pulse source are 50ps and 2 volt, respectively, as shown in Fig.2. Figure 4 compares NEXT and FEXT for various vias numbers of guard trace. The simulated NEXT and FEXT waveforms on victim line are microstrip coupled line without the guard trace (called case 1), microstrip coupled line inserted a guard trace with two shorting vias (ℓ_d=60mm, ℓ_{stub}=0mm; called case 2), and the microstrip coupled line inserted a guard trace with one shorting via and stub (ℓ_d=0mm, ℓ_{stub}=60mm; called case 3). In case 2, when the signal is a ramp pulse applied at the aggressor line, the guard trace effectively suppresses crosstalk noise, additional ringing noise appears on the crosstalk waveforms of victim line; this is likely owing to that FEXT propagates on guard trace back and forth since both ends are

---

The work was supported in part by the National Science Council, Republic of China, under Grant NSC 97-2221-E-033-061.

grounded [3]. However, in case 3, i.e. the guard trace stub case, the noise interference worsens. The extra noise interference by the guard trace consists of a peak voltage and flat voltage noise waveforms; this differs from two shorting vias case, as shown in Fig. 4. In the guard trace stub case, a ramp pulse with amplitude V_s is considered, in which rise time t_r propagates along the aggressor line; the NEXT and FEXT noise interference then appear at both guard trace and victim line, respectively. As the FEXT noise interference propagates at the left-end of guard trace stub, the NEXT noise interference is cancelled because the reflection coefficient is - 1. The NEXT noise interference then inverts the voltage polarity and propagates towards the right end portion of the guard trace stub during the time $2T_d \sim 3T_d$ (T_d: time delay). Because the right end portion of guard trace stub is open, the NEXT noise interference maintains the voltage polarity and propagates towards the left end portion of the guard trace during the time $3T_d \sim 4T_d$. The following time, the NEXT noise interference repeats the same action back and forth on the guard trace stub. For the same time duration, the FEXT noise interference on the guard trace stub keeps the voltage polarity due to an open right end portion and propagates towards the left end portion of the guard trace stub and, then, inverts the voltage polarity due to short left end portion and propagates towards the right end portion of the guard trace stub. However, the NEXT and FEXT noise interference on the guard trace stub both induce further far-end and near-end crosstalk noise interference on the victim line.

The flat voltage, V_{flat}, on the far-end or near-end crosstalk waveforms of the victim line belongs to the near-end type crosstalk noise induced by the NEXT as the main signal propagates back and forth on the guard trace stub. Moreover, the maximum peak voltage, $V_{n\_max\,peak}$, on the near-end crosstalk waveform of victim line belongs to the far-end type crosstalk noise induced by the NEXT and FEXT noise interference as the main signal propagates towards the left end of guard trace stub. Therefore, the magnitude of $V_{n\_max\,peak}$ is the sun of FEXT induced by NEXT and FEXT noise interference of the guard trace stub at the time $2T_d$. Hence, the magnitudes of $V_{n\_max\,peak}$ [3] and V_{flat} [4] can be derived as

$$V_{n\_\max\ peak} \cong \left(\frac{1}{\sqrt{2}} + \frac{1}{2} \cdot \left| k_{near,g} \right| \right) \cdot \frac{V_s}{8} \cdot \frac{T_{d\_stub}}{t_r} \cdot \left(\frac{L_{m,g}}{L_{s,g}} - \frac{C_{m,g}}{C_{s,g}} \right) = \left(\frac{1}{\sqrt{2}} + \frac{1}{2} \cdot \left| k_{near,g} \right| \right) \cdot \frac{V_s}{8} \cdot \frac{T_{d\_stub}}{t_r} \cdot \left| k_{far,g} \right| \quad (1)$$

$$V_{flat} \cong \frac{V_s}{2} \cdot \frac{1}{8} \cdot \left(\frac{L_{m,g}}{L_{s,g}} + \frac{C_{m,g}}{C_{s,g}} \right)^2 = \frac{V_s}{16} \cdot \left| k_{near,g} \right|^2 \quad (2)$$

where V_s represents the ramp pulse input, T_{d_stub} represents the delay time of stub length, $k_{far,g}$ and $k_{near,g}$ are the far-end and near-end crosstalk coefficients and subscript g denotes the mutual term between guard trace and the aggressor line or victim line. According to Table I, the extra noise voltage $V_{n\_max\,peak}$ and V_{flat} on near-end and far-end crosstalk waveforms of victim line between approximation formula (1)-(2) and HSPICE [6] simulations are in good agreement.

Figure 5 shows the near-end crosstalk and far-end crosstalk noise waveforms on the victim line when the number of grounded vias is two, i.e. the spacing of via is 20mm, and the length of guard trace stub is 40mm. In this case, the guard trace comprises shorting vias section and open stub section. According to above analyses, ringing noise $V_{ringing}$ [3], flat voltage V_{flat} and maximum peak voltage, $V_{n\_max\,peak}$ all occur, as shown in Fig. 5. Obviously, the value of peak voltage $V_{n\_max\,peak}$ is larger than the ringing noise $V_{ringing}$ on the NEXT waveform, and flat voltage V_{flat} appears on NEXT and FEXT waveforms of the victim line. Consequently, the guard trace with a long stub significantly degrades the performance of SI.

III. Performance of eye diagram

The effects of crosstalk interference on eye diagram due to coupled line inserted guard trace with/without the open stub are examined. In HSPICE simulation, the pseudorandom incident signals are launched with rise/fall time 50ps, data bit rate 2Gb/s, and voltage swing of 2volt. Recall the routing scheme in Fig. 1(a) with the same cross sectional parameters in Fig. 1(b). The simulated eye diagrams at the receiver are case 1, 2 and 3 as shown in Fig. 6(a), 6(b), and 6(c), respectively. Table II compares the parameters of the eye diagram between the three cases. Comparing Figs. 6(a) and 6(b) reveals that the microstrip coupled line additional guard trace with both shorting ends helps to improve the width and jitter of eye diagram. According to Fig. 6(c), the microstrip coupled line additional guard trace stub obviously degrades the performance of eye diagram significantly. Additionally, while no far-end crosstalk noise occurs in the stripline scheme, the guard trace still reduces the near-end crosstalk noise. However, the fact that the near-end crosstalk still propagates back and forth on the guard trace stub explains why far-end crosstalk interference is induced on the victim line, as shown in Fig. 7. Furthermore, this crosstalk interference worsens eye diagram, as shown in Fig. 8.

978-1-4244-4447-2/09 $25.00 © 2009 IEEE   218

IV. Conclusion

As the rise time of digital signal goes higher, the crosstalk noise is become more seriously. Inserting the guard trace is a conventional means of preventing crosstalk noise interference, although it must avoid the guard trace stub. Otherwise, the signal integrity performs the worst. In general, no far-end crosstalk noise occurs in the stripline scheme. However, due to the inserted guard trace with an open stub the crosstalk noise interference still appears on far end of coupled line for either the microstrip or strip line structure.

References

[1] D. N. Ladd and G. I. Costache, "SPICE simulation used to characterize the cross-talk reduction effect of additional tracks grounded with vias on printed circuit boards," *IEEE Trans. Circuits Syst. II,* vol. 39, pp. 342-347, June 1992.

[2] A. Suntives, A. Khajooeizadeh, and R. Abhari, "Using via fences for crosstalk reduction in PCB circuits," in *2006 IEEE Int. Symp. Electromagnetic Compat.*, pp. 34–37, 14-18 Aug. 2006, Portand, Orgeon, USA.

[3] Y. S. Cheng, W. D. Guo, G. H. Shiue, H. H. Cheng, C. C. Wang, and R. B. Wu, "Fewest vias design for microstrip guard trace by using overlying dielectric," in *2008 IEEE-EPEP*, pp. 321-324, 27-29 Oct. 2008, San Jose, CA.

[4] S. H. Hall, G. W. Hall, and J. A. McCall, *High-Speed Digital System Design*, Hoboken, NJ: Wiley, 2000. ch. 3, pp. 45-51.

[5] HFSS v11.1, ANSOFT Corporation (www.ansoft.com)

[6] HSPICE v2006.09, SYNOPSYS Corporation (www.synopsys.com)

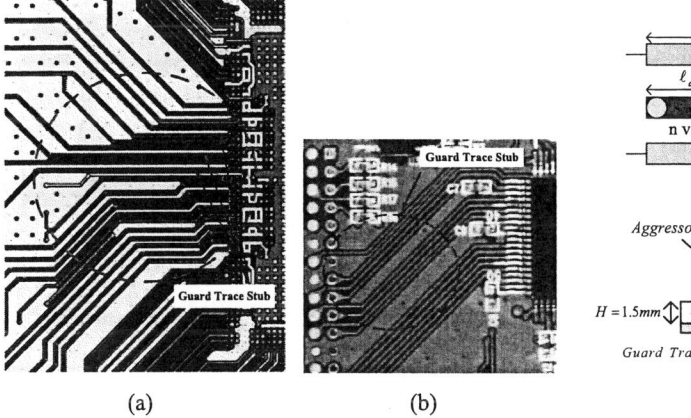

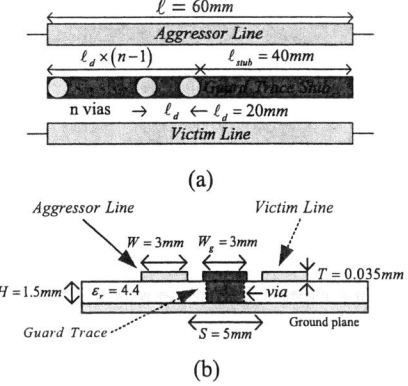

(a)　　　　　　　　　(b)

Fig. 1. Guard trace stub in a practical (a) package and (b) printed circuit board layouts.

Fig. 2. Conventional routing scheme for coupled microstrip line with a guard trace stub (a) top view (b) cross-sectional view.

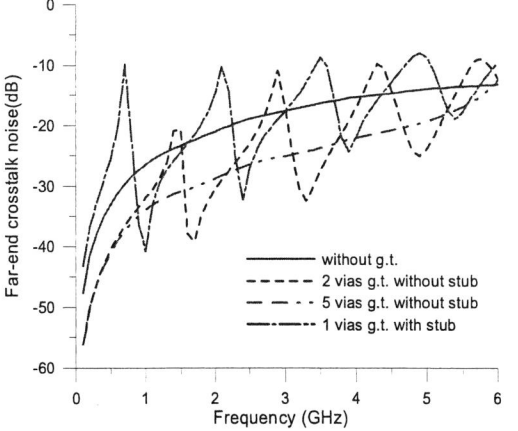

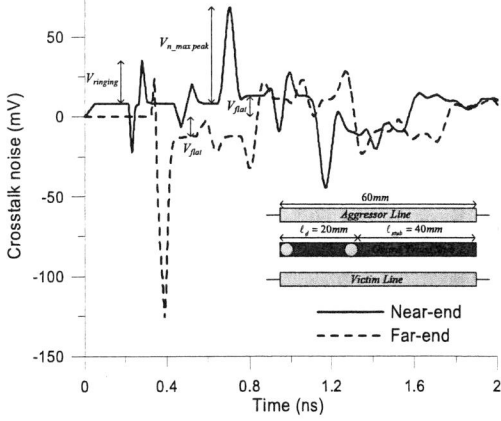

Fig. 3. Frequency response of far-end crosstalk for coupled microstrip line.

Fig. 5. Crosstalk noise waveforms for microstrip line inserted guard trace with shorting via section and open stub section.

978-1-4244-4447-2/09 $25.00 © 2009 IEEE

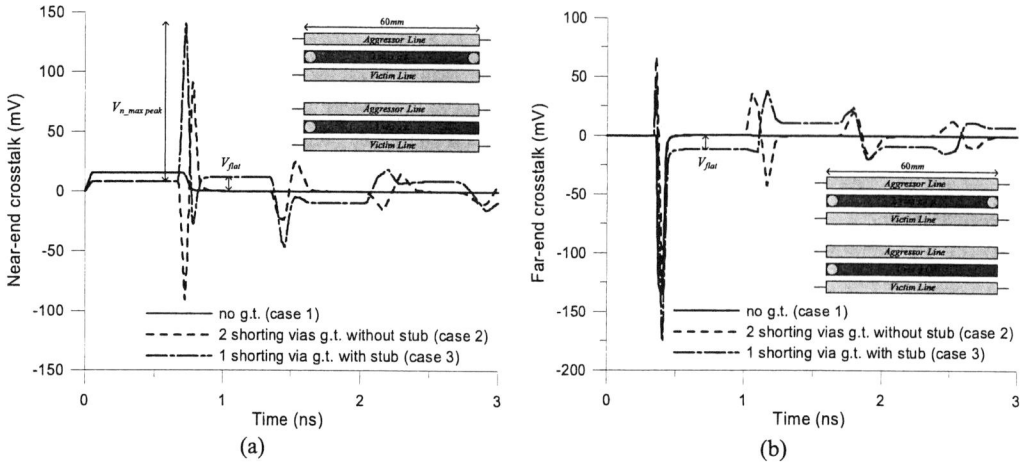

Fig. 4. Comparisons of (a) near-end and (b) far-end crosstalk noise time-domain waveforms between different cases.

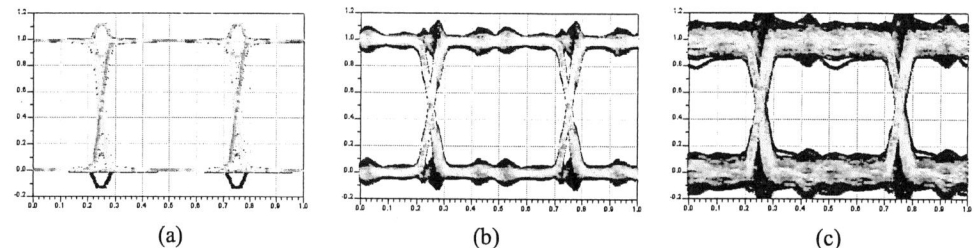

Fig. 6. Comparisons of eye diagrams between (a) case 1, (b) case 2, and (c) case 3.

Table I. Comparison of the maximum peak voltage and flat voltage on crosstalk noise waveforms between approximation formula and HSPICE simulation.

Extra Crosstalk Noise	Mircostrip line		Stripline
	$V_{n\_max\,peak}$	V_{flat}	V_{flat}
Approach Formula	142 mV	14.1 mV	21.4 mV
HSPICE Simulation	141 mV	11.8 mV	20.8 mV

Table II. Comparisons of eye diagram parameters between case1, case2, and case3.

	Csae1 (no g.t.)	Case2 (2 vias g.t.)	Case3 (g.t. stub)
Eye high (volt)	0.98	0.90	0.60
Eye width (psec)	482	489	471
Jitter (psec)	17.4	13.3	30.4

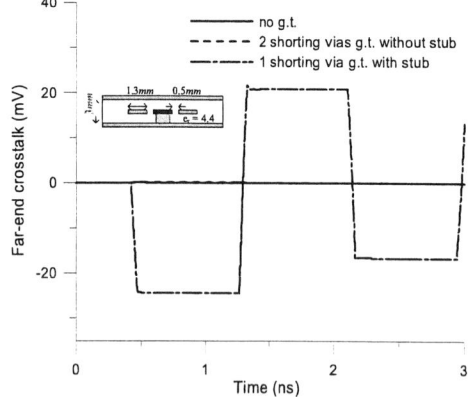

Fig. 7. Far-end crosstalk noise waveform for couple stripline inserted guard trace stub.

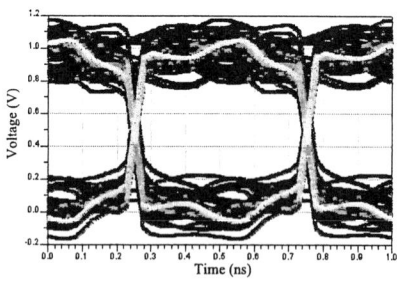

Fig. 8. Eye diagram for coupled stripline inserted guard trace stub.

978-1-4244-4447-2/09 $25.00 © 2009 IEEE

Numerical Acceleration Of Spectral Domain Approach For Shielded Microstrip Lines By Approximating Summation With Corrected Integral

Sidharath Jain and Jiming Song
Department of Electrical and Computer Engineering
Iowa State University, 2215 Coover Hall, Ames,IA-50011
Phone:(515)294-8396, Fax:(515)294-8432
Email: sjain@iastate.edu, jisong@iastate.edu

Abstract: In this paper, a technique for numerical acceleration of the spectral domain approach (SDA) for shielded microstrip lines has been presented. This is achieved by speeding up the convergence of series summation in the elements of the Galerkin matrix by leading term extraction and approximating the series summation of the tail as an integral with a correction. This technique results in highly accurate results for the effective permittivity without the need for numerical integration and incurring a very low computational cost.

I. INTRODUCTION

The spectral domain approach (SDA) is most widely used technique for the analysis of planer microstrip structures. It gives accurate results for the propagation constant using a few basis functions. However, as the required accuracy increases the computation time increases very rapidly because of the slow convergence of the spectral series summation making this approach computationally intensive for computer aided design (CAD) purposes. In order to accelerate the convergence of spectral series summation several techniques have been used [1], [2] and [3]. Also for open microstrip with double negative materials an approach based on leading term extraction has been demonstrated [4]. As shown in [2] that the time intensive part in the SDA is the calculation of the matrix elements and not finding the determinant as the matrix size is small in most cases. By using appropriate basis functions and adding the asymptotic tails of the series involved drastic improvement in accuracy and CPU time can be obtained [5]. In this paper, we propose a novel technique for numerical acceleration of the SDA for shielded microstrip by accelerating the convergence of series summation in the elements of the Galerkin matrix. This is done by subtracting the leading term of the matrix elements from the finite summation and adding the closed form for the infinite summation of the leading terms obtained from the theory of Riemann Zeta function. The series summation of the tail is approximated by an integral corresponding to the mid point summation approximation with a correction.

II. SPECTRAL DOMAIN APPROACH (SDA)

Figure 1 shows a cross section of a shielded microstrip. Region 2 is air and region 1 consists of a lossless dielectric material with relative permittivity and permeability ϵ_r and μ_r respectively. The structure is uniform and infinite along the z axis and the thin metal casing and the thin metal strip are assumed to be perfect electric conductors (PECs). The solutions for microstrip lines are hybrid modes that can be expressed in terms of superposition of infinite transverse electric (TE) and transverse magnetic (TM) modes which can inturn be expressed in terms of the scalar vector potentials [6]. Then by taking the Fourier transform of all the field components and applying the boundary conditions in the Fourier domain including the fact that for the dominant quasi-transverse electromagnetic (TEM) wave E_y should be an even function of x we obtain [6]:

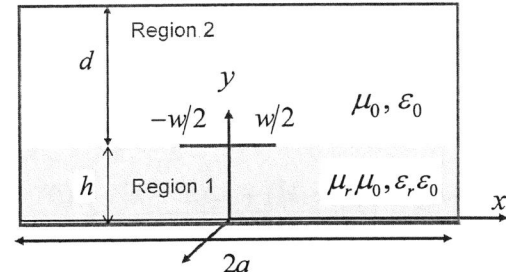

Fig. 1. Shielded microstrip

$$\tilde{E}_{x2}(\alpha_n, h) = G_{xx}(\alpha_n, \beta)\tilde{J}_x(\alpha_n) + G_{xz}(\alpha_n, \beta)\tilde{J}_z(\alpha_n) \tag{1}$$

$$\tilde{E}_{z2}(\alpha_n, h) = G_{zx}(\alpha_n, \beta)\tilde{J}_x(\alpha_n) + G_{zz}(\alpha_n, \beta)\tilde{J}_z(\alpha_n) \tag{2}$$

where $\alpha_n = (n - 1/2)\pi/a$. The z dependency of the electric and magnetic field has the form of $e^{-j\beta z}$. The expressions for the Green's functions are reported in [6].

The Fourier transforms of the unknown current $\tilde{J}_x(\alpha_n)$ and $\tilde{J}_z(\alpha_n)$ are expanded in terms of basis functions $\tilde{J}_{xi}$ and $\tilde{J}_{zi}$ as follows:

$$\tilde{J}_x(\alpha_n) = \sum_{i=1}^{N_x} a_i \tilde{J}_{xi}(\alpha_n) \text{ and } \tilde{J}_z(\alpha_n) = \sum_{i=1}^{N_z} b_i \tilde{J}_{zi}(\alpha_n) \tag{3}$$

The basis currents are chosen such that their inverse Fourier transforms $J_{xi}(x)$ and $J_{zi}(x)$ are nonzero only on the strip $|x| < w/2$, and $J_{xi}(x)$ is a real odd function, $J_{zi}(x)$ is a real even function. So from the properties of Fourier transforms, $\tilde{J}_{xi}(\alpha_n)$ is a purely imaginary and odd function, $\tilde{J}_{zi}(\alpha_n)$ is a purely real and even function. Further, using the Galerkin method, followed by the Parseval's theorem and the fact that the product of the tangential field component and the surface current is always zero on the line we obtain the following matrix equation:

$$\begin{bmatrix} K^{xx} & K^{xz} \\ K^{zx} & K^{zz} \end{bmatrix} \begin{bmatrix} A \\ B \end{bmatrix} = \begin{bmatrix} 0 \\ 0 \end{bmatrix}$$

where

$$K_{ij}^{pq} = \sum_{n=1}^{\infty} \tilde{J}_{pi}(\alpha_n) G_{pq}(\alpha_n, \beta) \tilde{J}_{qj}(\alpha_n) \quad p = x, z \text{ and } q = x, z \tag{4}$$

Finally, the propagation constant β can be obtained by solving $\det[K] = 0$. If Chebyshev polynomials are chosen as the basis function there is a possibility of speeding up the convergence of the elements of the K matrix hence accelerating the SDA as discussed in the next section. $J_z(x)$ and $J_x(x)$ are even and odd functions, respectively and can be expanded with the Chebyshev polynomials of the first and second kind, respectively [7]. Their Fourier transforms are given by

$$\frac{\tilde{J}_z(\alpha_n)}{w} = \frac{\pi}{2} \sum_{i=1}^{N_z} I_{z(i-1)}(-1)^{i-1} J_{2(i-1)}(\delta_n) \text{ and } \frac{\tilde{J}_x(\alpha_n)}{w} = \frac{\pi}{\delta_n} \sum_{i=1}^{N_x} I_{xi} i (-1)^i J_{2i}(\delta_n) \tag{5}$$

where $\delta_n = \alpha_n w/2$ and $J_n(z)$ is the Bessel function.

III. Leading Term Extraction and Approximating Summation with Integral

As $\alpha_n \to \infty$ keeping the first two terms in Taylor expansion the Green's functions are approximated as [4]:

$$G_{xx} \approx \frac{\alpha_n^3}{(1+\epsilon_r)(\alpha_n^2 + y_{xx}^2)}, \ G_{xz} \approx \frac{\beta \alpha_n^2}{(1+\epsilon_r)(\alpha_n^2 + y_{xz}^2)} \text{ and } G_{zz} \approx \frac{[(\beta^2 - k_2^2) + \mu_r(\beta^2 - k_1^2)]\alpha_n}{(1+\epsilon_r)(1+\mu_r)(\alpha_n^2 + y_{zz}^2)} \tag{6}$$

where

$$y_{xx}^2 = \frac{\beta^2}{2} + \frac{k_1^2 + \epsilon_r k_2^2}{2(1+\epsilon_r)}, \ y_{xz}^2 = \frac{\beta^2}{2} + \frac{(k_1^2 - k_2^2)(1-\mu_r)}{2(1+\mu_r)} - \frac{(k_2^2 + \epsilon_r k_1^2)}{2(1+\epsilon_r)} \text{ and}$$

$$y_{zz}^2 = \frac{1}{2}\left[\frac{(\beta^2 - k_2^2) + \mu_r(\beta^2 - k_1^2)}{1+\mu_r} + \frac{(\beta^2 - k_2^2) + \epsilon_r(\beta^2 - k_1^2)}{1+\epsilon_r} - \frac{(\beta^2 - k_1^2)(\beta^2 - k_2^2)(1+\mu_r)}{(\beta^2 - k_2^2) + \mu_r(\beta^2 - k_1^2)}\right] \tag{7}$$

Using the asymptotic forms of the Bessel function [8] and Green's functions for large α_n, we have

$$\tilde{J}_{pi}(\alpha_n) G_{pq}(\alpha_n, \beta) \tilde{J}_{qj}(\alpha_n) \approx C_{ij}^{pq} \frac{1 + \sin(\alpha_n w)}{\alpha_n^2 + y_{pq}^2} \tag{8}$$

where $p = x, z$, $q = x, z$, $C_{ij}^{xx} = \frac{8\pi ij}{w(1+\epsilon_r)}$, $C_{ij}^{xz} = \frac{2\pi i\beta}{w(1+\epsilon_r)}$ and $C_{ij}^{zz} = \frac{\pi}{2}\frac{(\beta^2-k_2^2)+\mu_r(\beta^2-k_1^2)}{w(1+\epsilon_r)(1+\mu_r)}$.

From the theory of Riemann Zeta function [9] for integer powers we can obtain

$$\sum_{n=1}^{\infty}\frac{1}{[2n-1]^2} = \frac{\pi^2}{8} \implies \sum_{n=1}^{\infty}\frac{1}{[n-1/2]^2} = \frac{\pi^2}{2} \tag{9}$$

Using equations (8) and (9), we have

$$K_{ij}^{pq} \approx \sum_{n=1}^{N}\left[\tilde{J}_{pi}(\alpha_n)G_{pq}(\alpha_n,\beta)\tilde{J}_{pj}(\alpha_n) - \frac{C_{ij}^{pq}}{\alpha_n^2}\right] + C_{ij}^{pq}\sum_{n=N+1}^{\infty}\left[\frac{1+\sin(\alpha_n w)}{\alpha_n^2+y_{pq}^2} - \frac{1}{\alpha_n^2}\right] + C_{ij}^{pq}\sum_{n=1}^{\infty}\frac{1}{\alpha_n^2} \tag{10}$$

In the above equation, the first term can be evaluated using direct summation. The third term is calculated using the Riemann Zeta function (9) analytically. Arranging the second term yields

$$\sum_{n=N+1}^{\infty}\left[\frac{1+\sin(\alpha_n w)}{\alpha_n^2+y_{pq}^2} - \frac{1}{\alpha_n^2}\right] = \sum_{n=N+1}^{\infty}\frac{\sin(\alpha_n w)}{\alpha_n^2+y_{pq}^2} - y_{pq}^2\sum_{n=N+1}^{\infty}\frac{1}{\alpha_n^2(\alpha_n^2+y_{pq}^2)} \tag{11}$$

The summation in the right hand side of the above equation can be approximated by the corresponding integral. Then the integral for the first term is approximated by the leading term of the corresponding integral obtained using integral by part with a correction.

$$\sum_{n=N+1}^{\infty}\frac{1}{\alpha_n^2(\alpha_n^2+y_{pq}^2)} \approx \frac{a}{\pi y_{pq}^3}\left[\frac{y_{pq}}{A} - \arctan\left(\frac{y_{pq}}{A}\right)\right] \qquad \sum_{n=N+1}^{\infty}\frac{\sin(\alpha_n w)}{\alpha_n^2+y_{pq}^2} \approx C_p\frac{a}{\pi w}\frac{\cos(wA)}{A^2+y_{pq}^2} \tag{12}$$

where $A = N\pi/a$ for mid point summation to integral and $A = (N+1/2)\pi/a$ for end point summation to integral, respectively. The correction coefficient C_p is derived as

$$C_p = \frac{w\pi/(2a)}{\sin[w\pi/(2a)]} \tag{13}$$

IV. NUMERICAL RESULTS

The approach was numerically validated using a shielded microstrip with parameters $\epsilon_r = 11.7, \mu_r = 1, f = 4$GHz, $h = 3.17$mm, $w = 3.04$mm, $2a = 34.74$mm, $d = 50$mm [2]. From Figure 2(a) it is observed that the convergence of Kzz changes from $1/N$ using the direct summation to $1/N^2$ using approximating the tail with end point integral and finally to $1/N^3$ using mid point integral. Here, the result using leading term with mid point integral at $N = 1,000,000$ is used as a reference. The direct summation is performed by truncating the infinite series summation in (4) at N. It is worth noting that using the mid point integral approximation and the end point summation differ by one order of magnitude. Figure 2(b) show the convergence of Kzz using the direct summation, using the first and third term in (10), using all the three terms with $C_p = 1$, using C_p equal to the derived value. For the case with $C_p=1$ the convergence is slower as $1/N^2$. But when the summation is approximated as the integral with a correction shown in (12), the error decreases as $1/N^3$.

Figures 3(a) and 3(b) show that the convergence of the determinant of K matrix and ϵ_r also is accelerated from $1/N$ without using leading term extraction to $1/N^3$ by using leading term extraction with the tail summation approximated by an integral with a correction.

V. CONCLUSION

Computation speed and accuracy are a major issue for CAD applications and it is very computationally intensive to use the spectral domain approach in such applications because of the slowly converging series in evaluation of the elements of the Galerkin matrix. With the development of the accelerated spectral domain approach using leading term extraction and integral to approximate the tail with a correction the convergence of the seris summation for shielded microstrips has been accelerated by several orders of magnitude.

Acknowledgements: This work was supported in part by the National Science Foundation CAREER Grant ECS-0547161.

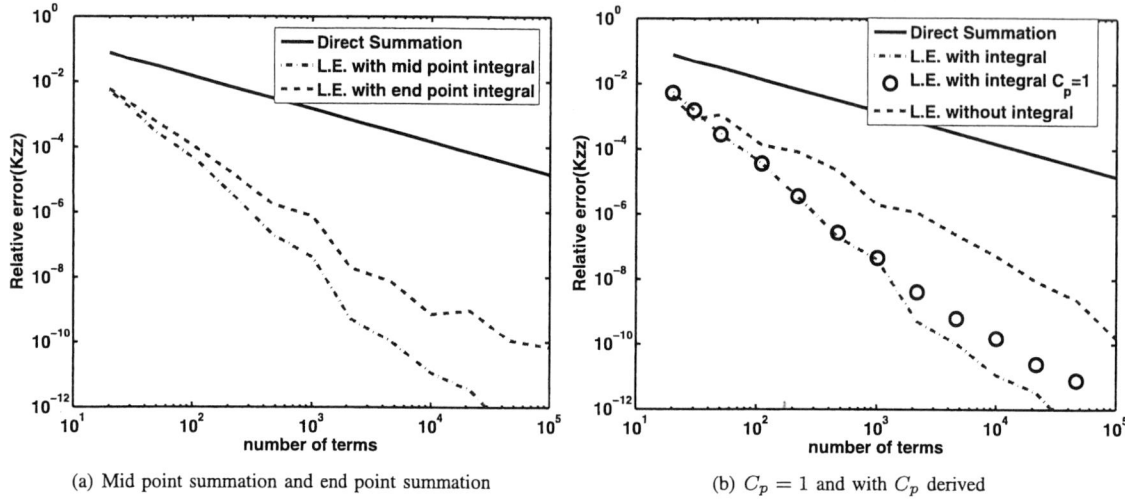

(a) Mid point summation and end point summation (b) $C_p = 1$ and with C_p derived

Fig. 2. The convergence of Kzz (a) using the direct summation, using approximating the tail with mid or end point integral. (b) using the direct summation, using the first and third term in (10), using all the three terms with $C_p = 1$, using C_p equal to the derived value.

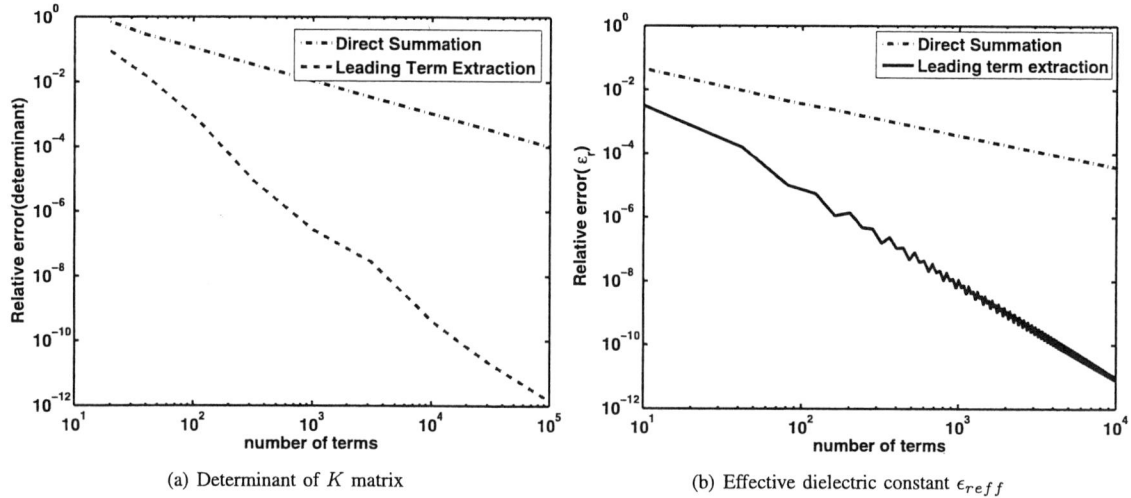

(a) Determinant of K matrix (b) Effective dielectric constant ϵ_{reff}

Fig. 3. Convergence of (a) determinant of K matrix and (b) ϵ_{reff} using the direction summation and the approach proposed.

REFERENCES

[1] G. Cano, F. Mesa, F. Medina, and M. Horno, "Systematic computation of the modal spectrum of boxed microstrip, finline, and coplanar waveguides via an efficient sda," *IEEE Transactions on Microwave Theory and Techniques*, vol. 43, no. 4 pt 1, pp. 866 – 872, 1995.

[2] G. Cano, F. Medina, and M. Horno, "On the efficient implementation of sda for boxed strip-like and slot-like structures," *IEEE Transactions on Microwave Theory and Techniques*, vol. 46, no. 11 pt 1, pp. 1801 – 1806, 1998.

[3] J. L. Tsalamengas and G. Fikioris, "Rapidly converging spectral-domain analysis of rectangularly shielded layered microstrip lines," *IEEE Transactions on Microwave Theory and Techniques*, vol. 51, no. 6, pp. 1729 – 1734, 2003.

[4] W. Shu, *Electromagnetic waves in double negative metamaterials and study on numerical resonances in the method of moments.* Phd thesis: Iowa State University, 2008.

[5] F. Medina and M. Horno, "Quasi-analytical static solution of the boxed microstrip line embedded in a layered medium," *IEEE Transactions on Microwave Theory and Techniques*, vol. 40, no. 9, pp. 1748 – 1756, 1992.

[6] T. Itoh and R. Mittra, "Technique for computing dispersion characteristics of shielded microstrip lines." *IEEE Transactions on Microwave Theory and Techniques*, vol. 22, no. 10, pp. 896 – 898, 1974.

[7] R. E. Collin, *Field Theory of Guided Waves, 2nd ed.* Piscataway ,New Jersey: IEEE Press, 1991.

[8] M. Abramowitz and I. A. S. (Eds), *Handbook of Mathematical Functions.* New York: Dover Publications, 1970.

[9] H. M. Edwards, *Riemann's Zeta Function.* N. Chemsford, MA: Courier Dover Publications, 2001.

A New Isolation Structure for Crosstalk Reduction of Pogo Pins in a Test Socket

Ruey-Bo Sun, Chang-Yi Wen, Yen-Chih Chang*, and Ruey-Beei Wu
Department of Electrical Engineering and Graduate Institute of Communication Engineering,
National Taiwan University, Taipei 10617, Taiwan
E-mail: rbwu@ew.ee.ntu.edu.tw; d93942011@ntu.edu.tw

*Network Interconnection Business Group, Hon Hai Precision Ind. Co., Ltd., Taipei 23699, Taiwan

Abstract

This paper proposes a new isolation structure integrated in the dielectric holder of a test socket for reducing the crosstalk noise. The isolation structure consists of inserted vias and metallic planes, connecting the inserted vias with ground pogo pins. It can provide better than -20dB crosstalk and reflection noises from DC to 10GHz, with 10dB improvement in crosstalk at 10 GHz.

I. Introduction

Crosstalk noise is primarily caused by the electromagnetic coupling between interconnects and is also one of the most critical signal integrity issues for high-speed digital systems operated at up to several gigabit data rate. For the pogo pin structures, the crosstalk noise can be easily reduced by just enlarging the pin pitch or decreasing the pin radius. However, for an advanced test system with several gigabit data rate, e.g., 5 Gb/s, the specification of the crosstalk noise will be more stringent, say below -20 dB from DC to 10 GHz. It is hard to be satisfied by only adjusting the pogo pin geometries.

As a result, an isolation structure for shielding the coupling between adjacent signal lines has been proposed in [1], where some vias are inserted in the board of automated test equipment (ATE). However, this structure needs the co-design between the ATE board and the pogo pin structures, and the cost may be high. Hence, a new isolation scheme, which is also an inserted-vias like structure, will be proposed in this work. It can be directly fabricated in the dielectric holder of the test socket, with smaller cost.

This paper is organized as follows. The three-ground pin pattern (S/G = 2) is taken as an example to investigate its associated crosstalk and reflection noises in Section II. The isolation structure is proposed and analyzed by full wave method in Section III, followed by brief conclusions in Section IV.

II. Simulation of Original Pogo Pin Structure

Some frequently-used pin patterns for single-ended pogo pin structures have been shown in [2]. The three-ground pin pattern is very common for the case with S/G = 2 and is chosen as the example to demonstrate the analysis and design in this work. To begin with, it is characterized by employing the full wave software HFSS [3]. The whole and lateral views of the simulation structure are shown in Figs. 1(a) and (b) with the geometric parameters: barrel radius $R_1 = 0.15$ mm, plunger radius $R_3 = 0.09$ mm, barrel length $L_1 = 3.1$ mm, upper plunger length $L_2 = 0.31$ mm, bottom plunger length $L_3 = 0.69$ mm, and pin pitch $D = 0.7$ mm, while the associated pin assignment is presented in Fig. 1(c), where "S" and "G" indicate the signal and ground pins, respectively.

The return loss of the central signal pin indexed by 1 versus frequency with the barrel radius R_1 as a parameter is expressed in Fig. 2(a). Besides, the near-end crosstalk (NEXT) and far-end crosstalk (FEXT) noises between signal 1 and other signal pins indexed by 2, 3, and 4, respectively are shown in Figs. 2(b) and (c), where the barrel radius $R_1 = 0.1$ mm. It is seen that the pogo pin structure has minimum reflection at certain barrel radius near $R_1 = 0.15$ mm which corresponds to a better impedance match to the reference impedance, i.e., 50Ω. In Fig. 2(b) and (c), it can be found that the adjacent pins, i.e., signal pins 1 and 2 have the worst crosstalk noise as expected. Consequently, only the crosstalk between signal 1 and 2 is considered in the following work.

978-1-4244-4447-2/09 $25.00 © 2009 IEEE

III. Analysis of New Isolation Structure

Due to the manufacturing limitation, the barrel radius R_1 should not be smaller than 0.1 mm. It is well known that the crosstalk is reduced for smaller R_1 provided that the pin pitch D remains fixed. However, it is evident from Fig. 2(b) that the NEXT exceed -20 dB at 2.5 GHz even if $R_1 = 0.1$ mm. Hence, a new isolation structure is proposed to assure sufficiently small crosstalk up to 10 GHz.

Figure 3 shows the preliminary scheme of this isolation structure. Four vias with the radius R_2 are inserted between pogo pins in the dielectric holder of the test socket. Different from the previous work [1], these inserted vias should not contact the upper and lower ground planes, which are practically located in the chip under test and ATE board, respectively. A new metallic plane is inserted in the socket to connect all ground pogo pins and the inverted-vias, with anti-pad for the signal pins. These inserted vias act as the additional return current paths, providing better isolation between signal pins and thus smaller crosstalk. However, these additional return current paths will reduce the equivalent impedance of the pogo pin structure and hence, the geometric parameters of the inserted vias and the pogo pins should be well designed to concurrently meet the specifications of crosstalk and reflection noise.

The high-frequency characterization of this one-layer isolation structure is simulated by the full wave software HFSS, with isolation plane placed in the center of the barrel and the radius of inserted via R_2 equal to R_1. The NEXT and FEXT as well as the reflection coefficient versus frequency with R_1 as a parameter are shown in Fig. 4. It is seen that the structure with $R_1 = 0.1$ mm has the lowest crosstalk noise, the NEXT keeps smaller than -20 dB until 9 GHz, but the FEXT exceeds -20 dB at frequency higher than 4.5 GHz. Also, the reflection is smaller than the original one because additional charges will be induced on the inserted vias, which makes the equivalent impedance closer to 50Ω.

Since the inserted vias are open circuited at both ends, the one-layer isolation structure is not effective at higher frequencies. More isolation planes can overcome this drawback and the structures are shown in Fig. 5. In case of two isolation planes, the two metallic planes are placed at the two ends of barrel, respectively. For the structures with three isolation planes, the third metallic plane is placed in the center of barrel. Four inserted vias are also employed, whose radius R_2 remains equal to R_1.

In case of two isolation planes, the NEXT and FEXT as well as the reflection coefficient versus frequency with R_1 as a parameter are shown in Fig. 6. It is seen that the NEXT and FEXT as well as the return loss are all smaller than -20dB from DC to 10GHz if R_1 and R_2 are both designed as 0.1 mm. In case of three isolation planes, the simulated results of crosstalk and reflection noises are quite similar to those of two isolation planes structure and not presented here. The additional central metallic plane can make the half wavelength resonance on the inserted via shift to a higher frequency. However, this resonant frequency is far beyond 10 GHz and cannot be observed from Fig. 6.

Finally, the NEXT and FEXT as well as the reflection noises of the original pogo pin structure and isolation structures with different layers of metallic planes are compared in Fig. 7. It is observed that one-layer isolation structure satisfies the noise specification only in the range from DC to 4GHz. With one more metallic plane, i.e., the two-layer isolation structure, the crosstalk and reflection noises will be smaller than -20 dB from DC to 10 GHz, which is sufficient to meet the requirement for the modern high-speed test socket. It also exhibits 10dB improvement in both crosstalk and reflection noise at 10 GHz as compared with the original structure.

IV. Conclusion

A new isolation structure, consisting of inserted vias and metallic planes, has been proposed and analyzed in this paper. This structure is highly pattern-dependent, since the assignments of signal and ground pins should be given initially. Nonetheless, it has the advantage of low cost and easy fabrication compared with the previous work [1]. The three-ground pin pattern is taken as an example to demonstrate the benefit of this structure. The crosstalk and reflection noises are both smaller than -20 dB from DC to 10 GHz, exhibiting 10dB improvement by employing two metallic planes with the radii of barrel and inserted vias equal to 0.1 mm.

References

[1] B. B. Szendrenyi, H. Barnes, J. Moreira, M. Wollitzer, T. Schmid, and M. Tsai, "Addressing the broadband crosstalk challenges of pogo pin type interfaces for high-density high-speed digital appications," in *IEEE MTT-S Int. Microw. Symp. Dig.*, pp. 2209-2212, June 2007.

[2] R.-B. Sun, R.-B. Wu, and S.-W. Hsiao, "Compromised impedance match design for signal integrity of pogo pins structures with different signal-ground patterns," in *IEEE Workshop Signal Propag.Intercon.*, pp. S2-3, May 2009.

[3] Ansoft Corp, *HFSS, Ver. 11*. (www.ansoft.com).

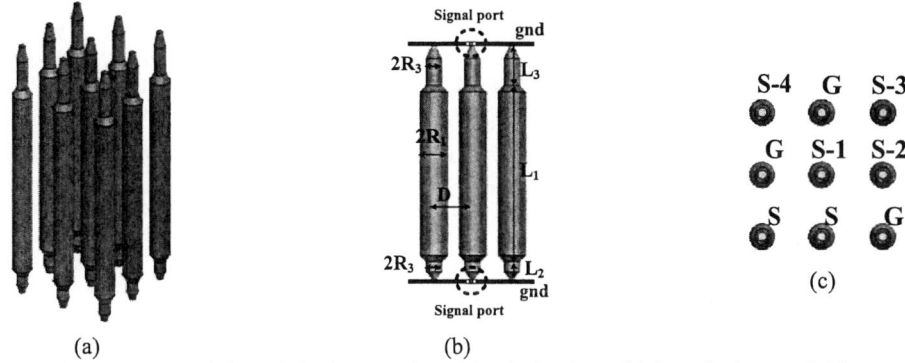

Fig. 1. Simulation structure of the original pogo pins: (a) whole view, (b) lateral view, and (c) associated pin assignment.

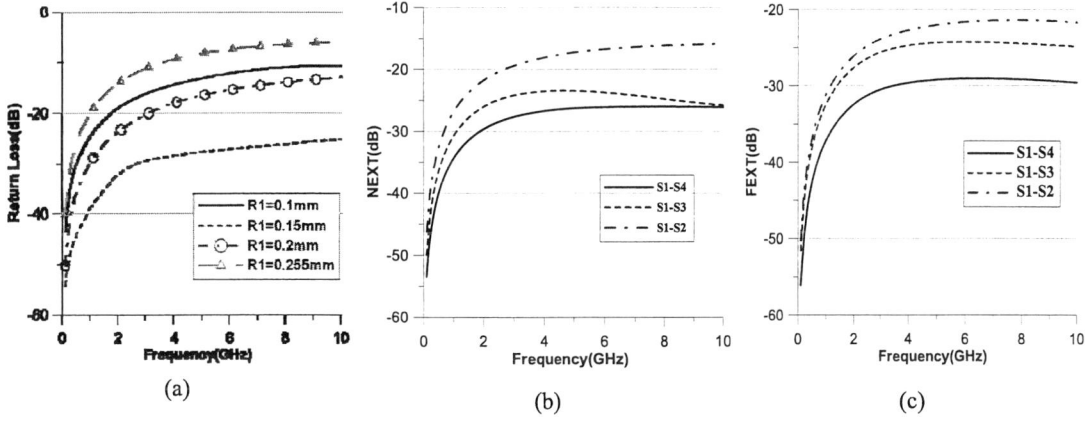

Fig. 2. Simulated results of the pogo pin structure shown in Fig. 1. (a) Return loss versus frequency with R_1 as a parameter. (b) NEXT and (c) FEXT versus frequency between different signal pins with $R_1 = 0.1$mm.

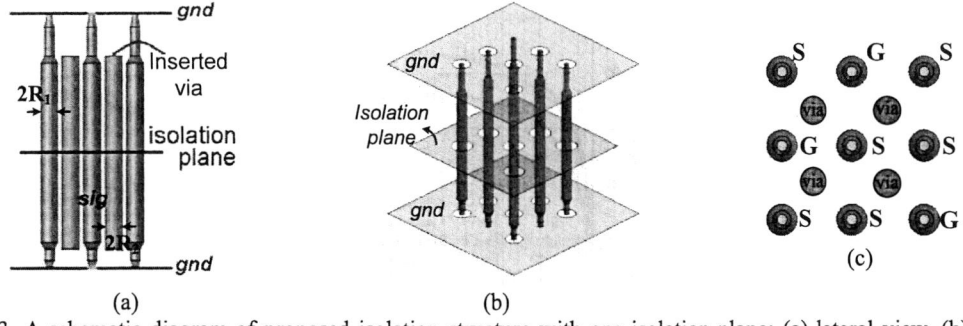

Fig. 3. A schematic diagram of proposed isolation structure with one isolation plane: (a) lateral view, (b) whole view, and (c) top view.

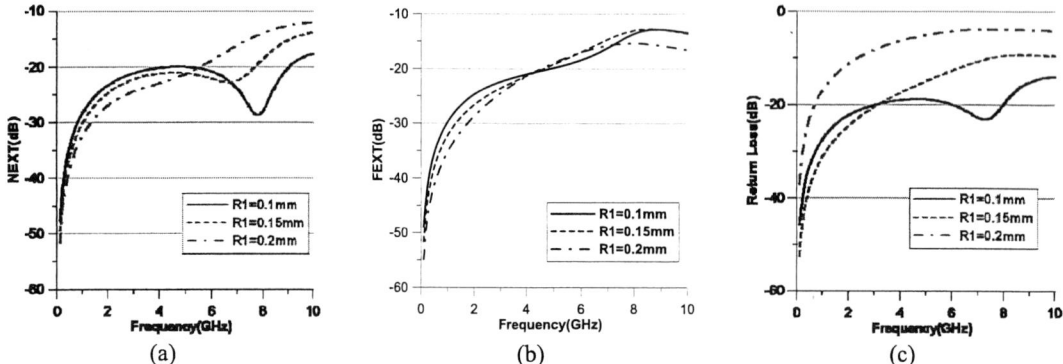

Fig. 4. Simulated results of the one-layer isolation structure shown in Fig. 3 versus frequency with R_1 as the parameter. (a) NEXT, (b) FEXT, and (c) return loss.

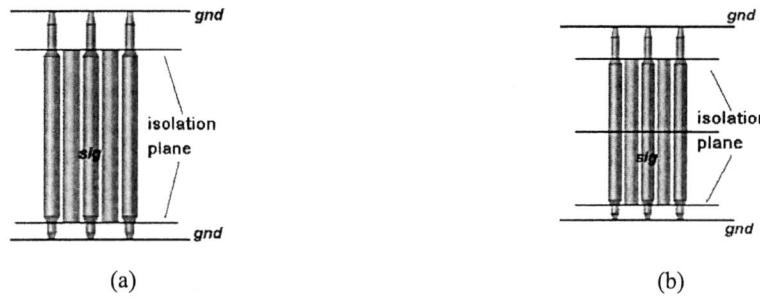

Fig. 5. A schematic diagram of isolation structure with (a) two and (b) three isolation planes.

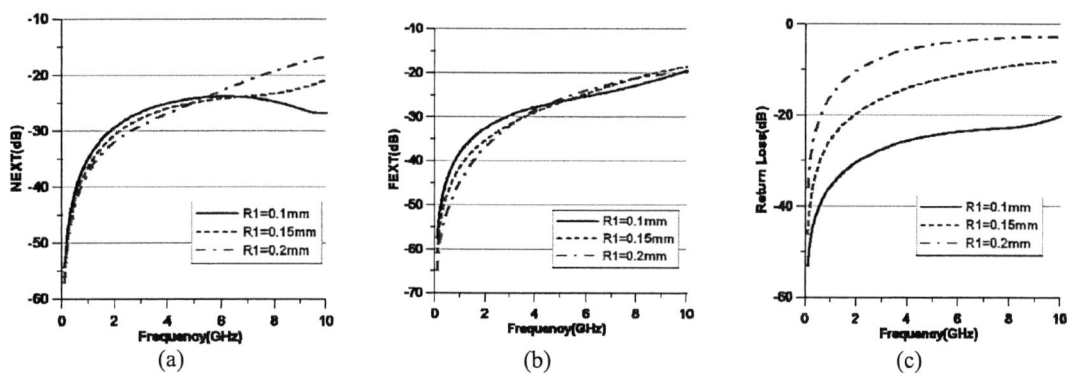

Fig. 6. Simulated results of isolation structure shown in Fig. 5(a) versus frequency with R_1 as a parameter. (a) NEXT, (b) FEXT, and (c) return loss.

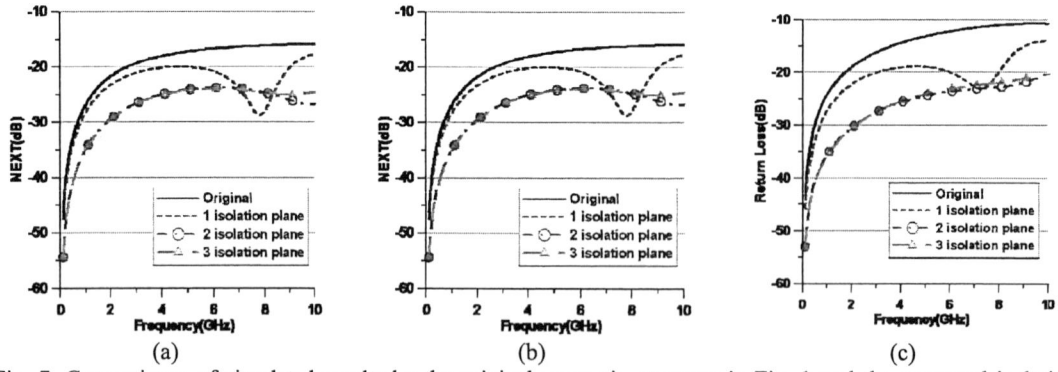

Fig. 7. Comparisons of simulated results by the original pogo pin structure in Fig. 1 and the proposed isolation structure with different numbers of isolation planes. (a) NEXT, (b) FEXT, and (c) return loss.

978-1-4244-4447-2/09 $25.00 © 2009 IEEE   228

Chip-Package Codesign with Redistribution Layer

Mahadevan Suryakumar, Yidnek Mekonnen, Ananda Sarangi

Intel Corporation, MS: CH5-157
5000 W. Chandler Blvd, Chandler, Arizona - 85226
Telephone: (480) 554 5193
Email: mahadevan suryakumar@intel.com

Abstract
The use of redistribution layers to connect I/O circuit to the I/O pad is introduced and the electrical performance of a conceptual design with and without signal redistribution was compared.

Introduction
In today's platform design environment, more and more functions are being integrated into each piece of silicon. For Intel components, this is seen with feature integration into the processor. The feature integration drives the need to increase the number of high speed Input/Output (I/O) C4 bumps for the processor. In flip chip packaging, the upper metal layers of the silicon chip are used for redistributing signal routes from the I/O circuit to the I/O pads. Even though the use of signal redistribution layer is not new and is widely used on many chip designs, careful design consideration is critical in minimizing the parasitics of the redistribution layer on high speed I/O signaling [1] [2]. The paper compares the electrical performance of a conceptual design with and without on chip redistribution layers.

Leadway wire – Electrical characteristics.
The bump to buffer signal routing on the redistribution layer is referred to as lead way routing. The use of redistribution layer even though allows the use of coarser first level interconnect features on the package presents two challenges as shown in Figure 1. The I/O circuit is no longer under the shadow of the I/O bump pad and the signal wire connecting the pad to the circuit adds parasitics to the signal path. The logic adjacent to the I/O circuit doesn't see a power/ground bump directly underneath as the I/O pads encroach into this region. The power delivery to the logic that borders the I/O is through the power/ground bumps and silicon interconnects near the encroached region. Careful design of the silicon upper metal layers interconnect is critical in minimizing the leadway wire parasitics to the I/O while providing robust power delivery to the logic in the encroached region.

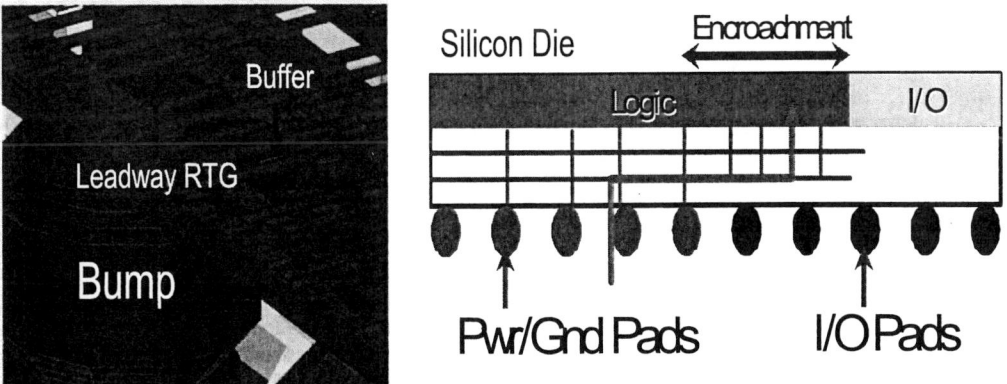

Figure1: Leadway wire and Pad Encroachment

The choice of interconnect layer used for signal redistribution significantly affects the propagation constant. An electromagnetic simulation was developed using Ansoft Q2D and two models were created to study the impact to propagation constant. In the first model, the leadway wire was placed on the upper most thick copper layer (Mx) closest to the pad with the wires on adjacent layers (Mx-1) being orthogonal to it. In the second model, the signal redistribution was placed on the thin metal layer (Mx-1) with wires on adjacent layers (Mx and Mx-2) orthogonal to it. Figure2 plots attenuation (alpha) and phase constant (Beta) of signal redistribution on thick vs thin metal layers. One can clearly observe that the redistribution wire on a thicker metal layer namely Mx is mostly LC dominated while the leadway wire on the Mx-1 layer transitions from LC to RC dominated below 1GHz.

978-1-4244-4447-2/09 $25.00 © 2009 IEEE

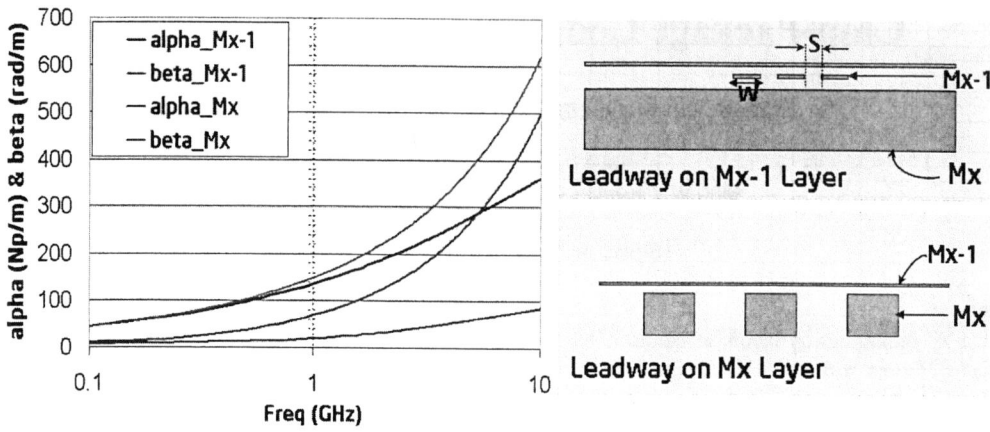

Figure 2: Attenuation/Phase constant plots

Figure3 shows the TDR plot of the conceptual designs with and without leadway wire and characteristic impedance (Zo) vs frequency for different leadway wire geometries. From the TDR plot, one can observe that the design with signal redistribution is more capacitive than the design without leadway routing. Even though the design without leadway routing has a capacitive discontinuity due to the bump pad, the leadway wire is more capacitive than the bump pad. It is also important to note that the impedance discontinuity of the package plated through hole via and the socket pin is much larger than the discontinuity of leadway wire. The Zo vs frequency plot shows that by modulating the wire width and spacing on the signal redistribution layer one can create controlled onchip transmission lines to tune the characteristic impedance of the leadway wire to the desired target value. On designs where leadway wire behaves as a transmission line, the choice of Mx-1 layer for redistribution is more conducive for tuning the leadway wire characteristic impedance to desired value than using the thick copper layer Mx for redistribution.

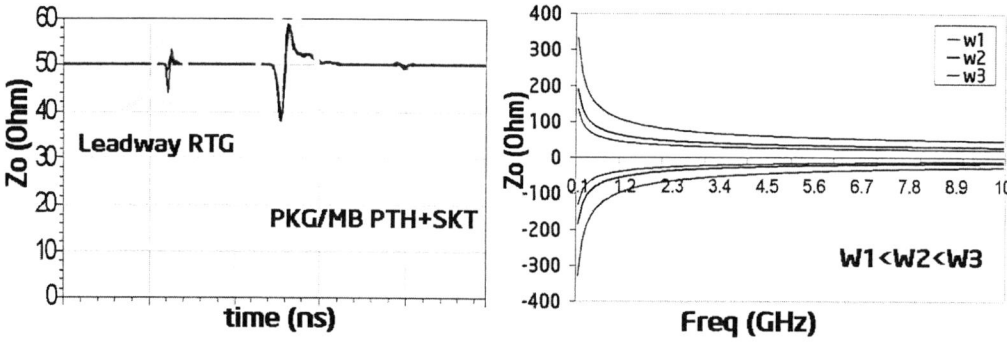

Figure 3: TDR plot and Zo vs frequency plot

Design Case Studies

Optimal arrangement of circuit elements within the I/O cell is an important consideration in minimizing the impact of leadway routing. To illustrate this, a conceptual differential I/O cell is analyzed with and without leadway routing. The I/O cell comprises of line driver, Digital to Analog Converter current source (iDAC), switches, equalizer and control signal buffer circuits. The line driver comprises of ESD diode array, serializer, pre-driver, iDAC current switches, link detect circuit and termination resistor (R_{term}). The iDAC block consists of a bias generator, cascode current source and control signal decoupling. The serializer mux is used to join the two half speed parallel data paths into one full speed serial data path. The output of the serializer mux is driven to the predriver circuit which is used to drive the current steering switches in the DAC. The multi phase clock path in the serializer circuit is matched relative to each other and are also matched between both line drivers to minimize common mode variation and eye height degradation at the transmitter. The predriver is broken into many segments with the strength of each segment binary weighted to drive the corresponding binary weighted DAC switch. The cascode current source array contains many instances of the base current source cell with all instances capable of supplying constant current to the pads. They are grouped in binary weights so that the current levels can be selected by the equalizer coefficients. The driver outputs are terminated to ground through a resistor such that the desired voltage level is achieved at the pad when the current sinks through the driver into the termination. The cascoded current sources achieve greater output impedance which in turn improves the linearity across de-emphasis setting and pad voltage. With lower output impedances, the step size at lower output voltage can be significantly higher than the step size at higher output

978-1-4244-4447-2/09 $25.00 © 2009 IEEE   230

voltages. These differences can show inconsistent differential signaling levels across pad voltages and cause potential common mode variations. The current steering switches are also binary weighted to match the current sources and are mapped one to one to each current source unit cell and routed accordingly. Careful considerations need to be given to the wire geometries as the minimum metal is dependent on the unit cell current while the maximum width is limited by the routing capacitance. At higher frequencies, the capacitance at the tail current node (C_{tail}) between the current source and switches will reduce the output impedance when Ctail increases.

From Figure 4, Design A has bump placed directly underneath the line driver with no leadway routing from driver to pad for optimal ESD performance. Although the current switches are part of the iDAC they are placed in the line driver. The iDAC is placed in the center to achieve balanced routing between each line driver and the iDAC current source. This configuration although minimizes the capacitance at the pad (C_{pad}), increases the capacitance at the tail current node (C_{tail}). In Design B, all circuits are centralized with only ESD diodes at the pads and R_{term} placed locally at the driver. The pad is connected to the driver through the leadway route as shown. This configuration eventhough marginally increases the C_{pad} considerably reduces C_{tail} due to local R_{term} conducting only half the current and the length of routes from the switches to the current source is considerably reduced. The increase in Cpad due to leadway wire can further be minimized by routing the leadway wire on upper metal layers and by minimizing the wire length. To understand the impact of leadway routing and to quantify the sensitivity of C_{tail}, C_{pad} on I/O performance, detailed simulations were done using Hspice.

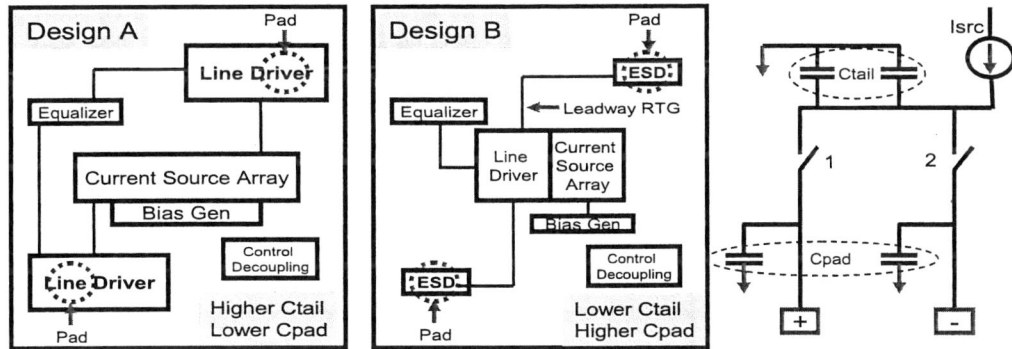

Figure 4: Differential I/O – Design A vs Design B

Figure 5 shows the voltage at + pad and sensitivity of tail capacitance on the rise and fall time. C_{tail} doesn't affect fall time however significantly affects rise time. Increasing C_{tail} reduces rise time while reducing C_{tail} slows down the rise time. This is because when switch 1 turns ON both the current source and C_{tail} sink current and the voltage at the + pad transitions from low to high faster. The voltage at the – pad however transitions from high to low and the fall time doesn't get affected as switch 2 is turned OFF.

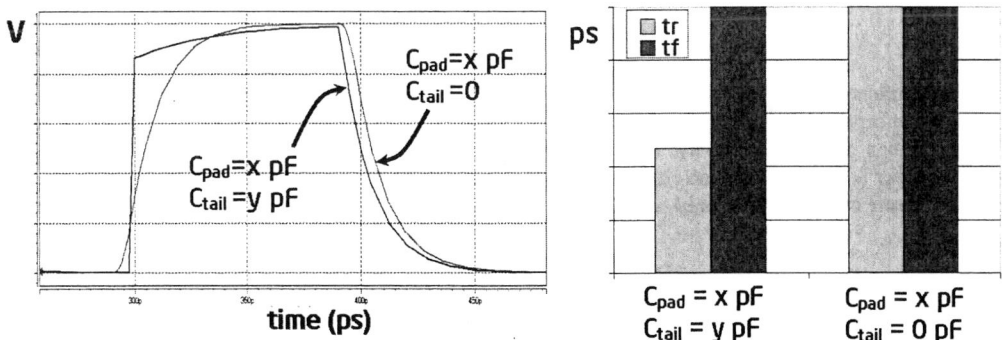

Figure 5: Ctail sensitivity to rise and fall times.

Figure 6 shows the AC Common Mode voltage (ACCM) for a random bit pattern and rise, fall time differences for Design A and B respectively. The rise and fall times for Design B become slightly slower due to the larger C_{pad} but they are more balanced than in Design A. This balance in rise and fall times considerably reduces ACCM for Design B when compared to Design A. Improvement in ACCM is expected to improve eye margins as differential to common mode conversion when the signal reach the receiver pad is reduced [3].

978-1-4244-4447-2/09 $25.00 © 2009 IEEE   231

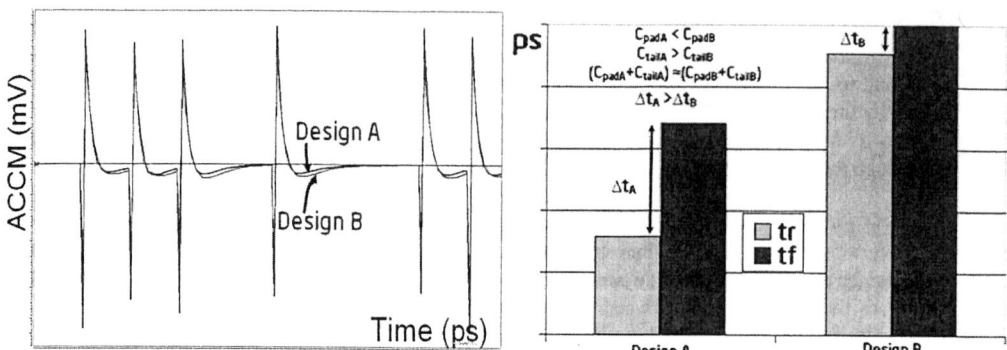

Figure 6: Design A vs. Design B – ACCM and rise, fall time comparison

Power Delivery to Encroached Region

To study the power delivery impact, a sample power grid layout was created. The power grid layouts on adjacent layers were made orthogonal to each other to reduce both resistance and inductance. In the intersections between power and ground metals, farms of vias are added to stitch the different the layers together. The parastics of the power grid was then extracted using a custom power grid simulation tool. The size of the functional unit block was made equal to the model size and the lower metal layer was partitioned into many microcells. The loads were then uniformly distributed with each DC current source attached to the center of each microcell between the lower metal power and ground wires. The max power workload was then exercised in the logic that borders the encroached region and voltage drop plotted on power and ground wires closest to the device. Figure 7 shows the voltage drop as a function of encroachement region. As the encroachment region widens the voltage drop increases. To mitigate the performance risks with the increased voltage drop, careful floor planning of the chip is essential so speed critical circuits that are voltage droop sensitive are not placed in the encroached region or thick interconnect metal layers are added to the silcon chip to minimize the voltage drop.

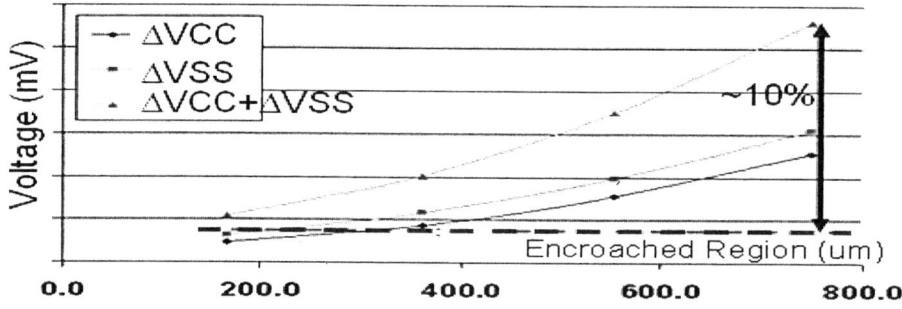

Figure 7: Voltage drop vs Encroached Region

Conclusion

Careful design tradeoffs required to realize a robust design with signal redistribution layers. The scalability limit of the redistribution concept will be limited by the electrical parasitics of the leadway wire. Even though the use of redistribution layer on the silicon chip allows the use of coarser first level interconnect features on the package requires the need for redistribution layer added to the silicon chip, hence cost tradeoff of adding layers to the silicon chip vs using finer first level interconnect feature on the package need to be analyzed in detail.

References

[1] Zhonghua Wu and Oscar Siguenza, "Flip Chip Redistribution Layer Electrical characterization and SSO Noise simulation," in Proceedings of IEEE 4th Tropical meeting on EPEP, pp. 117-120, Oct 1997

[2] Ganesh Balamurugan, Joseph Kennedy, Gaurab Banerjee, James Jaussi, Mozhgan Mansuri, Frank O'Mahony, Bryan Casper and Randy Mooney, "A Scalable 5–15 Gbps, 14–75 mW Low-Power I/O Transceiver in 65 nm CMOS", "IEEE Journal of Solid-State Circuits, Vol. 43, NO. 4, April 2008

[3] H. Heck, S. Hall, B. Horine, T. Liang, "Modeling and Mitigating AC Common Mode Conversion in multi-Gb/s Differential Printed Circuit Boards," in Proceedings of IEEE 13th Topical Meeting on EPEP, pp. 29-32, Oct 2004

An Approach for Quantifying the Conductor and Dielectric Losses in PCB Transmission Lines

Reydezel Torres-Torres, Víctor H. Vega-González

Instituto Nacional de Astrofísica, Óptica y Electrónica, Tonantzintla, Puebla 72480, Mexico.

reydezel@inaoep.mx, hvega@inaoep.mx

Abstract — **An experimental method for separately determining the contribution of the conductor and dielectric losses to the signal attenuation occurring in PCB transmission lines is presented in this paper. This method is based on an analytical extraction of the model parameters for the propagation constant and takes into consideration the dominant effects influencing the signal attenuation. It is demonstrated through a careful model-experiment correlation that the application of this method allows the accurate representation of the loss mechanisms in PCB technology. Furthermore, the frequency beyond which the dielectric losses surpass the skin effect losses is also easily determined in a simple and direct way using the proposed method.**

I. INTRODUCTION

In current technologies, the attenuation (α) is becoming a critical parameter due to the accentuation of the loss mechanisms as the electronic systems move toward higher densities [1]. For this reason, several methods for obtaining α have been proposed in the literature [2], [3]. This has originated many studies where the impact of the transmission line (TL) downscaling on α is analyzed on chip and on PCB [4]-[7]. In these studies it is pointed out that different effects influence the trend of the α versus frequency curve depending on the dimensions, fabrication process and materials, and the operation frequency of the TLs. In fact, using the models presented in [8], the impact of the conductor and dielectric losses in the overall attenuation of a TL can be assessed at different frequency ranges. This provides information about the viability of using a given dielectric or metal for fabricating an interconnect working within a specific frequency range for a particular application. In order to determine the separate contribution of the conductor and dielectric losses to the attenuation of a signal, different approaches have been proposed. However, these approaches either require a priori knowledge of some parameters [9],[10], are only valid for particular structures (e.g. using low loss materials) [11], or use optimization procedures [12]. Owing to this, a method that allows the assessment of the dielectric and conductor losses in practical TLs directly from experimental data is desirable.

In this paper, a simple and physically based method to experimentally determine the separate contribution of the dominant losses associated with TLs fabricated in PCB technology is presented. This method allows to directly obtaining the losses associated with the conductor and dielectric materials as a function of frequency using the experimentally determined propagation constant (γ) and not requiring any additional data. Moreover, the proposed method allows to analytically determining the frequency beyond which the dielectric losses surpass the conductor losses.

II. THEORY

The α versus frequency curve associated with any TL presents five different regions when a single propagation mode is occurring [8]. Depending on the region in which the TL is being operated, the corresponding model for the line can be simplified without significantly losing accuracy. These regions in strict increasing frequency order are:

i) *Lumped-element region*, where the TL can be represented by means of a lumped equivalent circuit.

ii) *RC region*, where the resistance associated with the conductor losses in the TL model is much bigger than the distributed inductive reactance of the TL. This allows to represent the line with a distributed RC network.

iii) *LC region*, where the distributed inductive reactance of the TL is much bigger than the resistance associated with the conductor losses. This allows to represent the line with a distributed LC network.

iv) *Skin-effect region*, where the resistive losses are frequency dependent due to the skin effect.

v) *Dielectric loss region*, where the dielectric losses surpass the conductor losses.

In accordance to the criteria established in [8], the region i) may span up to a couple of gigahertz for interconnects with a few millimeters in length, whereas the regions ii) and iii) are distinguishable when the trace is so thin that the onset frequency of the skin effect presents a relatively high value (e.g. for on-chip interconnects). Thus, since PCB interconnects are usually thick enough so that the onset frequency of the skin effect is well below 1 GHz, the regions i), ii), and iii) take place in very reduced frequency ranges and below the frequency at which the lines are normally operated. This makes possible to use the equations associated with the regions iv) and v) to describe α for PCB TLs as shown afterwards.

The model that describes γ for any homogeneous TL in terms of the RLGC parameters for the regions iv) and v) is [8]:

$$\gamma = \sqrt{j\omega L + R} \cdot \sqrt{j\omega C + G} = \sqrt{1 + R/j\omega L} \cdot \sqrt{j\omega L(j\omega C + G)} \approx (1 + R/2j\omega L) \cdot \sqrt{j\omega L(j\omega C + G)} \qquad (1)$$

In this equation, the factor $(j\omega C + G)$ is replaced by a frequency dependent capacitance C' that takes into account the dielectric losses occurring in the transmission line. This complex capacitance is expressed as:

$$C' = C_0(j\omega/\omega_0)^{-2\theta_0/\pi} \approx C_0(\omega/\omega_0)^{-2\theta_0/\pi}(1 - j\theta_0/2)^2 \qquad (2)$$

where C_0 is the capacitance of the TL, and θ_0 is determined from $\tan\theta_0$, which is the loss tangent of the dielectric. The sub-index '0' specifies that the parameters are defined at frequency $\omega = \omega_0$.

Once that the onset frequency of the skin effect has been surpassed, and neglecting the DC resistance of the TL, the resistance of the line can be defined using the following frequency dependent equation:

$$R = R_0(1 + j)\sqrt{\omega/\omega_0} \qquad (3)$$

To obtain an expression for α, $\mathrm{Re}(\gamma)$ is determined after substituting (2), (3), $L = L_0$, and $\omega = 2\pi f$ in (1); this yields:

$$\alpha = \mathrm{Re}(\gamma) = K_1\Theta\sqrt{f} + K_2\Theta f \qquad (4)$$

where $K_1 = \sqrt{\pi}R_0(\theta_0 + 2)/(\sqrt{8\omega_0 L_0/C_0}) \approx \sqrt{\pi}R_0/(\sqrt{2\omega_0 L_0/C_0})$, $K_2 = \theta_0\sqrt{L_0 C_0}/2$, and $\Theta = (\omega/\omega_0)^{-\theta_0/\pi}$. Notice that, for typical PCB substrates, $\theta_0 \ll 1$, which allows to assume $\Theta \approx 1$. In this case, (4) can be simplified to:

$$\alpha = K_1\sqrt{f} + K_2 f \qquad (5)$$

In the right hand side of this equation, the first and second term represent the attenuation associated with the conductor and dielectric losses (i.e. α_c and α_d) respectively. Thus, once K_1 and K_2 are determined using the method proposed in this paper, α_c and α_d can be separately obtained as explained hereafter.

Dividing both sides of (5) by $\sqrt{f}$ yields:

$$\alpha/\sqrt{f} = K_1 + K_2\sqrt{f} \rightarrow y = b + mx \qquad (6)$$

Since K_1 and K_2 are assumed to be independent of frequency, (6) represents the equation of a line. Moreover, the experimental α can be easily obtained using TRL-based methods [3]. Thus, K_1 can be determined from the intercept with the abscises b of the linear regression of the experimental $\alpha/\sqrt{f}$ versus $\sqrt{f}$ data, whereas K_2 can be obtained from the corresponding slope m. Afterwards, α_c and α_d can be separately obtained from:

$$\alpha_c = K_1\sqrt{f} \qquad (7a) \qquad\qquad \alpha_d = K_2 f \qquad (7b)$$

In order to determine the frequency at which the dielectric losses surpass the skin effect losses, the so-called crossover frequency f_θ can be determined by equating α_c and α_d as defined in equations (7) and solving for f. In this case, f_θ can be experimentally obtained by applying:

$$f_\theta = (K_1/K_2)^2 \qquad (8)$$

Notice that, other than the experimental α, no additional data is required for determining α_c, α_d, and f_θ.

III. EXPERIMENTS

Copper microstrip lines with fixed width of 150 μm, an approximated thickness of 50 μm, and different lengths were fabricated in the top layer of a PCB made of FR4 material. The thickness of this dielectric layer (i.e. the separation between the traces and the ground plane) is 100 μm and the material presents a nominal permittivity and loss tangent of ε_r=4.4 and $\tan\theta$ =0.02 respectively. In addition, the lines are terminated with 1.85-mm female coaxial connectors so that two-port S-parameters can be measured to these structures using coaxial cables and a Precision Network Analyzer (PNA). Once the measurements are performed, the line-line method reported in [3] is used to extract the experimental complex propagation constant of the microstrip lines, which allows to apply the characterization methodology presented in this paper. Fig. 1 shows a sketch of the fabricated lines.

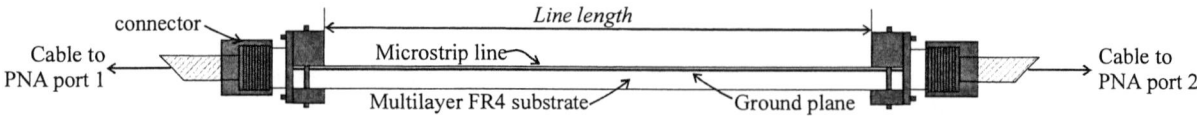

Fig. 1. Sketch of the side view of the fabricated board illustrating a microstrip terminated with coaxial connectors (the figure is not to scale).

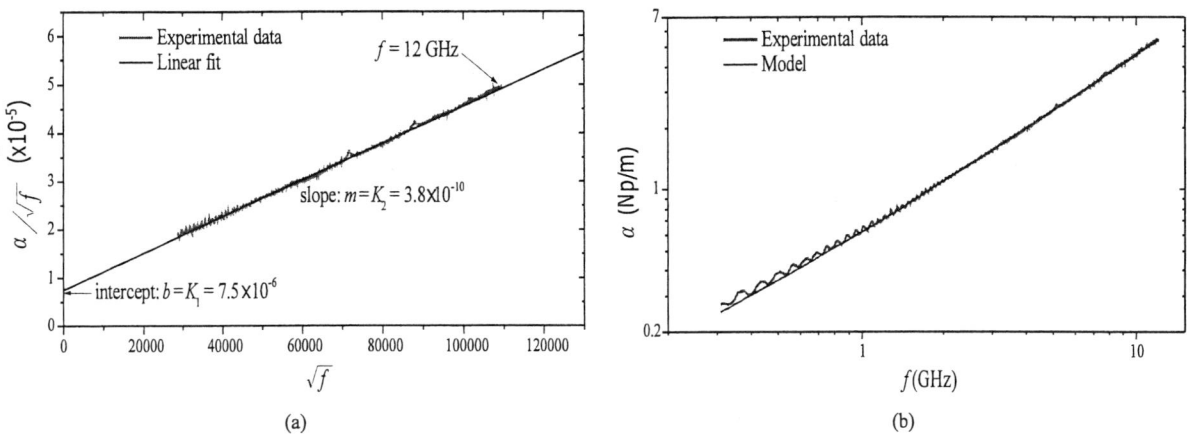

(a)   (b)

Fig. 2. a) Regression used to determine the parameters in the model given in (5). b) Plot showing the model-experiment correlation in log-log scale.

IV. RESULTS AND DISCUSSION

Once that the propagation constant of the fabricated lines was determined, the attenuation was obtained as a function of frequency using $\alpha = \mathrm{Re}(\gamma)$. Afterwards, in accordance with (6), K_1 and K_2 were determined from the intercept and slope of the linear regression shown in Fig. 2a. Notice the good agreement between the experimental data and the linear fit, which allows to obtain the model parameters in a very simple way. Fig. 2b shows the comparison between the model given by (6) and experimental data in a full logarithmic plot. As can be seen, the model accurately predicts the experimental curve even at frequencies below 1 GHz, which validates the parameter extraction technique. The reason why the model describes well the characteristics of these lines within the measured frequency range is related to the skin depth (δ) of copper, which is the material used to form the lines. As the operation frequency increases, δ eventually becomes as small as the line thickness; this point can be used to provide a first-order definition of the onset frequency of the *skin-effect region* (f_δ). Thus, once that the operation frequency is well above f_δ, the conductor losses become frequency dependent and can be accurately modeled using equation (7a). For microstrip lines fabricated on PCB technology, f_δ takes values in the range of some megahertz due to the relatively thick metal traces used to form the lines. For instance, as can be seen in Fig. 3, the fabricated copper lines present a thickness of 50 μm, which equals δ at a frequency $f_\delta \approx 1.7$ MHz. In this case, regions i), ii), and iii) (as defined in Section II) will be present below this frequency, which suggests that only regions iv) and v) are of interest for practical high-speed PCB interconnects.

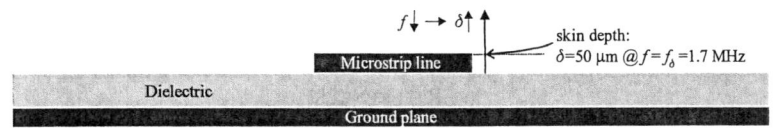

Fig. 3. Cross section view of the fabricated copper microstrip line showing that the skin depth equals the thickness of the line at a frequency lower than 2 MHz.

As demonstrated in Section II, the model parameters K_1 and K_2 have a physical significance, being the coefficients of the functions describing the conductor and dielectric losses respectively. Hence, when these parameters are known (7a) and (7b) can be used to calculate the attenuation associated with the conductor and dielectric losses respectively. Fig. 4 shows the curves corresponding to α_c and α_d as well as the comparison between the simulated $\alpha_c + \alpha_d$ with experimental data. As expected, the dielectric losses represent the dominant mechanism contributing to the signal attenuation in these lines due to the relatively high value of $\tan\theta$ for FR4, whereas the conductor losses remain in a lower level since in PCB technology thick traces are used, reducing the corresponding resistance. In this particular case, f_θ is found to be 390 MHz after applying (8), and this value can be graphically seen in Fig. 4b at the frequency point where $\alpha_c = \alpha_d$. Notice that f_θ will highly depend on the type of materials and dimensions of the lines. For instance, in packages, low loss dielectrics and much thinner lines are used, which yields higher conductor losses and lower dielectric losses when compared with regular PCB lines. Thus, in package interconnects it is expected that f_θ will easily be within the gigahertz range. Finally, in order to explicitly present the impact of the conductor and dielectric loss mechanisms on the total line attenuation, the corresponding contribution (in percentage) of these losses is plotted as a function of frequency in Fig. 5. Notice that,

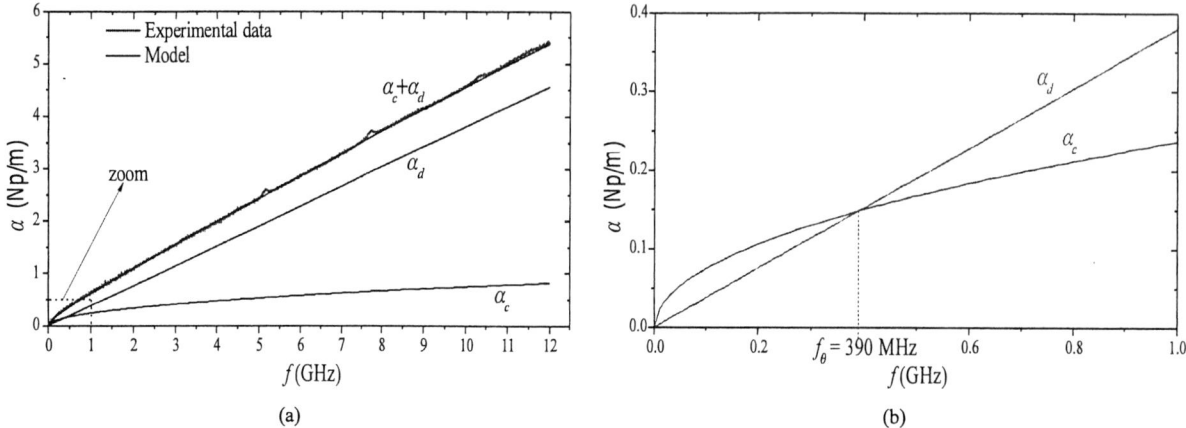

(a)   (b)

Fig. 4. a) Plot illustrating a model-experiment correlation in linear scale and including the curves that represent the conductor and dielectric losses. b) Zoomed-in plot showing in detail the region where α_c equals α_d, and f_θ can be obtained.

beyond 1 GHz, more than 60% of the line losses are associated with the dielectric for the studied structures.

V. CONCLUSIONS

A new method for separately determining the conductor and dielectric losses in PCB transmission lines has been theoretically formulated and experimentally verified. The method allows the determination of the model parameters directly from experimental data and not requiring any additional data. This simple method can be used to compare transmission lines fabricated in different technologies and to identify the effects limiting the performance of interconnects. It is necessary to mention that additional effects have to be taken into consideration for extending frequency ranges of applicability for the method (e.g. surface roughness). Work in this direction is currently ongoing.

ACKNOWLEDGEMENT

This work was partially supported by Intel Co. and CONACyT-Mexico.

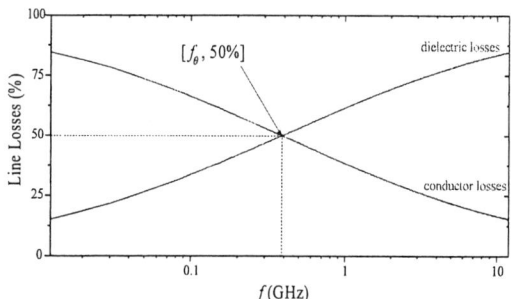

Fig. 5. Percentage of losses as a function of frequency. Notice that the dielectric losses represent the main loss mechanism in current FR4 PCB substrates within the microwave range.

REFERENCES

[1] R. Tummala, "Moore's law meets its match", *IEEE Spectrum*, vol. 43, pp. 44-49, Jun. 2006.

[2] G.F. Engen and C.A. Hoer, "Thru-reflect-line: an improved technique for calibrating the dual six-port automatic network analyzer," *IEEE Trans. Microw. Theory Tech.*, vol. 37, pp. 987-993, Dec. 1979.

[3] J.A. Reynoso-Hernández, "Unified method for determining the complex propagation constant of reflecting and nonreflecting transmission lines," *IEEE Microwave Wireless Comp. Lett.*, vol. 13, pp. 351-353, Aug. 2003.

[4] L.P. Vakanas, A.C. Cangellaris, and J.L. Prince, "A parametric study of the attenuation constant of lossy microstrip lines," *IEEE Trans. Microw. Theory Tech.*, vol. 38, pp. 1136-1139, Aug. 1990.

[5] C. Schöllhorn, W. Zhao, M. Morschbach, and E. Kasper, "Attenuation mechanisms of aluminum millimeter-wave coplanar waveguides on silicon," *IEEE Trans. Electron Devices*, vol. 50, pp. 740-746, Mar. 2003.

[6] J.C. Rautio and V. Demir, "Microstrip conductor loss models for electromagnetic analysis," *IEEE Trans. Microw. Theory Tech.*, vol. 51, pp. 915-921, Mar. 2003.

[7] A.K. Rastogi and S. Hardikar, "Attenuation characteristics and sensitivity analysis for coplanar waveguide," *Int. Jour. Infrared Millim. Waves*, vol. 22, pp. 703-714, May 2001.

[8] H. Johnson and M. Graham, High Speed Signal Propagation: Advanced Black Magic. Englewood Cliffs, NJ: Prentice-Hall, 2003.

[9] H. Yue, K.L. Virga, and J.L. Prince, "Dielectric constant and loss tangent measurement using a stripline fixture," *IEEE Trans. Compon., Packag., Manuf. Technol.*, vol. 21, pp. 441-446, Nov. 1998.

[10] J.E. Chan, K. Sivaprasad, and K.A. Chamberlin, "High-frequency modeling of frequency-dependent dielectric and conductor losses in transmission lines," *IEEE Trans. Compon. Packag. Technol.*, vol. 30, pp. 86-91, Mar. 2007.

[11] M. Cauwe and J. De Baets, "Broadband material parameter characterization for practical high-speed interconnects on printed circuit board," *IEEE Trans. Advanced Packag.*, vol. 31, pp. 649-656, Aug. 2008.

[12] Q. Yu and O. Wing, "Computational models of transmission lines with skin effects and dielectric loss," *IEEE Trans. Circ. Syst. – I*, vol. 41, pp. 107-119, Feb. 1994.

GPGPU-FDTD Method for 2-Dimensional Electromagnetic Field Simulation and Its Estimation

Masaki Unno[†] Yuta Inoue[‡] and Hideki Asai[†‡*]

†Graduate School of Engineering, Dept. of Systems Eng., Shizuoka University

‡Graduate School of Science and Technology, Shizuoka University

* Dept. of Systems Eng., Shizuoka University

3-5-1 Johoku, Naka-ku, Hamamatsu-shi, 432-8561 Japan

Phone: +81-53-478-1237, Fax: +81-53-478-1269

Email: {kouy01umk, inoue}@tzasai7.sys.eng.shizuoka.ac.jp, hideasai@sys.eng.shizuoka.ac.jp

Abstract: For signal/power integrity analysis of the high density packages and printed circuit boards, the FDTD (Finite-Difference Time-Domain) method has been widely used. In order to apply to large-scale problems, a variety of acceleration techniques are required. This paper describes a GPGPU-FDTD (General Purpose computing on GPU (Graphic Processing Unit)-Finite-Difference Time-Domain) method for massively parallel electromagnetic field simulation. Finally, it is confirmed that GPGPU-FDTD method shows the high-performance when the computational algorithm is programmed suitably for the architecture of GPU.

Keywords: FDTD method, GPU, GPGPU

1. Introduction

With the progress of the high-density integration circuit technology, the power/signal integrity and EMI problems in the high-speed signal transmission have become important, and the simulation techniques for them have also attracted a great deal of attention [1].

The FDTD (Finite-Difference Time-Domain) method is one of the useful electrical simulation techniques to solve the Maxwell's equation [2]. This method has been widely used for a variety of problems such as the antenna analysis and the PCB analysis. In the FDTD method, the space to be analyzed is discretized by an enormous number of cells, and the values of electric and magnetic fields are updated alternately in the time domain. However, in this method, the time step size is restricted by the Courant stability condition. Therefore, the very large amount of CPU time is required for the large problem. In order to cope with these problems, the parallel processing by the PC cluster machine and the multi-core CPU [3][4][5] and the acceleration algorithm such as ADI(Alternating-Direction Implicit)-FDTD method [6], which is an unconditionally stable method independent of the Courant stability condition, have been discussed.

In this research, we focus attention on the use of GPU (Graphics Processing Unit) and GPGPU (General Purpose GPU). In this paper, two-dimensional (2-D) GPGPU-FDTD (GPGPU-based FDTD) method is described. Finally, it is shown that the FDTD simulation method with GPU is more than 30 times faster than the single CPU-based FDTD method without losing accuracy.

2. The FDTD method

In the FDTD method, the space to be analyzed is discretized by a lot of fine cells and the Maxwell's equation is differentiated and solved according to the Yee's algorithm. Applying the central difference method in time and space, electrical and magnetic field unknown vectors are located alternately. The update formulas for the TE wave in 2-dimensional space are given by:

$$E_x^{\,n}(i+1/2, j) = C_{EX}(i+1/2, j)E_x^{\,n-1}(i+1/2, j)$$
$$+C_{EXLY}(i+1/2, j)\left\{H_z^{\,n-1/2}(i+1/2, j+1/2) - H_z^{\,n-1/2}(i+1/2, j-1/2)\right\}, \tag{1}$$

$$E_y^{\,n}(i, j+1/2) = C_{EY}(i, j+1/2)E_y^{\,n-1}(i, j+1/2)$$
$$+C_{EYLX}(i, j+1/2)\left\{H_z^{\,n-1/2}(i+1/2, j+1/2) - H_z^{\,n-1/2}(i-1/2, j+1/2)\right\}, \tag{2}$$

978-1-4244-4447-2/09 $25.00 © 2009 IEEE   237

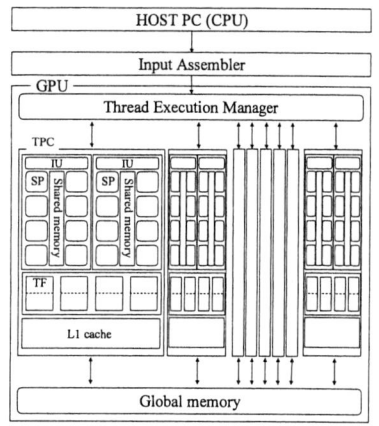
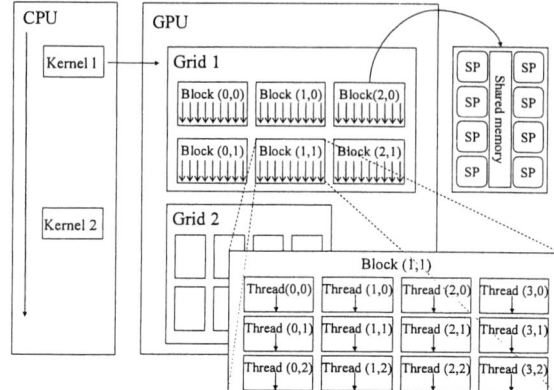

Fig. 1: The diagram of GeForce 8800 GTX. Fig. 2: The relationship between kernel and multi-threads.

$$H_z^{n+1/2}(i+1/2, j+1/2) = H_z^{n-1/2}(i+1/2, j+1/2)$$
$$-C_{HZLX}(i+1/2, j+1/2)\{E_y^{n}(i+1, j+1/2) - E_y^{n}(i, j+1/2)\} \tag{3}$$
$$+C_{HZLY}(i+1/2, j+1/2)\{E_x^{n}(i+1/2, j+1) - E_x^{n}(i+1/2, j)\},$$

where C's are the coefficients which depend on the material properties and n is time step. In addition, i and j denote the coordinate in 2-dimensional space.

3. GPGPU and CUDA [7][8]

In a few years, applications of general purpose GPU to the physical simulations have been explored. As an example, GeForce 8800 GTX which is one of the GPUs is shown in Fig.1. GeForce 8800 GTX is composed of 8 TPCs (Thread Processor Clusters) and 512MB global memory. Each TPC is composed of two SMs (Streaming Multi-processors), 8 TF (Texture Filtering) units and an L1 cache. And each SM consists of 8 SPs (Streaming processors) which can carry out single precision floating point calculations, IU (Instruction Unit) and 16KB shared memory. The frequency of operation is 1.35GHz.

CUDA is an environment for the development with API (Application Program Interface) and specific compiler. GPU includes the structure to operate the threads in parallel, where the thread is the smallest element of the process and it is executed by each SP. In CUDA, a group of threads is denoted by block. Each block is assigned to each SM. A group of 32 threads, which is defined as warp, is executed by one SM in parallel. In addition, the cluster of two or more blocks is called a grid. And then, the number of blocks and threads can be arbitrarily specified. All of the threads in the grid execute the same program which is called kernel. The relationships between them are shown in Fig. 2.

All the operations among the threads are synchronized in a block, but the operations among the blocks in a grid are not always synchronized. Therefore, an algorithm which dose not require the synchronization among the blocks is needed.

4. 2-D GPGPU-FDTD method with CUDA

In this section, 2-D GPGPU-FDTD method with CUDA is described. In GPGPU-FDTD method, (1)-(3) are calculated on GPU. In order to compute these equations, the data which are allocated and initialized on CPU have to be transferred to GPU. Also, at each time step in FDTD method, the data transfer from GPU to CPU is required in order to obtain the computation results.

On GPGPU with CUDA, every thread should be operated in parallel in the block for acceleration of the simulation. Therefore, in the FDTD method, the space to be analyzed is divided into sub-domains and each sub-domain is assigned to each block. In addition, in each block, the electric and magnetic field components assigned to each edge and each surface should be correspondent to each thread.

978-1-4244-4447-2/09 $25.00 © 2009 IEEE   238

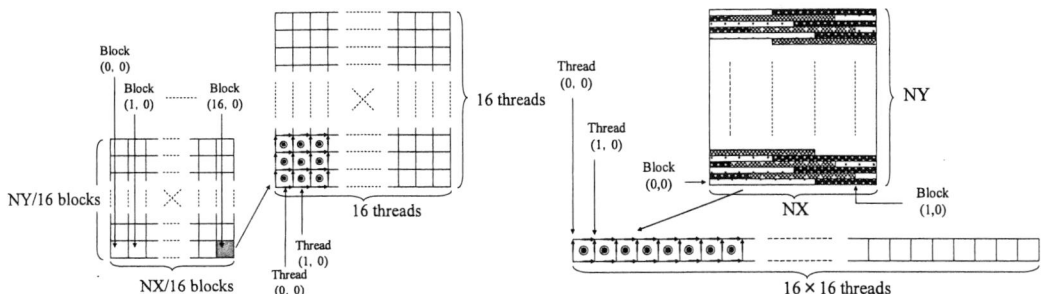

Fig. 3: Domain partitioning and block assignment – 1 (Scheme-1).

Fig. 4: Domain partitioning and block assignment -2 (Scheme-2).

Here, two schemes of domain partitioning and block assignment have been tried. One is the scheme-1 where the domain is divided as shown in Fig. 3. In general, it is well-known that the suitable number of threads in a block is 196 or 256 [8]. In Fig. 3, the number of threads in a block is fixed to 256(= 16 × 16). As a result, the number of blocks become NX/16 × NY/16. The other is scheme-2 where the domain is divided as shown in Fig. 4. As well as the case in Fig. 3, the number of threads is 256. The difference between two schemes is in the process of memory access. In Fig. 4, each thread accesses the data in the global memory which are stored at the continuous physical memory addresses.

5. Numerical results

In order to confirm the accuracy and the speed of the FDTD methods with GPGPU and single CPU, the simulations have been done for the 2-D free space in Fig.5. The cell size is set as follows:

$$\Delta(k)_{(mm)} = \begin{bmatrix} 100.0 & (k = 1 - 32) & \Delta(k) : \text{constant} \\ 68.4 - 6.0 & (k = 33 - 39) & \Delta(k) = 1.5\Delta(k+1) \\ 4.0 & (k = 40 - 62) & \Delta(k) : \text{constant} \\ 6.0 - 68.4 & (k = 63 - 69) & 1.5\Delta(k) = \Delta(k+1) \\ 100 & (k = 70 - \text{NX}) & \Delta(k) : \text{constant} \end{bmatrix} \tag{4}$$

where $\Delta(k) = \Delta x(i)$ for the x-direction and $\Delta(k) = \Delta y(j)$ for the y-direction. Additionally, it is assumed that NX = NY where both of NX and NY are the numbers which are just dividable by 16. The excitation is given by the H_z component at the point ($i = j = 50$), and the following waveform was input as the excitation.

$$Hz = Hz + \sin^2\left(\frac{\pi t}{T}\right) \quad T = 9.4ns \tag{5}$$

The time step size is set to 9.43 ps according to Courant stability condition. The total simulation interval is [0 ns, 40 ns]. In the simulation, GeForce GTX295 was used. This graphics card is composed of two GPUs. Each unit has 240SPs and the 895MB global memory. The frequency of operation is 1242MHz. In this research, only one GPU was used for the simulations, and the shared memory was not used. In addition, Intel Xeon X5482 3.2GHz was used as the single CPU PC.

In Fig. 6 and Fig. 7, the waveform at the observation point ($i = j = 90$) and the speed-up ratio for the problem composed of 1696 × 1696 cells are shown, respectively. From Fig. 6, it is confirmed that good agreement between the waveforms obtained by single CPU and GPGPU is shown. In addition, from Fig. 7, it can be seen that GPGPU-FDTD method for the scheme-2 is more than 30 times faster than the FDTD method with single CPU.

6. Conclusion

In this paper, we verified the computational performance of 2-D GPGPU-FDTD method with CUDA. From the numerical results, it has been confirmed that the simulation with GPU is more than 30 times faster than the simulation

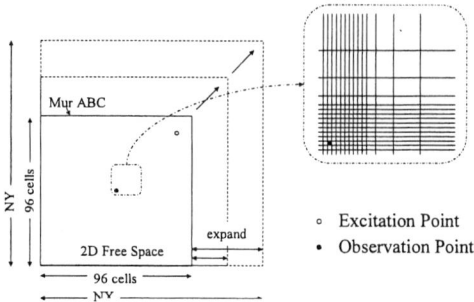

Fig. 5: An example space.

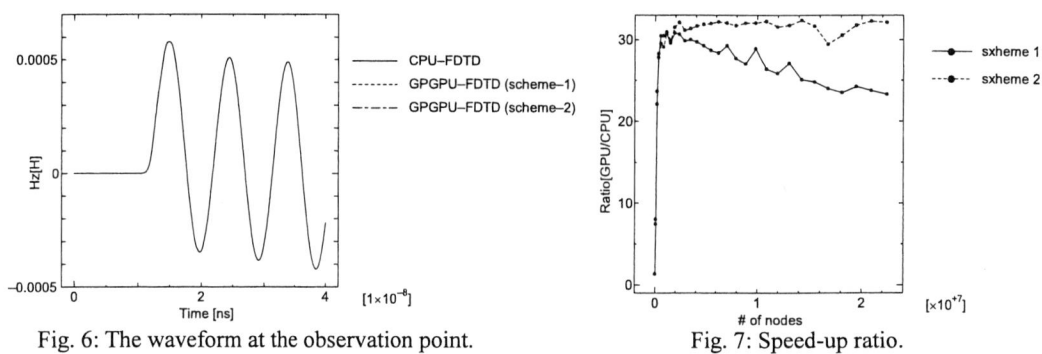

Fig. 6: The waveform at the observation point.

Fig. 7: Speed-up ratio.

with single CPU without losing accuracy. Additionally, the difference of the performance due to the scheme of domain partitioning and block assignment was shown.

Reference

[1]. H. Asai, "Present Status and Future Trend of Electrical Simulation Technology for High-Speed Electronic Design," EDAPS 2006, Shanghai, China, Dec 2006.

[2]. K. S. Yee, "Numerical solution of initial boundary value problems involving Maxwell's equations in isotropic media," *IEEE Trans. Antennas Propagat.*, vol. AP-14, pp. 302-307, May 1966.

[3]. H. Asai, T. Watanabe, T. Sasaki, and K. Araki, "An EMI Simulator Based on the Parallel-Distributed FDTD Method for Large-Scale Printed Wiring Boards", Proc. SPI2002, pp.141-144, May 2002.

[4]. K. Araki, T. Murayama, M. Suzuki, T. Watanabe and H. Asai, "With the Development of Tools to Conquer Electromagnetic Noise, PCB's Can be Analyzed in 4 Hours", Nikkei Electronics, No.892, pp.117-130, Jan 2005.

[5]. N. Oguni, H. Asai, "Estimation of parallel FDTD-based electromagnetic field solver on PC cluster with multi-core CPUs," EDAPS 2008, pp. 167-170, Dec. 2008.

[6]. T. Namiki, "A New FDTD Algorithm Based on Alternating-Direction Implicit Method," *IEEE Trans. Microwave Theory Tech*, vol. 47, pp. 2003-2007, Oct 1999.

[7]. J. D. Owens, D. Luebke, N. Govindaraju, M. Harris, J. Krüger, A. E. Lefohn, and T. J. Purcell, "A Survey of General-Purpose Computation on Graphics Hardware," In Eurographics 2005, State of the Art Reports, pp.21-51, Aug 2005.

[8]. CUDA 2.2 Programing Guide, http://www.nvidia.com/object/cuda_develop.html.

Distributed Via Connectivity in High Resolution Package Power Delivery Modeling

Omer Vikinski
Intel Corp, MTM Park, Haifa, Israel
Phone: 972-4-8651418, FAX: 972-4-8655055
omer.vikinski@intel.com

Key words: Distributed Models, Via Connectivity, Power Delivery

Abstract

High resolution electrical models can be generated for chip packaging power delivery networks. The lumped versus distributed stitching of vertical path connections such as micro-via and plated-through-holes can introduce various effects on the model predictions.

Introduction

Many internal and external tools [1-3] are being used to aid the package designer with modeling of the various power supply rails of his product. Some of the tools are oriented to DC electrical solution, some are oriented to AC sweep, and some to transient simulations. Most of the engines treat the on-die chip, package, and board structures as distributed electrical phenomena which require structural partitioning and model complexity limited by reasonable run time and computing resources.

The most basic analysis of a DC solution can contain quite a large research and innovation ground for tool complexity and innovation. Modeling size, scope, and efficient solving are among the computed aided design challenges which lie in this field. In addition, most designers would find DC solution data results very valuable in assisting them to establish intuition and understanding of the electrical behavior and level of robustness for their power supply design. In a second stage, the DC solution's results may assist with design tuning to communicate characteristic numbers of the design with the various project stake holders and to tradeoff between performance and cost pressures and constraints acting on the design.

As the available computer resources and solver engine algorithms become more powerful and efficient, and some of them specialize for the specific case of DC solution, the opportunity to explore high resolution modeling capabilities and limitations becomes possible.

In this work the focus was devoted to explore the effects of incorporating distributed vertical path resistive array modeling. Vertical path stitch between power planes and connection line structures located at different layers of a conventional package or board's multi conductive layer substrates. Among the common vertical path structures are the fine resolution micro-via connections and the grosser plated-through-holes connections. In the first section, high resolution DC model generation will be described. Next, the test case chosen for illustration and its baseline modeling results would be discussed. In the results section, the effect of treating vertical path stitching as lumped versus distributed array connection on the modeling results will be shown.

Model Generation

In order to explore various modeling options, simple program scripts were coded to generate a high resolution resistive mesh netlist that represent the power delivery structures. The input information for the multi conductive layer package or board design and their vertical connection are collected as text stream from the editing tool. The text stream represents power bussing planes by their external and internal (void) contour lines and arcs. Basic point-in-polygon graphic processing algorithm [4] is used to digitize those planes into a discrete pixel array. The digitization resolution is set by the user. Figure 1 shows a power plane layout drawing and its digitized pixel array generated by the aforementioned code. The pixel arrays representing the power planes digitized image are stitched by resistive mesh according to the conductive layer thickness and conductivity.

Vertical paths are being stitched according to the structural feature (micro-via or plated-through-hole as example) central location. Single resistance element captures the lumped electrical contact of that vertical path. Once the complete electrical resistive mesh is translated to a netlist including all the lateral mesh and

978-1-4244-4447-2/09 $25.00 © 2009 IEEE

vertical stitching components, terminals are assigned. Current and voltage sources are applied according to the expected usage model or analysis of interest. Once all the described content is included in the spice netlist, a specialized DC solver engine - internal tool equivalent in performance to Hypre [5] - is applied. Its output is the model solution. This solution includes all the model element's currents and all of the model node's voltages. Few auxiliary simple applications assist for graphical display of the results. Figure 2 illustrates the visual user interface, which shows pseudo color images the current distribution among an array of vias.

Test Case and Distributed Model Exploration

Intel's new generation microprocessor quad-core desktop product would serve as a test case. The investigated power supply rail would be the IA (Intel Architecture) core supply. The current stimuli applied at each die bump will be taken from a representative TDP (Thermal Design Power) virus application power map of the chip. The voltage termination is applied at the board side, sufficiently further away from the package pin field to avoid boundary conditions artifacts for the package electrical results.

For establishing results baseline, vertical connections for both micro-vias and plated-through-holes were treated as a lumped singular stitch. This means that a single resistive element was stitched between two conductive layers to represent the vertical path connectivity. The location for the stitch was at the center of the vertical structure position, ignoring its specific structural details. Next, the netlist generating script was enhanced to stitch an array of resistive elements to better capture the vertical path distributed electrical connectivity as implied by its geometrical structural details. Filled micro-vias were stitched by an array of resistive connections, capturing their full contact footprint according to their characteristic contact diameter. Plated-through-holes were stitched by an array of resistive connections to capture their conductive plated wall on the perimeter of the through-hole. Figure 3 (left pane) visualizes the pixel based lumped and distributed via connectivity array stitching.

Results

From each simulation deck, all the micro-via and plated-through-holes currents information can be captured and stored. In the distributed vertical model cases, the total current flowing through the vertical path stitched array components is back annotated into single value. The ratio between the numerical results for micro-via's current when the model is distributed versus the cases where it is lumped (the baseline) is plotted as a histogram (figure 3 upper right pane). The histogram mean is unity, since the total sum current as injected from the chip in both cases is equal. It can be seen that the deviations between the two modeling cases is quite low and in the order of 5% to 10%. The base side micro-vias show slightly lower error spread as they are located further away from the chip stimuli application spots, making the current spread reaching them more uniformly and steadily.

In addition, the chip bump's voltage drop is probed, at all the model nodes, and the ratio between the results in both modeling cases (lumped and distributed) is plotted as a bar histogram. The results show a small gap of 4% pessimistic predicted values for the lumped mode at uniform chip stimuli, and a slightly larger 7% gap when the chip side stimuli is more localized and singular in nature. These results are seen in figure 3 lower right pane.

The DC solver returns all the data over the model resistive mesh element's passing through currents. This allows very easy post processing to obtain the power loss dissipated over the modeled structure by simple sum of the I^2R contributions from each element in the model. The summing script also featured categorizing the loss partial contributions per each package layer and micro-via cluster type. The results show sensitivity of the modeling type for the power loss calculus. Planes and micro-via display pessimistic losses in the case of lumped vertical path modeling, and distributed models can result in 10% less losses. The equivalent loss resistance of the full path (stimuli to source) which can be implied by those results would be similarly reduced in the distributed model prediction. In some less common package structures and for some planes the current density observed was 50% larger in the lumped model, due to the singular sinks created by the single entry for each vertical conductive element on the digitized plane's mesh. Figure 4 shows current distribution pseudo color maps for two substrate layers (core layer in the upper panes, buildup layer in the lower panes). The maps clearly indicate the location of via singular spots for the lumped via models (left panes) and much more smooth maps for the distributed via cases (right panes).

Summary and Conclusions

Strong dedicated DC solvers allow the investigation of high resolution resistive meshes for large scale fine elements structures such as chip package substrates. Graphic algorithms and basic scripting skills allow the infrastructure coding of a custom netlist tool to allow exploration of stitching features. The feature of distributed via (micro-vias, plated through holes) vertical path connectivity was used to explore the sensitivity of the results to the selected modeling style. Distributed via models demonstrate an observable, yet not dramatic change over the simulation results. These results justify the use of simpler vertical paths stitching methods, yet highlight the cases where distributed via connectivity effects can grow in significance. Among those situations are proximity to stimuli (substrate face layers), locality of stimuli and level of required accuracy of power losses results.

References

[1] J. Auernheimer, "New methods for power distribution system design and analysis", ECTC 2004

[2] M. Suryakumar, H. Jiangqi, "Power delivery validation of processor front side bus", ECTC 2005

[3] H. Yi-feng, Y. Zhao-wen, "The simulation and pre-design on the PCB of the simulator", APEMC 2008

[4] L.T. Chen, L.S. Davis, C.P. Kruskal, "Efficient parallel processing of image contours", PAMI 1993

[5] F. Bacchus, J. Winter, "Effective Preprocessing with Hyper Resolution...", SAT 2003

Figures

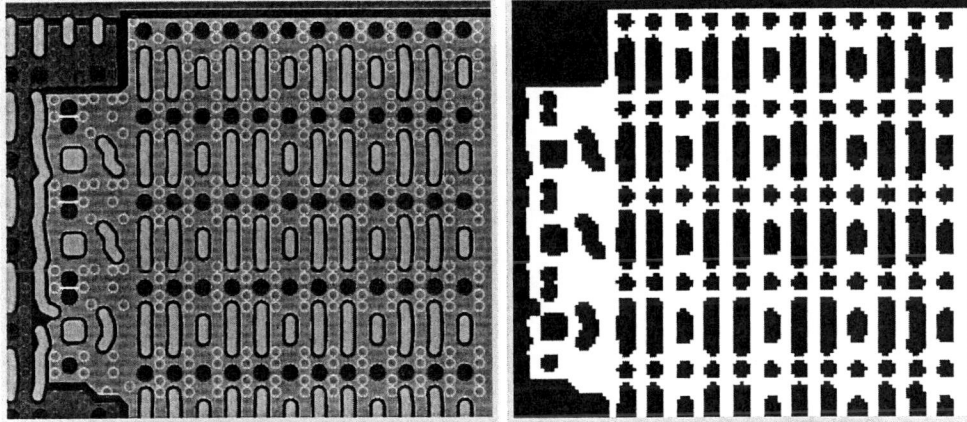

Figure 1. IA core supply power plane at one of the package substrate layers and its digitized pixel image

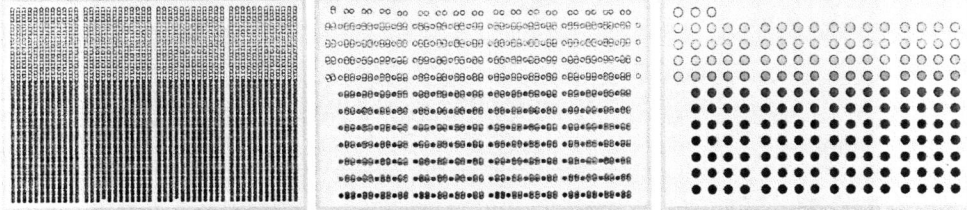

Figure 2. Grayscale current map through (left to right):
micro-via under the chip bumps, deeper face layers micro-via and plated-through-holes

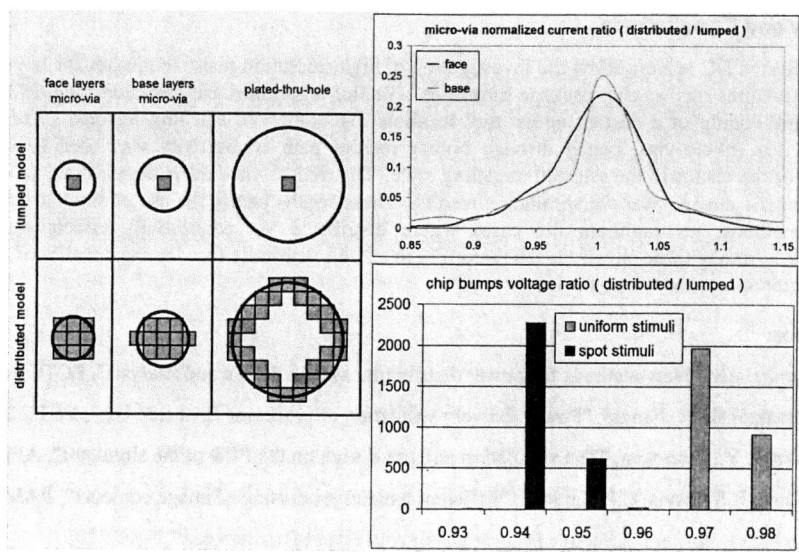

Figure 3. Via lumped and distributed stitching arrays,
current ratio normalized histograms and bar plot for chip bump voltage ratio
between distributed and lumped micro-via models

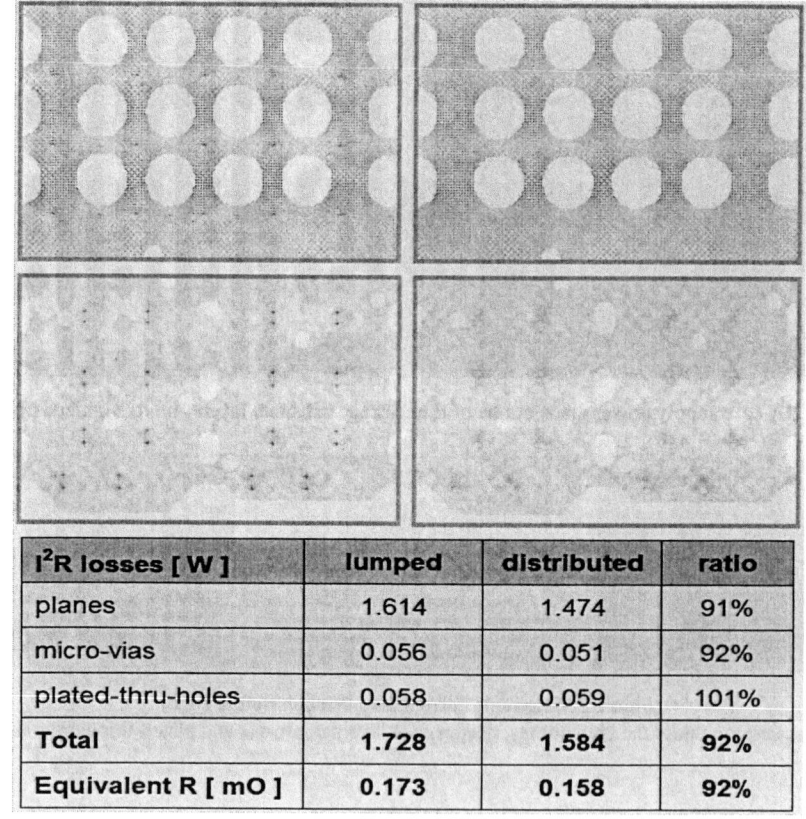

I^2R losses [W]	lumped	distributed	ratio
planes	1.614	1.474	91%
micro-vias	0.056	0.051	92%
plated-thru-holes	0.058	0.059	101%
Total	1.728	1.584	92%
Equivalent R [mO]	0.173	0.158	92%

Figure 4. Conductive losses under lumped or distributed micro-via model

Design of Shorting Vias in Alternative PCB Planes for Suppressing Ground-Bounce Induced Electromagnetic Emission

Kai-Bin Wu, Fu-Sheng Chang, and Ruey-Beei Wu

Department of Electrical Engineering and Graduate Institute of Communication Engineering,
National Taiwan University, Taipei, Taiwan, 10617, R.O.C.
Tel: +886-2-33663630   Fax: +886-2-23638247, E-mail: rbwu@ew.ee.ntu.edu.tw

Abstract

The stacked ground planes with vias stitching is employed to suppress the ground-bounce induced electromagnetic emission in multilayer printed circuit boards. This paper developed an analytic normal-mode analysis and used it to design the optimized positions of the vias in the board periphery. Comparison with full-wave simulation validates that the proposed systematic design procedure can easily achieve the desired EMI suppression from dc up to several GHz.

I. Introduction

Throughout decades of continuous advances in semiconductor technology, there have always been concerns about the ability of components to operate safely in an increasingly disruptive electromagnetic environment. Among them the reduction of electromagnetic emission from integrated circuits is of great significance [1]. To meet the EMI requirement such as IEC or FCC standard, the first priority of EMC engineers is to mitigate or block the severe radiated noise. Considerate efforts have been presented in the literature, using spread spectrum clock generation [2], reduced package inductance [3], de-coupling capacitors [4], [5], absorbent materials [6], asynchronous circuits [7], electromagnetic bandgap structures [8], and so on.

In multilayer printed circuit boards (PCB), the ground bounce noise will be excited when a sudden change of current passes through a through-hole via. Significant fringing electric field built on the board edges radiates to the surrounding space and results in EMI concerns. A popular approach is using decoupling capacitors to suppress not only the ground bounce but also the radiated emission [4], [5], but the applicable range is limited to about 1GHz due to the significant ESR and ESL at higher frequencies. The structures of stacked power/ground planes with vias stitching was presented to reduce the fringing electric field, but the analysis relies on FDTD and the design was mostly based on case studies [9].

In this paper, a three-layer structure with ground-power-ground (GPG) stack-up is chosen as the example. Two radiation modes are identified and their excitation and suppression mechanisms are investigated separately. After that, a canonical problem model is proposed to effectively simplify the multi-layer structure with shorting vias along the periphery. An analytical solution of the field distribution is obtained by applying the normal-mode analysis. Based on it, the geometric parameters of the vias are determined to achieve the maximum EMI suppression, followed by the validation with the full-wave simulation.

II. Mode Mechanism in EMI Suppression

Consider a current source excitation between the power and ground planes in the two-layer GP and three-layer GPG stack-up structures shown in Figs. 1(a) and (b). The top layer in Fig. 1(a) and the upper and bottom layers in Fig. 1(b) are ground planes. By superposition theory, the GPG stack-up with one source between the first and second layers can be decomposed into two structures of even- and odd-mode excitation as depicted in Figs. 1(c) and (d), respectively.

The radiated emission of the two structures can be attributed to the radiation from the equivalent magnetic surface current $\vec{M}_s = -\hat{n} \times \vec{E}$ due to the fringing electric filed $\vec{E}$ along the board edges. Note that the two equivalent magnetic current sources corresponding to the even-mode excitation in Fig. 1(c) are of opposite polarity, while those for odd- mode excitation in Fig. 1(d) are in phase. As a result, the even-mode radiation is much smaller than that of the odd-mode excitation by

$$A = A_{ideal} \cos\theta; \quad A_{ideal} = \tfrac{1}{2} j k_0 h \qquad (1)$$

due to the array factor, where k_0 is the propagation constant in free space, h is the height of each layer, and θ is the zenith angle of the observation points.

978-1-4244-4447-2/09 $25.00 © 2009 IEEE

Without the shorting vias, the radiation due to the odd-mode excitation is similar to that in the original problem Fig. 1(a), while the even-mode radiation is negligible due to (1). With shorting vias to connect the upper and lower planes, it is noted that the even-mode excitation in Fig. 1(c) is hardly affected. However, provided that the vias are sufficiently closely spaced, the fringing electric field on the board edges due to the odd-mode excitation in Fig. 1(d) is strongly reduced and the even-mode radiation becomes the dominant term. Hence, A_{ideal} in (1) becomes a good approximate to the theoretical limit of the EMI suppression by the present technique of stacked planes with vias stitching. For example, if an EMI suppression of 20dB from dc up to 3GHz is desired, one can easily choose layer height $h < 3.1$mm so that $A < 0.1$ for all θ.

III. Design of Shorting Vias

The above reasoning relies on the negligence of odd-mode radiation, which can be justified only if the vias are spaced densely enough so that the fringing electric field on the board edge is very small. More vias can yield better shielding, but also implies larger manufacture cost and less freedom for the routing near the board edges. An optimal design of the vias is that the electric field on the board edges can be reduced by an attenuation factor given by (1), thereby assuring that the odd-mode radiation remains smaller than that of the even-mode excitation.

Consider the via stitching in a three-layer structure with an odd-mode excitation. The problem can be simplified to a two-layer structure with shorting vias as depicted in Fig. 2. The major design parameters are the spacing P between two adjacent vias, distance D from via center to board edge, and via radius r. Without loss of generality for the design of these parameters, consider a rectangular two-layer structure with shorting vias at all the four sides as shown in Fig. 3(a). Furthermore, provided that size of the plane is large as compared with a wavelength, the field distribution in the region near the vias and board edge can be approximated as a canonical problem shown in Fig. 3(b) due to the equivalence by image theory. It is a wave scattering problem in a waveguide with two perfectly magnetic conductors (PMC) as the side walls; while the bottom and upper walls are metal planes as shown in Fig. 3(c). An incident field E_{inc} is excited from the left side and scattered by the shorting via in the right side (the board edge is chosen at $x = 0$ and assumed to be a PML, while the two PML walls of the waveguide are located at $y = 0$ and P).

The problem can be solved by applying the normal-mode waveguide theory [10]. The mathematical details are omitted due to the limit of space. The electric field can be expressed analytically by

$$E_z^{tot}(x,y) = 2E_{inc}\cos kx - j\omega\mu I_{via}\cdot\left[g\left(x,-D;y,\tfrac{1}{2}P\right) + g\left(x,D;y,\tfrac{1}{2}P\right)\right] \tag{2}$$

where the Green's function is given by

$$g(x,x';y,y') = \frac{1}{2jkP}e^{-jk|x-x'|} + \sum_{m=1}^{\infty}\frac{1}{k_mP}e^{-k_m|x-x'|}\cos\left(\frac{m\pi y}{P}\right)\cos\left(\frac{m\pi y'}{P}\right), \quad k_m = \sqrt{\left(\frac{m\pi}{P}\right)^2 - k^2} \tag{3}$$

k is the propagation constant in the material, and I_{via} is the current on the shorting via. Since the total electric field in (2) should equal zero at the via boundary, say $x = -(D-r)$ and $y = P/2$, it can be solved as

$$I_{via} = \frac{2\cos(kD-kr)\cdot E_{inc}}{j\omega\mu\cdot\left[g(r,0;\tfrac{1}{2}P,\tfrac{1}{2}P) + g(2D-r,0;\tfrac{1}{2}P,\tfrac{1}{2}P)\right]} \tag{4}$$

The efficiency of field reduction due to the via fence can be given by the transmission coefficient which is defined by the ratio of the maximum field on the board edge, i.e., $E_z^{tot}(0,P)$, to the incident field. A good design calls for suitable parameters P, D, and r to make the transmission coefficient smaller than A_{ideal} in (1).

IV. Simulation Results

Consider a test power/ground plane pair as shown in inset of Fig. 4 with length, width, height, and dielectric constant being 140mm, 140mm, 1mm, and 4.4 respectively, and the location of the current source at (70mm, 70mm). Figure 4 shows the total radiated power of two-layer, three-layer, three-layer with even-mode excitation, and three-layer with odd-mode excitation calculated by a full-wave software HFSS. It can be seen that all the curves are almost the same expect the three-layer case with even-mode excitation, whose radiation is much smaller.

A MatLab program has been tailored to calculate the transmitted field on the board edge. The parameter study on the transmission coefficient versus the parameters D, P, and r have been performed and the results are summarized here. Roughly speaking, the via spacing P should be no larger than a quarter wavelength to reduce the transmitted field. The transmission coefficient decreases for smaller P, and varies approximately linearly versus P if $P < \lambda/6$. Larger via radius r gives better field reduction, but its value is limited by the manufacturing and routing consideration. The choice of D is also important in reducing the transmitted field. It should be smaller than a quarter wavelength, else the structure may become resonant and the transmitted field at board edge becomes very large. The transmission coefficient becomes small in certain range of D, which can be determined by numerical calculation. Although not shown here, design charts are constructed to obtain the values of these parameters for satisfying the desired reduction in the transmitted field.

For example, consider a design of EMI suppression by 20dB from dc up to 3GHz for the three-layer structure in Fig. 4. The requires that the transmission coefficient should be smaller than 0.1. A possible choice is $kP = 0.208$, $kD = 0.179$, and $kr = 0.01$ at the highest frequency from the design chart, or $P = 1.58$mm, $D = 1.36$mm, and $r = 3$mil. Figure 5 shows the total radiated power from 0 to 3GHz calculated by HFSS. It can be seen that the total radiated power is suppressed in general by 20dB or more as compared with the structure without vias, and larger suppression for lower frequency.

V. Conclusions

The radiated emission due to the ground bounce in a multi-layer PCB causes EMI concerns and can be reduced by using the stacked alternate ground planes with vias stitching. The radiation mechanism can be decomposed into the even- and odd-mode parts. The radiation due to the even-mode excitation is much smaller than that of the odd-mode excitation, while the odd-mode radiation can be greatly suppressed by placing dense enough vias to reduce the fringing electric field on the PCB edges. By this way, the EMI due to the ground bounce in multi-layer PCB can be greatly suppressed.

For the design of shoring vias to achieve the EMI suppression, the problem can be simplified to a canonical problem which is then solved analytically by the waveguide normal-mode theory. A systematic procedure is established for the placement design of shorting vias. Based on this, a design example to achieve 20dB EMI suppression over dc up to 3GHz is demonstrated and validated by a full-wave simulation.

References

[1] M. Ramdani, E. Sicard, A. Boyer, S. Ben Dhia, J. J. Whalen, T. H. Hubing, M. Coenen, and O. Wada, "The electromagnetic compatibility of integrated circuits - past, present, and future," *IEEE Trans. Electromagn. Compat*, vol. 51, pp. 78-100, Feb. 2009.

[2] K. B. Hardin, J. T. Fessler, and D. R. Bush, "Spread spectrum clock generation for the reduction of radiated emission," *Proc. IEEE Int. Symp. Electromagn. Compat.*, 1994, pp. 227-231.

[3] T. Sudo, "Behavior of switching noise and electromagnetic radiation in relation to package properties and on-chip decoupling capacitance," *Proc. 17th Int. Zurich Symp. Electromagn. Compat.*, 2006, pp. 568-573.

[4] J. Kim, H. Kim, W. Ryu, and J. Kim, "Effects of on-chip and off-chip decoupling capacitors on electro-magnetic radiation emission," *Proc. Electron. Comp. Technol. Conf.*, 1998, pp. 610-616.

[5] L. van Wershoven, "Characterization of an EMC test-chip," *Proc. IEEE Int. Symp. Electromagn. Compat.*, 2000, pp. 117-121.

[6] S.-H. Kim, S.-B. Lee, K.-I. Ouh, C.-B. Rim, K.-S. Moon, H.-G. Yoon, and T.-J. Moon, "Reduction of radiated emissions from semiconductor by using absorbent materials," *Proc. IEEE Int. Symp. Electromagn. Compat.*, 2000, pp. 153-156.

[7] D. Panyasak, G. Sicard, and M. Renaudin, "A current shaping methodology for lowering EM disturbances in asynchronous circuits," *Microelectron. J.*, vol. 35, pp. 531-540, Jan. 2004.

[8] T.-L. Wu, Y.-H. Lin, T.-K Wang, C.-C. Wang, and S.-T. Chen, "Electromagnetic bandgap power/ground planes for wideband suppression of ground bounce noise and radiated emission in high-speed circuits," *IEEE Trans. Microw. Theory Tech.*, vol. 53, pp. 2935-2941, Sept. 2005.

[9] X. Ye, D. M. Hockanson, M. Li, Y. Ren, W. Cui, J. L. Drewniak, and R. E. DuBroff, "EMI mitigation with multilayer power-bus stacks and via stitching of reference planes," *IEEE Trans. Electromagn. Compat.*, vol. 43, pp. 538-548, Nov. 2001.

[10] R. E. Collins, *Field Theory of Guided Waves*, 2nd ed., IEEE Press, 1991, Chap. 7.

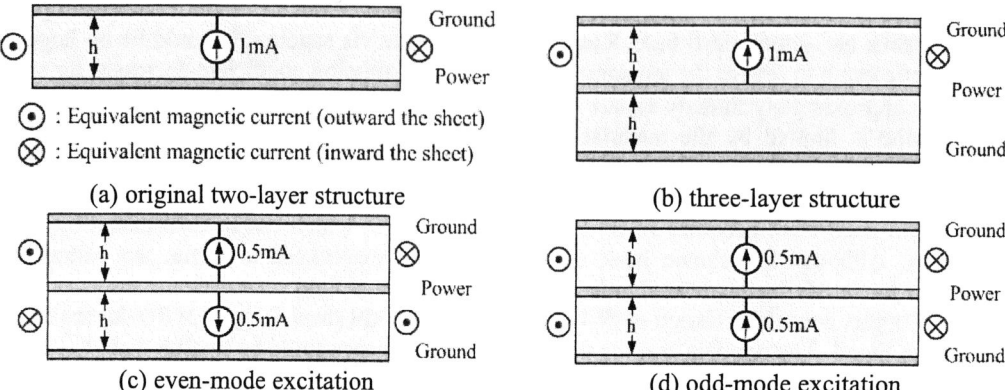

(a) original two-layer structure

(b) three-layer structure

⊙ : Equivalent magnetic current (outward the sheet)

⊗ : Equivalent magnetic current (inward the sheet)

(c) even-mode excitation

(d) odd-mode excitation

Fig. 1 Two- and three-layer power/ground planes with even- and odd-mode current sources.

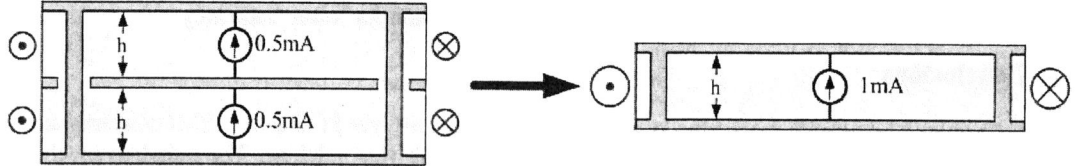

Fig. 2 Simplification from the three-layer structure with odd-mode excitation to a two-layer structure

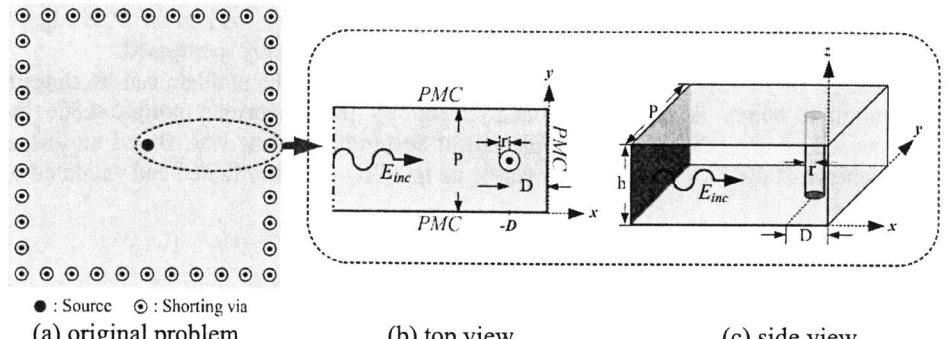

● : Source   ⊙ : Shorting via

(a) original problem   (b) top view   (c) side view

Fig. 3 Original structure with shorting vias and its canonical problem.

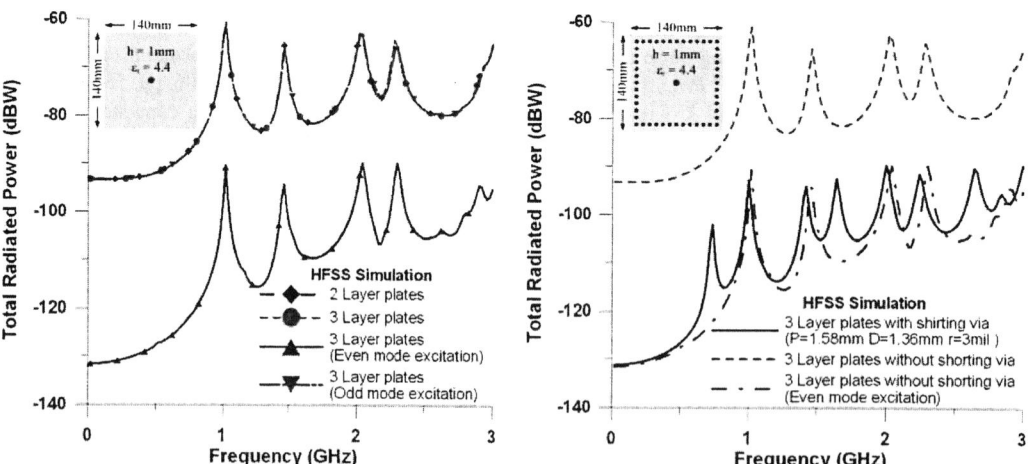

Fig. 4 Total radiated power from three- layer structure without shorting vias

Fig. 5 Suppression of total radiated power using shorting vias in the board periphery

On-Chip Global Clock Distribution Using Directional Rotary Traveling-Wave Oscillator

Yulei Zhang[1], James F. Buckwalter[1], Chung-Kuan Cheng[2]

[1]ECE Dept., [2]CSE Dept., University of California, San Diego, 9500 Gilman Drive, La Jolla, CA 92093

Tel:1-858-534-8174, Fax:1-858-534-7029, Email:{ylzhang, jfbuckwalter, ckcheng}@ucsd.edu

Abstract: We revisit the rotary clock scheme in [1], and firstly propose and analyze the multiple-mode operations of rotary clock structure, which can be utilized to generate multiple clock frequencies within one loop. By replacing original inverter pairs with feed-forward differential buffers and adding U-shape T-line segments, a novel rotary clock scheme is presented to guarantee the directionality of traveling wave in the closed-loop. Design tradeoffs and methodologies are discussed for this novel directional rotary clock and a $3.3GHz$, $30pF$ loading rotary clock design example is given. The simulation results show that, the proposed directional rotary clock can at least reduce 34% power compared with fCV^2, and achieve lower power dissipation and sharper rising-edge compared with the conventional rotary clock in [1].

1. Introduction

Distributing multi-GHz global clock signals with low jitter/skew and low power consumption has become a challenging design task for microprocessors and other high-performance VLSI chips. As the clock frequency goes up to gigahertz range, it becomes more difficult to maintain the clock skew and jitter within a certain limit, which is normally 10% of the cycle time for previous designs [2]. To reduce the clock skew, much efforts have been done to balance and buffer clock trees, bringing additional design efforts and power dissipation [3]. Another approach that uses clock grids/meshes could reduce the skew but consuming significantly more power [4]. In [5], it has been shown that up to 40% of total dynamic power in a 0.13μm microprocessor is dissipated by clocks.

The conventional clock distribution networks, which include trees, grids(or meshes), spines, and other hybrid structures, need a large amount of buffers/repeaters to improve the clock quality and reduce the skew in order to meet certain skew and jitter requirement. It becomes difficult to design such buffered clock distribution networks as clock frequency approaches $10GHz$ [6]. Meanwhile, since all the load capacitances, mostly from the buffers and driven registers, are charged and discharged all the time at the highest frequency within the whole chip, conventional clock distribution networks consume significant power, described in the form of $P = fCV^2$. If considering the technology scaling, the clock power dissipation will keep increasing due to the increasing total capacitance and clock frequency.

As an alternative clocking approach, wave-based oscillators have been intensely studied recently. According to the type of wave that propagates along the transmission line (T-line), these oscillators are categorized into standing-wave oscillators (SWOs) and traveling-wave oscillators (TWOs) [7]. Standing waves can be generated by sending incident wave along the T-line and reflect it back with a short termination. To eliminate the residual traveling-wave component, different compensation methods are proposed to reduce the wire loss [8, 9, 10]. The clock skew can be greatly reduced by adopting SWOs due to the uniform-phase of standing wave, but the signal amplitude varies with position. Besides, SWOs can only provide a sinusoidal signal, sine-to-square converters are needed to generate useable clock, making the whole scheme consume similar power compared with the conventional clock trees [8]. On the other hand, rotary clock distribution using TWOs has been proposed and implemented in [1]. The full-swing, multi-phase square waves can be generated along a $M\ddot{o}bius$-connected closed-loop built by differential T-lines, with much lower power dissipation because of the energy is recirculated (The main power loss mechanism here is I^2R, where R is the wire resistance.) instead of charging and discharging the large capacitances. However, how to control the wave propagation direction, which determines the phase of clock signal at a specific position, remains a problem.

In this work, we revisit the rotary clock scheme in [1] and propose an approach to guide the traveling wave along the desired direction inspired by the idea in [11]. By replacing the original inverter pairs with well-designed feed-forward buffers and U-shape T-line segments, the traveling waves that propagate along the feed-forward direction are reinforced, forming a **directional rotary clock scheme** which can generate signals with stable phases. The main contributions of this work are summarized below: 1) By analyzing the fundamental principle of rotary clock scheme, we firstly point out the **multiple-mode operation** which can be utilized to generate multiple clock frequencies within one loop; 2) We propose a directional rotary clock scheme to control the wave propagation direction by adopting controllable differential buffers and U-shape T-lines; 3) We explore the tradeoff relation between rise time and power dissipation in this novel clock scheme and provide the insights on how to approach the optimal design.

2. Directional Rotary Clock Scheme

Fig. 1 shows the proposed directional rotary clock scheme with the illustrations of how this rotary clock scheme works. Different from the conventional rotary clock proposed in [1], this scheme compensates the wire loss and generate the full-swing square-wave clock by introducing feed-forward buffers, as the inverter pairs shown in Fig. 1(a). The inputs and outputs of the buffers are connected by the U-shape T-line segments which form the whole closed-loop. Traveling waves along the desired direction (as indicated in Fig. 1(a)) are reinforced by the feed-forward buffers, guaranteeing the directionality of this rotary clock scheme. Buffer size and length of U-shape T-line segment should be carefully designed to achieve the lowest power dissipation as discussed in later sections. To further understand the proposed directional rotary clock, some fundamentals and multiple-mode operations of such rotary structures are firstly illustrated as follows.

978-1-4244-4447-2/09 $25.00 © 2009 IEEE

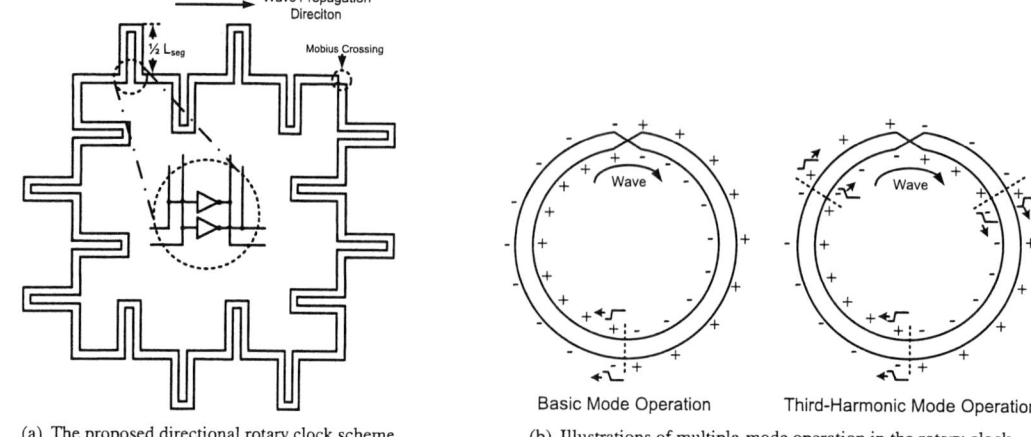

(a) The proposed directional rotary clock scheme.

(b) Illustrations of multiple-mode operation in the rotary clock.

Figure 1. The overall directional rotary clock scheme studied in this work.

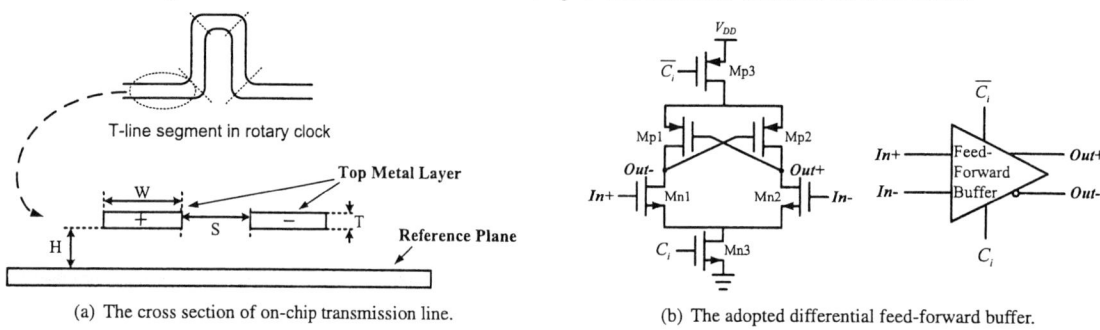

(a) The cross section of on-chip transmission line.

(b) The adopted differential feed-forward buffer.

Figure 2. The implementation of transmission line and feed-forward buffer in the proposed clock scheme.

2.1 Fundamentals and Multiple-Mode Operations

Fig. 1(b) illustrates the working principle of the idea rotary clock by the examples of basic mode and third-harmonic mode operations. The basic mode operation is briefly reviewed here, as explained in [1]. Assuming a pair of voltage waves with opposite polarity is generated (by internal noise or external biasing) in the closed-loop, as shown at the bottom of the left loop in Fig. 1(b), the waves will travel down the differential T-lines, pass the $M\ddot{o}bius$-crossing, and finally return to the bottom position but with switched polarity, causing the signal inversion at this position. If not considering wire losses, the waves will travel continuously, resulting in a toggling clock signal at every position of the closed-loop with the frequency,

$$f_{basic} = \frac{v_p}{2 \times \text{loop length}} = \frac{1}{2\sqrt{L_{total}C_{total}}} \tag{1}$$

where $v_p = 1/\sqrt{L_{perlen}C_{perlen}}$ is the velocity of the traveling wave, and L_{total}, C_{total} indicate the total inductance and capacitance of the T-line loop.

Inspired by the basic mode behavior of rotary clock, we found that the $M\ddot{o}bius$-connected loop can also support higher-order harmonic mode operations by intentionally starting several pairs of waves with specific polarities at equally-spaced points of the closed-loop. As an example, third-harmonic mode operation is illustrated in the right loop of Fig. 1(b). In this scenario, three pairs of voltage waves with alternate polarities are generated at three different positions (indicated by dash lines in the figure) that divide the whole loop into three equal parts. These waves switch polarities after traveling 1/3 length of the whole loop, causing the signal inversions at those positions. In the stable state, a toggling clock signal appears at every position of the loop, with the frequency that equals to three times of f_{basic} described in (1). It is interesting to notice that, only odd-order harmonic mode can be supported in this kind of rotary clock structure, due to the $M\ddot{o}bius$ characteristic of the loop. In summary, the rotary clock structure can generate clock signal with frequencies,

$$f_{clk} = n \times f_{basic} = \frac{n}{2\sqrt{L_{total}C_{total}}} (n = 1, 3, 5...) \tag{2}$$

2.2 Circuit Implementation

The proposed directional rotary clock can be implemented using typical CMOS process. The differential T-lines are usually fabricated on the top metal layer of CMOS chip with the largest dimension to reduce wire loss and achieve high performance. U-shape T-line layout has been adopted and implemented in [11] without adding complexity. Fig. 2(a) shows the cross section of a micro-strip configuration to model one segment of top-layer on-chip T-lines. The reference plane shown there can represent the adjacent or sub-adjacent metal layer below the top metal layer, depending on the different parameters that need to be extracted. W, S, T, H indicate width, spacing, thickness, and dielectric height of this interconnect structure.

978-1-4244-4447-2/09 $25.00 © 2009 IEEE   250

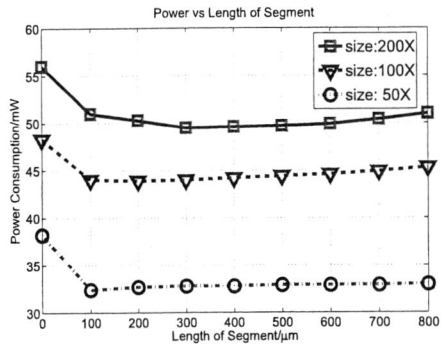

(a) The influence of U-shape length on the power dissipation.

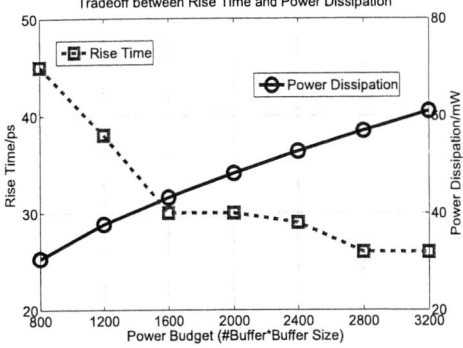

(b) The tradeoff relation between rise time and power dissipation.

Figure 3. Exploring the optimal design for the proposed directional rotary clock scheme.

Inverter pairs shown in Fig. 1(a) are replaced by a differential buffer in the real design, as shown in Fig. 2(b). This differential buffer adopts a pair of cross-coupled PMOSFETs as the loading to speed up the output transition and achieve better common-mode rejection compared with two separate inverters. Transistor size of Mp1/Mp2 can be adjusted to reduce the buffer latency and short-circuit power dissipation. To further ensure the ***directionality*** of traveling wave, we add two MOSFET switches at the head and the foot of the differential buffer to control the time at which buffer starts to work. In order to switch the ***operation mode*** of the loop, a simple bias circuit (a single NMOS or PMOS transistor with control signal) is added to the buffer inputs (not shown in Fig. 2(b)) to set the initial voltage to some specific values ($1V$ or gnd). Finally, a top-level control block can be designed to manipulate the working timing of each buffer (e.g., a start-up sequence according to the relative position of buffers) and initial conditions of internal nodes, resulting in a clock signal with the desired phase and frequency.

3. Simulation Results and Discussion

In this section, experiments are performed to verify the previous analysis and to study the design methodology of the novel directional clock scheme. To model on-chip T-lines, U-element from HSPICE [12] with geometric description is adopted in the following experiments. As a case study, we use the following dimensions for the micro-strip model shown in Fig. 2(a): $W=S=30\mu m$, $T=2\mu m$, and $H=6\mu m$. In terms of technology, we use TSMC $90nm$ 1P9M process with BSIM3 SPICE model card. All the following simulations are performed in HSPICE.

3.1 Approaching the Optimal Design

In the proposed rotary clock scheme, number of inserted buffers, buffer size, total loop length, and length of the U-shape T-line segment need to be determined to approach the optimal design. Fig. 3(a) shows the impact of U-shape length on the total power dissipation for different buffer sizes. There are totally 16 buffers evenly distributed along a $16mm$ T-line loop, and the clock frequency is fixed to be $3GHz$ in this experiment. It is shown that, there exists an optimal U-shape length in terms of the lowest power dissipation, and this optimal value varies with the buffer size. For longer U-shape segment, the wire loss becomes larger, therefore larger buffers are needed in order to inject stronger current in the transition edge. On the other hand, the zero-segment-length case, which corresponds to the conventional rotary clock ([1]) in which input and output of the buffers are placed at the same position along the loop, will dissipate about 10% extra power compared with the proposed feed-forward buffer scheme. This difference is mostly attributed to the larger short-circuit current of conventional buffers due to the simultaneous switching of input and output. As a result, the proposed directional rotary clock further reduces the power dissipation by reducing the short-circuit current.

Tradeoff between rise time and power dissipation in this clock design is explored in Fig. 3(b). For the same clock frequency ($3.3GHz$) and the same loading capacitance ($10pF$), we study the best rise time that can be achieved for the given power budget, which is defined as the product of number of buffers and buffer size. It can be seen that, total power dissipation increases approximately linearly with the power budget, whereas the rise time decreases accordingly. For a given power budget, which means the total buffer size, the way to distribute these buffers affects the rise time of generated clock signal. Typically, placing more buffers with smaller size can improve the rise time because of shorter U-shape segment, however, in the extreme case, rise time will be limited by the technology due to the poor driving capability of the smallest buffer. In the real design, total power dissipation needs to be reduced with a limitation on the upper bound of the rise time, providing us a methodology on how to approach the optimal design.

3.2 A Design Example

A directional rotary clock design example is presented here by applying the methodology discussed in Subsection 3.1. The total load capacitance is assumed to be $30pF$ and is evenly distributed along the whole loop with $50\mu m$ separation to represent the loaded on-chip registers. The target clock frequency is set to be $3.3GHz$ ($300ps$ cycle time). To reduce the power dissipation as much as possible without sacrificing the rise time, finally we choose to place 32 buffers with $50\times$ size along the whole loop. The total loop length is $8mm$ which is adjusted to satisfy the cycle time requirement. Fig. 4(a) shows the simulated clock waveforms probed from three equally-spaced positions of the loop. As a result, each clock signal should shift phase $60°$ from the previous one as shown in the figure. The rise time and power dissipation of this clock design are summarized in the third row of Table 1. It is clear to see, rotary clock scheme can at least reduce the power dissipation by 34% compared with fCV^2, which is $100mW$ in this case[1].

To verify the multiple-mode operation of rotary clock scheme, we also simulate the third-harmonic mode waveforms in Fig. 4(b). Waveforms are probed from the same positions as in Fig. 4(a) in order to make a comparison. It can be seen that, all the clock signals change frequency

---

[1] Only consider load capacitance in this simple power calculation, if considering wire and buffer capacitance, this power reduction can become larger.

978-1-4244-4447-2/09 $25.00 © 2009 IEEE

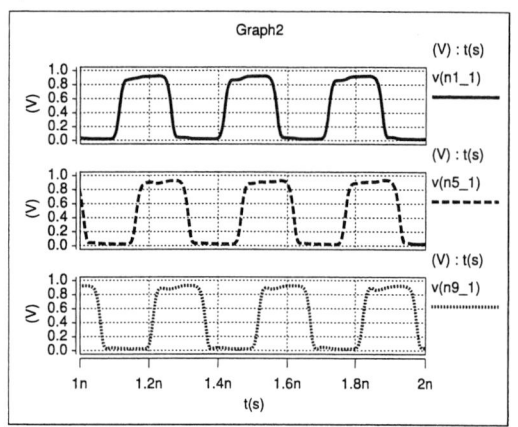

(a) Simulated waveforms for basic mode operation.

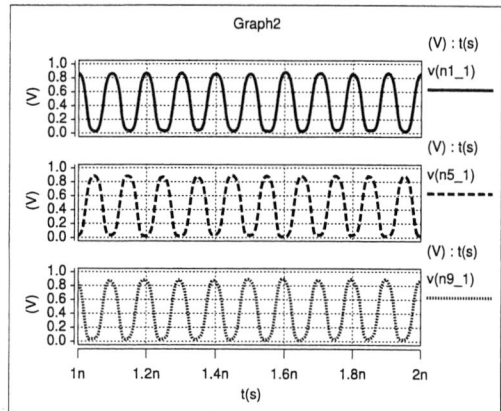

(b) Simulated waveforms for third-harmonic mode operation.

Figure 4. A directional rotary clock example: $C_{load} = 30pF$, $f_{basic} = 3.3GHz(T_C = 300ps)$.

Table 1. Performance comparison between conventional rotary clock [1] and the proposed scheme

Scheme	Power Dissipation/mW	Rise Time/ps	Loop Length/mm	# Buffers	Buffer Size
Conventional Rotary Clock	74.3	36	6.4	32	50×
Directional Rotary Clock	66.3	30	8.0	32	50×

to $10GHz$ ($100ps$ cycle time), which corresponds to three times of the basic frequency, and each signal shifts phase $180°$ from the previous one (the $180°$ phase shift between the last and the first signal is due to $M\ddot{o}bius$-crossing). The results here verify our previous analysis in Subsection 2.1. Because the whole clock scheme remains the same, the rise time of third-harmonic mode clocks is still $30ps$, making the waveforms look like sinusoidal signals.

Finally, we compare the performance of the proposed directional rotary clock with conventional rotary clock scheme ([1]) by setting the same target frequency ($3.3GHz$) to drive the same amount of load capacitance ($30pF$). Table 1 gives the comparison results between these two schemes. To make a fair comparison, we choose the same number of buffers and buffer size. As a result, the loop length of conventional rotary clock is $6.4mm$, which is shorter than that of the proposed scheme because of 25% larger equivalent buffer capacitance loaded to the T-lines. In terms of the power consumption, as discussed in Subsection 3.1, conventional rotary clock consumes 12% more power due to the larger short-circuit current. Also, the conventional rotary clock shows a little bit larger rise time in this design. In summary, the proposed clock scheme could cover larger chip area with sharper rising-edge and lower power dissipation while driving the same load capacitance at the same clock frequency.

4. Conclusion and Future Works

In this paper, a novel rotary clock scheme, which is composed of inserted feed-forward differential buffers and U-shape T-line segments, is presented to guarantee the directionality of traveling wave along the T-line loop. The design methodology of this novel clock scheme is explored by studying the tradeoff of rise time and power dissipation. A $3.3GHz$ clock design example shows that, the proposed clock scheme can reduce total power 34% compared with traditional clock tree structure and about 10% compared with conventional rotary clock.

Future works include more detailed considerations to implement the proposed directional rotary clock. A high-level control block is needed to set initial voltages for internal nodes along the loop and to control the timing of feed-forward buffers, in order to make the loop work at desired mode. The layout of T-lines (geometry, twisted, ...) also should be studied with the consideration of crosstalk and other coupling noise.

References

[1] J. Wood, T. C. Edwards, and S. Lipa, "Rotary Traveling-Wave Oscillator Arrays: A New Clock Technology," *IEEE Journal of Solid-State Circuits*, vol. 36, no. 11, pp. 1654–1665, Nov 2001.

[2] J. Nurmi, H. Tenhunen, J. Isoaho, and A. Jantsch, Eds., *Interconnect-Centric Design for Advanced SoC and NoC*. Kluwer Academic Publishers, 2004.

[3] E. G. Friedman, "Clock Distribution Networks in Synchronous Digital Integrated Circuits," *Proceeding of the IEEE*, vol. 89, no. 5, pp. 665–692, May 2001.

[4] A. Sobczyk, A. Luczyk, and W. A. Pleskacz, "Power Dissipation in Basic Global Clock Distribution Networks," in *Design and Diagnostics of Electronic Circuits and Systems*, April 2007, pp. 1–4.

[5] N. Magen, A. Kolodny, U. Weiser, and N. Shamir, "Interconnect Power Dissipation in a Microprocessor," in *IEEE/ACM Int. Workshop on System Level Interconnect Prediction*, Feb 2004, pp. 7–13.

[6] S. C. Chan, K. L. Shepard, and P. J. Restle, "Uniform-Phase Uniform-Amplitude Resonant-Load Global Clock Distributions," *IEEE Journal of Solid-State Circuits*, vol. 40, no. 1, pp. 102–109, Jan 2005.

[7] W. Andress and D. Ham, "Recent Developments in Standing-Wave Oscillator Design: Review," in *Radio Frequency Integrated Circuits (RFIC) Symposium*, Jun 2004, pp. 119–122.

[8] F. O'Mahony, C. P. Yue, M. Horowitz, and S. S. Wong, "10GHz Clock Distribution Using Coupled Standing-Wave Oscillators," in *IEEE Int. Solid-State Circuits Conf.*, Feb 2003, pp. 428–504.

[9] W. Andress and D. Ham, "Standing Wave Oscillators Utilizing Wave-Adaptive Tapered Transmission Lines," *IEEE Journal of Solid-State Circuits*, vol. 40, no. 3, pp. 638–651, Mar 2005.

[10] V. H. Cordero and S. P. Khatri, "Clock Distribution Scheme Using Coplanar Transmission Lines," in *Design Automation and Test in Europe*, March 2008, pp. 985–990.

[11] J. F. Buckwalter and J. Kim, "A 26dB-Gain 100GHz Si/SiGe Cascaded Constructive-Wave Amplifier," in *IEEE Int. Solid-State Circuits Conf.*, Feb 2009, pp. 488–450.

[12] Synopsys, "HSPICE Signal Integrity Guide," March 2005.

Minimizing crosstalk in high-speed differential buses by optimizing power/ground and signal assignment

Yang Yi[*], Yifang Liu[*], Yaping Zhou[+], and Wiren Dale Becker[°]

[*] Department of Electrical and Computer Engineering, Texas A&M University, College Station, TX 77843
[+] IBM Systems&Technology Group, 11400 Burnet Rd., Austin, TX 78758
[°] IBM Systems&Technology Group, 2455 South Rd., Poughkeepsie, NY 12601
Email:yaping@us.ibm.com, Tel: (512)286-9401

Abstract

Vias in packages and boards, land-grid-array pins, and connector pins introduce significant crosstalk noise in high-speed differential buses. In this paper, we propose an algorithm that minimizes crosstalk noise by optimizing power/ground distribution as well as differential signal assignment. Experimental results demonstrate its high efficiency and significant electrical benefit.

I. INTRODUCTION

With the continuing increase on the speed of differential buses, crosstalk noise has become one of the most significant factors affecting the performance of differential buses [1]. Crosstalk noise degrades signal waveforms and introduces jitters, which leads to eye closure to high-speed buses. In order to reduce crosstalk noise, high-speed differential signals are typically routed in packages and boards with very good reference planes. However, significant crosstalk noise can arise in package and board features that lack sound referencing. Examples of such features include vias in packages and boards, land-grid-array pins, and connector pins. Fig. 1 shows an example with high-speed differential signal pins/vias surrounded by power/ground pins. When differential buses go through those regions, signal pins can be paired vertically, horizontally, or diagonally to form a differential signal assignment. Each of such assignments has different electromagnetic field distribution which leads to different electrical performance.

The conventional simulation flow to determine the optimal differential signal assignment is very time consuming [2] [3]. Multi-port S-parameters of 3D pin/vias are usually extracted by tools such as Ansoft HFSS. An assumption for the differential signal assignment is needed to run time domain simulation. Eye-diagram or the worst coupling noise is then used to compare different differential signal assignments. Since time domain simulation is required, the number of possible assignments that can be examined is very limited. It is impractical to find the real optimal assignment.

Recently, we made an advance in minimizing the crosstalk noise by optimizing the differential signal assignment [4]. The method finds many possible differential signal assignments based on available pins, and uses a benchmark parameter to find the optimal one. Circuit simulation of practical examples demonstrates that an optimal differential signal assignment can have as little as half the coupling noise compared to a non-optimal assignment. Studies also show that the benchmark parameter can be correlated to actual hardware performance. However, the method assumes a pre-determined power/ground distribution, and it replies on searching a huge amount of (e.g., millions of), possibly redundant, assignments to find the best one. Therefore, it cannot guaranteed that all possibility is covered, and it still takes long time to cover those assignments for practical cases.

In this paper, we further improve the performance of our method by developing a fast algorithm that optimizes power/ground distribution by force-directed optimization and enumerates all possible differential pairings with each power/ground distribution by linear perturbation. An optimal assignment is obtained by comparing the electrical performance of all differential signal assignments. Compared with the previous method, the new algorithm makes significant improvements. First of all, it optimizes power/ground distribution as well as differential signal assignment. Secondly, it guarantees finding all possible differential signal assignments, which is not realized in the previous method. Thirdly, it is very fast due to our speed up method. The experimental results validate the accuracy and efficiency of the new algorithm.

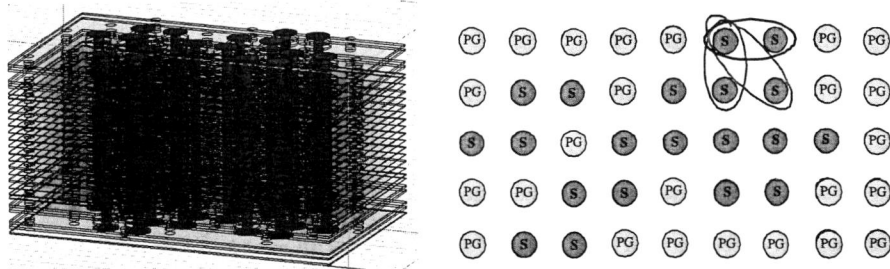

Fig. 1. A pin/via configuration example.

II. ALGORITHM

The high-speed differential signal pins/vias shown in Fig. 1 can be treated as a target $m \times n$ grid. Every point in the grid can be categorized as power/ground, signal, or unused, and it is identified by its coordinate (x, y). The performance of a bus is usually gated by the performance of the worst differential signal in the bus. It is expected, therefore, that the power/ground should be distributed nearly uniformly in the whole bus region. We propose using force-directed optimization algorithm to generate nearly uniform power/ground distributions. Note that the number of nearly uniform power and ground distributions is defined by users. With each power and ground distribution, our linear perturbation

978-1-4244-4447-2/09 $25.00 © 2009 IEEE

algorithm enumerates all possible differential signal pairing combinations, each of which consists of the pairing assignments of all signal pins on the grid. These differential signal assignments are then assessed by applying a benchmark parameter [4] that describes the relative performance of each assignment without doing time domain simulations. At the end, the best configuration with the optimal power/ground distribution and differential signal pairing assignment can be found.

A. Power/Ground Distribution by Force-Directed Optimization

In the power/ground distribution problem, a given number of power/ground pins need to be scattered over the grid on the usable points, so that the pins are uniformly spaced from one another. Let the distance between two power/ground pins P_i and P_j be d_{ij}. The attempt of spacing a pair of pins away from each other can be translated into resisting force between them, which we defined to be inversely proportional to the square of distance between the two pins, i.e.,

$$f_{ij} = \gamma \frac{1}{d_{ij}^2} \mathbf{u}_{ij}, \tag{1}$$

where γ is the coefficient of the resisting force, and $\mathbf{u}_{ij}$ is the unit vector pointing from pin i to j. In our problem, the uniform distribution of power/ground pins is expressed in the way of balancing the resisting force between all pins. Let F_j be the summation of all the vector forces the jth pin is subject to, i.e., $F_j = \sum_i f_{ij}$, which include the responsive force from the fixed parts (boundaries and unused spots) right next to the power/ground pin. The responsive force from fix components always neutralizes the force applied on the power/ground pin on the direction against the fix component, which makes the fix components act like a smooth wall to the pins. When the overall amount of force $\sum_i |F_i|$ is minimized, the pins tend to be equally spaced from each other, due to the posynomial resisting force function. Therefore, the problem of power/ground distribution is turned into a pin placement problem with a force balancing objective. The formal problem formulation of our power/ground distribution problem is:

$$\text{minimize:} \quad \sum_{i \in PG} |F_i|$$
$$\text{subject to:} \quad 0 \leq x \leq m, \quad 0 \leq y \leq n, \quad \forall P_i(x,y) \in PG,$$

where PG is the set of power/ground pins.

According to the problem formulation, we take gradient-based optimization approach to solve the continuous problem and then project the continuous solution to the closest available locations on the grid. Our gradient-based optimization is force directed, which is also employed in some of the cell placement methods, such as [5], [6]. Our method uses the total force applied on a pin as the derivative of the overall force with respect to the pin's coordinates. The individual forces on all the pins update the pin coordinates in the steepest descent direction to reduce the summation of force strength. The optimization procedure comprises gradient-based iterations following an initial placement of power/ground pins on available (usable) points on the grid. At the beginning of each iteration, total force on every PG pin is calculated. Then, all PG pins are moved in the direction of the total force on it. Specifically, the coordinates (x^k, y^k) of a PG pin P_i is updated to (x^{k+1}, y^{k+1}) according to equation:

$$(x^{k+1}, y^{k+1}) - (x^k, y^k) = \alpha F_i, \tag{2}$$

where α is the step size of update in an iteration. The step size α is determined by standard line search in each iteration. The method explained above is outline with the pseudo code in Alg. 1. When multiple uniform power/ground distributions are needed, different placement results can be obtained by starting the above procedure with different initial placement setups.

Input : Grid with Unusable Spots
Output : Coordinates of Power/Ground Pins

1 make an initial legal placement of all PG pins;
2 **repeat**
3   calculate $F_i, \forall i \in PG$;
4   calculate stepsize α;
5   update the coordinate of each PG pin $(x^{k+1}, y^{k+1}) \leftarrow \alpha F_i + (x^k, y^k)$;
6 **until** *no improvement*;
7 round up all x_i and y_i to the closest integers;
8 return $P_i(x,y), \forall P_i \in PG$;

Algorithm 1: $Force\_Directed\_Optimization$

B. Signal Pairing by Linear Perturbation

Once the power/ground are placed, all signals pins are also settled on specific locations on the grid. To search for the optimal differential assignment, an algorithm is needed to find all possible pairing combinations. Here, we propose an algorithm that guarantees to find all possible pairing combinations and runs in $O(K)$ time, where K is the number of all possible pairing combinations. We call our method a linear perturbation approach, which is basically a depth-first search in the pair assignment space. In our method, we first line up all the signals in a sequence by some order, where the order itself does not matter. Then, starting from the first signal in the sequence, we recursively go through the pairing options of all signals in the sequence. At each signal we check all its available pairing options. Once an available pairing option is picked for a signal, the same operation continues on the next signal in the sequence. This recursive procedure enumerates all pairing combinations of all the signals after the current signal, given a specific pairing combination of the current signal and the ones before it in the sequence. The recursive enumeration of available pairing options makes sure all possible combinations are discovered. And, because of the forward pairing checking operation on each signal, uniqueness of every pairing combination is guaranteed. The runtime of this algorithm is linear, because the operation cost is constant on each pairing combination. Alg. 2 gives a sketch of the method above.

Input : a signal sequence $Q[1..z]$, the set of possible pairing neighbor A_i for every signal s_i
Output : pairing combinations

1 **if** *current_signal_index ==0* **then**
2   current_combination $\leftarrow \emptyset$;
3   signal_array $\leftarrow \emptyset$;
4   current_signal_index $\leftarrow 0$;

5 **if** *current_signal_index ==z* **then**
6   output current_combination;
7   current_signal_index $\leftarrow$ current_signal_index-1;

8 **if** *current_signal_index $\notin$ current_combination* **then**
9   **forall** $s_j \in A_i$ **do**
10   current_combination $\leftarrow$ current_combination + (current_signal_index, s_j);
11   signal_array $\leftarrow$ signal_array + current_signal_index;
12   current_signal_index $\leftarrow$ current_signal_index + 1;
13   Pairing(current_signal_index);

14 **else**
15   signal_array $\leftarrow$ signal_array + current_signal_index;
16   current_signal_index $\leftarrow$ current_signal_index + 1;

17 signal_array $\leftarrow$ signal_array - current_signal_index;
18 current_combination $\leftarrow$ current_combination - the pair with the current_signal_index being its first element;
19 current_signal_index $\leftarrow$ current_signal_index-1;

Algorithm 2: Pairing(current_signal_index)

C. Benchmark Different Differential Signal Assignments

[7] shows that the crosstalk noise is due mostly to inductive coupling in crosstalk hotspot areas such as vias in packages and boards, BGA/LGA pins, connectors. Therefore, the mutual inductance can be used to asses the crosstalk noise of each differential signal assignment.

To consider the power and ground distribution, power/ground pins will be extracted in electromagnetic tools just like signals. Assume S-parameters are available from electromagnetic tools. For each power/ground distribution, the power/ground pins can be grounded by removing those pins in the Y-matrix. We can then define an inductance matrix L_{se}, which describes inductive coupling between N single-ended signals with a certain power/ground distribution. To obtain inductance matrix $N/2 \times N/2$ L_{diff} for N differential signals, we need to construct an $N/2 \times N$ matrix Q. Each row of Q matrix corresponds to a differential pair with 1 and -1 for the true and complement signals and 0 for all other signals. L_{diff} can then be calculated as $QL_{se}Q^T$. .

Assume the victim differential pair is i and all other pairs are acting as aggressors, the total coupling to i from all the aggressors is proportional to $L_{i,total} = \sum_{j \neq i} |L_{diff}[i,j]|$. Absolute values of $L_{diff}[i,j]$ are used to consider the worst case when all aggressors are acting in phase. In a differential bus, the worst victim differential pair, the one with $L_{max} = max(L_{i,total})$, dictates the performance of the whole bus. L_{max} can then be used to compare different differential pair assignments. If an assignment has a lower L_{max}, it will have less crosstalk in the via/pin areas and better electrical performance. Consequently, an optimal assignment with nearly uniform power/ground distribution and the lowest inductive coupling differential signal assignment can be found. Alg. 3 shows detailed procedures.

---

1 Optimize power/ground distributions by force-directed optimization;
2 **for** *all power/ground distributions* **do**
3   Ground the power/ground pins;
4   Get all possible signal pair assignments by linear perturbation;
5   **for** *all possible signal pair assignments* **do**
6   Extract a $2N \times 2N$ S-parameter matrix;
7   Transform the S-parameter matrix to a $2N \times 2N$ Y-matrix by shorting one side of all signals together to form a current loop;
8   Reduce the size of Y-matrix to $N \times N$ by assuming right hand side ports of all signals are shorted together;
9   Invert Y-matrix to Z-matrix;
10   Transform the $N \times N$ Z-matrix to matrix $L_{SE} = \frac{imag(Z)}{2\pi f}$ at a user-defined frequency f;
11   Transform L_{se} matrix into matrix $L_{diff} = QL_{se}Q^T$;
12   Compute $L_{i,total}$ of each pair and get L_{max};

13 **return** Optimal power/ground distribution and signal pairings with lowest L_{max};

Algorithm 3: Benchmark different differential signal assignments

III. EXPERIMENTAL RESULTS

To demonstrate the accuracy and efficiency of the new algorithm, several classes of tests are implemented. The grid size of our test cases is 4×11. With a pre-determined power/ground distribution as shown in Fig. 2, previous method [4] takes almost one day to get a confident coverage of all possible differential assignments and it is still not guaranteed that all possibilities are covered. Our method enumerates all 78400 cases (diagonal paring allowed) in less than two seconds. The new algorithm is much more computationally efficient and guarantees that all possibilities are covered.

Another advantage of our algorithm is that it can optimize the power/ground distribution. Users can specify the number of power/ground pins. For the test case shown in Fig. 2, we demonstrate the first eight power/ground distributions (Fig. 3) extracted from our algorithm. For each power/ground configuration, we will list all possible differential signal assignments using linear perturbation. For example, with the first power/ground distribution as shown in Fig. 3, there are 2824 possible differential signal assignments, while with the second configuration, there are 10789 possibilities. All these possibilities can be assessed to find the optimal case within a few minutes.

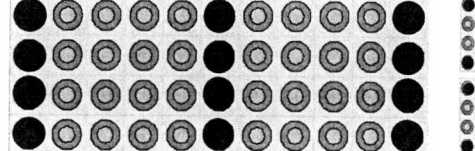

Fig. 2.   Pre-determined power/ground distribution.

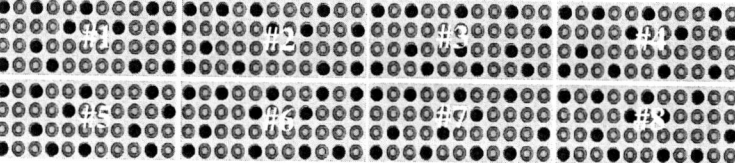

Fig. 3.   First eight power/ground distribution by force-directed optimization.

With a pre-determined power/ground distribution, the easy differential signal assignment has a benchmark coupling factor of 1.097. The previous method [4] gives an optimal one with a benchmark factor of 0.550, almost halved. Our method optimizes power/ground distribution as well as differential signal assignments. Fig. 4(c) shows the optimal case. The coupling benchmark factor is 0.475, a further improvement of 14%. A channel simulation tool is used to assess the time domain crosstalk noise in these three cases. The simulated signal-to-crosstalk noise ratio is 29.5dB, 33.9dB and 34.5dB, respectively. Eye diagrams (as shown in Fig. 5 with a baud rate of 6.4Gbps) confirm that electrical performance is enhanced by optimizing power/ground distribution and differential signal assignment in a high-speed differential bus.

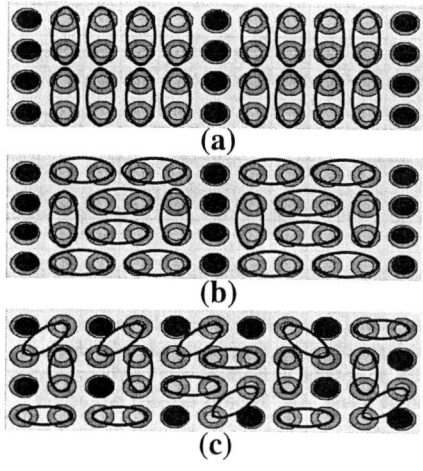

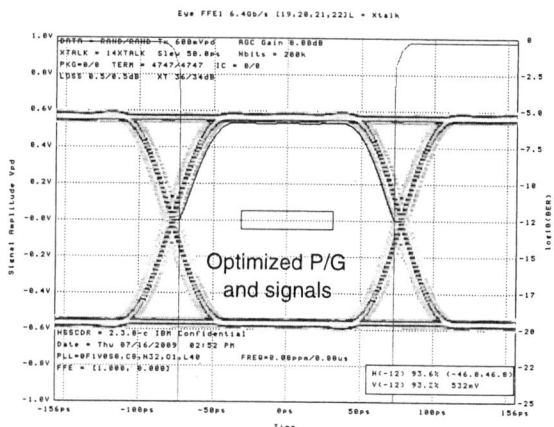

Fig. 4.   Differential signal assignment of non-optimized (a), optimized with pre-determined power/ground distribution (b), and optimized with optimal power/ground distribution (c).

Fig. 5.   Eye-diagrams of optimal assignment.

REFERENCES

[1] F. Mbairi, W. Siebert, and H. Hesselbom, "On The Problem of Using Guard Traces for High Frequency Differential Lines Crosstalk Reduction," *IEEE Transactions on Components and Packaging Technologies*, vol. 30, no. 1, pp. 67-74, 2007.

[2] B. Szendrenyi, H. Barnes, J. Moreira, M. Wollitzer, T. Schmid, and M. Tsai, "Addressing the Broadband Crosstalk Challenges of Pogo Pin Type Interfaces for High-Density High-Speed Digital Applications," in *proceedings of IEEE/MTT-S International Microwave Symposium*, vol. 3, no. 8, pp. 2209-2212, 2007.

[3] W. Yao, Y. Shi, L. He, and S. Pamarti, "Worst Case Timing Jitter and Amplitude Noise in Differential Signaling," in *proceedings of International Symposium on Quality Electronic Design*, pp. 40-46, 2009.

[4] Y. Zhou, R. Mandrekar, T. Zhou, S. Harvey, P. Weekly, and R. Weekly, "Minimizing Crosstalk Noise in Vias or Pins by Optimizing Signal Assignment in a High-speed Differential Bus," in *Proceedings of IEEE Conference on Electrical Performance of Electronic Packaging*, pp. 313-316, 2008.

[5] J. Cong, G. Luo, and E. Radke, "Highly Efficient Gradient Computation for Density-Constrained Analytical Placement," *IEEE Transactions on Computer-Aided Design of Integrated Circuits and Systems*, vol. 27, no. 12, pp. 2133-2144, 2008.

[6] T. F. Chan, J. Cong, J. R. Shinnerl, K. Sze, and M. Xie, "mPL6: Enhanced Multilevel Mixed-size Placement," in *Proceedings of ACM International Symposium on Physical Design*, pp. 212-214, 2006.

[7] J. Quine, H. Webster, H. Glascock, and R. Carlson, "Characterization of Via Connections in Silicon Circuit Boards," *IEEE Transactions on Microwave Theory and Techniques*, vol. 36, no. 1, pp. 21-27, 1988.

978-1-4244-4447-2/09 $25.00 © 2009 IEEE

AUTHOR INDEX

Abbasfar, Ali 21
Abdulhadi, Abdulhadi E..............149
Abhari, Ramesh 125, 149
Achar, Ram 65
Ahmad, Waqar 105
Ahn, Dal 185
Amirkhany, Amir 21
Aoyagi, Masahiro 25
Asai, Hideki 81, 237
Baba, Kazuhiro 25
Baks, Christian 173
Bandyopadhyay, Tapobrata..............117
Becker, Wiren D. 1, 253
Best, Scott 93
Beyene, Wendemagegnehu..........189, 21
Beygi, Amir 169
Bhattacharya, Swapan K................201
Bowles, Kevin 29
Braunisch, Henning 133
Broydé, Frédéric 5
Buckwalter, James F...................249
Cangellaris, A. C. 69, 161
Chada, Arun Reddy 153
Chan, Marcus 209
Chandrasekar, Karthik 121
Chang, Fu-Sheng........................245
Chang, Sam 33, 49
Chang, Yen-Chih 225
Charest, Andrew 65
Chatterjee, Ritwik 117
Chen, Qiang 105
Chen, Xiaoming 29
Cheng, Chung-Kuan 249
Chinea, A. 61
Ching, Michael 93
Chiu, Po-Wei 217
Cho, Jeonghyeon 13, 53, 157
Cho, Jonghyun 97
Choi, Kwansun 185
Chung, Daehyun................17, 113, 117
Chung, J. H. 161
Cressler, John D. 201
Daniel, Luca 177
Démoulin, Bernard 5
Deutsch, Alina 57, 113
Doblar, Drew 9
Dobre, Sorin 29
Dounavis, Anestis 169
Dreps, Daniel M. 1
Drewniak, James L. 109
El Sabbagh, Mahmoud A. 129
El-Ghazaly, Samir M. 129
El-Moselhy, Tarek 177
Erickson, Evan 121
Fan, Jun 109, 153

Frans, Yohan 33
Franzon, Paul D...................37, 77, 121
Gadfort, Peter 37
Gan, Houle 145
Gaskill, Steven G. 101
Giovannini, Thomas 93
Girardi, A. 85
Gomyo, Toshio 25
Grivet-Talocia, S. 61
Gu, Xiaoxiong 153, 173, 205
Gunupudi, P. 137
Gwon, Chilhyeun 185
Hayes, Jerry 89
Hsiao, Yu-Chung 177
Huang, Yu-Wen 181
Huh, Suzanne 17
Inoue, Yuta 237
Ito, Toshiyasu 189
Izzi, R. 85
Jain, Sidharath 221
Jakubczyk, J. 137
Jandhyala, Vikram 133
Jang, Daehoon 185
Ji, Steven Yun 197
Jiang, Lijun 113
Jiao, Dan 145
Kang, Minwoo 185
Kikuchi, Katsuya 25
Kim, Jaemin 109
Kim, Jingook 109
Kim, Jiseong 157
Kim, Joohee 13
Kim, Joong-Ho 33, 93
Kim, Joungho 13, 53, 97, 109, 157
Kim, Kwisoo 185
Klein, J. 137
Kollipara, Ravi 93, 189
Kusamitsu, Hideki 189
Kwark, Young H. 153
Lalgudi, Subramanian 73
Lee, Hyungdong 13, 97
Lee, Junho 13, 97
Lee, Juyoung 9
Leibowitz, Brian 33
Lessio, T. 85
Li, Ming 33, 189
Lim, Jongsik 185
Liu, Xiaoping 193
Liu, Yifang 253
Liu, Yi-Feng 193
Madden, Chris 33, 189
Mahani, Mohammad S....................125
Maio, I. A. 85
McAllister, Michael 113
Mekonnen, Yidnek 229

AUTHOR INDEX

Melde, Kathleen L....................57
Milosevic, Pavle 45
Min, Sung-Hwan 141
Mintarno, Evelyn 197
Moon, Se-Jung....................69
Mullen, Don 189
Nakhla, Michel 65
Narasimha, Rajan Lakshmi 41
Narayanan, T. V....................141
Nguyen, Nhat 33
Oh, Dan 33, 49, 93
Okhmatovski, V. 161
Okubo, Toshikazu 25
Otsuka, Kanji 25
Pak, Jun So 13, 97
Pan, Christopher 29
Papapolymerou, John 201
Park, Kunwoo....................13, 97
Park, Kwangsik 185
Patterson, Chad E. 201
Poh, Chung Hang John 201
Popovich, Mikhail 29
Ren, Liehui 109
Rigazio, L. 85
Ritter, Mark B. 205
Romo, G. 165
Rubin, Barry J. 113
Ruehli, Albert E. 89, 205
Sarangi, Ananda....................229
Sathanur, Arun V. 133
Schmitt, Ralf....................33
Schutt-Ainé, José E. 45
Scogna, A. Ciccomancini 165
Sekine, Tadatoshi 81
Shaeffer, Ian 93
Shanbhag, Naresh 41, 45
Shayan, Amirali 29
Shi, Hao....................213
Shilimkar, Vikas S. 101
Shim, Jongjoo....................97, 157
Shimada, Osamu 25
Shiue, Guang-Hwa 217
Smy, T. 137
Song, Eakhwan....................13, 53, 97, 157
Song, Jiming 221
Stievano, I. S. 85
Sudo, Toshio 25
Sun, Ruey-Bo 225
Suntives, Asanee....................125, 149
Suryakumar, Mahadevan....................229
Swaminathan, Madhavan....................17, 113, 117, 141
Takemura, Koichi 25
Takeuchi, Yukiharu 25
Tau, Yee Hung See 209
Tenhunen, Hannu 105

Thrivikraman, Tushar K. 201
Torres-Torres, Reydezel 233
Triverio, P. 61
Tsang, Leung 173
Tsuk, Michael 73
Tummala, Rao 117
Ueda, Chihiro 25
Unno, Masaki....................237
Vega-González, Víctor H. 233
Vikinski, Omer 241
Vitale, F. 85
Wang, Ting-Kuang 181
Ware, Fred 33
Weerasekera, Roshan 105
Weisshaar, Andreas 101
Weldezion, Awet Yemane 105
Wells, Lionelle F. 57
Wen, Chang-Yi 225
Wilson, John 93, 121
Wu, Boping 173
Wu, Kai-Bin 245
Wu, Ruey-Beei 225, 245
Wu, Tzong-Lin 181
Xie, Jianyong 113
Yi, Yang....................253
Yuan, Chuck 33, 93, 189
Zhang, Yulei 249
Zheng, Li-Rong 105
Zhou, Tingdong 1
Zhou, Yaping 253
Zhou, Zhen 57
Zhu, Ting....................77

9781424444472